PHYSIOLOGY OF HALOPHYTES

Signaling, Omics, and Tolerance Mechanisms

Nivas Manohar Desai, PhD

Ganesh Chandrakant Nikalje, PhD

Editors

AAP | APPLE ACADEMIC PRESS

First edition published 2025

Apple Academic Press Inc.
1265 Goldenrod Circle, NE,
Palm Bay, FL 32905 USA

760 Laurentian Drive, Unit 19,
Burlington, ON L7N 0A4, CANADA

CRC Press
2385 NW Executive Center Drive,
Suite 320, Boca Raton FL 33431

4 Park Square, Milton Park,
Abingdon, Oxon, OX14 4RN UK

Library and Archives Canada Cataloguing in Publication

CIP data on file with Canada Library and Archives

Library of Congress Cataloging-in-Publication Data

CIP data on file with US Library of Congress

ISBN: 978-1-77491-735-0 (hbk)
ISBN: 978-1-77491-782-4 (pbk)
ISBN: 978-1-00350-408-5 (ebk)

About the Editors

Nivas Manohar Desai, PhD
Assistant Professor, Department of Botany, Dapoli Urban Bank Senior Science College, Dapoli, Ratnagiri District, Maharashtra, India

Nivas Manohar Desai, PhD, is a fellow of the Association of Plant Science Research (FAPSR) as well as Assistant Professor in the Department of Botany, Dapoli Urban Bank Senior Science College, Dapoli, Ratnagiri District, Maharashtra, India. With over 15 years of experience in the field of plant science research, Dr. Desai has a unique identity that shines through his more than 37 international and national publications. His research areas are marine botany, plant physiology, functional foods, and photochemistry. He is a recipient of a DST Fastrack Young Scientist Fellowship and has successfully completed a project on developing biofuel from marine cyanobacteria. He holds three Australian and six Indian patents. He has attended more than 30 international and national conferences. He is also the recipient of several young scientist awards and one research excellence award. He is the author of many book chapters and has edited eight books.

Ganesh Chandrakant Nikalje, PhD
Assistant Professor in Botany, R. K. Talreja College, Ulhasnagar (affiliated to University of Mumbai), India

Ganesh Chandrakant Nikalje, PhD, is currently working as an Assistant Professor in Botany at R. K. Talreja College, Ulhasnagar (affiliated to University of Mumbai), India. His major areas of research are unravelling the salt adaptation mechanism of halophytes, development of bio-saline agriculture, phytoremediation, endophyte-assisted growth promotion and stress tolerance in plants, and assessment of wild vegetables for nutritional and value-added products. He has received grants from the University of Mumbai for research projects and travel grants from DST-SERB, India. He earned his doctorate in Botany from Savitribai Phule Pune University and the Bhabha Atomic Research Centre, Mumbai, India, under the BARC-SPPU Collaborative PhD programme. He has completed his postdoctoral research at Gyongsang National University, South Korea, as a SERB-SIRE Fellow of Science and Engineering Research Board, Govt. of India, India. He has received an MS Swaminathan Research Excellence Award 2021 and a Young Plant Scientist Award 2021 from the Association of Plant Science Researchers (APSR) and the Plantica Foundation Dehradun, India. He is a recognized adviser for PhD and MSc students at the University of Mumbai. He is serving as an editor and reviewer for several international journals. He has been invited as a resource person for different conferences and symposia. To date, he has published three books, 23 research/review papers in international journals, and 16 book chapters in edited books for international publishers.

Contents

Contributors

Sunil P. Bhagat
Department of Botany, ADT's Shardabai Pawar Mahila Arts, Commerce and Science College Shardanagar, Baramati, Maharashtra, India

Bhushan P. Bhusare
Ideal Institute of Biology (IIB)-PCB, Nanded, Maharashtra, India

Shreyasi Biswas
Department of Microbiology, Lady Brabourne College, Kolkata, West Bengal, India

Mahadev R. Chambhare
Department of Botany, Amruteshwar Art's, Commerce and Science College, Vinzar, Velha, Pune, Maharashtra, India

Sudarshan Charaple
Department of Botany and Plant Protection, Sadguru Gadage Maharaj College, Karad, Dist. Satara, Maharashtra

Laxmishree S. Chengala
Department of Botany, G.M. Momin Women's college, Bhiwandi, Thane, Maharashtra, India

Madhulika Gupta
Department of Microbiology, Lady Brabourne College, Kolkata, West Bengal, India

Priyanka Gupta
Department of Biotechnology, B. K. Birla College of Arts, Science and Commerce (Autonomous), Kalyan (West), Maharashtra, India

Vibha Gupta
Department of Botany, G. N. Khalsa College (Autonomous), Matunga, Mumbai, Maharashtra, India

Mali Ram Hindala
Department of Botany, University of Rajasthan, Jaipur, Rajasthan, India

Ritu L Jain
Department of Botany, G.M. Momin Women's College Bhiwandi University of Mumbai, India

Ridhi Joshi
Department of Botany, University of Rajasthan, Jaipur, Rajasthan, India

Amit D. Kale
Department of Botany, S.S.K.A.C.S. College, Kolpewadi, Maharashtra, India

Vidya R. Kothari
Department of Microbiology, ADT's Shardabai Pawar Mahila Arts, Commerce and Science College Shardanagar, Baramati, Maharashtra, India

Asha Kumari
Plant Omics Division, CSIR- Central Salt and Marine Chemicals Research Institute (CSIR-CSMCRI), Gijubhai Badheka Marg, Bhavnagar, Gujarat, India
Academy of Scientific and Innovative Research (AcSIR), Ghaziabad, Uttar Pradesh, India

Vinayak H. Lokhande
Department of Botany, J.T.S.S.P.M., Shri Shiv Chhatrapati College of Arts, Commerce and Science, Bodkenagar, Junnar, Pune, Maharashtra, India

Kamlakar More
PG Research Centre in Botany, Prof. Ramkrishna More Arts, Commerce and Science College, Akurdi, Pune, Maharashtra, India

Mahuya Mukhopadhyay
Department of Microbiology, Lady Brabourne College, Kolkata, West Bengal, India

Ganesh C. Nikalje
PG Department of Botany, Seva Sadan's R. K. Talreja College of Arts, Science and Commerce, Ulhasnagar, Thane, Maharashtra, India

Harichandra A. Nikule
Department of Botany, ADT's Shardabai Pawar Mahila Arts, Commerce and Science College Shardanagar, Baramati, Maharashtra, India

Asish Kumar Parida
Plant Omics Division, CSIR – Central Salt and Marine Chemicals Research Institute (CSIR-CSMCRI), Gijubhai Badheka Marg, Bhavnagar, Gujarat, India
Academy of Scientific and Innovative Research (AcSIR), Ghaziabad, Uttar Pradesh, India

Monika Patel
CSIR – Central Salt and Marine Chemicals Research Institute (CSIR-CSMCRI), Gijubhai Badheka Marg, Bhavnagar, Gujarat, India
Academy of Scientific and Innovative Research (AcSIR), Ghaziabad, Uttar Pradesh, India

Avinash Patil
Department of Biotechnology, B. K. Birla College of Arts, Science and Commerce (Autonomous), Kalyan (West), Maharashtra, India

Darshana Patil
Department of Botany, Smt. C.H.M. College, Ulhasnagar, Maharashtra, India

J. R. Patil
Department of Botany, SevaSadan's R. K. Talreja College of Arts, Science and Commerce, University of Mumbai, Ulhasnagar, Maharashtra, India

Manasi Patil
Department of Botany and Plant Protection, Sadguru Gadage Maharaj College, Karad, Dist. Satara, Maharashtra

Vidya Patni
Department of Botany, University of Rajasthan, Jaipur, Rajasthan, India

Ganesh Pawar
Post-Graduate Research Centre, Department of Botany, Tuljaram Chaturchand College of Arts, Science and Commerce, Baramati, Pune, Maharashtra, India

Satyam D. Pawar
Department of Botany, ADT's Shardabai Pawar Mahila Arts, Commerce and Science College Shardanagar, Baramati, Maharashtra, India

Umesh Pawar
Department of Botany, Shri Pancham Khemraj College, Sawantwadi, Maharashtra

Jayant P. Rathod
Department of Microbiology, ADT's Shardabai Pawar Mahila Arts, Commerce and Science College Shardanagar, Baramati, Maharashtra, India

Divya Raul
PG Department of Botany, Seva Sadan's R. K. Talreja College of Arts, Science and Commerce, Ulhasnagar, Maharashtra, India

Joshi. A. Samiksha
University of Pune, Department of Botany, Pune, Maharashtra, India

Ramesh R. Sanap
Department of Botany, S.S.G.M. College, Kopargaon, Maharashtra, India

Deepak B. Shelke
Department of Botany, Amruteshwar Art's, Commerce and Science College, Vinzar, Velha, Pune, Maharashtra, India

Hiralal B. Sonawane
Department of Botany, Sant Gadge Baba Amravati University, Amravati, Maharashtra, India

A B Telave
Post Graduate Research Centre, Department of Botany, AES's T. C. College of Arts, Science and Commerce (Autonomous), Baramati, Maharashtra, India

Minal Trivedi
Department of Biotechnology, B.K Birla College of Arts, Science and Commerce (Autonomous), Kalyan, Maharashtra, India

Foreword

Food security has become a prime concern as there is need to enhance food production by about 70% by 2050 to meet the demands of a growing population, especially in the context of dwindling land, water, and eco-resources. Plant biodiversity is one of the basic tenets of our sustenance on this planet, and hence, conservation and improvement are necessarily the prime focus of global research.

Stress both biotic and abiotic, have been a challenge to food production, especially in the environmentally challenged land and water ecosystems. While most plants cannot defend well against abiotic stresses like salinity, a group of plant halophytes are endowed with unique abilities to thrive under saline regions and also in a range of climatic conditions. These "specially abled" plant species include trees, grasses, and saltbushes that grow in salt marshes, saline deserts, and coastal areas are a small fraction of the vast majority of flowering plants.

The study of halophytes has intrigued many researchers and understanding is fundamental to plant biology as the knowledge can be used to adopt non-halophytes to grow under saline areas. It is very important to note the relevance of this halophyte research in "biosaline agriculture." Globally, several institutions are now working on developing knowledge on halophytes with considerable scope for the improvement of stress-tolerance in crop plants.

In this context, it is very pleasing to see a book highlighting the important research outcomes in halophytes. The editors of this book have done an excellent job by compiling chapters from contributing authors on a broad range of topics, including physiological, biochemical mechanisms, compatible solutes, signaling, and regulatory components, in addition to halophilic cyanobacteria and microalgae, halophyte-associated microbes, endophytes, adaptive pathways. The book has also relevant chapters on the role of halophytes in eco-friendly environmental and agricultural practices and environmental clean-up.

I congratulate the editors for compiling such a book, and I am hopeful that this will become a valuable knowledge resource for researchers, students, and policymakers looking to explore halophytes in sustainable agriculture.

—Prof Suprasanna Penna
Director, Amity Institute of Biotechnology (AIB),
Amity University, Maharashtra
Former Scientist, & Head, Nuclear Agriculture and Biotechnology Division
Bhabha Atomic Research Centre, Mumbai, India

Preface

Although halophytes have long been known, it was not until the 20th century that they were thoroughly investigated. Despite the fact that there is still much to understand about these salt-tolerant plants, it is obvious that multiple mechanisms contribute to tolerance, which explains the challenges associated with engineering tolerance in more salt-sensitive species. Because of this, it is critical to comprehend the multiple levels of tolerance—molecular, physiological, and ecological—in order to comprehend what it means to be a halophyte. The long-term goal of this research is to be able to use our understanding of halophytes to enhance the productivity of crops in salt-affected soils.

In order to exploit some of these fascinating plants in land re-vegetation, as fodder for cattle, and to develop salt-tolerant crops, it is critical to have a better understanding of halophytes. The evolution of salt tolerance in halophytes, their life-history features, and developments in our understanding of the molecular, biochemical, and physiological mechanisms behind salt tolerance are outlined in this book on halophytes and saline adaptations. Highlighted are the cellular mechanisms that enable halophytes to withstand high tissue concentrations of Na+ and Cl, including the control of membrane transport, the production of suitable solutes, and the management of reactive oxygen species.

This book, *Physiology of Halophytes: Signaling, Omics, and Tolerance Mechanisms,* provides scientific knowledge and expertise on the mechanisms of plant responses to various environmental stresses. Chapter 1 provides an overview of the metabolomic studies of halophytes under salinity stress. Chapter 2 focuses on the role of the halophyte–microbes duo in environmental clean-up. Chapter 3 presents an overview of endophytes-assisted salinity tolerance in halophytes. Chapter 4 gives an idea about salt signaling in halophytes. In Chapter 5, the author discusses the metabolomic, proteomic and ionomic adjustment and adaptive mechanisms of halophytes. Chapter 6 covers the habitat, growth response and adaptation under salinity in *Sonneratia* sp. Chapter 7 highlights the physiology and potential applications of sand dune halophytes. Chapter 8 gives a systematic review of physiological responses of halophytes to salinity. Chapter 9 discusses adaptive pathways of halophytes and the prospects of increasing plant salinity tolerance. Chapter

10 comprehensively deals with the halophilic cyanobacteria and microalgae, focusing on their role in environmental clean-up and value-added products. Chapter 11 presents structural hierarchy and discusses salinity stress tolerance improvement in the halophyte *Sesuvium portulacastrum* L. Chapter 12 deals with prospective role of halophytes and perspectives for sustainable, eco-friendly environmental and agricultural development. Chapter 13 deals with physiological, proteomics, and metabolomics approaches for elucidation of salinity and alkalinity tolerance in halophytes. Chapter 14 covers reactive oxygen species and oxidative damage in halophytes under salt stress. In Chapter 15, the author discusses that halophyte adaptation is nature's prodigy for environmental changes. Chapter 16 throws light on adaptive strategies and the role of endophytes in halophytes for amelioration of salt stress. Chapter 17 signifies the potential role of halophytes in environmental clean-up. In the last chapter, stress mitigating proteins and osmolytes of halophytes are elaborated.

We express our thanks to all the contributors. We would like to thank Prof. Suprasanna Penna for writing the foreword. Finally, it is a profound pleasure to thank AAP for taking up the publication of this book. We hope that this book will provide current knowledge on environmental stress in plants and will lead new discussions and efforts to deals with various abiotic and biotic stress factors.

Introduction

Soil salinity, a substantial abiotic stress that has an extensive impact on plant growth and productivity, is one of the main topics of research in the domains of biochemistry and plant physiology. Salinity has a lot of negative effects on plant growth and development as well as germination. In addition to nutritional imbalance, ion toxicity, the induction and modification of plant hormones, and various metabolic changes that result in molecular damage because of the generation of reactive oxygen species (ROS), its impact also includes water shortages linked to lower photosynthetic activity.

Plants known as halophytes may survive in salty conditions. They represent 1–2% of the world's flora and include both dicots and monocots. Based on their tolerance and need for salts, halophytes can be divided into facultative and obligatory halophytes. While obligatory halophytes need salt for growth and development, facultative halophytes can thrive in non-saline environments and under restricted freshwater circumstances. Contrarily, the bulk of agricultural crops are salt-sensitive glycophytes.

Based on the conditions in which they thrive, halophytes are further divided into hydro-halophytes and xerohalophytes. Mangroves and salt-marsh species can be found inland or along beaches; hydro-halophytes need water or moist soil to survive. Halophytes are known as extremophiles when they sprout and breed in high-salinity settings (500 mM NaCl) and may survive in dry habitats with salty soils (i.e., arid, semi-arid inlands with irregular rainfall).

Special consideration should be given to the widespread use of halophytes as alternatives for glycophytic crops, fodder and animal feeds, oilseeds and protein crops, energy crops (biofuels and fuelwood), phytoremediation, medicinal plants, and other commercial items. Halophytes (ephemeral, shrubs, and trees) have developed a variety of mechanisms for regulating growth and development in order to maintain their survival in high-salt settings (inland or coastal areas, salt marshes, dunes, and deserts). Halophytes need to undergo morphological and anatomical changes, such as the development of succulence (dilution of the ion concentration in the plant tissue) or salt glands or salt bladders (for the selective exclusion or accumulation of ions).

It is important to emphasize the role endogenous hormones and the seed coat play in controlling dormancy and germination. The genetics and developmental stage in particular have an effect on salt tolerance. Ion poisoning or an osmotic action may affect the delicate stages of germination and flowering. The type of salt in the soil where the seeds are placed (such as NaCl, $NaHCO^3$, and Na^2SO^4, etc.) also has an impact on how the plant responds. While some Amaranthaceae (formerly Chenopodiaceae) plants are salt tolerant, others actively avoid salt throughout the germination process.

High germination percentages and/or rates in high salinity (such as in *Haloxylon ammodendron, Suaeda physophora, Anabasis salsa,* and *Suaeda*) are indicative of typical salt tolerance, whereas a high percentage of recovery germination following the reduction in salinity is indicative of a salt-avoidance strategy (e.g., *Borszczowia aralocaspica, Ceratoides latens, Bassia dasyphylla*). According to the type of salt, some seeds may have either salt tolerance, salt avoidance, or both features.

This book discusses the biology of halophytes in great detail, with a particular emphasis on their ecology, physiology, omics, proteomics, and potential as crops.

CHAPTER 1

Metabolomic Studies of Halophytes Under Salinity Stress

VIBHA GUPTA

Department of Botany, G. N. Khalsa College (Autonomous), Matunga, Mumbai, India

ABSTRACT

Abiotic stresses give rise to morphological, biochemical, and molecular changes which affect the overall growth and development of plants. Among these stresses, salinity stress is the major threat for the plants. To combat this stress, halophytic plants which are also called physiological drought plants, have evolved mechanisms to deal with high levels of salinity. These adaptations at physiological, biochemical, and molecular level have been studied extensively with reference to Halophytic plants but study of a network of metabolites synthesized during salinity stress is yet to be fully explored. The study can provide a deeper insight about specific mechanisms which stress driven plants undergo when compared with normal plants and how they adapt to acclimatize the changed environment. A high-throughput technique named as "Metabolomics" also called metabolic profiling, is one of the fastest developing disciplines, enabling the scientists to use multivariate analysis from different versions of chromatography together with nuclear magnetic resonance (NMR), Mass spectroscopy (MS), liquid chromatography-MS, reverse-phase high-performance liquid chromatography (RP-HPLC), next-generation sequencing (NGS) and gene editing using CRISPR technologies. The data obtained with the help of these techniques are highly advantageous over the conventional methods (gel-based assays) in providing a great insight into new strategies in halophytic plants at cellular and molecular levels. The

Physiology of Halophytes: Signaling, Omics, and Tolerance Mechanisms.
Nivas Manohar Desai & Ganesh Chandrakant Nikalje (Eds.)

primary and secondary metabolites which are salinity-induced expressed in galophytes are proline, glycine, betaine, mannitol, polyamines, intermediates in ATP biosynthesis, metabolites of photosynthetic machinery, phytohormones, phenolics, alkaloids, etc. The present chapter summarizes the overview of halophytes showing different adaptations at biochemical level, techniques used in plant metabolomics with brief description of steps involved and various primary and secondary metabolites of salt-tolerant plants.

1.1 INTRODUCTION

Halophytes describe the plant species adapted to grow and reproduce in high saline conditions making them model plants for studying the physiological, biochemical basis of salt-tolerant plants and generating novel plants for deteriorated salinized lands which are otherwise unsuitable for agricultural use (Glenn et al., 1999; Saddhe et al., 2020). It is also well documented that some halophytes are not even affected at salt levels above sea water; salinity has led to their widespread use for sustainable development and reclamation of saline soils (Naikoo et al., 2020; Bueno and Lopes, 2020). For example, *Atriplex nummularia* accumulates considerable amounts of salts (20–40% on dry mass basis). Another species called *Suaeda fruticosa* can remove 2.5 mg of salt in a single harvest (Ghnaya et al., 2005). Based on their ability to maintain growth under saline conditions, Sengbusch (2003) proposed the grouping of halophytes into obligate halophytes, facultative halophytes, hydro halophytes, xerohalophytes, marine phanerogams. Initially, salt-stressed plants suffer from osmotic stress followed by ion effects. This retards the water absorption and cell expansion, and to adjust osmotically, most plant species grow slowly at the latter stages. Nutritional imbalance is another toxic effect of excess salinity further reducing the plant growth. Thus, high accumulation of Na^+ marks the onset of ion toxicity disrupting various metabolic activities once the cellular salt concentrations exceed the threshold level. Excess accumulation of Na^+ and or Cl^- affects photosynthetic activity by declining the carbon metabolism and photophosphorylation. Plant leaves show adverse effects such as decreased activities of cell organelles and metabolites; chlorosis and necrosis when they are facing osmotic and salt stress simultaneously.

Plant productivity is highly affected by salinity stress worldwide. It has been estimated that one-third of farmland lands are affected by soil salinity (Bhargava and Srivastava, 2020). Though several advances have been made in increasing agricultural production, improvement in the production of salt tolerance crop plants remains always a difficult task as it affects plant physiological mechanisms severely. Excessive salinity causes water deficit, ion toxicity, and nutrient deficiency, affecting the overall growth of halophytes (Roy et al., 2022; Aboelsoud et al., 2022). Reduction in the photosynthetic activity, production of reactive oxygen species (ROS), and accumulation of inorganic and organic metabolites along with changes at the molecular level (He et al., 2021; El-Badri et al., 2021; Qin et al., 2013) is also well documented.

Halophyte plants are physiologically drought, also known as salt-loving plants, they can combat salinity stress, and several genes are expressed, which are involved in conferring their resistance to high saline conditions (Alnusairi et al., 2021; Yu et al., 2011). The physiological, biochemical, and molecular level changes are subjected to intense research in halophyte studies. Thus, they have evolved several mechanisms to tolerate high salt conditions (Fig. 1.1).

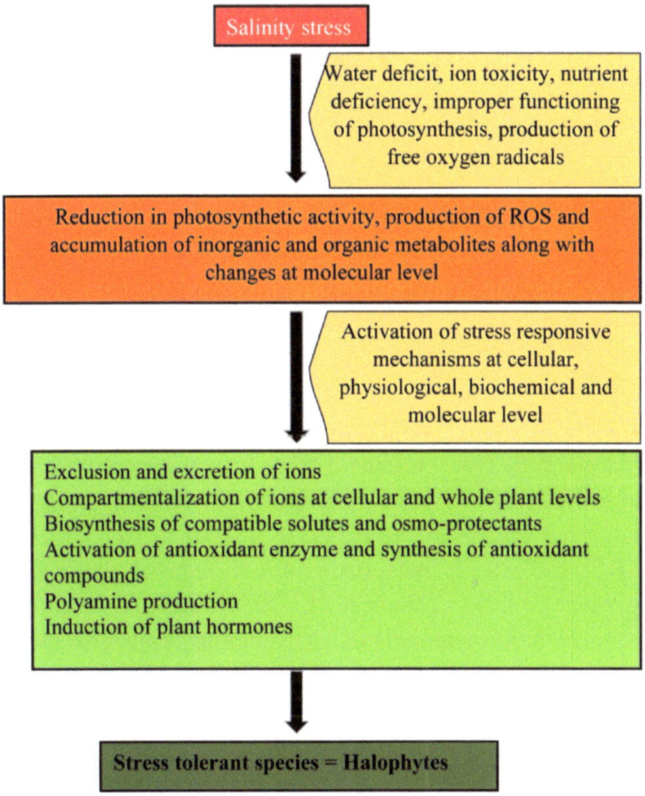

FIGURE 1.1 Generalized scheme for halophytes to combat salinity stress.

In this view, it is required to harness this knowledge for the improvement of halophytic plants. Genetically engineered plants are one of the ways to combat these challenges. This requires exploring the gene pools and various metabolic pathways of halophytic plants. eHALOPH database presently represents the information of more than 1500 species as salt-tolerant species. These are highly evolved and well-adapted to high salt concentrations (Flowers and Colmer, 2008).

Thus, this chapter aims at understanding the biochemical adaptations of halophytes, techniques commonly employed in the plant metabolomic field followed by understanding the nature of various metabolites produced under salinity stress.

1.2 BIOCHEMICAL ADAPTATIONS TO SALINITY STRESS

Comprehensive literature available on halophytes on their understanding of salt tolerance mechanisms. They show marked physiological and molecular adaptations in addition to morphological and anatomical changes. This increases their survival rate by completing their life cycle under abiotic stress conditions. There are three main biochemical adaptation mechanisms which are mentioned as:

1.2.1 ION HOMEOSTASIS

Many ion channels and transporters are involved in accumulating various inorganic ions mainly Na^+, Cl^-, K^+ in halophytic plants. Non-selected cation channels (NSCCs) including cyclic nucleotide-gated channels (CNGCs) and glutamate receptor-like channels (GLRs), high-affinity K^+ transporters (HKT2 and HAK5), and Na^+ influx aquaporins (plasma membrane intrinsic protein isoforms PIP) are symplastic transporters responsible for uptaking of Na^+. The influx of Na^+ is also assisted by low-affinity cation transporter (LCT1) and *Arabidopsis* K^+ transporter 1 (AKT1) (Arif et al., 2020). HKT2, HAK5, and AKT1 transporters are shown to uptake competitively K^+ in the presence of Na^+ (Zhao et al., 2020). Also, in addition to cation transporters, slow anion channels (SLAC channels) are also involved in the passive transportation of Cl^- during high salinity (Wu and Li, 2019). It was suggested by Himabindu et al. (2016) that the accumulation of excess salt in the cytosol of halophytes can be avoided by sequestrating ions toward the vacuole. The SOS system is also crucial for recognizing salt-mediated process which includes the exclusion of excess Na^+ by roots, loading of Na^+

into the xylem, and sequestration of excess Na^+ into the vacuoles using Na^+/H^+ exchangers (NHXs) (Sadhhe et al., 2020). Tran et al. (2019) established that in salt-tolerant plant *Mesembryanthemum crystallinum* the expression of genes is enhanced involved in encoding tonoplast antiporters H^+/Cl^-, Na^+/H^- exchangers and V-ATPase pumps.

1.2.2 OSMOTIC ADJUSTMENT

Osmotic adaptation plays a significant role in halophytes to adapt to high levels of salt conditions (Flowers and Colmer, 2008). Osmolytes, low molecular weight organic compounds, are highly soluble and harmless even at higher concentrations without disturbing the normal process of the cell (Slama et al., 2015). Examples of osmolytes are proline, trehalose, sucrose, polyols, glycine betaine (GB), etc. They are involved in the protection of subcellular structures, mitigation of oxidative damage caused by ROS, and maintenance of enzyme activities under stress conditions (Bose et al., 2014; Lokhande et al., 2013; Zaid and Wani, et al., 2019). Nikaljeetal (2017) indicated that the basal level of osmolytes are significantly higher in halophytes under normal conditions. Apart from osmolytes, halophytes also exhibit the synthesis of diverse metabolites. For example, metabolic profiling of *Prosopis strombulifera* confirms the activation of metabolites such as lysophosphatidylcoline, and 13- hydroxyoctadecadienoic acid (Llanes et al., 2016).

There are generally two strategies to attain and maintain turgor pressure and organellar volume within the growing plant cells (Zhao et al., 2020): (i) Enhanced uptake of inorganic ions, and (ii) Synthesis of organic osmolytes. Accumulation of Na^+, Cl^-, K^+ in plant vacuoles, as well as compatible organic solutes (proline and sugars) in the cytoplasm, are crucial to ensure osmotic adjustment under salinity stress (Munns et al., 2019). Zhao et al (2020) observed that the K+ efflux in the salinity-induced state takes place via ion channels GORK (guard cell outward potassium channels) and/or ROS-activated NSCCs. Although GORK channels play a central role, extensive research into its molecular studies remains unexplored.

It is significant to mention here that the metabolite profiling of halophytes has gained momentum, especially in the comprehensive analysis of the accumulation of osmolytes. For instance, Panda et al. (2020) and Kumari and Parida (2018) revealed the salt-responsive metabolites and their biosynthetic pathways in *Haloxylonsali cornicum* and *Salvadora persica* plants. Nevertheless, further exploration should be undertaken to strengthen our existing knowledge of salt tolerance mechanisms in halophytes.

1.2.3 ROS-DETOXIFICATION

It is well known that halophytic plants have efficient antioxidant defense mechanisms to protect themselves from various damages caused by oxidative stress. Oxidative stress is generally induced by the excessive accumulation of ROS under high saline conditions (Castro et al., 2021). Thus, maintaining ROS homeostasis by vigorous antioxidant mechanism helps halophytes to survive in high saline conditions (Bose et al., 2013; Wang et al., 2020). It consists of non-enzymatic antioxidants (flavonoids, aminosalicylic acid AsA, Glutathione GSH, and tocopherols) and enzymatic antioxidants (superoxidase SOD, catalase CAT, ascorbate peroxidase APX, glutathione reductase (GR), monodehydroascorbate reductase (MDHAR), dehydroascorbate reductase DHAR, glutathione peroxidase GPX and glutathione S transferase GST) (Czarnocka and Karpinski, 2018).

For example, in *Haloxylonsali cornicum* plant, the elevated activity of SOD and GSH and AsA/dehydroascorbate (DHA) and GSH/oxidized glutathione (GSSG) plays a pivotal role in scavenging free radicals (Panda et al., 2019).

Thus, ROS generation is an inherent mechanism to defend against salinity stress in plants and plays a key role in plant development and hormone regulation (Xia et al., 2015). Therefore, it is important to focus on the regulation of ROS production which can provide useful information about the signaling pathways mediated by hormone-modulation in stress-tolerant plant species.

1.3 PLANT METABOLOMICS

A metabolome refers to a whole set of metabolites accumulated at a precise stage of growth in the cell by an organism (Hollywood et al., 2006). These metabolites are known to play an essential role in several biochemical mechanisms (Patti et al., 2012). Metabolomics has been a rapidly growing field in biological sciences in the last decade involving metabolic fingerprinting, profiling, and targeted analysis (Fiehn, 2002; Shulaev, 2006). Currently, plant metabolomics is one of the upcoming disciplines of the omics field, having huge potential in crop improvement programs. It is one of the most rapid analytical tools that aims at the exhaustive and extensive analysis of large numbers of metabolites (Liu et al., 2011). It is widely used in the study of plant sciences as these primary producers are natural reservoirs of several metabolites, which do not exist in any other life forms. Brosché et al. (2005) referred to metabolomic research as the complete profiling of small metabolites (intermediates, hormones, signaling molecules, and secondary metabolites) found in biological samples. They are used as a key tool to understand the various responses due to abiotic and biotic responses in the plant system. Metabolomics is more adaptable to changes as compared with transcriptomes, genomics, and proteomics. The research has contributed to our valuable understanding of the importance and mechanism of primary metabolites playing a pivotal role in various biosynthetic pathways (Razzaq et al., 2019; Obata and Fernie 2012). There are several secondary metabolites synthesized by the regulatory genes during environmental stresses, which are the by-products of biochemical pathways (Hurmat and Bansal, 2020; Wink, 1988). In stress biology, metabolomic analysis helps in the identification of chemical compounds which include stress metabolism and signaling molecules which are synthesized on exposure to various abiotic and biotic stresses. These molecules can result in changes at transcript, proteomic, or expression levels (Razzaq et al., 2019). When plants sense such changes in environmental conditions, the signal transduction pathways are initiated and activated to perform homeostasis (Mittler et al., 2004). Though it is a difficult and costly affair to carry out metabolomic analysis for each plant, metabolic profiling of a subclass of metabolites can be analyzed at any time or growth stage of the plant (Kumari et al., 2015).

Salt stress is the second largest abiotic stress that affects physiological activities resulting in a decrease in photosynthetic rate, respiration, and protein synthesis, and disturbs the nucleic acid metabolism. Plants respond to these disturbances by certain modifications in primary metabolism, the most apparent are alterations in the levels of sugars/sugar alcohols, amino acids, and tricarboxylic acid cycle intermediates.

Thus, metabolomic studies offer a crucial role in exploring and studying the complex biological regulatory networks that play an important role in plant growth and development. The study relies on its methodologies and instruments to comprehensively identify and measure each metabolite (Hong et al., 2016). A brief discussion about the various techniques along with step-wise sample analysis is mentioned below:

1.3.1 TECHNIQUES USED IN PLANT METABOLOMICS

- NMR is a common method of choice to study the structure of chemical compounds. It is non-destructive and beneficial to molecules in solution and solid state (Berendt et al., 2006). NMR-based metabolomics studies have been effectively used in plant biotechnology research (Kim et al., 2007; Palama et al., 2010).

- MS is another widely used technique in Plant Metabolomics. It involves the recognition and quantification of compounds in a very precise manner and provides information about the arrangement and nature of molecules (Blaum, 2006). The compounds can be quantified and the nature of the molecules can be studied thoroughly. Numerous mass analyzers are used in MS: time-of-flight (TOF) analyzers, quadrupole mass analyzers, and ion trap analyzers (Feihn et al., 2000).
- Liquid chromatography-MS is the most advanced technique used by many researchers. It is the combination of different analytical techniques that leads to identifying and characterizing the metabolites of the test sample (Uslu and Özkan, 2002). RP-HPLC (reverse-phase–high-performance liquid chromatography) is the most frequently used technique.
- NGS is a powerful tool to explore novel metabolites at the cellular level. In the 21st century, it has provided opportunities for the discovery of new metabolites (Kim and Buell, 2015) including genes involved in biosynthetic pathways, regulatory pathways, and networks, etc. through genomics and proteomic-enabled technologies. The NGS technologies have been an amazing tool to revolutionize the field of plant metabolomics.
- Gene editing using CRISPR/Cas9 has recently developed an efficient nuclease-based method for genome engineering. Shan et al. (2014) suggested that CRISPR technology a cheap, safe, and efficient technology and can be considered in crop gene editing for better tolerance to both types of stresses, that is, biotic and abiotic stresses. They described a step-wise protocol for the selection of target sites, construction, and use of single-guide RNAs for site-specific mutagenesis in target genes of rice and wheat. This technology is considered safe in gene editing for the production of better tolerance to biotic stress quality (Kharabian-Masouleh and Henry, 2015) and for creating new metabolites.

The unequaled potential of genome editing tools, namely, CRISPR/Cas9-based and NGS for the generation of well-defined sequence with target-specific approach in plants was reviewed by Dey (2021) and Waseem et al. (2022). These technologies contributed exceptionally unparalleled achievements in the areas of metabolic engineering (genetic and regulatory mechanism) in plants. Through NGS and CRISPR/Cas systems, millions of SNPs can be discovered in a genome in which many of them belong to secondary metabolites and will help in fine-tuning of genes. Cheng et al. (2014) sequenced 529 rice accessions and generated 6.4 million SNPs, responsible for 840 metabolites which finally led to the identification of 36 candidate metabolite-modulating genes with potential physiological and nutritional importance. Cheng et al. (2010) found more than 200,000 SNPs in 96 *Arabidopsis* accessions to assess the natural diversity of 43 glucosinolate phenotypes out of 172 total genes. After the recognition of genes containing SNPs with direct effects on metabolites, they can be edited with CRISPR/Cas9 technology to produce novel metabolic compounds, with more or even less abundances.

1.3.2 TARGETED AND UNTARGETED METABOLOMICS

The tolerance ability of plants against salinity stress is based on their capacity to maintain a proper level of primary and secondary metabolic responses. During the stress, plants produce metabolites as excretory products which are secreted from shoots, roots, and leaves during different stages of development. Plant metabolomics is considered thus a bridge between plant's genotype and phenotype (Ni et al., 2015). It helps in investigating key differences between tolerant and sensitive plant species. Mainly two strategies are used for metabolic studies: targeted and non-targeted metabolomics.

Targeted metabolomics is a tool for the identification, estimation, and interpretation of specific or known metabolites in plants under stress (Zhang et al., 2018; Yu et al., 2020). Non-targeted metabolomics can produce an overview of abundant metabolites under stress (Patel et al., 2019, 2020). They are a method of choice for examining the physiology of stressed plants which enhances the identification and sensing of the differentially regulated metabolites. Though the non-targeted approach is time-consuming, semi-quantitative, and cannot be used for large numbers of biological samples, its ability to cover the entire metabolome under study helps in the identification of novel metabolites as biomarkers. Targeted profiling uses specific metabolite analysis and is thus commonly used to trail the dynamics of a limited number of metabolites under stress and can be used for large numbers of samples.

1.3.3 STEPS INVOLVED IN PLANT METABOLOMICS

The major steps involved in the experimental planning for metabolomic research are sample preparation, data mining, and statistical analysis. Since the production of metabolites is influenced by several factors (Aoki-Kinoshita,

2006; Kim and Verpoorte, 2010), an adequate methodology is required to categorize them into different metabolites. The steps are described briefly below and are schematically represented in Figure 1.2:

1.3.3.1 *SAMPLE PREPARATION*

Sample preparation is the first and an important step as it has a massive impact on the final results (Kim and Verpoorte, 2010). The aerial parts of the plant tissues are preferred as a sample material. The sample preparation is done in order to isolate the specific metabolites from the mixture of undesired metabolites. The techniques which are preferred for the preparation should be simple, economical, rapid, and maintain the sample intact without any denaturation (Causon and Hann, 2016).

1.3.3.2 *DATA MINING AND STATISTICAL ANALYSIS*

The current developments in the field of plant metabolomics have generated huge data sets because numerous metabolites are involved at different stages of development in cells or tissues or organs. Thus, the analysis of such enormous metabolites is more complicated (Kuhn et al., 2008) and requires very powerful automated tools to manage and annotate these complex data sets. The major challenge in the field of plant metabolomics is to gather and study the information about the specific metabolites obtained from the massive data sets generated. Data mining involves three basic steps (Sun and Weckwerth, 2012; Liland, 2011):

(i) Pre-processing, (ii) Pre-treatment, and (iii) Statistical analysis

Statistical analysis: Statistical tools are used to interpret the information obtained from high-throughput metabolomics data. Generally, the planning and execution of experimental procedures decide the selection of statistical tools to be used for the analysis (Ren et al., 2015). For example, to study the relationship between treated and untreated samples for metabolite analysis, the following predictive models can be exercised for the analysis:

- Pairwise Pearson's correlation matrix is used for the identification of a specific biomarker (metabolite) to the desired phenotypic trait.
- Canonical correlation analysis (CCA) is used to study the relationship between two sets of variables and helps in reducing the size of computational data (Song et al., 2016).

- Univariate analysis is a preferred choice tool at the initial level of omics studies in the identification of specific metabolites, studying one variable at a specific time of development.
- Multivariate analysis is used for efficient comparative evaluation among different genotypes, screening plant cultivars, and in metabolic marker discovery (Foito and Stewart, 2018).

1.3.3.3 BIOINFORMATICS TOOLS

Computational informatics thus becomes a prerequisite for metabolomic studies due to the elaborative datasets generated (Wishart, 2007). The various tools used for the analysis of the data mining search are mentioned in Figure 1.2.

1.4 PRIMARY METABOLITES

Primary metabolites are the metabolites which are essential for the growth and development of plants due to their involvement in several biochemical and physiological processes. They are classified into carbohydrates, fatty acids, proteins, and enzymes; serving as vital functions in physiology in abiotic or biotic stresses (Shulaev et al., 2008; Patel et al., 2020). Through amino acid profiling, it was revealed that the high levels of amino acids (except asparagine) showed an increased trend in plants subjected to salinity stress as compared with control plants (Pandey et al., 2015). Levels of sugar alcohols (galactitol, erythritol, arabitol, xylitol, mannitol, and inositol) increased during salinity stress (Dias et al., 2015).

Metabolomic analysis of *Nitraria sibrica* Pall. (Tang et al., 2021) reported that different metabolites between treated and untreated with NaCl were mainly amino acids (proline, aspartic acid, methionine); organic acids (oxaloacetic acid, fumaric acid, nicotinic acid); polyhydric alcohols (inositol, ribitol). Similarly, metabolomic profiling of *Haloxylon salicornicum* was carried out in a shoot of control and salinity treated. It was observed that out of 56 metabolites, 47 metabolites significantly changed in response to salinity as reported by Panda et al. (2021). These metabolites were mainly amino acids, organic acids, amines, sugar alcohols, alkaloids, and phytohormones. In response to salinity, most of the amino acids were downregulated except alanine, phenylalanine, and lysine which were upregulated. The metabolic response of *Cakile maritima* (Arbelet-Bonnin et al., 2020) was analyzed on exposure to the different salt concentrations and showed the change in

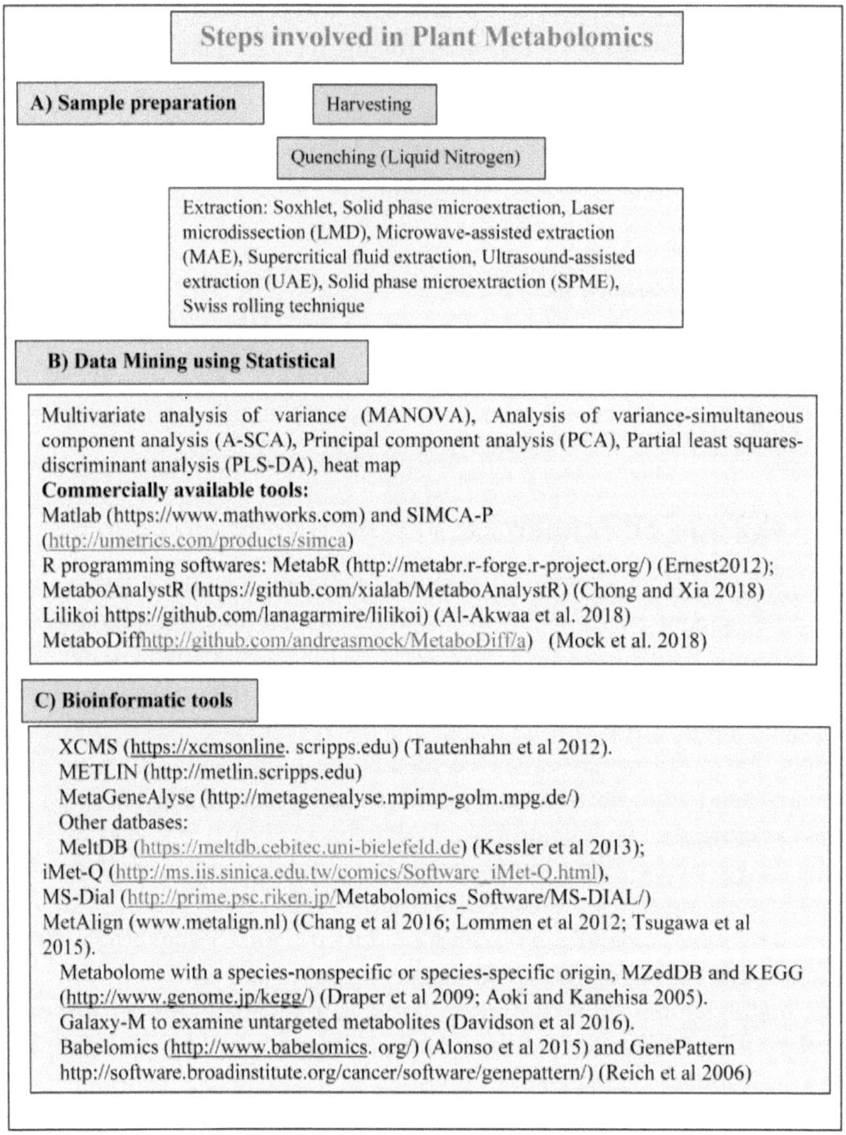

FIGURE 1.2 Schematic representation of steps involved in metabolomic analysis.

concentrations of different metabolites. For example, with 400 mM, there was an increase in amino acids and a depletion of organic acids and sugars.

Plant cells synthesize these organic solutes in response to salinity stress (Flowers and Colmer, 2008) which helps in maintaining the water balance.

Chemically, these can be broadly classified into amino acids (proline and tyrosine); polyols (myo-inositol, pinitol, and other methylated inositols) and Betaines (Rhodes and Hanson, 1993) with low molecular weight, nontoxic at high cellular concentrations without disturbing the cellular functions (Cushman, 2001). In plant cells, these organic solutes (osmoprotectants) are mainly restricted to the large central vacuole, cytosol, and chloroplast. They carry out the following functions by protecting (Ashraf and Foolad, 2007):

- Protects subcellular structures
- Protects from free radicals
- Maintains the enzyme activities
- Protection from dehydration injury

The most common metabolites are described (Table 1.1) in the following section:

TABLE 1.1 Identification of Secondary Metabolites in Halophytes Using Different Metabolomic Tools Under Salinity Stress.

Sr no.	Halophyte	Secondary metabolites	Analytical technique	Reference
1	*Amaranthus tricolor (Amaranthaceae)*	Polyphenol, flavonoid, antioxidant	HPLC-LC-MS	Sarker et al., 2018
2	*Atriplex mollis*	Rutin, t-ferulic acid, gallic acid	HPLC-DAD	Boutaoui et al., 2018
3	*Avicennia marina* subsp. *australasica*	A-Ketoglutarate, galactose, putrescine, p-toluic acid, syrengic acid, nicotinic acid, tridecane	GC-MS	Ravi et al., 2020
4	*Euphorbia peplus*	Simiarenol, 1-hexacosanol, β-sitosterol, β-sitosterol-3-O-glucoside	1D and 2D NMR	Amin et al., 2017
5	*Limonium species*	Total phenolics, flavonoids	UHPLC-QTOF-MS	Senizza et al.,2021
6	*Nitrariasibirica*	Proline, aspartic acid, methionine, oxaloacetic acid, fumaric acid, nicotinic acid, inositol, ribitol,	GC-TOF–MS	Li et al., 2021
7	*Prosopis Strombulifera*	Lysophosphatidylcholine, 13-hydroxyoctadecadienoic acid	UPLC-ESI-QTOF-MS	Llanes et al., 2016
8	*Salicornia ramosissima*	Ethyl linolenoate, sitostanol, octadecyl (E)-ferulate, eicosanyl (E)-ferulate, ethyl (E)-2-hydroxycinnamate, scopoletin, triacylglycerol of tetracosanoic acid, saliramoester, saliramophenone, and saliramopyrrole	1D and 2D NMR and MS	Ferreira et al., 2018

TABLE 1.1 *(Continued)*

Sr no.	Halophyte	Secondary metabolites	Analytical technique	Reference
9	*Sarcocornia ambigua*	Galangin, kaempferol, quercetin, isoquercetin, p-Coumaric acid, vanillic acid, caffeic acid, ferulic acid, syringic acid	ICP-MS	Bertin et al., 2016
10	*Sesuvium portulacastrum*	22,23-Dihydrostigmasterol, benzoic acid, 3,4,5-trihydroxy-(Gallic acid), capsaicin, (2R,3R)-(-)-epicatechin, ecdysones, flavonol 3-O-robinobiosides, 3-O-(2″-O-a-L-rhamnopyranosyl)-robinobiosides, sesuviosides AeF, ecdysterone, adenosine, 2′-O-methyladenosine, and L-tryptophan	FT-IR, NMR. GC-MS	Nikalje et al., 2017
11	*Suaeda salsa*	α-linolenic acid, quercetin, different organic acids, amino acids, nucleotide and its derivates	UPLC	Li and Song 2019
12	*Suaeda salsa and Salicornia europaea*	Glucosamine, maltose, D-(+)-sucrose, D-(+)-glucose, 2-propenyl (sinigrin), fructose 1-phosphate	LC-ESI-MS/MS, ESI-Q TRAP-Ms/Ms,	Wang et al., 2021

Source: Adapted and modified from Nikalje et al. (2018)

1.4.1 AMINO ACID AND ORGANIC ACID METABOLITE PROFILING

Amino acids such as proline, tyrosine, alanine, cysteine, arginine, glycine, amides and non -protein amino acids such as c-aminobutyric acid, pipecolic acid, citrulline and ornithine are accumulated in halophytes under high saline conditions. 55% of total free amino acids such as cysteine, arginine and methionine decreased when exposed to salinity stress whereas proline accumulation alleviated during salinity stress (Panda et al., 2021; Benjamin et al., 2019). Proline is a low molecular weight chaperone which accumulates during salt stress (Kumar et al., 2003). It reduces the inhibitory effects of sodium ions on various enzymatic activities. It is also involved in maintaining the NADP+/NADPH levels providing a sink for excess reductants to maintain the photosynthetic and respiratory process. This amino acid is predominantly synthesized from glutamate (primary precursor in stressed plants) using two important enzymes such as pyrroline 5 carboxylate synthetase (P5CS) and pyrroline 5 carboxylate reductase (P5CR). In some cases, Proline analogs (Munir et al., 2021; Furlan et al., 2020) are also reported to be synthesized such as N-methyl proline, thiazolidine-4-carboxylic acid (T4C).

Such analogs help in regulation and compartmentalization, thus aiding in the ability of plants to survive during salinity stress. A positive correlation between proline accumulation and salt stress was shown by Slama et al. (2008) in the obligatory halophyte *Sesuvium portulacastrum*, though the levels of soluble sugars and Na^+ remained unchanged. Similarly, the proline concentration was doubled when exposed to salt stress in the Alfalfa roots as compared with sensitive species (Petrusa and Winicov, 1997). It performs the following functions (Ashraf and Foolad, 2007; Lee et al., 2008; Babiychuk et al., 1995):

- Osmolyte to provide adjustment during osmosis
- Stabilizing membranes and proteins
- Ion homeostasis
- Scavengers for free radicals
- Sustaining $NADP^+$/NADPH ratios at required levels

Proline usually accumulates in the cytosol under saline conditions and is correlated with osmotic adjustment to improve the adaptation of a plant to saline conditions (Vinocur and Altman, 2005). High proline levels in the halophyte plant *Pancratium maritimum* increase the stabilization of detoxifying enzymes and the stimulation of stress proteins involved in protection (Khedr et al., 2003). Another example of proline accumulation in *Portulaca oleracea* L. plants may represent an adaptive feature to protect the plant's water balance in short-term saline conditions persuaded by osmotic stress (Yazici et al., 2007). However, the increased proline was not enough to maintain the water balance in this plant at the next long-term exposure to salinity stress.

Using transgenic technology proline production was enhanced by overexpressing the P5CS gene such as in transgenic tobacco, by 10- to 18-folds and provided better tolerance under salt stress (Kishor et al., 2005). Likewise, in *Arabidopsis* also, the overexpression of an antisense proline dehydrogenase cDNA resulted in the accumulation of the osmoprotectant which correlated with high salt tolerance (Nanjo et al., 2003). Unlike proline, amides accumulate to a lesser extent in salt-tolerant plants (Mansour et al., 2005), but the concentration of asparagine is found to increase in response to stress (Fougere et al., 1991).

Tyrosine is also reported to be abundant in certain halophytes which are produced as an intermediate in the shikimate pathway called prephenate. It is oxidatively decarboxylated followed by a transaminated process using glutamate as a nitrogen source and yields tyrosine and ketoglutarate amino acids.

These aromatic amino acids are crucial in photosynthesis (PSII) by acting as an electron donor in the reduction of oxidized chlorophyll. Along with amino acids, organic acids, such as citric, malic, malonic, succinic, fumaric, glycolic, and aconitic acids are also found to be important metabolites synthesized in halophytes (Panda et al., 2021; Arbelet-Bonnin et al., 2020). They are either found to be elevated or decreased in response to salinity.

1.4.2 QUATERNARY AMMONIUM COMPOUNDS (QACS)

These are an important class of osmoprotectants and the most common type of QAC is GB found in plants under salinity stress. It also includes β-alanine betaine, proline betaine, choline O-sulfate, etc. as described by Rhodes and Hanson (1993), Chen and Murata (2008). Betaine serves as an organic osmolyte which is synthesized from choline or glycine via two distinct pathways, that is, dehydrogenation of choline or N-methylation of glycine. Ferredoxin-dependent choline mono-oxygenase (CMO) mediates the first oxidation and the second oxidation is preceded by NAD-linked betaine aldehyde dehydrogenase (BADH) (Rathinasabapathi et al., 1997). Both plant enzymes are chloroplastic and have been used to engineer plants that lack GB.

GB is well documented to be one of the most abundant metabolites accumulated in halophytic species. The major roles played by GB are as follows:

- Stabilize the protein quaternary structure and association of extrinsic PSII complexes
- Reduce lipid peroxidation
- Osmotic adjustment (in the chloroplast), protectant (thylakoid membrane), and maintaining the photosynthetic machinery in the active state.

In plants belonging to Plumbaginaceae, well known for their stress tolerance, other QACs are preferred over GB like choline o-sulfate and betaines of alanine, proline nature, using the common substrate choline under different stress conditions. For example, choline-o-sulfate is suggested to be involved in sulfate detoxification along with its role as an osmoprotectant. This is explained because of the secretion of chloride ions from salt glands and not sulfate ions. The presence of sulfate ions leads to sulfate toxication but the conjugation of sulfate with choline such as choline-o-sulfate, helps in its removal by converting into a useful compatible solute (Hanson and Burnet, 1994). Also, β-alanine betaine has also been preferred over GB in the course of evolution. The reason might be the independence from choline availability in the synthesis of this QAC. Moreover, the synthesis of β-alanine betaine does not require oxygen which can be used to be suitable for osmoprotection under oxygen-depleted conditions like saline and hypoxic conditions (Rathinasabapathi et al., 2000). Proline betaine, also known as stachydrine, is found to be accumulated in minimal amounts in a few salt-tolerant plant species.

In developing transgenic plants by transferring bacterial choline oxidase cDNA, the betaine level was improved in *Arabidopsis* plants which resulted in enhanced seed yield production (Waditee et al., 2005). Similarly, choline-fed transgenic plants such as *Brassica napus* and *Nicotiana tabacum* synthesized substantially more amount of GB as compared with normal plants. The results indicated that the plants require an endogenous amount of choline for the synthesis of GB in minimal amounts under stress (Sairam and Tyagi, 2004).

1.4.3 CARBOHYDRATE METABOLISM

Accumulation of sugars is linked with stress-related mechanisms as it increases cellular osmolarity (Parida et al., 2002). The major roles played by the sugar metabolites are osmoprotection and scavenging ROS. Higher levels of the common metabolites found in plants during salt stress are sucrose, fructose, and glucose, which are stored even when there is a significant decrease in the net CO^2 assimilation rate (Murakeözy et al., 2003). An increase in reducing and nonreducing sugars along with polyphenol levels have been reported in the leaves of *Bruguiera parviflora* (Parida et al., 2002) though there was a decrease in starch content. Another metabolite named trehalose, a disaccharide molecule, is also observed to be accumulated under water deficit by providing protection to membranes and reducing aggregation of denatured proteins (Singer and Lindquist, 1998). According to Yamada et al. (2005), trehalose has a repressive effect on cells undergoing cell death.

Sugar alcohols: Also called polyols (polyhydric alcohols). The most common polyols in plants include acyclic forms, such as mannitol, glycerol, sorbitol, and cyclic forms ononitol and pinitol. During stress conditions, polyhydric alcohols, such as pinitol, myo-inositol, mannitol, or inositol production are enhanced for the existence of the plants (Bohnert et al., 1995). The various actions performed by polyols are:

- They act as compatible solutes and remove free radicals.

- They protect the cellular structures by interacting with enzymes and play a key role in signaling mechanisms (Valluru and Van den Ende, 2011).
- They overcome osmotic interferences by accumulating in the cytoplasm.

Mannitol, synthesized by the enzyme mannose-6-phosphate reductase (M6PR) (Zhifang and Loescher, 2003), shows enhanced production when plants are exposed to low water potential. The accumulation can be prevented by inhibiting metabolic pathways (Stoop et al. 1996). Transgenic studies of *Arabidopsis* have shown an improved level of mannitol accumulation among plants under stress (Thomas et al., 1995). It improves tolerance through the scavenging of hydroxyl radicals (OH^-) and stabilization of macromolecular structures. However, Abebe et al. (2003) observed negligible mannitol accumulation and its minimal effect on osmotic adjustment as compared with other carbohydrates during salt stress. These results demonstrate the major role of M6PR in developing salt-tolerant transgenic plants by manipulating mannitol biosynthesis (Zhifang and Loescher, 2003). Myo-inositol, a product of the enzyme MIPS (1-myo-inositol 1-phosphate synthase), plays a key role in the movement of sodium ions, and is found to be a substrate for the production of pinitol. The genes for the MIPS and IMT (inositol O-methyltransferase) which are involved in pinitol synthesis, have been characterized in certain halophytic species (Sengupta and Majumder, 2010).

1.4.4 POLYAMINES

These are polyvalent compounds with two or more amino groups which are majorly produced during osmotic stress, low pH, nutrient deficiency, etc. (Sairam and Tyagi, 2004). They play a crucial role in signaling mechanisms and ROS homeostasis. The common forms of polyamines reported during salt stress are putrescine, spermidine, and spermine. The conjugated polyamine accumulation assists as a scavenger for free radicals thus activating antioxidant enzyme activity indirectly. They are reported to be essential for efficient vacuolar sequestration of sodium ions. They efficiently block vacuolar ion channels preventing back leakage of sodium into cytosol. Based on their biological roles, Polyamines are divided into two groups (Bueno and Cordovilla, 2019):

i. The first group includes putrescine and cadaverine involved in cell elongation and root formation (similar to auxins and gibberellins)

ii. The second group comprises spermidine and spermine which regulate cell division and plant senescence like cytokinins.

Polyamines are polycations which have the ability to bind to nucleic acids and phospholipids at neutral pH, thereby providing stability of these macromolecules. It is also reported that these polyamines play distinguishing roles in promoting cell division during embryogenesis and delaying senescence (Kuznetsov and Shevyakova, 2007). However, their role in maintaining membrane stability has been questioned and the contribution of polyamines to osmotic adjustment appears to be small compared with other nitrogenous compounds.

Polyamines and the activities of their major metabolic enzymes were studied in *Brassica campestris* in response to stress (ElSayed et al., 2022). It was observed that if stress was prolonged there were minimal changes in the level of activities of arginine, ornithine decarboxylases, and polyamine oxidase enzymes, whereas stress of short periods caused a drastic increase in the levels of enzyme activities. Very few studies reported the accumulation of polyamines in salt-tolerant species or cultivars thus confirming the accumulation of polyamine as a non-specific response to salt stress.

1.4.5 ATP BIOSYNTHESIS METABOLITES

Since ATP is an important energy source for several biochemical pathways, thus metabolites associated with ATP generation are reported to be involved in regulatory mechanisms in halophytic species (Kumari et al., 2015). It is well documented in the literature about the enhanced expression of adenine, adenosine, and adenosine-5-monophosphate requiring energy sources under salinity stress. Enhancements of pentose phosphate pathway metabolites to produce large amounts of ATP in halophytes is observed to be maintained during high salt concentrations.

1.4.6 METABOLITES OF CELLULAR RESPIRATION

Metabolites involved in the cellular respiration process, such as photosystem I and II, carbon fixation, ATP generating system, and electron transport are severely hampered due to abiotic stresses (Nouri et al., 2015). Under salinity stress, the metabolites of tricarboxylic acid cycle are used up for the production of higher amounts of proline, tyrosine amino acids. Sobhanian et al. (2010) described the decrease in the expression of citrate, aconitate, 2-oxoglutarate, succinate, and fumarate levels in the citric acid cycle. Similar trend was observed with Calvin cycle metabolites (Ribose-5-phosphate, Ribulose-5-phosphate and glucose-6-phosphate) in the declining state, which decreased the photosynthetic rate and thereby increased the energy demand. Rubisco, a rate-limiting enzyme in the Calvin cycle, was reported to be present in abundance during salt stress. The level of Rubisco large subunit (LSU) and Rubisco small subunit (SSU) showed an increase in salt-induced state in *Physcomitrella patens* (Wang et al., 2008b) but a declining trend was observed in *Aeluropus lagopoides* (Sobhanian et al., 2010). A decrease in pyruvate, a metabolite of glycolytic pathway was also observed as a demand for proline increase in salt stress.

The expression of chlorophyll a/b binding (CAB) protein, ATP synthase, and oxidoreductases were enhanced when subjected to salt stress (Saibo et al., 2009; Nouri et al., 2015). It is also found that due to the conversion of Chlorophyll b to Chlorophyll a pigment, a marked increase in Chl a level was noted in a few salt-tolerant species (Akram and Ashraf, 2011; Ashraf and Harris, 2013). Likewise, Muchate et al. (2016) estimated the chlorophyll content and found a significantly higher content of chlorophyll a and carotenoids in *Sesuvium* plants when exposed to salinity stress.

1.4.7 PHOTOSYNTHETIC MACHINERY METABOLITES

Photosynthetic proteins are reported to be expressed differentially when plants are exposed to salt stress. Studies have reported that oxygen evolving enhancer protein (OEE1, OEE2, OEE3) plays an important role in light-induced oxidation of water in Photosystem II (PSII) of plants (Wungrampha et al., 2018; Sugihara et al., 2000). Tao et al. (2015) reported the effect of salt stress on stomatal and metabolic changes in perennial ryegrass genotypes. In *Bruguiera gymnorrhiza*, protein OEE2 is overexpressed in response to salt stress which is required to repair the injury of PSII complexes (Tada and Kashimura, 2009). But on the contrary, in *Puccinellia tenuiflora* (Yu et al., 2011), OEE1 expression was reported to be underexpressed. Parallelly, thylakoid proteins TL 19, 18.3 are reported to decrease in *Thellungiella halophila* (Pang et al., 2010) which indicates that salinity causes damage to thylakoid which might be responsible for reducing the efficiency of overall photosynthesis (Niewiadomska and Wiciarz, 2015).

1.4.8 ANTIOXIDANTS

Enzymatic and non-enzymatic antioxidant reactions play an important role in the detoxification of ROS produced during salinity stress. Surówka and Hura (2020), Akyol et al. (2022) mentioned the positive correlation of salinity tolerance with the activity of antioxidant enzymes such as superoxide dismutase, catalase, APX, glutathione peroxidase, and gutathione reductase. Non-enzymatic metabolites, such as proline, polyamines, polyols, and GB are also accumulated and show antioxidant properties due to their ability to scavenge hydroxyl radicals (Ben Hamed et al., 2020). Antioxidative defense systems such as the ascorbate–glutathione pathway (Parida et al., 2004; Groß et al., 2013; Asada, 1999) involve a series of enzymes including MDHAR, glutathione transferases, APXs, and superoxide dismutases, which play important roles in protecting against salt-induced oxidative stress in higher plants. The genes controlling these enzymes are overexpressed and lead to an increase in NaCl tolerance when subjected to salinity along with remarkable improvement in plant antioxidant activity. In addition to these enzymes, some sulfhydryl-rich proteins called metallothioneins (MTs) are able to detoxify and maintain homeostasis by binding to heavy metals. Similarly, S-adenosylmethionine synthetase, glycosyltransferase is also shown to participate in salt tolerance by limiting oxidative stress and helping maintain osmotic balance (Kim et al., 2013; Nisar et al., 2021).

1.4.9 PHYTOHORMONES

The phytohormones, also called chemical messengers, play crucial roles in regulatory mechanisms during plant stress. Halophytic species were detected with changes in the levels of phytohormones, such as jasmonic acid (JA), gibberellin (GA), ethylene (ET), abscisic acid (ABA), and enzymes related to their biosynthesis (Abobatta, 2020; Szepesi, 2020). ABA phytohormone is expressed in abundance (Xiong and Zhu, 2002) and is a major internal signal induced during salt stress. The level of ABA though increased was due to water deficit rather than the salt effect. An increase in enzyme NCED (9-cis-epoxycarotenoid dioxygenase) level shows an increase in ABA biosynthesis which is well mentioned in the literature. It plays a crucial role in the salt stress mechanism by decreasing pH causing stomatal closure and induction of early transcription factors which bind to ABA-responsive promoter elements (ABRE) of genes. It is reported that the products of these genes are accumulated in high amounts conferring their role in high salt tolerance. Jasmonates which include methyl jasmonate (MeJA) and JA are important growth regulators activating innate plant defense mechanisms in response to injury caused by insects/pathogens and other environmental stresses including salinity (Creelman and Mullet, 1997). Similarly, the study of the interaction of salt stress with gaseous hormone ethylene and the understanding of its signaling pathway is reported by Ryu and Cho (2015). On the similar lines, *Cakile maritima*, and *Thellungiella salsuginea* halophytes showed noteworthy accumulation of acetyl coenzyme A (ACC enzyme, ethylene precursor) in different parts of the plant when compared with the glycophytes (Ellouzi et al., 2014).

1.4.10 IONOMICS STUDIES

The study of the composition of minerals and trace elements at the cellular level of an organism called Ionomic studies (El-Esawi et al., 2020) involves the analysis of changes in mineral composition in response to physiological stimuli such as abiotic stress. It is a powerful tool for functional analysis of genes and their networks that control the mineral level in different physiological processes. All plant ionic profile information is stored in the database Purdue Ionomic (Mukherjee et al., 2022) which helps in the following ways:

- Understanding the composition of the inorganic ions and their continuous monitoring in different regions of plant organs/tissues/cell types/single cells affected by abiotic stress.
- Providing a framework for various biological pathways involved in regulating the ionic concentration.
- Molecular analysis to uncover the mechanism of various plant responses to harsh environmental conditions.

1.5 SECONDARY METABOLITES

Halophytes synthesize metabolites and a wide range of secondary metabolites to maintain the normal metabolism. Different halophytes synthesize these metabolites which increase in response to abiotic stress. These compounds include alkaloids, flavonoids, phenolics, etc. which act as antioxidants and are involved in scavenging toxic ROS (Slama et al., 2015). These halophytes are termed biofactories of secondary metabolites as there are several-fold increase in antioxidant level (Bose et al., 2014). Through metabolomic studies, it is revealed that plants respond differently to salinity stress such as alteration in plant growth and reduction in cytosolic and vacuolar volume via cellular dehydration with an increase in the concentration of secondary metabolites significantly. Salinity stress triggers the accumulation of osmolytes like proline and GB as mentioned above along with stimulating the production of ROS. These primary metabolites (precursors of secondary metabolites) affect the functional stability of biological membranes, nutrient balance, and redox homeostasis, which in turn affects the secondary metabolites (Slama et al., 2015). These changes act as elicitors for the enhanced production of antioxidant phenolic compounds, alkaloids, and tannins (Sytar et al., 2018). There is an increase in polyphenol content under salt stress such as *Cakile maritima* (Asteraceae) and *Mentha pulegium* (Labiatae) (Oueslati et al., 2010). Numerous studies have been carried out in the 21st Century to study the effect of salinity stress on plant growth and the production of secondary metabolites as summarized in Table 1.1. It is evident from the literature available that environmental factors have an effect on the development and biosynthesis of secondary metabolites.

1.5.1 ALKALOIDS

These are naturally occurring nitrogenous organic compounds, more than 3000 different types are identified from nearly about 4000 plant species. Several plant families are known to be rich in alkaloids such as Papaveraceae, Ranunculaceae, Solanaceae, and Amaryllidaceae. The effect of salinity in *Catharanthus roseus* (Hassan et al., 2021) showed elevated levels of desired alkaloid accumulation by using chitosan nanoparticles (CSNPs). These particles lowered the levels of malondialdehyde and H^2O^2, thereby retaining the membrane function and enhancing their tolerance to salinity. GC-MS analysis of *Aeluropus lagopoides* (Al-Moshaddak et al., 2020) conducted and found 50 compounds identified which included 8 alkaloids, out of which benzothiazole disulfane, cinnamido ethyl methoxyindole, and colchifoleine were abundant.

1.5.2 PHENOLICS

Phenolics are aromatic benzene ring compounds produced by plants mainly for protection against various types of stress. Phenolics are reported to play important roles in the development of plants especially in pigment and lignin biosynthesis (Bhattacharya et al., 2010). However, phenolic compounds show a positive correlation with antioxidant activity (Fu et al., 2011). For instance, it was shown that the polyphenol accumulation (especially anthocyanins) resulted in delimiting the damage caused by oxidative stress due to the generation of H^2O^2 (Hajlaoui et al., 2009; Lopes et al., 2021). It was suggested that antioxidative mechanisms in response to salt may result in a corresponding variability for the growth sustainability of some plants. The polyphenol content and antioxidant activity have been estimated to be high in leaves of the halophyte *Cakile maritima* (Ksouri et al., 2007). It was detected that the soil salinity caused a significant increase in concentrations of total chlorogenic acid in halophyte honeysuckle flower buds (Yan et al., 2016). Flavonoids are one of the largest classes of plant phenolics that play a role in plant defense mechanisms (Kondo et al., 1992). Isoflavonoids are derivatives of flavonone intermediate, for example, naringenin from legumes plays a crucial role in plant defense mechanisms and in supporting the configuration of nitrogen-fixing nodules by symbiotic rhizobia (Sreevidya et al., 2006). Posmyk et al. (2009) observed the role of flavonoids against ROS. Using LC-MS analysis, 18 phenolic compounds were identified in sea fennel extract (*Crithmum maritimum* L.) and the amount of total phenolic content exceeds 3% DW (Souid et al., 2021).

1.5.3 GLUCOSINOLATES

Glucosinolates are a large group of sulfur- and nitrogen-containing glucosides. Glucosinolates, a class of secondary metabolites, mainly found in Brassicaceae, coexist with an endogenous enzyme Myrosinase. These metabolites and enzymes though are stored separately in vacuoles but on tissue disruption, glucosinolates are released immediately and are hydrolyzed by Myrosinase enzyme. The products released chemically may be different due to the structure of the glucosinolate side chains, plant species, and reaction conditions (Mondal and Chakraborty et al., 2020; Redovnikovi et al., 2008). The mechanism of glucosinolate turnover regulation under salinity still is not completely clarified. However, a detailed analysis is required to understand the responses of glucosinolates at different physiological stages of the plant (in relation to glucosinolate—Myrosinase system) in response to salt stress. It was suggested that the increase in glucosinolates was related to the synthesis of osmoprotective compounds (Del Carmen Martínez-Ballesta et al., 2013; Liu et al., 2022).

1.5.4 *ESSENTIAL OILS*

Essential oils are secondary metabolites reported from plants which are volatile with strong odor. They can be synthesized by all plant organs (Castro et al., 2008). Terpenoids, which are a primary constituent of essential oils, represent a diverse group, derived from basic isoprene units (5-carbon nature) and are modified in several ways. In fact, almost all essential oils are extremely complex in composition, with respect to the large heterogeneity of chemical entities with functionalized groups, such as monoterpenoids, sesquiterpenoids, and phenylpropanoids. *Artemisia campestris* subsp. *maritima,* a halophytic plant was studied to characterize the essential oil (Rocha et al., 2021) along with antioxidant activities. *β*-pinene (54.5%) was reported to be present in high amounts followed by cadin-4-en-7-ol (9.5%), *Z*-*β*-ocimene (6.0%), and *Y*-terpinene (4.6%).

1.6 CONCLUSION

In response to salinity stress, a large number of proteins are induced but it must be acknowledged that in most cases, an integrated interpretation is still difficult. Few studies have shown that metabolomic studies allow better interpretation of physiological responses to environmental changes. A careful characterization of defense and resistance mechanisms in terms of metabolites is important for the understanding of various modifications in halophytic plants. Metabolomics will completely revolutionize our understanding of stress biology in plants by recognizing by-products of stress metabolism, signal molecules that are part of the acclimation response at different stages of development of plants. To further understand the intricacy of plant responses to environmental stresses we have to strengthen genomics and proteomic studies along with metabolite profiling in having deep insight knowledge about abiotic stresses.

Abbreviations used: Ultra performance liquid Chromatography electrospray quadrupole TOF mass spectrometry (UPLC-ESI-QTOF-MS); Inductively coupled plasma mass spectrometry (ICP-MS); High-performance liquid chromatography coupled with a photodiode array detector (HPLC-DAD); Gas chromatography mass spectrometry (GC-MS); High-performance liquid chromatography liquid chromatography mass spectrometry (HPLC-LC-MS); One-dimensional and two-dimensional nuclear magnetic resonance and mass spectrometry (1D and 2D NMR and MS); Fourier Transform InfraRed Spectroscopy (FTIR)

KEYWORDS

- **halophytes**
- **metabolomics**
- **primary**
- **secondary metabolites**

REFERENCES

Abebe, T.; Guenzi, A. C.; Martin, B.; Cushman, J. C. Tolerance of Mannitol-Accumulating Transgenic Wheat to Water Stress and Salinity. *Plant Physiol.* **2003,** *131* (4), 1748–1755.

Abobatta, W. F. Plant Responses and Tolerance to Extreme Salinity: Learning from Halophyte Tolerance to Extreme Salinity. In *Salt and Drought Stress Tolerance in Plants*; Springer: Cham, 2020; pp 177–210.

Aboelsoud, H. M.; Abdelrahman, M. A.; Kheir, A. M.; Eid, M. S.; Ammar, K. A.; Khalifa, T. H.; Scopa, A. Quantitative Estimation of Saline-Soil Amelioration Using Remote-Sensing Indices in Arid Land for Better Management. *Land* **2022,** *11* (7), 1041.

Akram, N. A.; Ashraf, M. Pattern of Accumulation of Inorganic Elements in Sunflower (Helianthus annuus L.) Plants Subjected to Salt Stress and Exogenous Application of 5-Aminolevulinic Acid. *Pak. J. Bot.* **2011**, *43* (1), 521–530.

Al-Akwaa, F.M.; Yunits, B.; Huang, S.; Alhajaji, H.; Garmire, L. X. Lilikoi: An R Package For Personalized Pathway-Based Classification Modeling Using Metabolomics Data. *Gigascience* **2018**, *7,* 136.

Al-Moshaddak A, Al Zahrani HS, El-Zohri M. Ecological and Phytochemical Evaluation of the Halophyte Aeluropus lagopoides Grows Wildly in Jeddah, Saudi Arabia. *Life Sci. J.* **2020,** *17* (4), 83–91.

Alnusairi, G. S.; Mazrou, Y. S.; Qari, S. H.; Elkelish, A. A.; Soliman, M. H.; Eweis, M.; ... Elnahhas, N. Exogenous Nitric Oxide Reinforces Photosynthetic Efficiency, Osmolyte, Mineral Uptake, Antioxidant, Expression of Stress-Responsive Genes and Ameliorates the Effects of Salinity Stress in Wheat. *Plants* **2021,** *10* (8), 1693.

Alonso, R.; Salavert, F.; Garcia-Garcia, F.; Carbonell-Caballero, J.; Bleda, M.; Garcia-Alonso, L.; Sanchis-Juan, A.; Perez-Gil, D.; Marin-Garcia, P.; Sanchez, R. Babelomics 5.0: Functional Interpretation for New Generations of Genomic Data. *Nucleic Acids Res.* **2015,** *43,* W117–W121.

Amin, E.; Moawad, A.; Hassan, H. Biologically-Guided Isolation of Leishmanicidal Secondary Metabolites from Euphorbia Peplus L. *Saudi Pharm. J.* **2017,** *25* (2), 236–240.

Aoki, K. F.; Kanehisa, M. Using the KEGG Database Resource. *Curr. Protoc. Bioinform.* **2005,** *11,* 1–12.

Aoki-Kinoshita, K. F. Overview of KEGG Applications to Omics-Related Research. *J. Pest. Sci.* **2006,** *31,* 296–299.

Arbelet-Bonnin, D.; Blasselle, C.; Palm, E. R.; Redwan, M.; Ponnaiah, M.; Laurenti, P.; ... Bouteau, F. Metabolism Regulation during Salt Exposure in the Halophyte Cakile Maritima. *Environ. Exp. Bot.* **2020,** *177,* 104075.
Arif, Y.; Singh, P.; Siddiqui, H.; Bajguz, A.; Hayat, S. Salinity Induced Physiological and Biochemical Changes in Plants: An Omic Approach Towards Salt Stress Tolerance. *Plant Physiol. Biochem.* **2020,** *156,* 64–77.
Asada, K. The Water-Water Cycle in Chloroplasts: Scavenging of Active Oxygens and Dissipation of Excess Photons. *Annu. Rev. Plant Physiol.* **1999,** *50* (1), 601–639.
Ashraf, M. F. M. R.; Foolad, M. R. Roles of Glycine Betaine and Proline in Improving Plant Abiotic Stress Resistance. *Environ. Exp. Bot.* **2007,** *59* (2), 206–216.
Ashraf, M. H. P. J. C.; Harris, P. J. Photosynthesis under Stressful Environments: An Overview. *Photosynthetica* **2013,** *51* (2), 163–190.
Babiychuk, E.; Kushnir, S.; Belles-Boix, E.; Van Montagu, M.; Inzé, D. Arabidopsis Thaliana NADPH Oxidoreductase Homologs Confer Tolerance of Yeasts toward the Thiol-Oxidizing Drug Diamide (∗). *J. Biol. Chem.* **1995,** *270* (44), 26224–26231.
Barreira, L.; Resek, E.; Rodrigues, M. J.; Rocha, M. I.; Pereira, H.; Bandarra, N.; ... Custódio, L. Halophytes: Gourmet Food with Nutritional Health Benefits?. *J. Food Compos. Anal.* 2017, *59,* 35–42.
Ben Hamed, K.; Dabbous, A.; Souid, A.; Abdelly, C. Antioxidant Molecules and Enzymes and Their Relevance to the Salt Adaptation of Halophytes. In *Handbook of Halophytes: From Molecules to Ecosystems Towards Biosaline Agriculture*; 2020; pp 1–17.
Benjamin, J. J.; Lucini, L.; Jothiramshekar, S.; Parida, A. Metabolomic Insights into the Mechanisms Underlying Tolerance to Salinity in Different Halophytes. *Plant Physiol. Biochem.* **2019,** *135,* 528–545.
Berendt, R. T.; Sperger, D. M.; Munson, E. J.; Isbester, P. K. Solid-State NMR Spectroscopy in Pharmaceutical Research and Analysis. *Trends Anal. Chem.* **2006,** *25* (10), 977–984.
Bertin, R. L.; Maltez, H. F.; De Gois, J. S.; Borges, D. L.; Borges, G. D. S. C.; Gonzaga, L. V.; Fett, R. Mineral Composition and Bioaccessibility in Sarcocornia ambigua Using ICP-MS. *J. Food Compos. Anal.* 2016, *47,* 45–51.
Bhargava, A.; Srivastava, S. Response of Amaranthus Sp. to Salinity Stress: A Review. In *Emerging Research in Alternative Crops*; 2020, pp 245–263.
Bhattacharya, P.; Samal, A. C.; Majumdar, J.; Santra, S. C. Uptake of Arsenic in Rice Plant Varieties Cultivated with Arsenic Rich Groundwater. *Environ. Asia* **2010,** *3* (2), 34–37.
Blaum, K. High-Accuracy Mass Spectrometry with Stored Ions. *Phys. Reports* **2006,** *425* (1), 1–78.
Bohnert, H. J.; Nelson, D. E.; Jensen, R. G. Adaptations to Environmental Stresses. *Plant Cell* **1995,** *7* (7), 1099.
Bose, J.; Rodrigo-Moreno, A.; Shabala, S. ROS Homeostasis in Halophytes in the Context of Salinity Stress Tolerance. *J. Exp. Bot.* **2014,** *65* (5), 1241–1257.
Boutaoui, N.; Zaiter, L.; Benayache, F.; Benayache, S.; Cacciagrano, F.; Cesa, S.; ... Locatelli, M. Atriplex Mollisdesf. Aerial Parts: Extraction Procedures, Secondary Metabolites and Color Analysis. *Molecules* **2018,** *23* (8), 1962.
Brosché, M.; Vinocur, B.; Alatalo, E. R.; Lamminmäki, A.; Teichmann, T.; Ottow, E. A.; ... Kangasjärvi, J. Gene Expression and Metabolite Profiling of Populus Euphratica Growing in the Negev Desert. *Genome Biol.* **2005,** *6* (12), 1–17.
Bueno, M.; Cordovilla, M. P. Polyamines in Halophytes. *Front. Plant Sci.* **2019,** *10,* 439.

Bueno, P. C. P.; Lopes, N. P. Metabolomics to Characterize Adaptive and Signaling Responses in Legume Crops under Abiotic Stresses. *ACS Omega* **2020,** *5,* 1752–1763.

Castro, B.; Citterico, M.; Kimura, S.; Stevens, D.M.; Wrzaczek, M.; Coaker, G. Stress-Induced Reactive Oxygen Species Compartmentalization, Perception and Signalling. *Nat. Plants* **2021,** *7,* 403–412.

Castro-Guerrero, N. A.; Rodríguez-Zavala, J. S.; Marín-Hernández, A.; Rodríguez-Enríquez, S.; Moreno-Sánchez, R. Enhanced Alternative Oxidase and Antioxidant Enzymes under Cd^{2+} Stress in Euglena. *J. Bioenerg. Biomembr.* **2008,** *40* (3), 227–235.

Causon, T. J.; Hann, S. Review of Sample Preparation Strategies for MS-based Metabolomic Studies in Industrial Biotechnology. *Anal. Chim.* **2016,** *938,* 18–32.

Chang, H. Y.; Chen, C. T.; Lih, T. M.; Lynn, K. S.; Juo, C. G.; Hsu, W. L.; Sung, T. Y.; Imet-Q: A User-Friendly Tool for Label-Free Metabolomics Quantitation Using Dynamic Peak-Width Determination. *PLoS ONE* **2016,** *11,* E0146112.

Chen, M.; Song, J.; Wang, B. S. Nacl Increases the Activity of the Plasma Membrane H+-Atpase in C3 Halophyte Suaeda Salsa Callus. *Acta Physiol. Plant.* **2010,** *32* (1), 27–36.

Chen, T. H.; Murata, N. Glycinebetaine: An Effective Protectant against Abiotic Stress in Plants. *Trends Plant Sci.* **2008,** *13* (9), 499–505.

Cheng, J.; Feng, J.; Sun, J.; Huang, Y.; Zhou, J.; Cen, K. Enhancing the Lipid Content of the Diatom Nitzschia sp. By 60Co-γ Irradiation Mutation and High-Salinity Domestication. *Energy* **2014,** *78,* 9–15.

Chong, J.; Xia, J. Metaboanalystr: An R Package for Flexible and Reproducible Analysis of Metabolomics Data. *Bioinformatics* **2018,** *34,* 4313–4314.

Creelman, R. A.; Mullet, J. E. Biosynthesis and Action of Jasmonates in Plants. *Annu. Rev. Plant Biol.* **1997,** *48* (1), 355–381.

Cushman, J. C. Crassulacean Acid Metabolism. A Plastic Photosynthetic Adaptation to Arid Environments. *Plant Physiol.* **2001,** *127* (4), 1439–1448.

Czarnocka, W.; Karpinski, S. Friend or Foe? Reactive Oxygen Species Production, Scavenging and Signaling in Plant Response to Environmental Stresses. *Free Radic. Biol. Med.* **2018,** *122,* 4–20.

Davidson, R. L.; Weber, R. J.; Liu, H.; Sharma-Oates, A.; Viant, M. R. Galaxy-M. A Galaxy Workflow for Processing and Analyzing Direct Infusion and Liquid Chromatography Mass Spectrometry-Based Metabolomics Data. *Gigascience* **2016,** *5,* 10.

Del Carmen Martínez-Ballesta, M.; Moreno, D. A.; Carvajal, M. The Physiological Importance of Glucosinolates on Plant Response to Abiotic Stress in Brassica. *Int. J. Mol. Sci.* **2013,** *14* (6), 11607–11625.

Dey, A. CRISPR/Cas Genome Editing to Optimize Pharmacologically Active Plant Natural Products. *Pharm. Res.* **2021,** *164,* 105359.

Dias, D. A.; Hill, C. B.; Jayasinghe, N. S.; Atieno, J.; Sutton, T.; Roessner, U. Quantitative Profiling of Polar Primary Metabolites of Two Chickpea Cultivars with Contrasting Responses to Salinity. *J. Chromatog. B* **2015,** *1000,* 1–13.

Draper, J.; Enot, D. P.; Parker, D.; Beckmann, M.; Snowdon, S.; Lin, W.; Zubair, H. Metabolite Signal Identification in Accurate Mass Metabolomics Data with MZedDB, An Interactive m/z Annotation Tool Utilising Predicted Ionisation Behaviour 'Rules'. *BMC Bioinform.* **2009,** *10,* 227.

El-Badri, A. M.; Batool, M.; Mohamed, I. A.; Khatab, A.; Sherif, A.; Wang, Z.; … Zhou, G. Modulation of Salinity Impact on Early Seedling Stage via Nano-Priming Application of Zinc Oxide on Rapeseed (Brassica napus L.). *Plant Physiol. Biochem.* **2021,** *166,* 376–392.

El-Esawi, M. A.; Sinha, R. P.; Chauhan, D. K.; Tripathi, D. K.; Pathak, J. Role of Ionomics in Plant Abiotic Stress Tolerance. In *Plant Life Under Changing Environment*; Academic Press, 2020; pp 835–860.

Ellouzi, H.; Hamed, K. B.; Hernández, I.; Cela, J.; Müller, M.; Magné, C.; … Munné-Bosch, S. A Comparative Study of the Early Osmotic, Ionic, Redox and Hormonal Signaling Response in Leaves and Roots of Two Halophytes and A Glycophyte to Salinity. *Planta* **2014,** *240* (6), 1299–1317.

Elsayed, A. I.; Mohamed, A. H.; Rafudeen, M. S.; Omar, A. A.; Awad, M. F.; Mansour, E. Polyamines Mitigate the Destructive Impacts of Salinity Stress by Enhancing Photosynthetic Capacity, Antioxidant Defense System and Upregulation of Calvin Cycle-Related Genes in Rapeseed (Brassica napus L.). *Saudi J. Biol. Sci.* **2022,** *29* (5), 3675–3686.

Ernest, B.; Gooding, J. R.; Campagna, S. R.; Saxton, A. M.; Voy, B. H. Metabr: An R Script for Linear Model Analysis of Quantitative Metabolomic Data. *BMC Res. Notes* **2012,** *5,* 596.

Ferreira, D.; Isca, V. M. S.; Leal, P.; Seca, A. M. L.; Silva, H.; Pereira, M. D.; … Pinto, D. Salicornia Ramosissima: Secondary Metabolites and Protective Effect Against Acute Testicular Toxicity. *Arabian J. Chem.* **2018,** *11* (1), 70–80.

Fiehn, O. Metabolomics—The Link Between Genotypes and Phenotypes. *Func. Genom.* **2002,** 155–171.

Fiehn, O.; Kopka, J.; Dörmann, P.; Altmann, T.; Trethewey, R. N.; Willmitzer, L. Metabolite Profiling for Plant Functional Genomics. *Nat. Biotechnol.* **2000,** *18* (11), 1157–1161.

Flowers, T. J.; Colmer, T. D. Salinity Tolerance in Halophytes. *New Phytol.* **2008,** 945–963.

Foito, A.; Stewart, D. Metabolomics: A High-Throughput Screen for Biochemical and Bioactivity Diversity in Plants and Crops. *Curr. Pharm.***2018,** *24,* 2043–2054.

Fougere, F.; Le Rudulier, D.; Streeter, J. G. Effects of Salt Stress On Amino Acid, Organic Acid, and Carbohydrate Composition of Roots, Bacteroids, and Cytosol of Alfalfa (Medicago Sativa L.). *Plant Physiol.* **1991,** *96* (4), 1228–1236.

Fu, Z. Y.; Zhang, Z. B.; Liu, Z. H.; Hu, X. J.; Xu, P. The Effects of Abiotic Stresses on the NADP-Dependent Malic Enzyme in the Leaves of the Hexaploid Wheat. *Biol. Plantarum* **2011,** *55* (1), 196–200.

Furlan, A. L.; Bianucci, E.; Giordano, W.; Castro, S.; Becker, D. F. Proline Metabolic Dynamics and Implications in Drought Tolerance of Peanut Plants. *Plant Physiol. Biochem.* **2020,** *151,* 566–578.

Ghnaya, T.; Nouairi, I.; Slama, I.; Messedi, D.; Grignon, C.; Abdelly, C.; Ghorbel, M. H. Cadmium Effects on Growth and Mineral Nutrition of Two Halophytes: Sesuvium portulacastrum and Mesembryanthemum Crystallinum. *J. Plant Physiol.* **2005,** *162* (10), 1133–1140.

Glenn, E. P.; Brown, J. J.; Blumwald, E. Salt Tolerance and Crop Potential of Halophytes. *Crit. Rev. Plant Sci.* **1999,** *18* (2), 227–255.

Groß, F.; Durner J. and Gaupels F. Nitric Oxide, Antioxidants and Prooxidants in Plant Defence Responses. *Front Plant Sci.***2013,** *4,* 419.

Hajlaoui, H.; Trabelsi, N.; Noumi, E.; Snoussi, M.; Fallah, H.; Ksouri, R.; Bakhrouf, A. Biological Activities of the Essential Oils and Methanol Extract of Tow Cultivated Mint Species (Mentha longifolia and Mentha pulegium) Used in the Tunisian Folkloric Medicine. *World J. Microbiol. Biotechnol.* **2009,** *25* (12), 2227–2238.

Hanson, A. D.; Burnet, M. Evolution and Metabolic Engineering of Osmoprotectant Accumulation in Higher Plants. In *Biochemical and Cellular Mechanisms of Stress Tolerance in Plants*; Springer: Berlin, Heidelberg, 1994; pp 291–302.

Hassan, F.; Ali, E.; Gaber, A.; Fetouh, M. I.; Mazrou, R. Ch Nanoparticles Effectively Combat Salinity Stress by Enhancing Antioxidant Activity and Alkaloid Biosynthesis in Catharanthus roseus (L.) G. Don. *Plant Physiol. Biochem.* **2021,** *162,* 291–300.

He, J.; You, X.; Qin, L. High Salinity Reduces Plant Growth and Photosynthetic Performance but Enhances Certain Nutritional Quality of C4 Halophyte Portulaca Oleracea L. Grown Hydroponically Under LED Lighting. *Front. Plant Sci.* **2021,** *12,* 651341.

Himabindu, Y.; Chakradhar, T.; Reddy, M.C.; Kanygin, A.; Redding, K. E.; Chandrasekhar, T. Salt-Tolerant Genes from Halophytes Are Potential Key Players of Salt Tolerance in Glycophytes. *Environ. Exp. Bot.* **2016,** *124,* 39–63.

Hollywood, K.; Brison, D. R.; Goodacre, R. Metabolomics: Current Technologies and Future Trends. *Proteomics* **2006,** *6* (17), 4716–4723.

Hong, Y.; Zhang, H.; Huang, L.; Li, D.; Song, F. Overexpression of A Stress-Responsive NAC Transcription Factor Gene ONAC022 Improves Drought and Salt Tolerance in Rice. *Front. Plant Sci.* **2016,** *7,* 4.

Hurmat, R. S.; Bansal, G. Does Abiotic Stresses Enhance the Production of Secondary Metabolites? A Review. *Pharma. Innov.* **2020,** *9* (1), 412–422.

Kessler, N.; Neuweger, H.; Bonte, A.; Langenkämper, G.; Niehaus, K.; Nattkemper, T. W.; Goesmann, A. Meltdb 2.0–Advances of the Metabolomics Software System. *Bioinformatics* **2013,** *29,* 2452–2459.

Kharabian-Masouleh, A.; Henry, R. J. Genome Editing can Boost Yield and Quality in Rice. *Trop Ag.* **2015,** 16–18.

Khedr, A. H. A.; Abbas, M. A.; Wahid, A. A. A.; Quick, W. P.; Abogadallah, G. M. Proline Induces the Expression of Salt-Stress-Responsive Proteins and may Improve the Adaptation of Pancratium maritimum L. to Salt-Stress. *J. Exp. Bot.* **2003,** *54* (392), 2553–2562.

Kim, H. K.; Verpoorte, R. Sample Preparation for Plant Metabolomics. *Phytochem. Anal. Int. J. Plant Chem. Biochem. Tech.* **2010,** *21,* 4–13.

Kim, J.; Buell, C. R. A Revolution in Plant Metabolism: Genome-Enabled Pathway Discovery. *Plant Physiol.* **2015,** *169* (3), 1532–1539.

Kim, S. G.; Kim, S. Y.; Park, C. M. A Membrane-Associated NAC Transcription Factor Regulates Salt-Responsive Flowering via FLOWERING LOCUS T in Arabidopsis. *Planta* **2007,** *226* (3), 647–654.

Kim, Y. H.; Khan, A. L.; Waqas, M.; Shim, J. K.; Kim, D. H.; Lee, K. Y.; Lee, I. J. Silicon Application to Rice Root Zone Influenced the Phytohormonal and Antioxidant Responses Under Salinity Stress. *J. Plant Growth Regul.* **2014,** *33* (2), 137–149.

Kishor, P. K.; Sangam, S.; Amrutha, R. N.; Laxmi, P. S.; Naidu, K. R.; Rao, K. S.; ... Sreenivasulu, N. Regulation of Proline Biosynthesis, Degradation, Uptake and Transport in Higher Plants: Its Implications in Plant Growth and Abiotic Stress Tolerance. *Curr. Sci.* **2005,** 424–438.

Kondo, F.; Ikai, Y.; Oka, H.; Okumura, M.; Ishikawa, N.; Harada, K.; ... Suzuki, M. Formation, Characterization, and Toxicity of The Glutathione and Cysteine Conjugates of Toxic Heptapeptide Microcystins. *Chem. Res. Toxicol.* **1992,** *5* (5), 591–596.

Ksouri, R.; Megdiche, W.; Debez, A.; Falleh, H.; Grignon, C.; Abdelly, C. Salinity Effects on Polyphenol Content and Antioxidant Activities in Leaves of The Halophyte Cakile Maritima. *Plant Physiol. Biochem.* **2007,** *45* (3–4), 244–249.

Kuhn, S.; Egert, B.; Neumann, S.; Steinbeck, C. Building Blocks for Automated Elucidation of Metabolites: Machine Learning Methods for NMR Prediction. *BMC Bioinform.* **2008,** *9,* 400.

Kumar, S. G.; Reddy, A. M.; Sudhakar, C. NaCl Effects on Proline Metabolism in Two High Yielding Genotypes of Mulberry (Morus alba L.) with Contrasting Salt Tolerance. *Plant Sci.* **2003,** *165* (6), 1245–1251.

Kumari, A.; Das, P.; Parida, A. K.; Agarwal, P. K. Proteomics, Metabolomics, and Ionomics Perspectives of Salinity Tolerance in Halophytes. *Front. Plant Sci.* **2015,** *6,* 537.

Kumari, A.; Parida, A. Metabolomics and Network Analysis Reveal the Potential Metabolites and Biological Pathways Involved in Salinity Tolerance of the Halophyte Salvadora Persica. *Environ. Exp. Bot.* **2018,** *148,* 85–99.

Kuznetsov, V. V.; Shevyakova, N. I. Polyamines and Stress Tolerance of Plants. *Plant Stress* **2007,** *1* (1), 50–71.

Lee, G.; Carrow, R. N.; Duncan, R. R. Identification of New Soluble Sugars Accumulated in a Halophytic Seashore Paspalum Ecotype under Salinity Stress. *Hortic. Environ. Biotechnol.* **2008,** *49* (1), 13–19.

Li, H.; Tang, X.; Yang, X. and Zhang, H. Comprehensive Transcriptome and Metabolome Profiling Reveal Metabolic Mechanisms of *Nitrariasibirica* Pall. To Salt Stress. *Sci. Rep.* **2021,** *11* (1), 1–19.

Liland, K. H. Multivariate Methods in Metabolomics–From Pre-Processing to Dimension Reduction and Statistical Analysis. *Trends Anal. Chem.* **2011,** *30,* 827–841.

Liu, X.; Yang, C.; Zhang, L.; Li, L.; Liu, S.; Yu, J.; … Wu, H. Metabolic Profiling of Cadmium-Induced Effects in One Pioneer Intertidal Halophyte Suaeda Salsa by NMR-based Metabolomics. *Ecotoxicology* **2011,** *20* (6), 1422–1431.

Liu, Z.; Wang, H.; Lv, J.; Luo, S.; Hu, L.; Wang, J.; … Yu, J. Effects of Plant Hormones, Metal Ions, Salinity, Sugar, and Chemicals Pollution on Glucosinolate Biosynthesis in Cruciferous Plant. *Front. Plant Sci.* **2022,** *13,* 856442–856442.

Li, Q.; Song, J. Analysis of Widely Targeted Metabolites of the Euhalophytesuaeda Salsa under Saline Conditions Provides New Insights into Salt Tolerance and Nutritional Value in Halophytic Species. *BMC Plant Biol.* **2019,** *19* (1), 1–11.

Llanes, A.; Arbona, V.; Gómez-Cadenas, A.; Luna, V. Metabolomic Profiling of the Halophyte Prosopis Strombulifera Shows Sodium Salt-specific Response. *Plant Physiol. Biochem.* **2016,** *108,* 145–157.

Lokhande, V. H.; Mulye, K.; Patkar, R.; Nikam, T. D.; Suprasanna, P. Biochemical and Physiological Adaptations of the Halophyte Sesuvium portulacastrum (L.) L.; (Aizoaceae) to Salinity. *Arch. Agron. Soil Sci.* **2013,** *59* (10), 1373–1391.

Lommen, A.; Kools, H.J. Metalign 3.0: Performance Enhancement by Efficient Use of Advances in Computer Hardware. *Metabolomics* **2012,** *8,* 719–726.

Lopes, M.; Sanches-Silva, A.; Castilho, M.; Cavaleiro, C.; Ramos, F. Halophytes as Source of Bioactive Phenolic Compounds and their Potential Applications. *Crit. Rev. Food Sci. Nutr.* **2021,** 1–24.

Mansour, M. M. F.; Salama, K. H. A.; Ali, F. Z. M.; Abou Hadid, A. F. Cell and Plant Responses to NaCl in Zea Mays L. Cultivars Differing in Salt Tolerance. *Gen. Appl. Plant Physiol.* **2005,** *31* (1–2), 29–41.

Mittler, R.; Vanderauwera, S.; Gollery, M.; Van Breusegem, F. Reactive Oxygen Gene Network of Plants. *Trends Plant Sci.* **2004,** *9* (10), 490–498.

Mock, A.; Warta, R.; Dettling, S.; Brors, B.; Jäger, D.; Herold-Mende, C. Metabodiff: An R Package for Differential Metabolomic Analysis. *Bioinformatics* **2018,** *34,* 3417–3418.
Mondal, S.; Chakraborty, K. Brassicaceae Plants Response and Tolerance to Salinity. In *the Plant Family Brassicaceae*; Springer: Singapore, 2020; pp 203–228.
Muchate, N. S.; Nikalje, G. C.; Rajurkar, N. S.; Suprasanna, P.; Nikam, T. D. Physiological Responses of the Halophyte Sesuvium portulacastrum to Salt Stress and their Relevance for Saline Soil Bio-Reclamation. *Flora* **2016,** *224,* 96–105.
Mukherjee, R.; Barwant, M.; Sinha, D. Ionomics vis À Vis Heavy Metals Stress and Amelioration. In *Heavy Metals in Plants Physiological to Molecular Approach*; CRC Press, 2022; pp 246–280.
Munir, N.; Hanif, M.; Dias, D. A.; Abideen, Z. The Role of Halophytic Nanoparticles Towards the Remediation of Degraded and Saline Agricultural Lands. *Environ. Sci. Pollut. Res.* **2021,** *28* (43), 60383–60405.
Munns, R.; Passioura, J. B.; Colmer, T. D.; Byrt, C.S. Osmotic Adjustment and Energy Limitations to Plant Growth in Saline Soil. *New Phytol.* **2019,** *225,* 1091–1096.
Murakeözy, É. P.; Nagy, Z.; Duhazé, C.; Bouchereau, A.; Tuba, Z. Seasonal Changes in the Levels of Compatible Osmolytes in Three Halophytic Species of Inland Saline Vegetation in Hungary. *J. Plant Physiol.* **2003,** *160* (4), 395–401.
Naikoo, M. I.; Kafeel, U.; Naushin, F.; Khan, F. A. Halophytes in India and their Role in Phytoremediation. In *Handbook of Halophytes: From Molecules to Ecosystems Towards Biosaline Agriculture*; 2020, 1–21.
Nanjo, T.; Fujita, M.; Seki, M.; Kato, T.; Tabata, S.; Shinozaki, K. Toxicity of Free Proline Revealed in an Arabidopsis T-DNA-Tagged Mutant Deficient in Proline Dehydrogenase. *Plant Cell Physiol.* **2003,** *44* (5), 541–548.
Ni, Z.; Liu, Z.; Huo, H.; Li, Z. L.; Nerry, F.; Wang, Q.; Li, X. Early Water Stress Detection Using Leaf-Level Measurements of Chlorophyll Fluorescence and Temperature Data. *Remote Sens.* **2015,** *7* (3), 3232–3249.
Niewiadomska, E.; Wiciarz, M. Adaptations of Chloroplastic Metabolism in Halophytic Plants. In *Progress in Botany*; Springer: Cham.; 2015, pp 177–193.
Nikalje, G. C.; Nikam, T.; Suprasanna, P. Looking at Halophytic Adaptation to High Salinity through Genomics Landscape. *Curr. Genom.* **2017,** *18* (6), 542–552.
Nikalje, G. C.; Srivastava, A. K.; Pandey, G. K.; Suprasanna, P. Halophytes in Biosaline Agriculture: Mechanism, Utilization, and Value Addition. *Land Degrad. Dev.* **2018,** *29* (4), 1081–1095.
Nisar, F.; Gul, B.; Aziz, I.; Hameed, A.; Egan, T. Increasing Salinity Leads to Differential Growth and H^2O^2 Homeostasis in Plants Produced from Heteromorphic Seeds of the Succulent Halophyte Arthrocnemum Indicum. *Plant Physiol. Biochem.* **2021,** *166,* 225–234.
Nouri, M. Z.; Moumeni, A.; Komatsu, S. Abiotic Stresses: Insight into Gene Regulation and Protein Expression in Photosynthetic Pathways of Plants. *Int. J. Mol. Sci.* **2015,** *16* (9), 20392–20416.
Obata, T.; Fernie, A. R. The Use of Metabolomics to Dissect Plant Responses to Abiotic Stresses. *Cell. Mol. Life Sci.* **2012,** *69* (19), 3225–3243.
Oucslati, S.; Karray-Bouraoui, N.; Attia, H.; Rabhi, M.; Ksouri, R.; Lachaal, M. Physiological and Antioxidant Responses of Mentha Pulegium (Pennyroyal) to Salt Stress. *Acta Physiol. Plant.* **2010,** *32* (2), 289–296.
Palama, T. L.; Fock, I.; Choi, Y. H.; Verpoorte, R.; Kodja, H. Biological Variation of Vanilla Planifolia Leaf Metabolome. *Phytochemistry* **2010,** *71* (5–6), 567–573.

Panda, A.; Rangani, J.; Parida, A. K. Cross Talk between ROS Homeostasis and Antioxidative Machinery Contributes to Salt Tolerance of the Xero-Halophyte Haloxylon salicornicum. *Environ. Exp. Bot.* **2019,** *166,* 103799.

Panda, A.; Rangani, J.; Parida, A.K. Unraveling Salt Responsive Metabolites and Metabolic Pathways Using Non-targeted Metabolomics Approach and Elucidation of Salt Tolerance Mechanisms in the Xero-Halophyte Haloxylon salicornicum. *Plant Physiol. Biochem.* **2020,** *158,* 284–296.

Pandey, P.; Ramegowda, V.; Senthil-Kumar, M. Shared and Unique Responses of Plants to Multiple Individual Stresses and Stress Combinations: Physiological and Molecular Mechanisms. *Front. Plant Sci.* **2015,** *6,* 723.

Pang, Q.; Zhang, A.; Zang, W.; Wei, L.; Yan, X. Integrated Proteomics and Metabolomics for Dissecting the Mechanism of Global Responses to Salt and Alkali Stress in Suaeda corniculata. *Plant Soil* **2016,** *402* (1), 379–394.

Parida, A. K.; Das, A. B.; Mohanty, P. Investigations on the Antioxidative Defence Responses to Nacl Stress in a Mangrove, Bruguiera Parviflora: Differential Regulations of Isoforms of Some Antioxidative Enzymes. *Plant Growth Regul.* **2004,** *42* (3), 213–226.

Parida, A.; Das, A. B.; Das, P. Nacl Stress Causes Changes in Photosynthetic Pigments, Proteins, and Other Metabolic Components in the Leaves of A True Mangrove, Bruguiera Parviflora, in Hydroponic Cultures. *J. Plant Biol.* **2002,** *45* (1), 28–36.

Patel, M. K.; Pandey, S.; Brahmbhatt, H. R.; Mishra, A.; Jha, B. Lipid Content and Fatty Acid Profile of Selected Halophytic Plants Reveal a Promising Source of Renewable Energy. *Biomass Bioenergy* **2019,** *124,* 25–32.

Patel, M.; Kumari, A.; Parida, A. K. Arsenic Tolerance Mechanisms in Plants and Potential Role of Arsenic Hyperaccumulating Plants for Phytoremediation of Arsenic-Contaminated Soil. In *Plant Ecophysiology and Adaptation under Climate Change: Mechanisms and Perspectives II*; Springer: Singapore, 2020; pp 137–162.

Patti, G. J.; Yanes, O.; Siuzdak, G. Metabolomics: the Apogee of the Omics Trilogy. *Nat. Rev. Mol. Cell Biol.* **2012,** *13* (4), 263–269.

Petrusa, L. M.; Winicov, I. Proline Status in Salt-Tolerant and Salt-Sensitive Alfalfa Cell Lines and Plants in Response to Nacl. *Plant Physiol. Biochem.* (*Paris)* **1997,** *35* (4), 303–310.

Posmyk, M. M.; Kontek, R.; Janas, K. M. Antioxidant Enzymes Activity and Phenolic Compounds Content in Red Cabbage Seedlings Exposed to Copper Stress. *Ecotoxicol. Environ. Safety* **2009,** *72* (2), 596–602.

Qin, S.; Bian, G. K.; Zhang, Y. J.; Xing, K.; Cao, C. L.; Liu, C. H.; ... Jiang, J. H. Modestobacter Roseus sp. Nov.; An Endophytic Actinomycete Isolated from the Coastal Halophyte Salicornia Europaea Linn.; and Emended Description of the Genus Modestobacter. *Int. J. Syst. Evol. Microbiol.* **2013,** *63* (Pt 6), 2197–2202.

Radojčićredovniković, I.; Glivetić, T.; Delonga, K.; Vorkapić-Furač, J. Glucosinolates and their Potential Role in Plant. *Periodicumbiologorum* **2008,** *110* (4), 297–309.

Rathinasabapathi, B.; Burnet, M.; Russell, B. L.; Gage, D. A.; Liao, P. C.; Nye, G. J.; ... Hanson, A. D. Choline Monooxygenase, an Unusual Iron-Sulfur Enzyme Catalyzing the First Step of Glycine Betaine Synthesis in Plants: Prosthetic Group Characterization and cDNA Cloning. *Proc. Natl Acad. Sci. 94* (7), 3454–3458, 1997.

Rathinasabapathi, B.; Sigua, C.; Ho, J.; Gage, D. A. Osmoprotectant β-Alanine Betaine Synthesis in the Plumbaginaceae: S-Adenosyl-l-Methionine Dependent N-Methylation of β-Alanine to Its Betaine is via N-Methyl and N, N-Dimethyl β-Alanines. *Physiol. Plant.* **2000,** *109* (3), 225–231.

Ravi, S.; Young, T.; Macinnis-Ng, C.; Nyugen, T. V.; Duxbury, M.; Alfaro, A. C.; Leuzinger, S. Untargeted Metabolomics in Halophytes: The Role of Different Metabolites in New Zealand Mangroves under Multi-Factorial Abiotic Stress Conditions. *Environ. Exp. Bot.* **2020,** *173,* 103993.

Razzaq, A.; Sadia, B.; Raza, A.; Khalid Hameed, M.; Saleem, F. Metabolomics: A Way Forward for Crop Improvement. *Metabolites* **2019,** *9* (12), 303.

Reich, M.; Liefeld, T.; Gould, J.; Lerner, J.; Tamayo, P.; Mesirov, J. P. Genepattern 2.0. *Nat. Genet.* **2006,** *38,* 500.

Ren, S.; Hinzman, A. A.; Kang, E. L.; Szczesniak, R. D.; Lu, L. J. Computational and Statistical Analysis of Metabolomics Data. *Metabolomics* **2015,** *11,* 1492–1513.

Rhodes, D.; Hanson, A. D. Quaternary Ammonium and Tertiary Sulfonium Compounds in Higher Plants. *Annu. Rev. Plant Biol.* **1993,** *44* (1), 357–384.

Rocha, M. I.; Gonçalves, M. J.; Cavaleiro, C.; Cruz, M. T.; Pereira, C.; Moreira, P.; ... Figueirinha, A. Chemical Characterization and Bioactive Potential of Artemisia Campestris L. Subsp. Maritima (DC) Arcang. Essential Oil and Hydrodistillation Residual Water. *J. Ethnopharmacol.* **2021,** *276,* 114146.

Roy, S.; Mishra, M.; Kaur, G.; Singh, S.; Rawat, N.; Singh, P.; ... Pareek, A. OsCyp2-P, an Auxin-Responsive Cyclophilin, Regulates Ca2+ Calmodulin Interaction for An Ion-Mediated Stress Response in Rice. *Physiol. Plant.* **2022,** *174* (2), E13631.

Ryu, H.; Cho, Y. G. Plant Hormones in Salt Stress Tolerance. *J. Plant Biol.* **2015,** *58* (3), 147–155.

Saddhe, A. A.; Jamdade, R.; Gairola, S. Recent Advances on Cellular Signaling Paradigm and Salt Stress Responsive Genes in Halophytes. *Handbook of Halophytes. From Molecules to Ecosystems towards Biosaline Agriculture*; 2020, 1–26.

Saibo, N. J.; Lourenço, T.; Oliveira, M. M. Transcription Factors and Regulation of Photosynthetic and Related Metabolism under Environmental Stresses. *Ann. Bot.* **2009,** *103* (4), 609–623.

Sairam, R. K.; Tyagi, A. Physiology and Molecular Biology of Salinity Stress Tolerance in Plants. *Curr. Sci.* **2004,** 407–421.

Sarkar, A. K.; Sadhukhan, S. Bioremediation of Salt-Affected Soil through Plant-Based Strategies. In *Advances in Bioremediation and Phytoremediation for Sustainable Soil Management*; **2022;** pp 81–100.

Sarker, U.; Oba, S. Augmentation of Leaf Color Parameters, Pigments, Vitamins, Phenolic Acids, Flavonoids and Antioxidant Activity in Selected Amaranthus Tricolor Under Salinity Stress. *Sci. Rep.* **2018,** *8* (1), 1–9.

Sengbusch, P. Halophytes. In *Botanik Online*; **2003,** pp 56–65.

Sengupta, S.; Majumder, A. L. Porteresia coarctata (Roxb.) Tateoka, a Wild Rice: A Potential Model for Studying Salt-Stress Biology in Rice. *Plant, Cell Environ.* **2010,** *33* (4), 526–542.

Senizza, B.; Zhang, L.; Rocchetti, G.; Zengin, G.; Ak, G.; Yıldıztugay, E.; ... Lucini, L. Metabolomic Profiling and Biological Properties of Six Limonium Species: Novel Perspectives for Nutraceutical Purposes. *Food Func.* **2021,** *12* (8), 3443–3454.

Shan, Q.; Wang, Y.; Li, J.; Gao, C. Genome Editing in Rice and Wheat Using the CRISPR/Cas System. *Nat. Protoc.* **2014,** *9* (10), 2395–2410.

Shulaev, V. Metabolomics Technology and Bioinformatics. *Brief. Bioinform.* **2006,** *7* (2), 128–139.

Shulaev, V.; Cortes, D.; Miller, G.; Mittler, R. Metabolomics for Plant Stress Response. *Physiol. Plant.* **2008,** *132* (2), 199–208.

Singer, M. A.; Lindquist, S. Multiple Effects of Trehalose on Protein Folding In Vitro and In Vivo. *Mol. Cell* **1998,** *1* (5), 639–648.
Slama, I.; Abdelly, C.; Bouchereau, A.; Flowers, T.; Savouré, A. Diversity, Distribution and Roles of Osmoprotective Compounds Accumulated in Halophytes under Abiotic Stress. *Ann. Bot.* **2015,** *115* (3), 433–447.
Slama, I.; Ghnaya, T.; Savouré, A.; Abdelly, C. Combined Effects of Long-Term Salinity and Soil Drying on Growth, Water Relations, Nutrient Status and Proline Accumulation of Sesuvium portulacastrum. *C R Biol* **2008,** *331* (6), 442–451.
Sobhanian, H.; Motamed, N.; Jazii, F. R.; Nakamura, T.; Komatsu, S. Salt Stress Induced Differential Proteome and Metabolome Response in the Shoots of Aeluropus lagopoides (Poaceae), A Halophyte C4 Plant. *J. Proteome Res.* **2010,** *9* (6), 2882–2897.
Song, Y.; Schreier, P. J.; Ramı'rez, D.; Hasija, T. Canonical Correlation Analysis of High-Dimensional Data with Very Small Sample Support. *Signal Process* **2016,** *128,* 449–458.
Souid, A.; Della Croce, C. M.; Frassinetti, S.; Gabriele, M.; Pozzo, L.; Ciardi, M.; ... Longo, V. Nutraceutical Potential of Leaf Hydro-Ethanolic Extract of the Edible Halophyte Crithmum maritimum L. *Molecules* **2021,** *26* (17), 5380.
Sreevidya, V. S.; Srinivasa Rao, C.; Sullia, S. B.; Ladha, J. K.; Reddy, P. M. Metabolic Engineering of Rice with Soybean Isoflavone Synthase for Promoting Nodulation Gene Expression in Rhizobia. *J. Exp. Bot.* **2006,** *57* (9), 1957–1969.
Stoop, J. M.; Williamson, J. D.; Pharr, D. M. Mannitol Metabolism in Plants: A Method for Coping with Stress. *Trends Plant Sci.* **1996,** *1* (5), 139–144.
Sugihara, K.; Hanagata, N.; Dubinsky, Z.; Baba, S.; Karube, I. Molecular Characterization of cDNA Encoding Oxygen Evolving Enhancer Protein 1 Increased by Salt Treatment in the Mangrove Bruguieragymnorrhiza. *Plant Cell Physiol.* **2000,** *41* (11), 1279–1285.
Sun, X.; Weckwerth, W. COVAIN: A Toolbox for Uni-And Multivariate Statistics, Time-Series and Correlation Network Analysis and Inverse Estimation of the Differential Jacobian from Metabolomics Covariance Data. *Metabolomics* **2012,** *8,* 81–93.
Surówka, E.; Hura, T. Osmoprotectants and Nonenzymatic Antioxidants in Halophytes. In *Handbook of Halophytes: From Molecules to Ecosystems Towards Biosaline Agriculture*; 2020, 1–31.
Sytar, O.; Hemmerich, I.; Zivcak, M.; Rauh, C.; Brestic, M. Comparative Analysis of Bioactive Phenolic Compounds Composition from 26 Medicinal Plants. *Saudi J. Biol. Sci.* **2018,** *25* (4), 631–641.
Szepesi, Á. Halotropism: Phytohormonal Aspects and Potential Applications. *Front. Plant Sci.* **2020,** *11,* 571025.
Tada, Y.; Kashimura, T. Proteomic Analysis of Salt-Responsive Proteins in the Mangrove Plant, Bruguiera gymnorhiza. *Plant Cell Physiol.* **2009,** *50* (3), 439–446.
Tang, X.; Zhang, H.; Shabala, S.; Li, H.; Yang, X.; Zhang, H. Tissue Tolerance Mechanisms Conferring Salinity Tolerance in a Halophytic Perennial Species Nitraria sibirica Pall. *Tree Physiol.* **2021,** *41* (7), 1264–1277.
Tao, J. J.; Chen, H. W.; Ma, B.; Zhang, W. K.; Chen, S. Y.; Zhang, J. S. the Role of Ethylene in Plants Under Salinity Stress. *Front. Plant Sci.* **2015,** *6,* 1059.
Tautenhahn, R.; Patti, G. J.; Rinehart, D.; Siuzdak, G. XCMS Online: A Web-based Platform to Process Untargeted Metabolomic Data. *Anal. Chem.* **2012,** *84,* 5035–5039.
Thomas, J. C.; Sepahi, M.; Arendall, B.; Bohnert, H. J. Enhancement of Seed Germination in High Salinity by Engineering Mannitol Expression in Arabidopsis Thaliana. *Plant, Cell Environ.* **1995,** *18* (7), 801–806.

Tran, D. Q.; Konishi, A.; Cushman, J. C.; Morokuma, M.; Toyota, M.; Agarie, S. Ion Accumulation and Expression of Ion Homeostasis-Related Genes Associated with Halophilism, NaCl-Promoted Growth in a Halophyte Mesembryanthemum Crystallinum L. *Plant Prod. Sci.* **2019,** *23,* 91–102.

Tsugawa, H.; Cajka, T.; Kind, T.; Ma, Y.; Higgins, B.; Ikeda, K.; Kanazawa, M.; Vandergheynst, J.; Fiehn, O.; Arita, M. MS-DIAL: Data-Independent MS/MS Deconvolution for Comprehensive Metabolome Analysis. *Nat. Methods* **2015,** *12,* 523.

Uslu, B.; Özkan, S. A. Determination of Lamivudine and Zidovudine in Binary Mixtures Using First Derivative Spectrophotometric, First Derivative of the Ratio-Spectra and High-Performance Liquid Chromatography–UV Methods. *Anal. Chim. Acta* **2002,** *466* (1), 175–185.

Valluru, R.; Van den Ende, W. Myo-Inositol and Beyond–Emerging Networks under Stress. *Plant Sci.* **2011,** *181* (4), 387–400.

Vinocur, B.; Altman, A. Recent Advances in Engineering Plant Tolerance to Abiotic Stress: Achievements and Limitations. *Curr. Opin. Biotechnol.* **2005,** *16* (2), 123–132.

Waditee, R.; Bhuiyan, M. N. H.; Rai, V.; Aoki, K.; Tanaka, Y.; Hibino, T.; ... Takabe, T. Genes for Direct Methylation of Glycine Provide High Levels of Glycine betaine and Abiotic-Stress Tolerance in Synechococcus and Arabidopsis. *Proc. Natl Acad. Sci. 102* (5), 1318–1323, 2005.

Wang, R.; Wang, X.; Liu, K.; Zhang, X.-J.; Zhang, L.-Y.; Fan, S.-J. Transcriptome Analysis of Halophyte Zoysia Macrostachya in Response to Salinity Stress. *Plants* **2020,** *9,* 458.

Wang, X.; Bai, J.; Wang, W.; Zhang, G.; Yin, S.; Wang, D. A Comparative Metabolomics Analysis of the Halophyte Suaeda Salsa and Salicornia Europaea. *Environ. Geochem. Health* **2021,** *43* (3), 1109–1122.

Wang, X.; Yang, P.; Gao, Q.; Liu, X.; Kuang, T.; Shen, S.; He, Y. Proteomic Analysis of the Response to High-Salinity Stress in Physcomitrella Patens. *Planta* **2008,** *228* (1), 167–177.

Waseem, M.; Nkurikiyimfura, O.; Niyitanga, S.; Nyimbo, W. J.; Shaheen, I.; Aslam, M. M. Role of Metabolomics and Next-Generation Sequencing for Sustainable Crop Production. In *Principles and Practices of OMICS and Genome Editing for Crop Improvement*; Springer: Cham.; 2022;pp 123–147.

Wink, M. Plant Breeding: Importance of Plant Secondary Metabolites for Protection against Pathogens and Herbivores. *Theor. Appl. Genet.* **1988,** *75* (2), 225–233.

Wishart, D. S. Current Progress in Computational Metabolomics. *Brief. Bioinform.* **2007,** *8,* 279–293.

Wu, H.; Li, Z. The Importance of Cl−Exclusion and Vacuolar Cl−Sequestration: Revisiting the Role of Cl−Transport in Plant Salt Tolerance. *Front. Plant Sci.* **2019,** *10,* 1418.

Wungrampha, S.; Joshi, R.; Singla-Pareek, S. L.; Pareek, A. Photosynthesis and Salinity: Are these Mutually Exclusive? *Photosynthetica* **2018,** *56* (1), 366–381.

Xia, X.-J.; Zhou, Y.-H.; Shi, K.; Zhou, J.; Foyer, C. H.; Yu, J.-Q. Interplay between Reactive Oxygen Species and Hormones in the Control of Plant Development and Stress Tolerance. *J. Exp. Bot. 66,* 2839–2856, 2015.

Xiong, L.; Zhu, J. K. Molecular and Genetic Aspects of Plant Responses to Osmotic Stress. *Plant, Cell Environ.* **2002,** *25* (2), 131–139.

Yamada, M.; Morishita, H.; Urano, K.; Shiozaki, N.; Yamaguchi-Shinozaki, K.; Shinozaki, K.; Yoshiba, Y. Effects of Free Proline Accumulation in Petunias Under Drought Stress. *J. Exp. Bot.* **2005,** *56* (417), 1975–1981.

Yan, K.; Xu, H.; Zhao, S.; Shan, J.; Chen, X. Saline Soil Desalination by Honeysuckle (Lonicera Japonica Thunb.) Depends On Salt Resistance Mechanism. *Ecol. Eng.* **2016,** *88,* 226–231.

Yazici, I.; Türkan, I.; Sekmen, A. H.; Demiral, T. Salinity Tolerance of Purslane (Portulaca Oleracea L.) is Achieved by Enhanced Antioxidative System, Lower Level of Lipid Peroxidation and Proline Accumulation. *Environ. Exp. Bot.* **2007,** *61* (1), 49–57.

Yu, J.; Chen, S.; Zhao, Q.; Wang, T.; Yang, C.; Diaz, C.; ... Dai, S. Physiological and Proteomic Analysis of Salinity Tolerance in Puccinellia tenuiflora. *J. Proteome Res.* **2011,** *10* (9), 3852–3870.

Yu, Z.; Duan, X.; Luo, L.; Dai, S.; Ding, Z.; Xia, G. How Plant Hormones Mediate Salt Stress Responses. *Trends Plant Sci.* **2020,** *25* (11), 1117–1130.

Zaid, A.; Wani, S. H. Reactive Oxygen Species Generation, Scavenging and Signaling in Plant Defense Responses. In *Bioactive Molecules in Plant Defense*; Springer Science and Business Media LLC: Berlin, Germany, 2019; pp 111–132.

Zhang, H.; Li, Y.; Zhu, J. K. Developing Naturally Stress-Resistant Crops for a Sustainable Agriculture. *Nat. Plants* **2018,** *4* (12), 989–996.

Zhao, C.; Zhang, H.; Song, C.; Zhu, J.-K.; Shabala, S. Mechanisms of Plant Responses and Adaptation to Soil Salinity. *Innovation*, 2020

Zhifang, G.; Loescher, W. H. Expression of A Celery Mannose 6-Phosphate Reductase in Arabidopsis Thaliana Enhances Salt Tolerance and Induces Biosynthesis of both Mannitol and A Glucosyl-Mannitol Dimer. *Plant Cell Environ.* **2003,** *26* (2), 275–283.

CHAPTER 2

Role of Halophyte–Microbes duo in Environmental Clean-Up

MINAL TRIVEDI[1], DIVYA RAUL[2], and GANESH C. NIKALJE[2]

[1]*Department of Biotechnology, B. K. Birla College of Arts, Science and Commerce (Autonomous), Kalyan, Maharashtra, India*

[2]*PG Department of Botany, Seva Sadan's R. K. Talreja College of Arts, Science and Commerce,, India*

ABSTRACT

The backbone of any country's economy is mainly strengthened by its industrial sector. However, due to industrialization, there has been huge contamination in the environment. In order to redevelop these polluted sites and restore the environment, it is important to remove the toxic compounds. Phytoremediation is a green, eco-friendly way to clean-up the polluted sites of the environment. Also, the use of salt-tolerant plants, that is, halophytes can aid to this problem. The halophytes not only resist the salt but they also accumulate and eliminate the salt in an effective way. The anatomical, morphological, and molecular adaptations help these plants to be a good option for environmental clean-up. In addition, soil microbes play a vital role in the development of halophytes and render some traits to the plants that can help to aid in microbial-assisted phytoremediation. This chapter focuses on the diversity of halophyte-associated microflora, sources of pollutants, types of pollution, applications of halophytes in phytoremediation, types of phytoremediation, and microbes-assisted phytoremediation potential in halophytes.

Physiology of Halophytes: Signaling, Omics, and Tolerance Mechanisms.
Nivas Manohar Desai & Ganesh Chandrakant Nikalje (Eds.)

2.1 INTRODUCTION

Halophytes refer to plants that grow in the saline habitat derived from "halos" meaning salt and "phyton" meaning plant. The plant is adapted to a saline environment to thrive in 8–10 dS/m of electrical conductivity of soil (Flowers, 2004). Depending on their levels to tolerate the salt, halophytes are classified into two groups: the first being obligate halophyte that can withstand, grow, and develop in high saline habitats just similar to seawater. The second class of halophytes includes facultative halophytes which are found in low saline habitats. They are generally found in low saline habitats but can grow in salt-devoid environments too (Flowers, 2004).

Halophytes have different adaptive mechanisms to withstand high salt concentrations, such as salt excretion, succulence and salt exclusion (Nikalje et al., 2018). Under this modification, halophytes secrete higher contents of salt ions onto the leaf surface. They have an epidermal structure called as salt glands which maintains cellular ion homeostasis (Roohi et al., 2011). These salt glands are surrounded by thick cuticle and have many plasmodesmata between the cells. Many halophytes show the presence of salt bladder. Plants belonging *to Limonium* genus (*L. aureum, L. bicolor, L. gmelinii, L. otolepis*) and Recretohalophyte, i.e. *Tamarix* genus tolerate salt by this mechanism (Rahman et al., 2021). Halophytes also adopt succulence mechanisms where they have thick leaves and stems, there is high turgor pressure, increased size of mesophyll cells. The salt tolerance is conferred by the accumulation of salts in leaves and stems without affecting the efficiency of photosynthesis. Examples are *Achras sapota, Sarcocornia fruticosa, Salsola drummondii*. In salt excretion mechanism, the plant root system-induced ultrafiltration works which exclude excessive salts via roots with the help of apoplastic barriers at the roots (Lokhande and Suprasanna, 2012; Nikalje et al., 2018).

2.2 DIVERSITY OF HALOPHYTE-ASSOCIATED MICROFLORA

Halophytes have the ability to grow in the salt concentration of more than 400 mM (Flowers et al., 2004; English and Colmer et al., 2011). This is possible because of the genes that limit the rate of salt uptake from the soil and affect the transport of salt through the plant. The plants are able to adjust the ionic and osmotic balance of cells in roots and shoots (Munns, 2005). The plant–microbiome interaction has been known for ages. The salt tolerance of the plants is associated with ecological processes within the rhizosphere and phyllosphere of the halophytes. Bacteria, archaea, and fungi are known to adapt to the change in external osmolarity. The various microbes isolated from halophytes are described in Table 2.1. One of the mechanisms for salt tolerance is the "salting out" processes wherein the cells adjust their intracellular environment by accumulating nontoxic organic osmolytes. These organisms avoid the accumulation of high salts in the cytoplasm to prevent water loss. Few properties in Archaea such as the presence of unique lipids and formation of tetra ether lipids, cause the cell membrane to become impermeable to the ions. The species from Halobacteriales contain S-layer on the cell wall which is approximately rich in 50% glycoprotein. In order to get stabilized, high concentrations of salts are needed, making these species tolerant to high salts (Oren, 2013).

Rhizobacteria are known to secrete exopolysaccharides which help to combat salt stress (Upadhyay et al., 2011). Apart from bacteria, fungi like *S. cerevisiae* and *Hortaea* sp. are known to tolerate high salt due to the presence of fatty acid modifying enzymes (Turk et al., 2004). Another method by which halophiles tolerate high salt concentration is by pumping ions out of the cell. Extra Na^+ is kept out of the cell by using Na^+ /H^+ antiporter, wherein in intracellular sodium ions are kept low. These antiporters are known in the *Synechococcus* sp. PCC7942, *A. amylolytica*, *Aphanothece halophytica*, *Alkalimonas amylolytica*. The fungal systems exclude ions if exposed to high salt concentration which is seen in *Aureobasidium pullulans* (Kogej et al., 2005) and some species of Saccharomyces (Kondrashov et al., 2002). Halophiles contain large amount of acidic amino acids like glutamic and aspartic acid which has highest water binding ability and the content of lysine and arginine which are basic in nature is low (Lanyi, 1974).

The prokaryotic community composition of halophytes has only rarely been investigated. Table 2.1 summarizes the microorganisms associated with halophytic plants and the role/function of microbes.

2.3 SOURCES OF POLLUTANTS

Environmental pollutants are those entities that directly or indirectly cause unfavorable alterations in the surrounding. These pollutants are the outcome of the industrialization and urban development where many natural resources are exploited. A pollutant can be of biological origin, physical in nature such as heat, sound, radiation, or chemical in nature such as metals, non-metals, or sediments which is intentionally introduced in the environment by humans.

TABLE 2.1 Summary of Different Microbes Associated with Halophytes.

Halophytes	Associated microbes	Role of microbes	References
Acacia sp.	*Bradyrhizobium* spp. *Ensifer* *Ochrobactrum* *Rhizobium* sp.	NA	Boukhatem et al. (2012) Thrall et al. (2009)
Aster tripolium	*Glomeromycota*	NA	Hildebrandt et al. (2001)
Avicennia marina	*P. maricaloris*	NA	El-Tarabily and Youssef (2010)
Salicornia bigelovii	*Bacillus* sp.	Diazotrophic, decomposition of organic matter	Rueda-Puente et al. (2010)
Salicornia brachiate	*Brevibacterium casei* *Haererehalobacter*	Diazotrophic	Gontia et al. (2011) Jha et al. (2012)
Salicornia europaea	*Actinobacteria, Firmicutes, Bacteroidetes Proteobacteria*	Helps to adapt to hypersaline conditions	Gao et al. (2022)
Seidlitzia rosmarinus	*Pseudomonas* *Staphylococcus* *Paenibacillus* *Brevibacterium*	Higher IAA content. Improves plant fitness under saline soils.	Shurigin et al. (2020)
Sesuvium portulacastrum	*Corynebacterium* *Pseudomonas* *P. maricaloris* *Avicennia marina* *Aeromonas*	Reduces the salinity stress effect on the activity of antioxidants	Anburaj et al. (2012)
Rhizophora apiculata	*Salinicola tamaricis* *Salinicola peritrichatus*	NA	Võ (2021)

**NA- Not available in the cited reference.

(Holdgate 1979). Depending on their nature, pollutants can be known to cause different types of pollution which are broadly classified as

1. Air pollution
2. Soil pollution
3. Water pollution

2.3.1 AIR POLLUTION

The major air pollutants as described by the World Health Organization (WHO) are carbon monoxide, ground-level ozone, lead, nitrogen oxide, particle pollution, and sulfur oxide. Air pollution has deleterious effects on the environment and also poses a threat to living beings. Some of such impacts include greenhouse effects, climatic change, and acid rain. Due to incomplete combustion of fossil fuels, carbon monoxide is released. The carbon monoxide has a greater affinity to hemoglobin as compared to oxygen which upon entering the blood may cause many health hazards, such as ischemia, cardiovascular disorders, nausea, headache, dizziness, and weakness. Carbon monoxide affects the greenhouse gases such as carbon dioxide and methane leading to climate changes indirectly (Grennfelt, 2020). When oxygen encounters high voltage, electric discharge leads to the formation of ozone gas. It is there in the stratosphere but can be there in the troposphere. Ozone is known to affect the plants and the associated microflora. It is a strong oxidant known to have antimicrobial properties. The reaction between volatile organic compounds (VOCs) and oxides of nitrogen emitted due to anthropogenic activities leads to the formation of ground-level ozone. High exposure to ozone leads to depletion in Vitamin C and E from the skin. It affects the tear ducts, the upper layer of the skin, and penetrates the lungs deeply (Ruidavets et al., 2005).

A commonly found heavy metal, lead is used in batteries, radiators, incinerators, and motor engines. The air is contaminated majorly by the lead released while mining and from the aircraft engines (Prüss-Üstün et al., 2004). Higher levels of lead are known to adversely affect plant growth. Lead can enter an organism via ingestion, inhalation, or skin absorption. It has shown neurological effects in humans which may cause disabilities in learning, hearing, and remembering. When inhaled in large amounts, it affects organs such as lungs, liver, bones, and even blood (Kumar et al., 2020). Another air pollutant which is emitted by automobiles is nitrogen oxide. This nitrogen oxide at a concentration above 2 ppm can have many deleterious effects on CD8+ cell, T-cells and NK cells that are known to provide immunity. It is a major cause of chronic lung disease. Apart from this, chronic exposure to nitrogen dioxide can lead to irritation of throat, eyes and nose (Glencross et al., 2020). Sulfur dioxide, another gas which is emitted from industrial activities leads to health hazards such as bronchitis, respiratory disorders, and may lead to cardiovascular diseases too. Upon entering the lungs, sulfur dioxide is converted to bisulfite which causes bronchoconstriction. The acid rains and acidification of soil are the consequence of sulfur dioxide emissions. In the atmosphere, particulate matter is formed due to a series of chemical reactions between pollutants. This includes particles of diameter 10 μm to less than 2.5 μm. Particulate contaminants include smoke, fly ash, smog, and soot, and they can also be biological, such as bacteria, fungi, viruses, pollens, allergens, and dust (Khan and Siddiqui, 2014). Air pollution not only harms the health of living beings but also impacts the environment. The acid rains that contain good amounts of sulfuric and nitric acids have the potential to acidify the soil and water, and damage the monuments (like the yellowing of Taj Mahal, India) (Bergin et al., 2015). Ozone from the stratosphere though is known to protect us from harmful UV rays, the ground-level ozone serves as a pollutant. Ozone penetrates through stomata leading to their closure and blocking the carbon dioxide exchange indirectly causing the reduction in photosynthesis (Grennfelt, 2020).

2.3.2 SOIL POLLUTION

The soil consists of inorganic and organic materials. The inorganic materials are obtained from the disintegration of various rocks leading to the availability of essential elements in the soil such as nitrogen, potassium, phosphorus, etc. These materials add bulk to the soil as well as serve as substrate for plant growth. The organic part of the soil includes animal and plant wastes like feces blood, skin, leaves, roots, etc. These are further processed as humus which is the natural soil fertilizer. Apart from this, soil is rich in microbes that aid in plant development by fixing nitrogen, solubilizing the phosphate and potassium, and synthesis of plant hormones. Hence, maintaining the soil health becomes very essential to be preserved (Chae and An, 2018). The soil gets contaminated due to industrial waste, household waste, and agrochemicals. The term "xenobiotic" has been derived from Greek origin where Xenos means roots and "Bios" means life. These xenobiotics that pollute the soil are known to be highly carcinogenic in nature. Heavy metals like arsenic, antimony, cadmium, chromium, lead, nickel, zinc also pollute the soils (Shaltami, 2014) and have an impact on human health (Table 2.2).

TABLE 2.2 Effect of Heavy Metals on human health.

Heavy metals	Impact on human health	Permissible limits (ppm)*	Reference
Arsenic	Causes malfunctioning of cell respiration and mitosis	0.05	Matta and Gjyli (2016)
Aluminium	Causes secondary hyperparathyroidism, osteomalacia, anemia	0.2	Ermias et al. (2013)
Cadmium	Absorbed more in the lungs. Causes morphological changes in kidneys, premature birth, and neurodegenerative diseases	0.005	Mudgal et al. (2010) Henson and Chedrese (2004) Branca et al. (2018)
Chromium	Leads to the formation of ulcers in the skin and it is carcinogenic	0.05	Matsumoto et al. (2006) Coetzee et al., (2020)
Iron	Causes gastrointestinal ulcerations, can initiate cancer	0.3	Albretsen (2006)
Lead	Increases reactive oxygen species and damage to cells and other biomolecules and causes mental retardation, birth defects, psychosis, and kidney damage	0.05	Jaishankar et al. (2014) Martin and Griswold (2009)
Mercury	It is carcinogenic, leads to tremors, memory problems, irritability, and changes in vision or hearing, affects renal functioning	0.001	Martin and Griswold (2009) Zhang et al. (2020)

*As per WHO (in drinking water)

The industries discharge their waste into water bodies wherein they encounter the soil also. These wastes include dioxins, chlorinated solvents, polychlorinated biphenyls, plasticizers. These wastes generate benzene and methylbenzene which are carcinogenic (Chen et al., 2020). For better development and yield of plants, different pesticides in the form of insecticides, herbicides, and fungicides are being used. These include triazines, carbamates, organophosphates, chlorinated hydrocarbons, etc. They affect the human immunity system and cause birth defects, and diseases of the central nervous system. Acid rains not only cause air pollution but also affect the soil. They change the soil chemistry which adversely affects the growth of plants and thus impacts the food chain (Chen et al., 2013). Another soil pollutant is electronic waste which contains many components. Upon disposal, these toxins may get leached (e.g., Salam and Varma, 2019). These soil contaminants change the soil chemistry which manifests the alteration in the metabolism of microbes. The persistence of DDT (pesticide) is harmful to avian leading to the weakening of egg shells. They are even known to affect the plant metabolism; hence the soil pollutant has harmful effects on the ecosystem too.

2.3.3 WATER POLLUTION

All organisms on this Earth use water for their survival. Pollution of water hence is threat to these life forms. Water pollutants can be classified as point source pollution where the source of pollution is known. The source of pollutants can be identified, for example, sewage treatment plants. The second classification of water pollutants includes non-point source pollution where the source of pollution cannot be identified, an example includes the entry of fertilizers and pesticides into water. The water gets contaminated due to poor sanitization wherein the fecal matter enters the water body. This contaminated water contains bacteria, viruses, and protozoans which leads to diseases such as cholera and diarrhea. Excessive use of nitrate fertilizers leads to leaching of nitrates in the water. The presence of nitrates in more than 10 ppm leads to blue baby syndrome. Similarly, pesticides and herbicides can also enter water bodies through run-offs disturbing the aquatic flora and fauna. These pesticides persist for long and are very hard to degrade (Chaudhary et al., 2017). These chemicals also contribute to eutrophication which encourages the growth of algae. As a result, the oxygen level in the water body depletes and makes it difficult for the survival of aquatic animals. Eutrophication is the excessive concentration of chemical nutrients in an environment to the extent that it encourages the dense growth of plant life and algae. Based on the degree of eutrophication, the effects are depletion of oxygen and extensive deterioration of water quality. As a result, it affects the survival of fish and other aquatic animals. Eutrophication is linked to an increased incidence of paralytic shellfish poisoning in humans, leading to death (Nazari-Sharabian et al., 2018). The entry of the water pollutant creates acidic lakes, depletes the oxygen level, and increases the temperature of water bodies. Due to this, the aquatic animals may move out to more adaptable zone or may get extinct creating a dead zone where the water is contaminated by pollutants (Simmonds and Isaac, 2007).

2.4 APPLICATIONS OF HALOPHYTES IN PHYTOREMEDIATION

2.4.1 PHYTOFILTRATION

It is a process through which plants absorb or adsorb heavy metals from the water streams. If the roots perform filtration it is known as rhizofiltration and if seedlings perform filtration, it is known as blastofiltration (Liang et al., 2017). In this process, the pollutants are adsorbed or absorbed from the water streams in order to reduce the passage of pollutants to underground water system (Mushtaq et al., 2021). Phytofiltration prevents the leaching of pollutants to the underground water. It is common in aquatic plants but it can be applied to terrestrial plants also. Microorganism associated with plants helps in the phytofiltration of contaminants from waste water. Plants are directly grown in the polluted water to check the phytofiltration ability of the plants (Usman et al., 2018). Duckweed halophytes are able to remove arsenic through phytofiltration. *Spirodela polyrhiza* and *Lemna gibba* shows the accumulation of As. A plant species should have rapid growth rate, it should be a high biomass species having high accumulation or tolerance of metals in it. Phytofiltration is a more effective alternative to traditional phytoremediation of metals in soil (Chen et al., 2013).

2.4.2 PHYTOEXTRACTION

It is a process of accumulation of toxic elements in plant tissues. It remediates heavy metals from soil and water having high salinity. The salt tolerance mechanism of halophytes enables them to accumulate high levels of toxic elements in them. Antioxidant systems present in halophytes make them tolerant towards toxic elements. Halophytes are considered ideal plants for phytoextraction of heavy metals from saline or non-saline soils. It performs the desalination of soil (Nikalje and Suprasanna, 2018). *Sesuvium portulacastrum* is an example of halophyte that sequestrate heavy metal and salt into the vacuole and salt glands. It removes the heavy metal from cytosol and performs the phytoextraction of heavy metal (Ayyappan et al., 2015).

The mechanism of phytoextraction involves the growth of plants which can tolerate the toxicity in contaminated soil. The plant absorbs the heavy metals from the soil and translocates them to the aerial parts of the plant. This reduces the target pollutant from the soil. After the accumulation of heavy metals in the plant part, it is harvested. These plant parts consist of a very high amount of heavy metal. Therefore, the harvested parts are incinerated and dumped. Incineration involves the release of energy that can be used in various processes. The accumulation of heavy metals in plants is higher than that of the amount of heavy metal present in the soil. Some plants can accumulate more than the desired amount of heavy metal. Due to hyperaccumulation, the metabolism of plants changes to tackle the stressful condition. This property of some halophytes can be studied further to understand the mechanism of tolerance (Ghori et al., 2016).

2.4.3 PHYTOSTABILIZATION

It is a technique in which plants stabilize contaminants like heavy metals in the soil instead of eliminating them from the soil. It primarily works on the principle of immobilization of toxic elements in the soil. This technique is also known as phytorestoration. It suppresses the vertical migration of contaminants into groundwater. It immobilizes contaminants by chemical fixation. Plants that perform phytostabilization are poor translocators of contaminants to the consumable plant parts present above the ground. It reduces the harmful effects caused by the consumption of plants by animals and humans. Halophytes like *Atriplex lentiformis and Atriplex undulate* are able to accumulate heavy metals in the root and maintain the concentration of metals like Cu, Pb, Cd, Zn, and Mn below the toxicity level in shoot tissues. They reduce the accumulation of contaminants in the aerial parts and perform phytostabilization (Eissa, 2014; Sruthi et al., 2016). The mechanism of phytostabilization includes accumulation of heavy metals in roots and reduces the translocation to the shoot by precipitation of metals (Eissa, 2014). It is mainly seen in areas with dense vegetation. Plants suitable for phytostabilization have dense root system which can restrict the accumulation of heavy metals near the roots and prevent transportation to aerial parts of the plants. Furthermore, studies are required to understand the uptake pattern of various halophytes in phytostabilization (Ghori et al., 2016).

2.4.4 PHYTODEGRADATION

In this technique, plants take up the contaminants and degrade them through metabolic reactions in order to perform the phytoremediation of soil (Newman and Reynolds, 2004). Different plant parts can remediate organic pollutants like pesticides by degradation or transformation. The plants may absorb pollutants and break them into less harmful, small compounds, and distribute them within plant tissue. The mangrove *Kandelia candel,* photodegrades/dissipates phenanthrene (Ph) and pyrene (Py) by 47.7% and 37.6% respectively in the rhizosphere (Lu et al., 2011).

2.4.5 PHYTOVOLATILIZATION

It is a technique for the conversion of pollutants into volatile form. Due to volatilization, pollutants are released into the air that reduces its concentration in the soil. Accumulation of heavy metals does not occur in the plant parts during phytovolatilization (Ghori et al., 2016). It occurs through transpiration.

It takes place for certain heavy metals and metalloids that are highly volatile. Mercury (Hg) and selenium are (Se) examples of some heavy metals which are eliminated from the soil using the method of phytovolatilization (Naikoo et al., 2021). In *Kosteletzkya pentacarpos*, soil salinity favors plants against Cd and Zn toxicity, and cytokinins help in increased Zn resistance (Zhou et al., 2019).

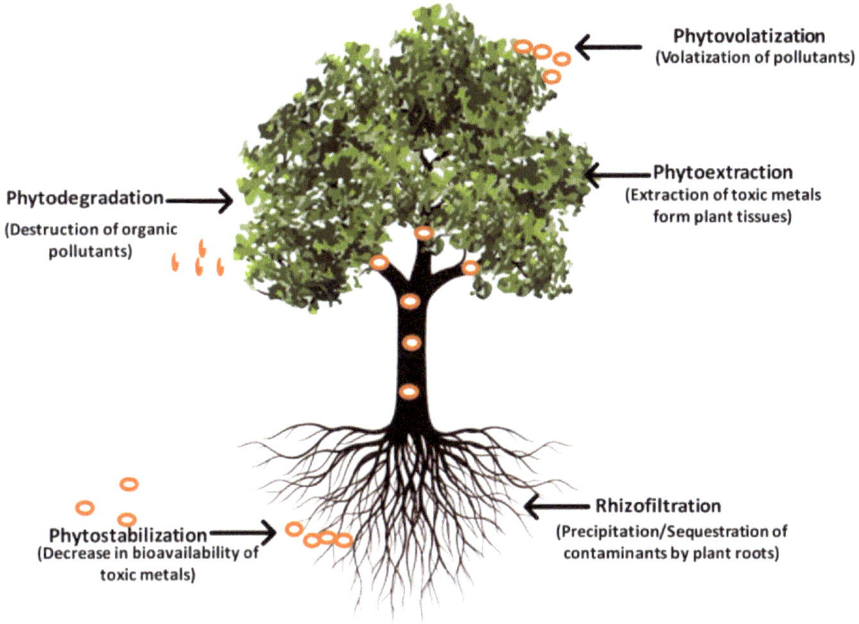

FIGURE 2.1 Types of phytoremediation.

TABLE 2.3 Summary of Different Halophytes Used in Phytoremediation of Heavy Metals.

Halophyte	Phytoremediation strategy	Plant part	Heavy metal decontaminated ($mg.kg^{-1}$)	References
Atriplex halimus	Phytoextraction	Roots	Cd (3174) and Zn (1423)	Lutts et al. (2004)
Atriplex tacamensis	Phytoextraction	Shoots and roots	Ar shoot (710) Ar root (8.3)	Vromman et al. (2011)
A. lagopoides	Phytoaccumulation	Shoots	Mn (62), Cr	Mujeeb et al. (2021)
A. macrostachyum	Phytoaccumulation	Shoots	Pb (1649), Zn (724)	Mujeeb et al. (2021)
Chenopodium quinoa	Phytoaccumulation	Leaves	Cd (59.76), Pb (44.63) and Ni (2.99)	Ghous et al. (2022)

TABLE 2.3 *(Continued)*

Halophyte	Phytoremediation strategy	Plant part	Heavy metal decontaminated (mg.kg^{-1})	References
Commelina communis	Phytostabilization	Roots	Cu (13,100)	He et al. (2021)
H. glomeratus	Phytoextraction	Roots	Ni (188.7), Cu (417.1), Zn (197.4)	Li Baochun et al. (2019)
Heliotropium bacciferum,	Phytoaccumulation	Shoot	Pb (11), Fe (200), Cu (9)	Mujeeb et al. (2020)
Halopyrum mucronatum	Phytoaccumulation	Shoot	Pb (8), Cr (9.5)	Mujeeb et al. (2020)
Juncus maritimus	Phytoextraction	belowground organs	Hg	Marques et al. (2011)
Moringa oleifera	Phytostabilization	Leaf	Cd	Howladar (2014)
Phragmites australis	Phytostabilization	Roots	Cd (90)	da Silva et al. (2014)
Portulaca oleracea	Phytoextraction	Roots and stems	Cr (Root- 4600, Shoot- 1067-1400)	Alyazouri et al. (2013)
Salicornia bigelovii	Phytovolatization	Leaf	Se	Huang et al. (2013)
Salicornia ramosissima	Phytostabilization	Roots	Cd (85.94)	Pedro et al. (2013)
Salsola imbricata	Phytoaccumulation	Shoot	Pb (4.5), Fe (180)	Mujeeb et al., (2020)
Sesuvium portulacastrum	Phytoaccumulation	Root, stem, and leaf	Cs (Root- 284.69, Stem-413.74, Leaves-536.1)	Nikalje et al. (2022)
Spartina alterniflora	Phytostabilization	Roots	Cu (1000)	Chai et al. (2014)
Spartina alterniflora	Phytostabilization	Roots	Cu (22) and Pb (20)	Nalla et al. (2012)
Suaeda salsa	Phytostabilization	Roots	Pb (4.2) and Zn (25.6)	Wu et al. (2013)
Triglochin maritima	Phytoextraction	belowground organs	Hg (0.184)	Castro et al. (2009)

2.5 MICROBES-ASSISTED PHYTOREMEDIATION

Microorganisms are known for influencing the process of phytoremediation. Microbes resist heavy metal toxicity by the synthesis of organic acids and polysaccharides which chelate metal ions, absorb metal ions within the cell wall, envelop, and sequester metal ions into vacuoles, transforming them into less toxic form (Gadd, 2010). They can be utilized in two ways: bioaugmentation and biostimulation. In bioaugmentation, natural or genetically modified microbes can be added to contaminated soil for the removal of toxic pollutants (Goswami et al., 2018). In biostimulation, the existing microflora of contaminated sites can be stimulated through the addition of electron donors and nutrients such as nitrogen and phosphorus for the degradation of toxic pollutants (Goswami et al., 2018). In a recent study, Lacalle et al. (2020) found that the addition of organic matter prior to bioaugmentation with a consortium of actinobacteria along with *Brassica napus* is highly effective for the transformation of lindane hexavalent chromium to trivalent chromium. Some fungi affect the bioavailability of certain heavy metals by modulating the composition of root exudates and soil properties (Bhantana et al., 2021). The AM fungi (Arbuscular mycorrhiza) help their host roots to increase the absorptive surface area which enhances water and nutrient uptake and bioavailability of metals (Göhre and Paszkowski, 2006). In addition, AM fungi produce phytohormones which facilitate phytoremediation and promote growth. For example, the inoculation of *Trifolium pratense* with AM fungi in Zn-contaminated soil resulted in higher yield and accumulation of Zn in plant roots (Chen et al., 2003). The interaction between microbes and plants in the rhizosphere region improves the phytoremediation of both organic and inorganic pollutants by synergistic effect (Rane et al., 2022). In a halophyte *Salicornia ramosissima*, bioaugmentation with consortium of *Vibrio neocaledonicus* SRT1, *Thalassospira australica* SRT8 and *Pseudarthrobacter oxydans* SRT15 resulted in the accumulation of arsenic (200 $mg.kg^{-1}$), cadmium (0.50 $mg.kg^{-1}$), copper (25 $mg.kg^{-1}$), nickel (25 $mg.kg^{-1}$), lead (300 $mg.kg^{-1}$), and zinc (300 $mg.kg^{-1}$) in roots. The bioaugmentation with microbial consortium enhanced the phytoremediation potential of *S. ramosissima* by increasing plant biomass and accumulation ability (Mesa-Main et al., 2020). The endophytes isolated from *Arthrocnemum meridionale* have improved its capacity to remediate Cd from contaminated soil (Navarro-Torre et al., 2017). The *Cyperus luzulae and C. laxus* showed a high accumulation of Hg in their roots (1.5–3.8 $mg.kg^{-1}$). The *C. luzulae harbors Ralstonia solanacearum, Burkholderia* sp and *C. laxus*, and possess bacteria belonging to the family Xanthomonadaceae which are highly Hg resistant. These endophytes may help *Cyperus* sp. for effective phytoremediation of Hg. The *Juncus acutus* harbors the members of Sphingomonadaceae as root endophytes. These bacteria showed significant activities in the removal of heavy metals such as Ni, Cd, and Zn from the soil. Thus, *J. acutus* in cooperation with endophytes efficiently remediates the polluted soil (Al-Ateeqi, 2014). A comparative study conducted by De Souza et al. (1999), revealed that the rhizospheric bacteria associated with *Polypogon monspeliensis* facilitate significant accumulation of Se and Hg. Interestingly, plants inoculated with these bacteria showed higher accumulation than non-inoculated plants (De Souza et al., 1999).

2.6 CONCLUSION

The world economy has developed due to industrialization but at the same time, it has affected the environment adversely. There are many pollutants in the form of smoke, effluents, and dust, etc. that are released into the environment and polluting it. These pollutants are known to impact the health of living beings; hence it becomes very essential to decontaminate these pollutants. Phytoremediation is commonly used to decontaminate these pollutants. Halophytes are known to be better adapted to cope with salt and heavy metal compounds. The occurrence of microorganisms as endophytes or rhizosphere microbes in halophytes helps them in enhanced phytoremediation of toxic metals and other pollutants. These halophyte–microbes duo are emerging as prospective candidates for environmental clean-up. Future research on halophytes and associated microbes will help to rehabilitate the contaminated soil for the cultivation of conventional crops.

ACKNOWLEDGMENT

GCN is thankful to the Science and Engineering Research Board (SERB), India for fellowship under the SERB- SIRE program (Grant code- SIR/2022/000391).

KEYWORDS

- **halophytes**
- **phytoremediation**
- **pollution**
- **soil microbes**

REFERENCES

Al-Ateeqi, S. S. Phytoremediation of Oil-Polluted Desert Soil in Kuwait Using Native Plant Species. Ph.D. Thesis, University of Glasgow, Glasgow, UK, 2014.

Albretsen, J. The Toxicity of Iron, an Essential Element. In *Veterinary Medicine*; 2006; pp 82–90.

Alyazouri, A. H.; Jewsbury, R. A.; Tayim, H. A.; Humphreys, P. N.; Al-Sayah, M. H. Phytoextraction of Cr (VI) from Soil Using *Portulaca Oleracea*. *Toxicol. Environ. Chem.* **2013,** *95* (8), 1338–1347.

Anburaj, R.; Nabeel, M. A.; Sivakumar, T.; Kathiresan, K. The Role of Rhizobacteria in Salinity Effects on Biochemical Constituents of the Halophyte Sesuvium Portulacastrum. *Russ. J. Plant Physiol.: A Compr. Russ. J. Mod. Phytophysiol.* **2012,** *59* (1), 115–119

Awa, S. H.; Hadibarata, T. Removal of Heavy Metals in Contaminated Soil by Phytoremediation Mechanism: A Review. *Water, Air, Soil Pollut.* **2020,** *231* (2), 1–15.

Ayyappan, D.; Sathiyaraj, G.; Ravindran, K. C. Phytoextraction of Heavy Metals By*sesuvium Portulacastruml*. A Salt Marsh Halophyte from Tannery Effluent. *Int. J. Phytoremediat.* **2015,** *18* (5), 453–459.

Backer, R.; Rokem, J. S.; Ilangumaran, G.; Lamont, J.; Praslickova, D.; Ricci, E.; ... Smith, D. L. Plant Growth-Promoting Rhizobacteria: Context, Mechanisms of Action, and Roadmap to Commercialization of Biostimulants for Sustainable Agriculture. *Front. Plant Sci.* **2018,** 1473.

Bergin, M. H.; Tripathi, S. N.; Jai Devi, J.; Gupta, T.; McKenzie, M.; Rana, K. S.; ... Schauer, J. J. The Discoloration of the Taj Mahal Due to Particulate Carbon and Dust Deposition. *Environ. Sci. Technol.* **2015,** *49* (2), 808–812.

Bhantana, P.; Rana, M. S.; Sun, X. C.; Moussa, M. G.; Saleem, M. H.; Syaifudin, M.; Shah, A.; Poudel, A.; Pun, A. B.; Bhat, M. A.; Mandal, D. L. Arbuscular Mycorrhizal Fungi and its Major Role in Plant Growth, Zinc Nutrition, Phosphorous Regulation and Phytoremediation. *Symbiosis* **2021,** *84* (1), 19–37.

Boukhatem, Z. F.; Domergue, O.; Bekki, A.; Merabet, C.; Sekkour, S.; Bouazza, F.; Duponnois, R.; de Lajudie, P.; Galiana, A. Symbiotic Characterization and Diversity of Rhizobia Associated with Native and Introduced Acacias in Arid and Semi-Arid Regions in Algeria. *FEMS Microbiol. Ecol.* **2012,** *80* (3), 534–547.

Branca, J. J. V.; Morucci, G.; Pacini, A. Cadmium-Induced Neurotoxicity: Still Much Ado. *Neural Regen. Res.* **2018,** *13,* 1879–1882.

Castro, R.; Pereira, S.; Lima, A.; Corticeiro, S.; Válega, M.; Pereira, E.; ... Figueira, E. Accumulation, Distribution and Cellular Partitioning of Mercury in Several Halophytes of a Contaminated Salt Marsh. *Chemosphere* **2009,** *76* (10), 1348–1355.

Chae, Y.; An, Y. J. Current Research Trends on Plastic Pollution and Ecological Impacts on the Soil Ecosystem: A review. *Environ. Pollut.* **2018,** *240,* 387–395

Chai, M.; Shi, F.; Li, R.; Qiu, G.; Liu, F.; Liu, L. Growth and Physiological Responses to Copper Stress in a Halophyte *Spartina alterniflora (Poaceae)*. *Acta Physiol. Plant.* **2014,** *36* (3), 745–754.

Chaudhry, F. N.; Malik, M. F. Factors Affecting Water Pollution: A Review. *J. Ecosyst. Ecography.* **2017,** *7* (225), 1–3.

Chen, B. D.; Li, X. L.; Tao, H. Q.; Christie, P.; Wong, M. H. The Role of Arbuscular Mycorrhiza in Zinc Uptake by Red Clover Growing in a Calcareous Soil Spiked with Various Quantities of Zinc. *Chemosphere* **2003,** *50* (6), 839–846.

Chen, G.; et al. A Short-Term Study to Evaluate the Uptake and Accumulation of Arsenic in Asian Willow (Salix Sp.) from Arsenic-Contaminated Water. *Environ. Sci. Pollut. Res.* **2013,** *21* (5), 3275–3284.

Chen, Y.; Ouyang, D.; Zhang, W.; Yan, J.; Qian, L.; Han, L.; Chen, M. Degradation of Benzene Derivatives in the CuMgFe-LDO/Persulfate System: The Role of the Interaction Between the Catalyst and Target Pollutants. *J. Environ. Sci.* **2020,** *90,* 87–97.

Coetzee, J. J.; Bansal, N.; Chirwa, E. Chromium in Environment, its Toxic Effect from Chromite-Mining and Ferrochrome Industries, and its Possible Bioremediation. *Exposure Health* **2020,** *12* (1), 51–62.

da Silva, M. N.; Mucha, A. P.; Rocha, A. C.; Silva, C.; Carli, C.; Gomes, C. R.; Almeida, C. M. R. Evaluation of the Ability of Two Plants for the Phytoremediation of Cd in Salt Marshes. *Estuarine, Coastal Shelf Sci.* **2014,** *141,* 78–84.

De Souza, M. P.; Huang, C. P. A.; Chee, N.; Terry, N. Rhizosphere Bacteria Enhance the Accumulation of Selenium and Mercury in Wetland Plants. *Planta* **1999,** *209,* 259–263.

El-Tarabily, K. A.; Youssef, T. Enhancement of Morphological, Anatomical and Physiological Characteristics of Seedlings of the Mangrove *Avicennia Marina* Inoculated with a Native Phosphate Solubilizing Isolate *of Oceanobacillus Picturae* Under Greenhouse Conditions. *Plant Soil* **2010,** *332* (1–2), 147–162.

English, J. P.; Colmer, T. D. Salinity and Waterlogging Tolerances in Three Stem-Succulent Halophytes (Te*cticorn*ia Species) from the Margins of Ephemeral Salt Lakes. *Plant Soil* **2011,** *348* (1), 379–396.

Ermias, A.; Shimelis, H.; Mark, L.; Fentahun, M. Aluminium Toxicity Tolerance in Cereals: Mechanisms, Genetic Control and Breeding Methods. *Afr. J. Agric. Res.* **2013,** *8* (9), 711–722.

Flowers, T. J. Improving Crop Salt Tolerance. *J. Exp. Bot.* **2004,** *55* (396), 307–319.

Gadd, G. M. Metals, Minerals and Microbes: Geomicrobiology and Bioremediation. *Microbiology* **2010,** *156,* 609–643.

Gao, L.; Huang, Y.; Liu, Y.; Mohamed, O. A. A.; Fan, X.; Wang, L.; ... Ma, J. Bacterial Community Structure and Potential Microbial Coexistence Mechanism Associated with Three Halophytes Adapting to the Extremely Hypersaline Environment. *Microorganisms* **2022,** *10* (6), 1124.

Ghori, Z.; et al. Phytoextraction. *Plant Metal Interaction*; 2016; pp 385–409.

Ghous, M.; Iqbal, S.; Bakhtavar, M. A.; Nawaz, F.; ul Haq, T.; Khan, S. *Asian J. Agric. Biol.* **2022,** *4,* 1–9.

Glencross, D. A.; Ho, T. R.; Camina, N.; Hawrylowicz, C. M.; Pfeffer, P. E. Air Pollution and its Effects on the Immune System. *Free Radic. Biol. Med.* **2020,** *151,* 56–68.

Göhre, V.; Paszkowski, U. Contribution of the Arbuscular Mycorrhizal Symbiosis to Heavy Metal Phytoremediation. *Planta* **2006,** *223* (6):1115–1122.

Gontia, I.; Kavita, K.; Schmid, M.; Hartmann, A.; Jha, B. *Brachybacterium Saurashtrense* Sp. Nov., a Halotolerant Root-Associated Bacterium with Plant Growth-Promoting Potential. *Int. J. Syst. Evol. Microbiol.* **2011,** *61,* 2799–2804

Goswami, M.; Chakraborty, P.; Mukherjee, K.; Mitra, G.; Bhattacharyya, P.; Dey, S.; Tribedi, P. Bioaugmentation and Biostimulation: A Potential Strategy for Environmental Remediation. *J. Microbiol. Exp.* **2018,** *6* (5), 223–231.

Grennfelt, P.; Engleryd, A.; Forsius, M.; Hov, Ø.; Rodhe, H.; Cowling, E. Acid Rain and Air Pollution: 50 Years of Progress in Environmental Science and Policy. *Ambio* **2020,** *49* (4), 849–864.

He, L.; Ren, Y.; Zeng, W.; Wu, X.; Shen, L.; Yu, R.; ... Li, J. Deciphering the Endophytic and Rhizospheric Microbial Communities of a Metallophyte *Commelina Communis* in Different Cu-Polluted Soils. *Microorganisms* **2021,** *9* (8), 1689.

Henson, M. C.; Chedrese, P. J. Endocrine Disruption by Cadmium, a Common Environmental Toxicant with Paradoxical Effects on Reproduction. *Exp. Biol. Med. (Maywood)* **2004,** *229* (5), 383–392

Hildebrandt, U.; Janetta, K.; Ouziad, F.; Renne, B.; Nawrath, K.; Bothe, H. Arbuscular Mycorrhizal Colonization of Halophytes in Central European Salt Marshes. *Mycorrhiza* **2001,** *10* (4), 175–183.

Holdgate, M. W. *A Perspective of Environmental Pollution*; Cambridge University Press, 1979.

Howladar, S. M. A Novel Moringa Oleifera Leaf Extract can Mitigate the Stress Effects of Salinity and Cadmium in Bean (*Phaseolus Vulgaris* L.) Plants. *Ecotoxicol. Environ. Saf.* **2014,** *100,* 69–75.

Huang, J. C.; Suárez, M. C.; Yang, S. I.; Lin, Z.-Q.; Terry, N. Development of a Constructed 883 Wetland Water Treatment System for Selenium Removal: Incorporation of an Algal Treatment 884 Component. *Environ. Sci. Technol.* **2013,** *47* (18), 10518–10525.

Jaishankar, M.; Tseten, T.; Anbalagan, N.; Mathew, B. B.; Beeregowda, K. N. Toxicity, Mechanism and Health Effects of Some Heavy Metals. *Interdiscip. Toxicol.* **2014,** *7* (2), 60.

Jha, B.; Gontia, I.; Hartmann, A. The Roots of the Halophyte *Salicornia Brachiata* are a Source of New Halotolerant Diazotrophic Bacteria with Plant Growth-Promoting Potential. *Plant Soil* **2012,** 356 (1–2), 265–277.

Karakas, S.; Dikilitas, M.; Tıpırdamaz, R. Phytoremediation of Salt-Affected Soils Using Halophytes. *Handb. Halophytes: From Mol. Ecosyst. Towards Biosaline Agric.* **2020,** 1–18.

Khan, R. R.; Siddiqui, M. J. Review on Effects of Particulates: Sulfur Dioxide and Nitrogen Dioxide on Human Health. *Int. Res. J. Environ. Sci.* **2014,** *3* (4), 70–73.

Khonsue, N.; Kittisuwan, K.; Kumsopa, A.; Tawinteung, N.; Prapagdee, B. Inoculation of Soil with Cadmium-Resistant Bacteria Enhances Cadmium Phytoextraction by Vetiveria nemoralis and Ocimum gratissimum. *Water, Air, Soil Pollut.* **2013,** *224* (10), 1–9.

Kogej, T.; Ramos, J.; Plemenitas, A.; Gunde-Cimerman, N. Halophilic Fungus *Hortaea werneckii* and the Halotolerant Fungus *Aureobasidium Pullulans* Maintain Low Intracellular Cation Concentrations in Hypersaline Environments. *Appl. Environ. Microbiol.* **2005,** 71 (11), 6600–6605

Kondrashov, F. A.; Rogozin, I. B.; Wolf, Y. I.; Koonin, E. V. Selection in the Evolution of Gene Duplications. *Genome Biol.* **2002,** *3* (2), 1–9.

Kumar, A.; Kumar, A.; MMS, C. P.; Chaturvedi, A. K.; Shabnam, A. A.; Subrahmanyam, G.; ... Yadav, K. K. Lead Toxicity: Health Hazards, Influence on Food Chain, and Sustainable Remediation Approaches. *Int. J. Environ. Res. Public Health* **2020,** *17* (7), 2179.

Kumar, P.; Kumar, A.; Kumar, R. Phytoremediation and Nanoremediation. In *New Frontiers of Nanomaterials in Environmental Science*; Springer, Singapore, 2021; pp 281–297.

Lacalle, R. G., Aparicio, J. D.; Artetxe, U.; Urionabarrenetxea, E.; Polti, M. A.; Soto, M.; Garbisu, C.; Becerril, J. M. Gentle Remediation Options for Soil with Mixed Chromium (VI) and Lindane Pollution: Biostimulation, Bioaugmentation, Phytoremediation and Vermiremediation. *Heliyon* **2020,** *6* (8), e04550.

Lanyi, J. K. Salt-Dependent Properties of Proteins from Extremely Halophilic Bacteria. *Bacteriol. Rev.* **1974,** *38* (3), 272–290.

Li, B.; et al. Halophyte *Halogeton glomeratus*, a Promising Candidate for Phytoremediation of Heavy Metal-Contaminated Saline Soils. *Plant Soil* **2019,** *442.1,* 323–331.

Liang, L.; Liu, W.; Sun, Y.; Huo, X.; Li, S.; Zhou, Q. Phytoremediation of Heavy Metal Contaminated Saline Soils Using Halophytes: Current Progress and Future Perspectives. *Environ. Rev.* **2017,** *25* (3), 269–281.

Lokhande, V. H.; Suprasanna, P. Prospects of Halophytes in Understanding and Managing Abiotic Stress Tolerance. In *Environmental Adaptations and Stress Tolerance of Plants in the Era of Climate Change*; Springer: Berlin, Germany, 2012; pp 29–56.

Lone, M. I.; He, Z. L.; Stoffella, P. J.; Yang, X. E. Phytoremediation of Heavy Metal Polluted Soils and Water: Progresses and Perspectives. *J. Zhejiang Univ. Sci. B* **2008,** *9* (3), 210–220.

Lu, H.; Zhang, Y.; Liu, B.; Liu, J.; Ye, J.; Yan, C. Rhizodegradation Gradients of Phenanthrene and Pyrene in Sediment of Mangrove (Kandelia Candel (L.) Druce). *J. Hazardous Mater.* **2011,** *196,* 263–269.

Lutts, S.; Lefevre, I.; Delpérée, C.; Kivits, S.; Dechamps, C.; Robledo, A.; Correal, E. Heavy Metal Accumulation by the Halophyte Species Mediterranean Saltbush. *J. Environ. Q.* **2004,** *33* (4), 1271–1279.

Marques, B.; Lillebø, A. I.; Pereira, E.; Duarte, A. C. Mercury Cycling and Sequestration in Salt Marshes Sediments: An Ecosystem Service Provided by *Juncus Maritimus* and *Scirpus Maritimus*. *Environ. Pollut.* **2011,** *159* (7), 1869–1876.

Martin, S.; Griswold, W. Human Health Effects of Heavy Metals. *Environ. Sci. Technol. Briefs Citizens* **2009,** *15,* 1–6.

Matsumoto, S. T.; Mantovani, M. S.; Malaguttii, M. I. A.; Dias, A. L.; Fonseca, I. C.; Marin-Morales, M. A. Genotoxicity and Mutagenicity of Water Contaminated with Tannery Effluents, as Evaluated by the Micronucleus Test and Comet Assay Using the Fish *Oreochromis niloticus* and Chromosome Aberrations in Onion Root-Tips. *Genet. Mol. Biol.* **2006,** *29,* 148–158.

Matta, G.; Gjyli, L. Mercury, Lead and Arsenic: Impact on Environment and Human Health. *J. Chem. Pharm. Sci.* **2016,** *9* (2), 718–725.

Mesa-Marín, J.; Pérez-Romero, J. A.; Redondo-Gómez, S.; Pajuelo, E.; Rodríguez-Llorente, I. D.; Mateos-Naranjo, E. Impact of Plant Growth Promoting Bacteria on Salicornia Ramosissima Ecophysiology and Heavy Metal Phytoremediation Capacity in Estuarine Soils. *Front. Microbiol.* **2020,** *11*, 553018.

Mudgal, V.; Madaan, N.; Mudgal, A.; Singh, R. B.; Mishra, S. Effect of Toxic Metals on Human Health. *Open Nutraceut. J.* **2010,** *3* (1).

Mujeeb, A.; Aziz, I.; Ahmed, M. Z.; Alvi, S. K.; Shafiq, S. Comparative Assessment of Heavy Metal Accumulation and Bio-Indication in Coastal Dune Halophytes. *Ecotoxicol. Environ. Saf.* **2020,** *195,* 110486.

Mujeeb, A.; Aziz, I.; Ahmed, M. Z.; Shafiq, S.; Fatima, S.; Alvi, S. K. Spatial and Seasonal Metal Variation, Bioaccumulation and Biomonitoring Potential of Halophytes from Littoral Zones of the Karachi Coast. *Sci. Total Environ.* **2021,** *781,* 146715.

Munns, R. Genes and Salt Tolerance: Bringing them Together. *New Phytol.* **2005,** *167* (3), 645–663.

Mushtaq, W.; Bedair, H.; Shakeel, A. Halophytes. *Handbook of Halophytes*, 2021; pp 2303–18.

Naikoo, M. I.; et al. Halophytes in India and their role in Phytoremediation. *Handbook of Halophytes*, 2021; pp. 2345–2365.

Nalla, S.; Hardaway, C. J.; Sneddon, J. Phytoextraction of Selected Metals by the First and Second Growth Seasons of *Spartina Alterniflora*. *Instrum. Sci. Technol.* **2012,** *40* (1), 17–28

Navarro-Torre, S.; Barcia-Piedras, J. M.; Caviedes, M. A.; Pajuelo, E.; Redondo-Gómez, S.; Rodríguez-Llorente, I. D.; Mateos-Naranjo, E. Bioaugmentation with Bacteria Selected from the Microbiome Enhances Arthrocnemum Macrostachyum Metal Accumulation and Tolerance. *Mar. Pollut. Bull.* **2017,** *117,* 340–347.

Nazari-Sharabian, M.; Ahmad, S.; Karakouzian, M. Climate Change and Eutrophication: A Short Review. *Eng. Technol. Appl. Sci. Res.* **2018,** *8* (6), 3668.

Newman, L. A.; Reynolds, C. M. Phytodegradation of Organic Compounds. *Curr. Opin. Biotechnol.* **2004,** *15* (3), 225–230.

Nikalje, G. C.; Srivastava, A. K.; Pandey, G. K.; Suprasanna, P. Halophytes in Biosaline Agriculture: Mechanism, Utilization, and Value Addition. *Land Degradat. Dev.* **2018,** *29* (4), 1081–1095.

Nikalje, G. C.; Suprasanna, P. Coping with Metal Toxicity – Cues from Halophytes. *Front. Plant Sci.* **2018,** *9.*

Oren, A. Salinibacter: An Extremely Halophilic Bacterium with Archaeal Properties. *FEMS Microbiol. Lett.* **2013,** *342* (1), 1–9.

Pandey, N.; Chandra, J.; Xalxo, R.; Sahu, K. Concept and Types of Phytoremediation. In *Approaches to the Remediation of Inorganic Pollutants*; Springer: Singapore, 2021; pp 281–302.

Pedro, C. A.; Santos, M. S.; Ferreira, S. M.; Gonçalves, S. C. The Influence of Cadmium Contamination and Salinity on the Survival, Growth and Phytoremediation Capacity of the Saltmarsh Plant *Salicornia Ramosissima. Mar. Environ. Res.* **2013,** *92,* 197–205.

Perez, A.; Martinaz, D.; Barraza, Z.; Marrugo-Negrete, J. L. Endophytic Bacteria Associated to Genus Cyperus and Paspalum in Soils with Mercury Contamination. *Rev. UDCA Actual. Divulg. Científica* **2016,** *19,* 67–76.

Prüss-Üstün, A.; Fewtrell, L.; Landrigan, P. J.; Ayuso-Mateos, J. L. Lead Exposure. *Comparative Quantification of Health Risks*; World Health Organization; Geneva, 2004; pp 1495–1542.

Rahman, M. M.; Mostofa, M. G.; Keya, S. S.; Siddiqui, M. N.; Ansary, M. M. U.; Das, A. K.; ... Tran, L. S. P. Adaptive Mechanisms of Halophytes and their Potential in Improving Salinity Tolerance in Plants. *Int. J. Mol. Sci.* **2021,** *22* (19), 10733.

Rane, N. R.; Tapase, S.; Kanojia, A.; Watharkar, A.; Salama, E. S.; Jang, M.; Yadav, K. K.; Amin, M. A.; Cabral-Pinto, M. M.; Jadhav, J. P.; Jeon, B. H. Molecular Insights into Plant–Microbe Interactions for Sustainable Remediation of Contaminated Environment. *Bioresour. Technol.* **2022,** *344,* 126246.

Roohi, A.; Nazish, B.; Maleeha, M.; Waseem, S. A Critical Review on Halophytes: Salt Tolerant Plants. *J. Med. Plants Res.* **2011,** *5* (33), 7108–7118.

Rueda-Puente, E.; Castellanos-Cervantes, T.; Diaz de Leon-Alvarez, J.; PreciadoRangel, P.; Almaguer-Vargas, G. Bacterial Community of Rhizosphere Associated to the Annual Halophyte *Salicornia Bigelovii* (Torr.). *Terra Latino americana* **2010,** *28* (4), 345–353.

Ruidavets, J. B.; Cournot, M.; Cassadou, S.; Giroux, M.; Meybeck, M.; Ferrières, J. Ozone Air Pollution is Associated with Acute Myocardial Infarction. *Circulation* **2005,** *111* (5), 563–569.

Shaltami, O. R. Characteristics of Heavy Metals in Soil at Al Qawarishah Village, NE Libya. *Trace Elements Environ.* **2014,** *13,* 16–19.

Shamsutdinov, N. Z.; Shamsutdinova, E. Z.; Orlovsky, N. S.; et al. Halophytes: Ecological Features, Global Resources, and Outlook for Multipurpose Use. *Her. Russ. Acad. Sci.* **2017,** *87,* 1–11.

Shurigin, V.; Egamberdieva, D.; Li, L.; Davranov, K.; Panosyan, H.; Birkeland, N. K.; ... Bellingrath-Kimura, S. D. Endophytic Bacteria Associated with Halophyte Seidlitzia Rosmarinus Ehrenb. Ex Boiss. from Saline Soil of Uzbekistan and their Plant Beneficial Traits. *J. Arid Land* **2020,** *12* (5), 730–740.

Simmonds, M. P.; Isaac, S. J. The Impacts of Climate Change on Marine Mammals: Early Signs of Significant Problems. *Oryx* **2007,** *41* (1), 19–26.

Singh, S. N.; Verma, A. Phytoremediation of Air Pollutants: A Review. *Environ. Bioremediat. Technol.* **2007,** 293-314.

Sruthi, P.; Shackira, A. M.; Puthur, J. T. Heavy Metal Detoxification Mechanisms in Halophytes: An Overview. *Wetlands Ecol. Manag.* **2016,** *25* (2), 129–148.

Thrall, P. H.; Broadhurst, L. M.; Hoque, M. S.; Bagnall, D. J. Diversity and Salt Tolerance of Native Acacia Rhizobia Isolated from Saline and Non-Saline Soils. *Austral Ecol.* **2009,** *34* (8), 950–963

Turk, M.; Mejanelle, L.; Sentjurc, M.; Grimalt, J. O.; Gunde-Cimerman, N.; Plemenitas, A. Salt-Induced Changes in Lipid Composition and Membrane Fluidity of Halophilic Yeast-Like Melanized Fungi. *Extremophiles* **2004,** *8* (1), 53–61

Upadhyay, S.; Singh, J.; Singh, D. Exopolysaccharide-Producing Plant Growth-Promoting Rhizobacteria Under Salinity Condition. *Pedosphere* **2011,** *21* (2), 214–222.

Usman, K.; Al-Ghouti, M. A.; Abu-Dieyeh, M. H. Phytoremediation: Halophytes as Promising Heavy Metal Hyperaccumulators. *Heavy Metals* [Preprint], 2018.

Võ, T. N. H. Morphological and Molecular Characterization of Plant Growth Promoting Salt-Tolerant Bacteria Associated with Halophytes in the Southeast Seaside of Vietnam. *Tạp. Chí Nông Nghiệp Phát Triển* **2021,** *20* (6), 17–26.

Vromman, D.; Flores-Bavestrello, A.; Šlejkovec, Z.; Lapaille, S.; Teixeira-Cardoso, C.; Briceño, M.; ... Lutts, S. Arsenic Accumulation and Distribution in Relation to Young Seedling Growth in *Atriplexata Camensis* Phil. *Sci. Total Environ.* **2011,** *412,* 286–295.

Wu, H.; Liu, X.; Zhao, J.; Yu, J. Regulation of Metabolites, Gene Expression, and Antioxidant 1196 Enzymes to Environmentally Relevant Lead and Zinc in the Halophyte Suaeda Salsa. *J. Plant 1197 Growth Regul.* **2013,** *32* (2), 353–361.

Zhang, C.; Gan, C.; Ding, L.; Xiong, M.; Zhang, A.; Li, P. Maternal Inorganic Mercury Exposure and Renal Effects in the Wanshan Mercury Mining Area, Southwest China. *Ecotoxicol. Environ. Saf.* **2020,** *189,* 109987.

Zhou, M.; Ghnaya, T.; Dailly, H.; Cui, G.; Vanpee, B.; Han, R.; Lutts, T. S. The Cytokinin Trans-Zeatine Riboside Increased Resistance to Heavy Metals in the Halophyte Plant Species Kosteletzkya Pentacarpos in the Absence But not in the Presence of NaCl Chemosphere. **2019,** 233, 954–965.

CHAPTER 3

Endophytes-Assisted Salinity Tolerance in Halophytes

DEEPAK B. SHELKE[1], MAHADEV R. CHAMBHARE[1], KAMLAKAR MORE[3], HIRALAL B. SONAWANE[2], and GANESH C. NIKALJE[4]

[1]*Department of Botany, Amruteshwar Art's, Commerce and Science College, Vinzar, Velha, Pune, MS, India*

[2]*Department of Botany, Sant Gadge Baba Amravati University, Amravati, Maharashtra, India*

[3]*PG Research Centre in Botany, Prof. Ramkrishna More Arts, Commerce and Science College, Akurdi, Pune, India*

[4]*PG Department of Botany, Seva Sadan's R. K. Talreja College of Arts, Science and Commerce, Ulhasnagar, Thane, Maharashtra, India*

ABSTRACT

Saline soil is a major constraint for crop productivity globally and its exponential rise leads to climate change. Saline habitats cover a wide area of Earth planet and halophytes are frequently utilized for human interests, next to their physiological, biochemical, and genetic adaptation to salinity, complex ecological routes have an effect on the salinity tolerance of halophytes. Recently, there is an increasing attention toward habitat-imposed plant stress tolerance carried by plant-associated microbes especially endophytes. Hence, endophytes inhabiting roots, stems, and leaves can contribute significantly to plant growth promotion and salt stress tolerance. Halophyte-associated endophytes are known to be able to adapt plants to a range of changes in external osmolarity contributing to salt stress tolerance.

Physiology of Halophytes: Signaling, Omics, and Tolerance Mechanisms.
Nivas Manohar Desai & Ganesh Chandrakant Nikalje (Eds.)

It is evident that endophytes influence host actions, such as improving antioxidant, osmolytes, and phytohormonal signaling, plant nutrient uptake efficiency, photosynthetic efficiency, and enhancing DNA repair mechanism. Therefore, the present chapter highlights the functionality of endophytes and their modes of action to enhance salinity tolerance and growth promotion in halophytes by modulating morphological, physiological, biochemical, and molecular mechanisms. Several reports on endophytes have helped us to know the mechanism of colonization of endophytes and their interactions with halophytes.

3.1 INTRODUCTION

As estimated that about 50% of the world's arable land will be salt-affected by the year 2050 and this could further increase due to climate change (Shelke et al., 2017). Salinity hampers crop production and there is an increase in the percentage of salinity-affected land thatexceeds 20% of the world's cultivated land (Fouda, 2019). Soil can be termed saline soil if its pH ranges from 7 to 8.5 and its electrical conductivity becomes 4 dS/m or more, which is equivalent to approximately 40 mM NaCl, with an osmotic potential of approximately 0.2 MPa (Muchate et al., 2016). Usually, saline soil contains a number of cations of soluble salts like Ca^{2+}, Mg^{2+}, Na^{+}, and K^{+}, and anions SO_4^{2-}, Cl^{-}, HCO_{3-} with CO_3^{2-}, and NO_{3-} (Ghosh et al., 2016). Salinity stress and heavy metals are the major and most destructive abiotic stresses (Zhu, 2001; Angulo-Bejarano et al., 2021) that limit the growth, development, and productivity of major crops; no toxic substances restrict the plant growth more than salt and heavy metals worldwide (Xiong and Zhu, 2002; Angulo-Bejarano et al., 2021). As a consequence, mainly due to Na^{+} accumulation, ion toxicity causes chlorosis and necrosis that interferes with many physiological processes in plants (Munns, 2002).

Salinity causes osmotic and ionic stresses in plants. Osmotic stress occurs immediately after exposure to salinity by the formation of hypertonic surroundings at cell's surface, while ionic stress increases due to Na^{+} and Cl^{-} ion accumulation inside the cell after several days of salt exposure (Shelke et al., 2019). These osmotic alterations affect the water balance inside the cell and hamper cell's turgor pressure, cell elongation, and cell division events (Shelke et al., 2017). The ionic stress modulates ion homeostasis inside the cell leading to alterations in hormonal regulation, nutrient translocation, photosynthesis, transpiration, and other metabolic routes (Munns, 2002). The mutual investigations incurred toward salt tolerance in several plant systems are targeted at sustaining the osmotic balance and ion homeostasis. Various breeding studies have been accomplished on diverse crop systems to recognize salt-tolerant genotypes; however, only a small number of genotypes has been recognized to date. Munns and Tester (2008) reported that these recognized salt-tolerant genotypes are region-specific and have failed to survive in diverse regions under stress environments. Therefore, conventional breeding programs are neglected by researchers due to time-consuming approaches and insignificant results. In this direction, genetic engineering strategies could be an effective tool for breeding programs but these are often cost-intensive techniques.

Provisionally, the administration of useful microbes employed in the improvement of salt tolerance in plants is a simple and feasible strategy to reclaim salinity-prone cultivated lands (Berg, 2009). Microbes occupy the region of plants and extensively contribute to growth promotion and salt tolerance in plants. Microbial incidence with plants improves soil–water–plant relationships and stimulates phytohormonal signaling mechanisms and triggers several other mechanisms that are implicated in salt stress tolerance in halophytic plants (Dodd and Pérez-Alfocea, 2012; Forni et al., 2017). A beneficial microbiota comprises plant growth promoting rhizobium (PGPR), arbuscular mycorrhizal (AM) fungi, and other bacterial and fungal endophytes. Yet, there is a huge gap in the perception of concerned microbe-assisted mechanisms linked with plant growth promotion that desires to be overcome. With this consideration, the authors attempted to solve the following issues: How do caring endophytes control salt stress in plants? What kind of physiological modulations occur in plants under saline environment? and what kind of biochemical and molecular modifications driven by endophytes will be useful in plant defense mechanisms under salinity stress? Recently, there has been rising interest in the identification of pathways associated with plant tolerance to stress is linked to plant-associated microorganisms (Hardoim et al., 2015; Qin et al., 2017; Shelke et al., 2022). Bacterial endophytes and arbuscular mycorrhiza sustain plant health and fitness under adverse environments. Due to increasing salt-affected area, currently, researchers are looking toward the halophyte for biosaline agriculture. Even though halophytes are salt tolerant, endophytes associated with them give additional benefits to them by improving their tolerance to some extent. Currently, various reports have confirmed an array of mechanisms of endophytes to enhance the growth of plants under diverse abiotic stress conditions (Brader et al., 2014; Santoyo et al., 2016; Teo et al., 2022). Microbes are unable to withstand extreme environments which becomes a challenging task for scientists to generate novel bioinoculants appropriate to sustain under such extreme environments. There is a need to develop strategies to investigate stress-tolerant microbes and their utilization in sustainable agriculture. Stress-tolerant microbes possess various strategies to overcome the causes induced by abiotic stresses and also to incidence with halophytes. In this chapter, we especially focused on endophytes and their application in plant salt tolerance with a brief outline of soil salinity and its tolerance strategy in plants. Therefore, the present chapter can be useful for the exploitation of halophytes-associated endophytes for the establishment of salt stress alleviation protocols.

3.2 ENDOPHYTES

A term "endophyte" originated from the Greek word "endon" which means inside or within and the word "phyton" means plant body. They are widely distributed almost in all climatic zones as in terrestrial as well as aquatic habitats. Both bacterial and fungal microbes are categorized as endophytes and possessed symbiotic associations with plants. Endophytes establish symbiosis with plants by producing metabolites which are employed in plant growth promotion, protection from pathogens, and improvement in plant defense (Shelke et al., 2022). The fungal endophytes, mostly belonging to group Ascomycota, are present internally (into the intercellular fluids and wall spaces) in the aerial plant parts (stem, leaves, flowers, etc.) or in underground parts like root, rhizome, and tubers without causing any damaging effect or without exhibiting any external symptoms of their presence. According to De Bary (1866), Endophyte is a microorganism that grows within/inside plant tissues symbiotically. Likewise, Petrini (1991) stated that any microorganism that can live some part of their life span within plant tissues without causing any damaging effect to the hosts. On the other hand, these descriptions resume changing as research on endophytes progresses (Bacon and White, 2000).

Endophytes have symbiotic or to some extent pathogenic relations with host plants and it is a sort of symbiotic association, and their enormous utilization has been emphasized by the widespread investigations driven by several workers (Sonawane et al., 2020; Sonawane et al., 2021). Generally, endophytes may be obligate or facultative and are associated with all known plant species. Fungal endophytes show numerous biological distinctiveness that is limited by multiple factors, such as host specificity, host-assisted development, inoculum density, and climatic aspects (Patle et al., 2018; Sonawane et al., 2020, 2021). They can synthesize a wide range of enzymes and secondary metabolites showing a variety of biological activities and are useful in stress tolerance such as heavy metal toxicity, drought resistance, and salt tolerance (Arnold et al., 2007; Rhoden et al., 2014; Sonawane et al., 2020, 2021). These are endosymbiotic microorganisms colonize intercellular and intracellular spaces of different plant compartments, although they can cause significant morphological changes (Hassan et al., 2019). Fungal endophytes such as ascomycetes or basidiomycetes exist within the roots or tissues of the hosts (Ganley, 2004; Arnold and Lutzoni, 2007). In mutualistic mode, endophytic fungi increase plant fitness by curtailing environmental stresses (Bacon, 1993; Waller et al., 2005) and in return, obtain access to the host plant's nutrients (Ali et al., 2012). However, endophytes are recorded as active biocontrol agents and plant growth promoters, and employed in the production of secondary metabolites of biotechnological or commercial pharmaceutical application (Sharma et al., 2017).

3.3 ENDOPHYTES AND SOIL ENVIRONMENT

The endophytes not only promote plant growth but also alleviate abiotic stresses (especially salt stress), and this is comprehending the functioning of plant–microbe–soil interactions to understand the underlying mechanisms which are beneficial in crop plant research for the management of salt stress (Compant et al., 2016). Halophytes and their progenitors grow well in saline soil with better colonization of endophytes in roots, but desert soil affects the colonization of endophytes in roots. Hence, the soil as a prominent factor influencing endophyte communities inhabiting the roots of halophytes. Soil conditions after measurement of plant root growth and endophyte colonization in roots need to take substantial care of soil health. Furthermore, such soils need large-scale applications of fertilizers to improve root health for better colonization of endophytes (Yasseen and Al-Thani, 2022). Kearl et al. (2019) reported that various bacterial endophytic strains were isolated from rhizospheric soil and roots of three different native Utah halophytic plants (*Allenrolfea occidentalis, Salicornia rubra,* and *Sarcocornia utahensis*). These isolated endophytes (*Bacillus*, *Enterobacter*, *Pantoea*, and *Sphingomonas*) are able to stimulate plant growth in saline environment which results in the enhancement of salinity tolerance of hybrid elephant grass (Li et al., 2016; Kearl et al., 2019).

3.4 DIVERSITY OF HALOPHYTE-ASSOCIATED ENDOPHYTES

The halophytic plants grow under saline environments. Furtado et al. (2019) reported that a bacterial endophyte (genus *Marinimicrobium)* was found in all roots of halophyte *Salicornia europaea.* As per previous reports, various halophilic bacterial and fungal endophytes associated with halophytes (in root, stem, and leaves) have potential application in salinity tolerance (Shelke et al., 2022). For instance, endophytes *Penicillium* and *Trichoderma* are reported from *Pinus thunbergii* halophyte (Min et al., 2014). Ruppel et al. (2013) have screened the novel bacteria endophytes including *Brachybacterium saurashtrense*, *Brevibacterium casei*, *Cronobacter sakazakii*, *Halomonas*, *Mesorhizobium*, *Pseudomonas*, and *Rhizobium radiobacter* strains were isolated from the roots of *Salicornia brachiata* which possessed enhanced ACC deaminase action. The bacterial endophyte genera like *Bacillus*, *Pseudomonas*, *Corynebacterium,* and *Escherichia* are reported from *Sesuvium portulacastrum* (Ruppel et al., 2013). Likewise, the bacterial endophytes, such as *Bacillus, Halobacillus, Oceanobacillus, Brevundimonas, Exiguobacterium, Serratia, Staphylococcus,* and *Vibrio* were isolated from the halophyte *Prosopis strombulifera* (Piccoli et al., 2011; Ruppel et al., 2013). The endophytes like *Aspergillus flavus, A. fumigates, A. japonicas, A. niger, A. ochraceous, A. terreus, Penicillium* sp. are isolated from *Portulaca portulacastrum* (Kannan et al., 2014). Mateu et al. (2020) have reported the dark septate endophytes (DSE) from halophyte *Phragmites australis.* Song et al. (2015) have reported that isolation of endophyte (*Epichloe* spp.) was recorded from *Hordeum brevisubulatum. Peyronellaea glomerata, Phoma macrostoma*, and *Alternaria tenuissima.* These fungal isolates were reported from *Phragmites australis* halophyte (Soares et al., 2016). *Neotyphodium coenophialum* endophytic isolate was reported from *Festuca arundinacea* (Sabzalian and Mirlohi, 2010). Kannan et al. (2014) have reported these different fungal isolates (*Alternaria alternata, A. fumigatus, A. glaucus, A. nidulans, A. niger, A. terreus, Chaetomium globosum, Chrysonilia sitophila, Curvularia brachyspora, Curvularia lunata, Fusarium culmorum, F. oxysporum, F. oxysporum,* and *Trichoderma* spp.) from halophytic plant *Spinifex littoreus*. A fungal endophyte *Alternaria* was isolated from several halophytic plant species, such as *Limonium tetragonum, Suaeda australis, Suaeda maritima, Suaeda glauca,* and *Phragmites australis* (Khalmuratova et al., 2020). The fungal endophytes such as *Alternaria* sp., *Aspergillus* sp., *Cladosporium* sp., *Fusarium* sp., and *Penicillium* sp. were isolated from halophyte *Climacoptera crassa* (Kondrasheva et al., 2022). Kondrasheva et al. (2022) has studied the endophytes like *Cladosporium* sp., *Aspergillus* sp., and *Penicillium* sp. in *Halocnemum strobilaceum* halophyte. Likewise, several studies have been performed on screening of endophytes from halophytes. Most of the bacterial strains categorized as *Bacillus* sp. are isolated endophytes from the roots of *P. strombulifera* and the leaves of salt-tolerant plants, such as *Cressa cretica, Salicornia brachiata, Spharantus indicus,* and *Suaeda nudiflora* (Szymańska et al., 2016).

3.5 ENDOPHYTES AND SALT TOLERANCE IN HALOPHYTES

The halophytes have improved mechanisms involved in salt stress through their morphological, physiological, biochemical, and molecular actions with the association of endophytes. The modulations of endophyte-associated salt stress-related mechanisms in halophytes are described below:

3.5.1 ENDOPHYTE-ASSISTED GROWTH RESPONSE OF HALOPHYTES UNDER SALT STRESS

The salinity tolerance in halophytes revolves around a complex morphological modulations. The plants can restrict influx and accumulation of salts, adjust cells' osmotic and ionic balance, and regulate leaf development; onsets of senescence are evident as salt-tolerance strategies (Shelke et al., 2019; Nikalje et al., 2019). For example, halophytic plants flourish and complete their life cycle under saline conditions.

For instance, a halophyte *Limonium sinense* inoculated with halotolerant endophytic actinomycete strain, KLBMP 5084 had greater fresh weight, root length, and leaf length (Qin et al., 2017). The endophytic strains *Penicillium* and *Trichoderma* in *Pinus thunbergii* exhibited morphological, physiological, and biochemical modulations that facilitate growth promotion, nutrient absorption, resistance to herbivores, and defense response against pathogen attacks (Min et al., 2014). Several endophytic fungal isolates such as *Alternaria alternata, Aspergillus flavus, Cladosporium oxysporum, C. tenuissimum, Fusarium chlamydosporum, F. oxysporum, F. proliferatum, Dioporthe phaseolorum, Lasiodiplodia pseudotheobromae*, and *Penicillium spinulosum* were screened from the tissues of various parts of *Ipomoea pes-caprae*. They showed significant salt tolerance in *Oryza sativa* (Rice). A fungal endophyte *F. oxysporum* (MH511104) isolated from *Ipomoea pes-caprae* halophyte, when inoculated in *O. sativa* resulted in 54% increase in growth in the presence of salt (Manasa et al., 2020) (Table 3.1).

Similarly, the halophilic *F. oxysporum* (MH511104) is capable of incidence with salt-sensitive IR-64 rice genotype, and if so, this fungal isolate is also capable of driving salt tolerance to the crop. The inoculation of *F. oxysporum* endophyte to the salt-sensitive rice (IR-64) seedlings exhibited considerably improved growth of seedlings under salinity stress in comparison to the control condition of seedlings (not treated with *F. oxysporum*). Upon fungal (*F. oxysporum*) infection, seedling growth promotion was recorded, while seedling growth remained unaltered under no infection treatment with fungi. On endophyte infection, root and shoot growth of rice genotype improved by 46% and 50%, respectively under saline environment in comparison to control conditions (Manasa et al., 2020). Likewise, the endophyte *Sphingomonas prati* inoculation potentially enhanced the morphological characters, such as mean plant height, root length, fresh weight, and dry weight by 45.43%, 9.91%, 82.00%, and 102.25%, respectively, compared with the control conditions at 20% NaCl (Guo et al., 2021).

3.5.2 ENDOPHYTE-ASSISTED PHYSIOLOGICAL RESPONSES OF HALOPHYTES UNDER SALT STRESS

The halophyte roots-associated endophytes exhibited physiological modulations that result in high growth promotion and salt stress tolerance in plants (Nanjundappa et al., 2021). The association of endophytes with halophytes may help the plant with stress alleviation by modulating plant hormone levels and nutrient uptake and/or controlling the reactive oxygen species (ROS) production (Rodríguez-Lorente et al., 2019).

Endophytes influence the host plants directly by modulating host nutrient uptake and by promoting growth, development, and stress-assisted phytohormones (cytokinins, ethylene, and gibberellins). For instance, ethylene controls different physiological routes such as root induction, abscission, leaf senescence, cell elongation, fruit ripening, and transport of auxin. Bacterial endophytes can trigger a mechanism such as induced systemic resistance which provides protection to halophytes against phytopathogens. They also exhibited antimicrobial activity against fungal pathogens by producing fungal cell-wall degrading enzymes, such as chitinases, glucanases, and proteases (Afzal et al., 2019). Mateu et al. (2020) reported that the DSE inoculation in *Phragmites australis* halophyte showed significantly increased seedling survival rate under salt stress through direct water and nutrient uptake into the seedling by inducing the synthesis of osmolytes. DSE are characterized

TABLE 3.1 Role of Endophytes in Improving Salinity Tolerance in Halophytes

Sr. no.	Endophytes	Halophytes	Studied modulation	Plant responses	Reference
1	Actinomycete strain KLBMP 5084	*Limonium sinense*	Molecular	Activation of PGP-related genes	Qin et al. (2017)
2	Actinomycete strain KLBMP 5084	*Limonium sinense*	Morphological	Greater fresh weight, root length, leaf length	Qin et al. (2017)
3	*Aspergillus flavus, A. fumigatus, A. japonicus, A. niger, A. ochraceous, A. terreus, Penicillium* sp.	*Portulaca portulacastrum*	Morphological	-	Kannan et al. (2014)
4	*Aspergillus flavus, A. fumigatus, A. japonicus, A. nidulans, A. niger, A. terreus, Chaetomium* sp., *Cladosporium* sp., *Emericella nidulans, Fusarium* sp., *Penicillium corylophilum*	*Salicornia* sp.	Morphological	Improvement in growth parameters of root, and stem	Kannan et al. (2014)
5	*Alternaria alternata, Aspergillus flavus, Cladosporium oxysporum, C. tenuissimum, Dioporthe phaseolorum, Fusarium chlamydosporum, F. oxysporum, F. proliferatum, Lasiodiplodia pseudotheobromae* and *Penicillium spinulosum*	*Ipomea pes-caprae* and *Oryza sativa*	Morphological	Significantly improved root, shoot and seedling growth under salinity stress with endophytes	Manasa et al. (2020)
6	*Alternaria alternata, A. flavus, A. niger, A. terreus, A. tamari, Chaetomium globosum, Curvularia brachyspora, Curvularia lunata, Fusarium culmorum, F. oxysporum, F. monoliformce, F. oxysporum, Fusarium* sp.	*Bauhinia* sp.	Morphological	Growth promotion in root, stem and leaves	Kannan et al. (2014)
7	*Alternaria*	*Limonium tetragonum, Suaeda australis, Suaeda maritima, Suaeda glauca, Phragmites australis*	Biochemical	*Alternaria alternata* and is a producer of gibberellins	Khalmuratova et al. (2020)

TABLE 3.1 *(Continued)*

Sr. no.	Endophytes	Halophytes	Studied modulation	Plant responses	Reference
8	*Alternaria* sp., *Aspergillus* sp.	*Suaeda heterophylla*	Biochemical	*Alternaria* produces the plant growth regulator indole-acetic acid for the host plant	Kondrasheva et al. (2022)
9	*Aspergillus* sp., *Penicillium* sp., *Alternaria* sp., *Fusarium* sp., *Cladosporium* sp.	*Climacoptera crassa*	Morphological	*Penicillium* and *Fusarium* could improve the survival and growth of their hosts by enhancing tolerance to environmental stress	Kondrasheva et al. (2022)
10	*Alternaria alternata, A. fumigatus, A. glaucus, A. nidulans, A. niger, A. terreus, Chaetomium globosum, Chrysonilia sitophila, Curvularia brachyspora, Curvularia lunata, Fusarium culmorum, F. oxysporum, F. oxysporum, Trichoderma* sp.	*Spinifex littoreus*	Morphological	Growth promotion in root, stem and leaves	Kannan et al. (2014)
11	*Alternaria alternata, Cladosporium* and *Fusarium*	*Limonium tetragonum, Suaeda australis, Suaeda maritima, Suaeda glauca* and *Phragmites australis*	Morphological and biochemical	Plant growth promoting capacity. A producer of gibberellins	Khalmuratova et al. (2020)
12	*Candida utilis*	*Halocnemum strobilaceum*	Morphological	Improvement in plant growth	Al-Mailem et al. (2010); Marasco et al. (2016)
13	*Cladosporium* sp., *Aspergillus* sp., *Penicillium* sp.	*Halocnemum strobilaceum*	Biochemical	*Aspergillus* and *Penicillium* GAs that confer biotic stress resistance to pathogenic attack	Kondrasheva et al. (2022)

TABLE 3.1 *(Continued)*

Sr. no.	Endophytes	Halophytes	Studied modulation	Plant responses	Reference
14	*Cladosporium* sp., *Alternaria* sp.	*Aeluropus litoralis*	Biochemical	The plant growth promoting effects of Alternaria Sm-3–7-5 and GA production	Kondrasheva et al. (2022)
15	Dark septate endophytes (DSE)	*Phragmites australis*	Physiological	Increased seedling survival rate	Mateu et al. (2020)
16	*Epichloë* sp.	*Hordeum brevisubulatum*	Biochemical	Increased photosynthesis, chlorophyll content, total antioxidant capacity and glycine betaine content and lower leaf malondialdehyde content	Taixiang et al. (2018)
17	*Epichloe* spp.	*Hordeum brevisubulatum*	Biochemical	Higher biomass as well as nitrogen: phosphorus ratio and lower carbon: nitrogen, carbon: potassium and sodium: potassium ratio	Song et al. (2014)
18	*Epicoccum*	*Phragmites australis*	Morphological	The highest plant growth promoting capacity	Khalmuratova et al. (2020)
19	Glomeromycota	*Aster tripolium* and Mangroves	Molecular	Up-regulation and down-regulation of several genes. acdS gene was detected	Hildebrandt et al. (2001); Juniper and Abbott (2006); Sengupta and Chaudhuri (2002); Wang et al. (2004); Yang et al. (2008)

TABLE 3.1 *(Continued)*

Sr. no.	Endophytes	Halophytes	Studied modulation	Plant responses	Reference
20	*Glomus geosporum* *Glomus caledonium*	*Atriplex* spp.	Morphological	Modifications in root and shoot characters	Aguilera et al. (1998); Landwehr et al. (2002); Sonjak et al. (2009)
21	*Khuskia*	*Suaeda australis*	Morphological	Plant growth promotion	Khalmuratova et al. (2020)
22	*Microascus pyramidus*	*Salsola tomentosa*	Biochemical	Antimicrobial properties,	Jalili et al. (2020)
23	*Neotyphodium coenophialum*	*Festuca arundinacea*	Biochemical	Greater sodium, potassium, and chloride ion concentrations with higher leaf survival rates and Root dry matter	Sabzalian and Mirlohi (2010)
24	*Peyronellaea glomerata, Phoma macrostoma,* and *Alternaria tenuissima*	*Phragmites australis*	Biochemical	Higher resistance to zinc and mercury	Soares et al. (2016)
25	*Penicillium* and *Trichoderma*	*Pinus thunbergii*	Morphological and physiological	Playing roles in plant growth, nutrient absorption, resistance to harsh environments, defense response against pathogens, and resistance from attack of herbivores	Min et al. (2014)
26	*Paraconiothyrium*	*Limonium tetragonum*	Morphological	Improved morphology of halophyte	Khalmuratova et al. (2020)
27	*Paraphaeosphaeria*	*Suaeda glauca*	Morphological	Improved plant growth	Khalmuratova et al. (2020)

TABLE 3.1 *(Continued)*

Sr. no.	Endophytes	Halophytes	Studied modulation	Plant responses	Reference
28	*Paecilomyces formosus*, *Monosporascus ibericus*	*Anabasis iranica*	Morphological, physiological and Biochemical	Showed superiority in terms of enzymes and antibacterial properties	Jalili et al. (2020)
29	*Piriformospora indica*	*Medicago truncatula*	Biochemical	Accumulation of osmoprotectant (Proline), stimulating antioxidant enzymes, reduced malondialdehyde (MDA) activity, Na+ content and relative electrolyte conductivity (REC)	Li et al. (2017)
30	*Piriformospora indica*	*Medicago truncatula*	Molecular	Over-expression of defense-related genes (PR1, PR10, P5CS2, MtAlfin1-like and MtZfp-c2h2)	Li et al. (2017)
31	*Piriformospora indica*	*E. coli*	Molecular	Up-regulation of six genes involved in metabolism, energy and biosynthetic processes, DNA repair, regulation of protein turnover, transport and salt stress tolerance	Gahlot et al. (2015)
32	*Piriformospora indica*	*E. coli*	Molecular	SiPHO80 (a homolog of *Saccharomyces cerevisiae* PHO80) gene is the first report of a cyclin involved in salt tolerance	Loha et al. (2018)
33	*Stemphylium*	*Suaeda maritima*	Morphological	Plant growth promotion and improved morphological characters	Khalmuratova et al. (2020)

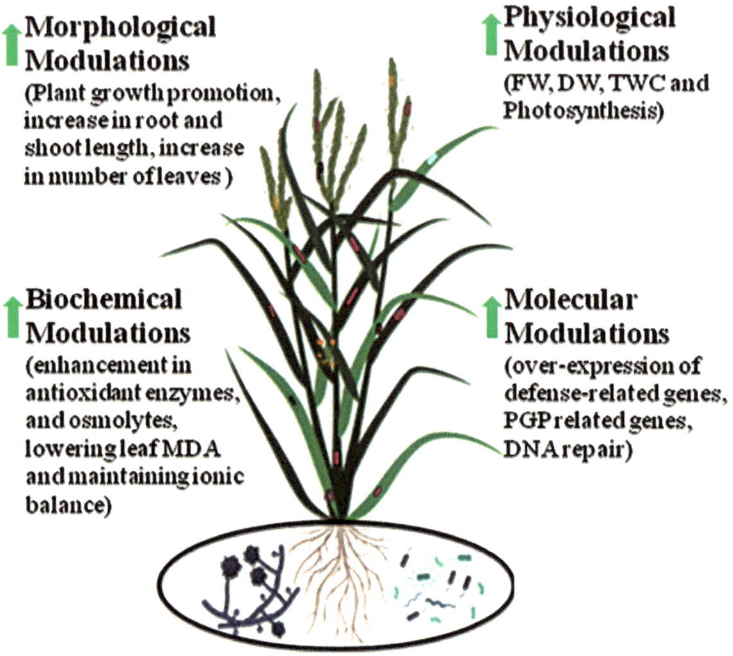

FIGURE 3.1 Alleviation of salt stress in endophytes-assisted halophytic plants.

as conidial, dark melanized hyphae, and microsclerotia, and are likely to play a crucial role in the ecophysiology of plants. Guo et al. (2021) have reported that the endophytes *Sphingomonas prati* and *Sphingomonas zeicaulis* exhibited antagonistic effects in halophyte *Suaeda salsa* as per inferences drawn from "interaction equation" (most G values were negative). Moreover, PLS model and PCA analysis revealed two processes in the regulation of plant salt tolerance, for example, endophyte *Sphingomonas prati* improved the growth of *Suaeda salsa* through (1) enhanced intracellular osmotic balance and (2) improved CAT production in antioxidant enzyme route and maintained the cell permeability.

3.5.3 ENDOPHYTES-ASSISTED BIOCHEMICAL RESPONSES OF HALOPHYTES UNDER SALT STRESS

The endophyte-assisted halophytes adapt themselves by different mechanisms and biochemical modulation is one of them. Under saline conditions, both ions, including sodium and chloride are essential for osmotic adjustment and many physio-biochemical modulations inside plant tissues. An increase of Cl^- content narrates to salinity tolerance which is linked with plant growth promotion, water use efficiency, transpiration, and defense system.

For instance, a halophyte *Limonium sinense* inoculated with halotolerant endophytic actinomycete strain (KLBMP 5084) exhibited higher total chlorophyll and proline content and also considerably lowered leaf malondialdehyde (MDA) content and higher antioxidant enzymes under both normal and saline environments. An endophyte-inoculated plant had significantly lower accumulation of Na^+ in leaves and roots through the compartmentalization of Na^+ in the vacuole of a cell (Qin, 2017). Endophytic community with 18 species mainly *Peyronellaea glomerata, Phoma macrostoma*, and *Alternaria tenuissima* associated with halophyte *Phragmites australis* showed higher resistance to zinc, mercury, and salt stress by promoting biochemical modulations (IAA production and phosphate solubilization). Bacterial endophytes such as *Achromobacter spanius, Bacillus amyloliquefaciens,* and *Microbacterium oxydans* recognized for growth promotional characteristics, auxin production, secretion of hydrolytic enzymes, and phosphate solubilization process (Soares et al., 2016). An endophytic ascomycete *Curvularia* sp. isolated from *Suaeda salsa* established a symbiosis with poplar (*Populus tomentosa*), which resulted in a significant increase in antioxidant enzymes, such as ascorbate peroxidase (APX) and superoxide dismutase (SOD), and modulations in chlorophyll a, chlorophyll b, proline content, PS-II-based electron transport rate, actual quantum yield (ϕPSII), and photochemical quenching events (qP) (Pan et al., 2018).

Piriformospora indica conferred high salt concentration in *Medicago truncatula* through the accumulation of osmoprotectant (proline), producing antioxidant enzymes, decline in MDA activity, Na^+, and relative electrolyte conductivity (REC) content under saline environment (Li et al., 2017).

An endophyte *Epichloe* spp. infected wild sea barley (*Hordeum brevisubulatum*) considerably improved chlorophyll content, photosynthetic rate, total antioxidants, and glycine betaine, while showing decline in leaf MDA content (Chen et al., 2018). Two tall fescue (*Festuca arundinacea*) genotypes (Fa75 and Fa83) infected with endophytic fungus *Neotyphodium coenophialum* revealed higher leaf survival rates, high root dry matter, greater Na^+ and Cl^- ion content in roots and accumulates more K^+, thereby the occurrence of higher K: Na ratio in the shoot (Sabzalian and Mirlohi, 2010).

Higher biomass as well as nitrogen:phosphorus ratio (N:P) and lower carbon and nitrogen (C:N), carbon and potassium (C:K) and sodium:potassium ratio (Na:K) was onserved in *Hordeum brevisubulatum* inoculated with endophyte *Epichloe* spp. (Song, et al., 2015). Likewise, inoculation of endophyte *Sphingomonas prati* declines MDA content by 23.78%, whereas enhancement of soluble sugar (15.08%) and soluble protein (12.57%) contents were recorded as compared with untreated endophyte conditions at a concentration of 20% NaCl. Increase in SOD and CAT content in the *Sphingomonas prati* inoculated *Suaeda salsa* halophyte were 1.03 and 1.47-fold greater, respectively, than in the endophyte *Sphingomonas zeicaulis* inoculation under 20% NaCl treatment (Guo et al., 2021).

3.5.4 ENDOPHYTES-ASSISTED MOLECULAR RESPONSES OF HALOPHYTES UNDER SALT STRESS

The bacterial endophytes express an array of parameters associated with cellular redox homeostasis, metabolism of amino acids, lipids, and nucleotides, transcription regulation, and energy production (Afzal et al., 2019). For instance, DREBs like transcription factors were employed in the regulation of gene expression in response to salt stress and other abiotic stresses through binding of DREBs to DRE (dehydration-responsive elements). Similarly, SERF1 belonging to group IIc ERFs of DREB sub-family, was found to be a positive regulator of short- and long-term salt stress tolerance, and upon salt exposure, it could augment the ROS-activated MAPK cascade in plant roots. In a genetic context, OsNAC genes might be employed in the crosstalk of various kinds of stress signaling-associated genes by positive regulation of stress-responsive genes exhibiting improved salt tolerance (Chen et al., 2021). An improved salt tolerance recorded in *Medicago truncatula* is bestowed by *Piriformospora indica* by employing over-expression of defense-responsive genes such as *PR1, PR10, P5CS2, MtAlfin1-like*, and *MtZfp-c2h2* (Li et al., 2017). Genomic analysis of halotolerant endophytic actinomycete strain (KLBMP 5084) inoculated with *Limonium sinense* which exhibited the activation of various PGP-related genes, among them some are implicated in adaptation to adverse environment and salt tolerance in all plant species (Qin et al., 2017). Likewise, SiPHO80 (a homolog of *Saccharomyces cerevisiae* PHO80) gene from *Piriformospora indica* endophyte is the first report of a cyclin which is involved in salt tolerance and other stress tolerance in several plants (Loha et al., 2018). For instance, six genes from *Piriformospora indica* are up-regulated under salt stress and are significant in salinity tolerance. These genes are involved in metabolic, energy and biosynthetic pathways, DNA repair mechanisms, regulation of protein turnover, translocation, and salt tolerance in plants (Gahlot et al., 2015).

The tested endophytes (8 strains) and more than half (14 strains) of the rhizospheric bacteria were recognized by the occurrence of acdS gene. This gene was identified in almost all known endophytes belonging to the phylum Proteobacteria; more than half of the strains belong to the phylum Actinobacteria. An occurrence of acdS gene in all strains of rhizospheric bacteria belonging to the phylum Proteobacteria while exactly half of the rhizopheric strains belong to the phylum Actinobacteria. Likewise, the acdS gene was identified in all endophytes and *Aster tripolium* rhizospheric strains (*Microbacterium, Pseudomonas,* and *Serratia*) (Szymańska et al., 2016).

3.6 CONCLUSION

The endophytes were isolated from root, stem, and leaf tissues of various halophytic plants and were reported as having high plant growth promoting activity and salt stress tolerance in plants growing under saline environment. It can be explained that salt-tolerant endophytes can live inside halophytic plants which significantly modulates morphological, physiological, biochemical, and molecular mechanisms. These endophytes were useful in nitrogen fixation, phosphate solubilization, synthesis of ACC deaminase, IAA, and siderophores which are known determinants for plant growth promotion. Modification at the molecular level to produce salt-tolerant cultivars is one of the promising fates, yet highly expensive techniques, whereas applications of endophytes on the other hand are very cheap. It is concluded that the selective best plant growth promoting endophytes can be used as potential bioinoculants for plant growth promotion in saline soil environment.

ACKNOWLEDGEMENT

GCN is thankful to Science and Engineering Research Board (SERB), India for fellowship under SERB- SIRE programme (Grant code-SIR/2022/000391).

KEYWORDS

- **endophytes**
- **halophytes**
- **osmotic stress**
- **salinity tolerance**

REFERENCES

Afzal, I.; Shinwari, Z. K.; Sikandar, S.; Shahzad, S. Plant Beneficial Endophytic Bacteria: Mechanisms, Diversity, Host Range and Genetic Determinants. *Microbiol. Res.* **2019,** *221,* 36–49.

Aguilera, L. E.; Gutierrez, J. R.; Moreno, R. J. Vesiculo Arbuscular Mycorrhizae Associated with Saltbushes *Atriplex* spp. (Chenopodiaceae) in the Chilean Arid Zone. *Rev. Chil. Hist. Nat.* **1998,** *71* (3), 291–302.

Ali, A.; Iqbal, N.; Ali, F.; Afzal, B. *Alternanthera bettzickiana* (Regel) G. Nicholson, a Potential Halophytic Ornamental Plant: Growth and Physiological Adaptations. *Flora Morphol. Distrib. Funct. Ecol. Plants* **2012,** *207* (4), 318–321.

Al-Mailem, D. M.; Sorkhoh, N. A.; Marafie, M.; Al-Awadhi, H.; Eliyas, M.; Radwan, S. S. Oil Phytoremediation Potential of Hypersaline Coasts of the Arabian Gulf Using Rhizosphere Technology. *Bioresour. Technol.* **2010,** *101* (15), 5786–5792. DOI: 10.1016/j.biortech.2010.02.082.

Angulo-Bejarano, P. I.; Puente-Rivera, J.; Cruz-Ortega, R. Metal and Metalloid Toxicity in Plants: an Overview on Molecular Aspects. *Plants* **2021,** *10* (4), 635.

Arnold, A. E.; Henk, D. A.; Eells, R. L.; Lutzoni, F.; Vilgalys, R. Diversity and Phylogenetic Affinities of Foliar Fungal Endophytes in Loblolly Pine Inferred by Culturing and Environmental PCR. *Mycologia* **2007,** *99* (2), 185–206.

Arnold, A. E.; Lutzoni, F. Diversity and Host Range of Foliar Fungal Endophytes: Are Tropical Leaves Biodiversity Hotspots? *Ecology* **2007,** *88* (3), 541–549.

Bacon, C. W. Abiotic Stress Tolerances (Moisture, Nutrients) and Photosynthesis in Endophyte-Infected Tall Fescue. *Agric. Ecosyst. Environ.* **1993,** *44* (1–4), 123–141.

Bacon, C, W.; White J. F. Jr. Physiological Adaptations in the Evolution of Endophytism in the Clavicipitaceae. In *Microbial Endophytes*; CRC Press, 2000; pp 251–276.

Berg, G. Plant–Microbe Interactions Promoting Plant Growth and Health: Perspectives for Controlled Use of Microorganisms in Agriculture. *Appl. Microbiol. Biotechnol.* **2009,** *84* (1), 11–18.

Brader, G.; Compant, S.; Mitter, B.; Trognitz, F.; Sessitsch, A. Metabolic Potential of Endophytic Bacteria. *Curr. Opin. Biotechnol.* **2014,** *27,* 30–37.

Chen, T.; Johnson, R.; Chen, S.; Lv, H.; Zhou, J.; Li, C. Infection by the Fungal Endophyte *Epichloë bromicola* Enhances the Tolerance of Wild Barley (*Hordeum Brevisubulatum*) to Salt and Alkali Stresses. *Plant Soil* **2018,** *428* (1), 353–370.

Chen, T.; Shabala, S.; Niu, Y.; Chen, Z.; Shabala, L.; Meinke, H.; Venkataraman, G.; Pareek, A.; Xu, J.; Zhou, M. Molecular Mechanisms of Salinity Tolerance in Rice. *Crop J.* **2021,** *9,* 506–520.

Compant, S.; Saikkonen, K.; Mitter, B.; Campisano, A.; Mercado-Blanco, J. Editorial Special Issue: Soil, Plants, and Endophytes. *Plant Soil* **2016,** *405* (1), 1–11.

de Bary, A. *Morphologie und Physiologie der Pilze, Flechten und Myxomyceten*; Engelmann, 1866; pp 1–315.

Dodd, I. C.; Pérez-Alfocea, F. Microbial Amelioration of Crop Salinity Stress. *J. Exp. Bot.* **2012,** *63* (9), 3415–3428.

Egamberdieva, D.; Lugtenberg, B. Use of Plant Growth-Promoting Rhizobacteria to Alleviate Salinity Stress in Plants. In *Use of Microbes for the Alleviation of Soil Stresses*; Springer: New York, NY, 2014; Vol. 1, pp 73–96.

Forni, C.; Duca, D.; Glick, B. R. Mechanisms of Plant Response to Salt and Drought Stress and their Alteration by Rhizobacteria. *Plant Soil* **2017,** *410* (1), 335–356.

Furtado, B. U.; Gołe̜biewski, M.; Skorupa, M.; Hulisz, P.; Hrynkiewicz, K. Bacterial and Fungal Endophytic Microbiomes of *Salicornia europaea*. *Appl. Environ. Microb.* **2019,** *85,* E00305–E00319. DOI: 10.1128/aem.00305-19.

Gahlot, S.; Joshi, A.; Singh, P.; Tuteja, R.; Dua, M.; Jogawat, A.; Tuteja, N. Isolation of Genes Conferring Salt Tolerance from *Piriformospora indica* by Random Overexpression in Escherichia Coli. *World J. Microbiol. Biotechnol.* **2015,** *31* (8), 1195–1209.

Ghosh, B.; Md, N. A.; Gantait, S. Response of Rice Under Salinity Stress: A Review Update. *Rice Res.* **2016,** 1–8.

Guo, J.; Chen, Y.; Lu, P.; Liu, M.; Sun, P.; Zhang, Z. Roles of Endophytic Bacteria in *Suaeda salsa* Grown in Coastal Wetlands: Plant Growth Characteristics and Salt Tolerance Mechanisms. *Environ. Pollut.* **2021,** *287,* 117641.

Hardoim, P. R.; Van Overbeek, L. S.; Berg, G.; Pirttilä, A. M.; Compant, S.; Campisano, A.; Sessitsch, A. The Hidden World within Plants: Ecological and Evolutionary Considerations for Defining Functioning of Microbial Endophytes. *Microbiol. Mol. Biol. Rev.* **2015,** *79* (3), 293–320.

Hassan, S. E. D.; Fouda, A.; Radwan, A. A.; Salem, S. S.; Barghoth, M. G.; Awad, M. A.; El-Gamal, M. S. Endophytic Actinomycetes *Streptomyces* spp Mediated Biosynthesis of Copper Oxide Nanoparticles as a Promising Tool for Biotechnological Applications. *J. Biol. Inorg. Chem.* **2019,** *24* (3), 377–393.

Hildebrandt, U.; Janetta, K.; Ouziad, F.; Renne, B.; Nawrath, K.; Bothe, H. Arbuscular Mycorrhizal Colonization of Halophytes in Central European Salt Marshes. *Mycorrhizamycorrhiza* **2001,** *10* (4), 175–183. DOI: 10.1007/s005720000074.

Jalili, B.; Bagheri, H.; Azadi, S.; Soltani, J. Identification and Salt Tolerance Evaluation of Endophyte Fungi Isolates from Halophyte Plants. *Int. J. Environ. Sci. Technol.* **2020,** *17* (7), 3459–3466. DOI: 10.1007/s13762-020-02626-y.

Juniper, S.; Abbott, L. K. Soil Salinity Delays Germination and Limits Growth of Hyphae from Propagules of Arbuscular Mycorrhizal Fungi. *Mycorrhiza* **2006,** *16* (5), 371–379. DOI: 10.1007/s00572-006-0046-9.

Kannan, K. P.; Madhan, K. D.; Ramya, P. R.; Madhu Nika, S.; Meenatchi, G.; Sowmya, A. N.; Bhuvaneswari, S. Diversity of Endophytic Fungi from Salt Tolerant Plants. *Int. J. Chem. Technol. Res.* **2014,** *6* (9), 4084–4088.

Kearl, J.; Mcnary, C.; Lowman, J. S.; Mei, C.; Aanderud, Z. T.; Smith, S. T.; West, J.; Colton, E.; Hamson, M.; Nielsen, B. L. Salt-Tolerant Halophyte Rhizosphere Bacteria Stimulate Growth of Alfalfa in Salty Soil. *Front Microbiol.* **2019,** *10,* 1849. DOI: 10.3389/fmicb.2019.01849.

Khalmuratova, I.; Choi, D. H.; Woo, J. R.; Jeong, M. J.; Oh, Y.; Kim, Y. G.; Kim, J. G. Diversity and Plant Growth-Promoting Effects of Fungal Endophytes Isolated from Salt-Tolerant Plants. *Korean Soc. Microbiol. Biotechnol.* **2020,** *30* (11), 1680–1687.

Kondrasheva, K. V.; Egamberdiev, F. B.; Suyarova, R. A.; Ruzieva, D. M.; Nasmetova, S. M.; Abdulmyanova, L. A.; Rasulova, G. A.; Gulyamova, T. G. In *Production of Indole-3-Acetic Acid by Endophytic Fungi of Halophyte Plants under Salt Stress.* IOP Conf. Series: Earth and Environmental Science, 1068, 012040, 2022. DOI: 10.1088/1755-1315/1068/1/012040.

Li, L.; Li, L.; Wang, X.; Zhu, P.; Wu, H.; Qi, S. Plant Growth-Promoting Endophyte *Piriformospora indica* Alleviates Salinity Stress in *Medicago truncatula*. *Plant Physiol. Biochem.* **2017,** *119,* 211–223.

Li, X.; Geng, X.; Xie, R.; Fu, L.; Jiang, J.; Gao, L.; et al. The Endophytic Bacteria Isolated from Elephant Grass (*Pennisetum purpureum* Schumach) Promote Plant Growth and Enhance Salt Tolerance of Hybrid *Pennisetum*. *Biotechnol. Biofuels.* **2016,** *9,* 190. DOI: 10.1186/s13068-016-0592-0.

Loha, A.; Kashyap, A. K.; Sharma, P. A Putative Cyclin, SiPHO80 From Root Endophytic Fungus *Serendipita indica* Regulates Phosphate Homeostasis, Salinity and Heavy Metal Toxicity Tolerance. *Biochem. Biophys. Res. Commun.* **2018,** *507* (1–4), 414–419.

Manasa, K. M.; Vasanthakumari, M. M.; Nataraja, K. N.; Uma Shaanker, R. Endophytic Fungi of Salt Adapted *Ipomea pes-caprae* L. R. Br: Their Possible Role in Inducing Salinity Tolerance in Paddy (*Oryza sativa* L.). *Curr. Sci.* **2020,** *118* (9), 1448.

Marasco, R.; Mapelli, F.; Rolli, E.; Mosqueira, M. J.; Fusi, M.; Bariselli, P.; Daffonchio, D. *Salicornia strobilacea* (Synonym of *Halocnemum strobilaceum*) Grown under Different Tidal Regimes Selects Rhizosphere Bacteria Capable of Promoting Plant Growth. *Front. Microbiol.* **2016,** *7,* 1286.

Mateu, G. M.; Baldwin, A. H.; Maul, J. E.; Yarwood, S. A. Dark Septate Endophyte Improves Salt Tolerance of Native and Invasive Lineages of *Phragmites australis*. *ISME J.* **2020,** *14* (8), 1943–1954.

Min, Y. J.; Park, M. S.; Fong, J. J.; Quan, Y.; Jung, S.; Lim, Y. W. Diversity and Saline Resistance of Endophytic Fungi Associated with *Pinus thunbergii* in Coastal Shelterbelts of Korea. *J. Microbiol. Biotechnol.* **2014,** *24* (3), 324–333. DOI: 10.4014/jmb.1310.10041.

Nanjundappa, M.; Manjunatha, N.; Li, H.; Sivasithamparam, K.; Jones, M. G.; Edwards, I.; Agarrwal, R. Fungal Endophytes from Salt-Adapted Plants Confer Salt Tolerance and Promote Growth in Wheat (*Triticum aestivum* L.) at Early Seedling Stage. *Biorxiv* **2021,** 1–18. DOI: 10.1101/2021.12.02.470908.

Nikalje, G. C.; Srivastava, A. K.; Sablok, G.; Pandey, G. K.; Nikam, T. D.; Suprasanna, P. Identification and Validation of Reference Genes for Quantitative Real-Time PCR Under Salt Stress in a Halophyte, *Sesuvium portulacastrum*. *Plant Gene* **2018,** *13,* 18–24.

Nikalje, G. C.; Bhaskar, S. D.; Yadav, K.; Penna, S. Halophytes: Prospective Plants for Future. In Hasanuzzaman, M., Nahar, K., Öztürk , M., Eds.; *Ecophysiology, Abiotic Stress Responses and Utilization of Halophytes*; Springer: Singapore, 2019. DOI: 10.1007/978-981-13-3762-8_10

Muchate, N. S.; Nikalje, G. C.; Rajurkar, N. S.; Suprasanna, P.; Nikam, T. D. Physiological Responses of the Halophyte Sesuvium Portulacastrum to Salt Stress and Their Relevance for Saline Soil Bio-reclamation. *Flora* **2016,** *224,* 96–105.

Munns, R. Comparative Physiology of Salt and Water Stress. *Plant Cell Environ.* **2002,** *25* (2), 239–250.

Munns, R.; Tester, M. Mechanisms of Salinity Tolerance. *Annu. Rev. Plant Biol.* **2008,** *59,* 651.

Pan, X.; Qin, Y.; Yuan, Z. Potential of a Halophyte-Associated Endophytic Fungus for Sustaining Chinese White Poplar Growth under Salinity. *Symbiosis* **2018,** *76* (2), 109–116.

Patle, B. K.; Pandey, A.; Jagadeesh, A.; Parhi, D. R. Path Planning in Uncertain Environment by Using Firefly Algorithm. *Defence Technol.* **2018,** *14* (6), 691–701.

Petrini, O. Fungal Endophytes of Tree Leaves. In *Microbial Ecology of Leaves*; Springer: New York, NY, 1991; pp 179–197.

Piccoli, P.; Travaglia, C.; Cohen, A.; Sosa, L.; Cornejo, P.; Masuelli, R.; Bottini, R. An Endophytic Bacterium Isolated From Roots of the Halophyte *Prosopis strombulifera*

Produces ABA, IAA, Gibberellins A(1) and A(3) and Jasmonic Acid in Chemically-Defined Culture Medium. *Plant Growth Regul.* **2011,** *64* (2), 207–210. DOI: 10.1007/s10725-010-9536-z.

Qin, Y.; Pan, X.; Kubicek, C.; Druzhinina, I.; Chenthamara, K.; Labbé, J.; Yuan, Z. Diverse Plant-Associated Pleosporalean Fungi From Saline Areas: Ecological Tolerance and Nitrogen-Status Dependent Effects on Plant Growth. *Front. Microbiol.* **2017,** *8,* 158.

Rhoden, S. A.; Corrêa, R. C. G.; Mota, T. R.; Azevedo, J. L.; Pamphile, J. A.; de Souza, C. G. M.; Peralta, R. M. Endophytic Fungi: Expanding the Arsenal of Industrial Enzyme Producers. *J. Ind. Microbiol. Biotechnol.* **2014,** *41* (10), 1467–1478.

Rodríguez-Lorente, I. D.; Pajuelo, E.; Navarro-Torre, S.; Mesa-Marín, J.; Caviedes, M. A. Bacterial Endophytes from Halophytes: How do they Help Plants to Alleviate Salt Stress? In *Saline Soil-Based Agriculture by Halotolerant Microorganisms*; Springer: Singapore, 2019; pp 147–160.

Ruppel, S.; Franken, P.; Witzel, K. Properties of the Halophyte Microbiome and their Implications for Plant Salt Tolerance. *Funct. Plant Biol.* **2013,** *40,* 940–951. DOI: 10.1071/fp12355.

Sabzalian, M.; Mirlohi, A. *Neotyphodium* Endophytes Trigger Salt Resistance in Tall and Meadow Fescues. *J. Plant Nutr. Soil Sci.* **2010,** *173* (6), 952–957.

Santoyo, G.; Moreno-Hagelsieb, G.; del C Orozco-Mosqueda, M.; Glick, B. R. Plant Growth-Promoting Bacterial Endophytes. *Microbiol. Res.* **2016,** *183,* 92–99.

Sengupta, A.; Chaudhuri, S. Arbuscular Mycorrhizal Relations of Mangrove Plant Community at the Ganges River Estuary in India. *Mycorrhiza* **2002,** *12* (4), 169–174. DOI: 10.1007/s00572-002-0164-Y.

Sharma, P.; Kharkwal, A. C.; Abdin, M. Z.; Varma, A. *Piriformospora indica*-Mediated Salinity Tolerance in *Aloe Vera* Plantlets. *Symbiosis* **2017,** *72* (2), 103–115.

Shelke, D. B.; Chambhare, M. R.; Sonawane, H. Fungal Endophytes: Potential Benefits of their Future Use in Plant Stress Tolerance and Agriculture. In *Beneficial Microorganisms in Agriculture*; Springer: Singapore, 2022; pp 177–209.

Shelke, D. B.; Nikalje, G. C.; Chambhare, M. R.; Zaware, B. N.; Penna, S.; Nikam, T. D. Na^+ and Cl^- Induce Differential Physiological, Biochemical Responses and Metabolite Modulations In Vitro in Contrasting Salt-Tolerant Soybean Genotypes. *3 Biotech* **2019,** *9* (3), 1–15.

Shelke, D. B.; Pandey, M.; Nikalje, G. C.; Zaware, B. N., Suprasanna, P.; Nikam, T. D. Salt Responsive Physiological, Photosynthetic and Biochemical Attributes at Early Seedling Stage for Screening Soybean Genotypes. *Plant Physiol. Biochem.* **2017,** *118,* 519–528. DOI: 10.1016/j.plaphy.2017.07.013.

Soares, M. A.; Li, H. Y.; Kowalski, K. P.; Bergen, M.; Torres, M. S.; White, J. F. Evaluation of the Functional Roles of Fungal Endophytes of *Phragmites australis* from High Saline and Low Saline Habitats. *Biol. Invasions* **2016,** *18* (9), 2689–2702.

Sonawane, H.; Ghule, S.; Math, S.; Shelke, D.; Nikalje, G. Rhizoctonia Bataticola: From Plant Pathogen to a Potential Source of Pharmaceutically Relevant Metabolites. *Curr. Res. Green Sustain. Chem.* **2021,** *4,* 100171. DOI: 10.1016/j.crgsc.2021.100171.

Sonawane, H. B.; Borde, M. Y.; Nikalje, G. C.; Terkar, A.; Math, S. K. HR-LC-MS based Metabolic Profiling of *Fusarium solani* a Fungal Endophyte Associated with *Avicennia officinalis*. *Curr. Res. Environ. Appl. Mycol.* **2020,** *10* (1), 262–273.

Song, M.; Chai, Q.; Li, X.; Yao, X.; Li, C.; Christensen, M. J.; Nan, Z. An Asexual Epichloë Endophyte Modifies the Nutrient Stoichiometry of Wild Barley (*Hordeum brevisubulatum*) Under Salt Stress. *Plant Soil.* **2015,** *387* (1), 153–165.

Sonjak, S.; Udovič, M.; Wraber, T.; Likar, M.; Regvar, M. Diversity of Halophytes and Identification of Arbuscular Mycorrhizal Fungi Colonising Their Roots in an Abandoned and Sustained Part of Sečovlje Salterns. *Soil Biol. Biochem.* **2009,** *41* (9), 1847–1856.

Szymańska, S.; Płociniczak, T.; Piotrowska-Seget, Z.; Złoch, M.; Ruppel, S.; Hrynkiewicz, K. Metabolic Potential and Community Structure of Endophytic and Rhizosphere Bacteria Associated with the Roots of the Halophyte *Aster Tripolium* L. *Microbiol. Res.* **2016,** *182,* 68–79.

Teo, H. M.; Bhubalan, K.; CI, M. S.; Ng, L. C. Setting a Plausible Route for Saline Soil-based Crop Cultivations by Application of Beneficial Halophyte-associated Bacteria: A Review. *Microorganisms* **2022,** *10* (3), 657.

Vaishnav, A.; Kumari, S.; Jain, S.; Varma, A.; Tuteja, N.; Choudhary, D. K. PGPR-Mediated Expression of Salt Tolerance Gene in Soybean through Volatiles Under Sodium Nitroprusside. *J. Basic Microbiol.* **2016,** *56* (11), 1274–1288.

Waller, F.; Achatz, B.; Baltruschat, H.; Fodor, J.; Becker, K.; Fischer, M.; Kogel, K. H. The Endophytic Fungus *Piriformospora indica* Reprograms Barley to Salt-Stress Tolerance, Disease Resistance, and Higher Yield. *Proc. Natl. Acad. Sci.* **2005,** *102* (38), 13386–13391.

Wang, F. Y.; Liu, R. J.; Lin, X. G.; Zhou, J. M. Arbuscular Mycorrhizal Status of Wild Plants in Saline-Alkaline Soils of the Yellow River Delta. *Mycorrhiza* **2004,** *14* (2), 133–137. DOI: 10.1007/s00572-003-0248-3.

Yang, Y. H.; Chen, Y. N.; Li, W. H. Arbuscular Mycorrhizal Fungi Infection in Desert Riparian Forest and Its Environmental Implications: A Case Study in the Lower Reach of Tarim River. *Prog. Nat. Sci.* **2008,** *18* (8), 983–991. DOI: 10.1016/J.Pnsc.2008.02.009.

Xiong, L.; Zhu, J. K. Molecular and Genetic Aspects of Plant Responses to Osmotic Stress. *Plant Cell Environ.* **2002,** *25* (2), 131–139.

Yasseen, B. T. Al-Thani, R. F. Endophytes and Halophytes to Remediate Industrial Wastewater and Saline Soils: Perspectives from Qatar. *Plants* **2022,** *11,* 1497. DOI 10.3390/PLANTS11111497.

Zhang, Y.; Li, X. M.; Wang, C. Y.; Wang, B. G. A New Naphthoquinoneimine Derivative from the Marine Algal-Derived Endophytic Fungus *Aspergillus niger* EN-13. *Chin. Chem. Lett.* **2007,** *18* (8), 951–953.

Zhu, J. K. Plant Salt Tolerance. *Trends Plant Sci.* **2001,** *6* (2), 66–71.

CHAPTER 4

Salt Signaling in Halophytes

JOSHI. A. SAMIKSHA

Department of Botany, University of Pune, Maharashtra, India

ABSTRACT

Salt stress is the major environmental stress that a plant has to undergo during the stages of growth and development. Excess salt in the soil results in salt stress. This makes the plants the/halophytes unable to take up water from roots. Halophytes get adversely affected by the osmotic stress and the stress by ions. NaCl is the most important salt to participate in salt tolerance. When under salt stress the halophytes have maximum concentration of K^+ ions and minimum conc. of Na^+, that is, they maintain high potassium ion and low sodium ion conc. This is done with the help of the regulation of the activity of K^+ and Na^+ transporters. The regulation and activity of the H^+ pump play an important role in driving the force for transporters. Various signaling of different protein kinases, salt overly sensitive (SOS2) proteins, and calcium-binding proteins SOS3, also get started with salt stress booster signals of calcium proteins. Also, the concentration of calcium increase in the cytosol is one of the early responses in halophytes. Three different types of calcium fluxes are identified in halophytes guanosine 3',5'-cyclic monophosphate (cGMP) is found to be one of the important proteins in halophytes which on increase and decrease in the cytosolic concentration induces changes in calcium import that further regulates Na^+ import and also reduces K^+ influx. In this chapter, the detailed salt stress signaling in halophytes is described.

Physiology of Halophytes: Signaling, Omics, and Tolerance Mechanisms.
Nivas Manohar Desai & Ganesh Chandrakant Nikalje (Eds.)

4.1 INTRODUCTION

Halophytes are those plants which are resistant to salt or salt-tolerant plants. Halophytes constitutes 1–2% of total world flora (Tester and Davenport, 2003). The salinity of soil is a global concern. It adversely affects the agriculture. There are many factors that contribute toward increased salinity. Some of them are anthropogenic activities, places nearby coastal areas, and more. Halophytes can complete their life cycle in 200 NaCl and are also capable of tolerating high concentration (Blumwald et al., 2000). Some also show maximum growth in such conditions. The saline soils are rich in salts such as $CaSO^4$ Na^2CO^3, and NaCl which majorly dominate the saline soil. This affects the soil in many ways such asthe osmotic stress is disturbed which leads to water deficiency due to high solute concentration in the soil (Munns and Tester, 2008). Disturbed ratio or decreased ratio of K^+ ion concentration (Tester and Davenport, 2003). Alterations of the K^+/Na^+ ratio at the maximum concentration exceeds the sodium ions inside the cell cytosol. In salt stress conditions, the influx of sodium is supported through a pathway that functions for K^+, since the radius of ions Na^+ and K^+ seems similar under the moistened state. Hence, a required balance needs to be maintained for the Na^+ level in the cytoplasm to keep the high concentration of K^+ ions resulting in high K^+/Na^+ ratio. In this chapter, different pathways and Na influx; efflux into the cytoplasm and out the cell; Na transport, the various transporters and different channels, vacuolar compartmentalization, and the SOS pathway are discussed. There are many more mechanisms such as osmolyte biosynthesis, alteration in ion homeostasis, intracellular compartmentalization of toxic ions, and ROS scavenging system against which the plants have developed complex defense to resist salt stress. Induction of these pathways for a brief exposure to salt stress develop more resistance to salinity. Based on the interaction with salt and salinity tolerance the halophytes are divided into pseudo halophytes, euhalophytes and recretohaophytes. Halophytes have also developed distinct morphological, structural, and physiological to sustain high salt concentrations. To investigate the molecular mechanism various salt–responsive genes have been studied.

4.2 PHYSIOLOGICAL EFFECTS

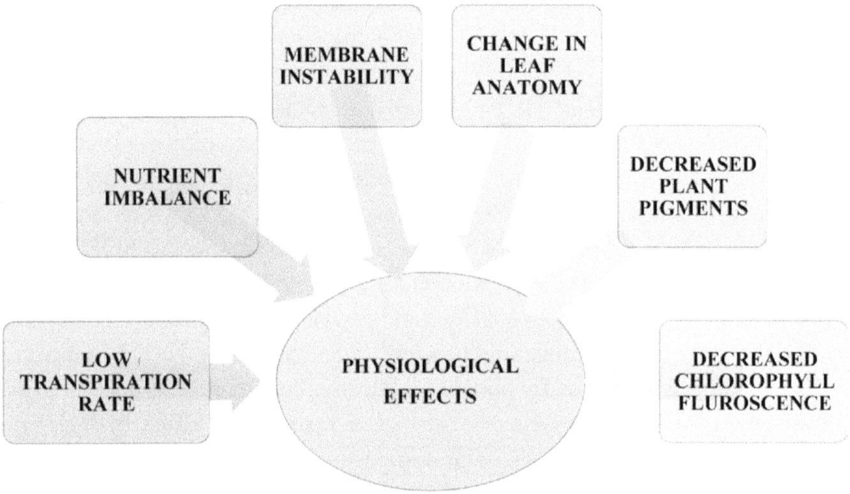

Salt stress directly or indirectly impacts the physiological and biological mechanism of plants thereby reducing the crop yield and production. Initially, the number of solutes in the soil is toward the higher side which increases the osmotic stress and this reduces water absorption from the roots (Hasegawa et al., 2000). This accelerates transpiration. Also, some ionic effect causes an increase of toxic sodium and chlorine ions Na^+ and Cl^- in the cells. This results in chlorosis and necrosis of leaves (Blumwald et al., 2000). Osmotic stress can be overcome when accompanied by the accumulation of internal solutes under low and moderate salt stress conditions. Due to the imbalance of solutes inside the cell the salt concentration stress increases resulting in deficiency symptoms similar to nutrient imbalance or nutrient deficiency symptoms. This occurs due to the ions interfering in the membrane function system which impacts the nutrient absorption and solute balance inside the cell. The degree of salt tolerance on the germplasms also varies from plant to plant of the domestic and the wild species limiting the cultivars, varieties, and species growth. The reduction in the photosynthesis and assimilation also occurs in the salt-sensitive plants as they fail to sequester salts at a higher rate in the vacuole than salt-tolerant plants. Other major changes that occur are a decrease in plant pigments such as chlorophyll, carotenoids, etc., a small aperture of stomatal pores, and a decrease in the transpiration rate.

4.3 SODIUM SENSING AND ENTRY IN HALOPHYTES

Entry of sodium is facilitated to the cell via channels and transporters. Here, the nonselective cation channels (NSCCs) and transporters such as high-affinity potassium transporter 5 (HAK5) and high-affinity potassium transporters (HKTs) are accompanied. There are several sensory mechanisms through which the molecules are sensed. The Na^+ is sensed by an unknown mechanism that activates the Ca^{2+} signaling cascade (Abideen et al., 2014). Through several steps. the Ca^{2+} increases in the cytosol, is responded by SOS3 which is a myristolated Ca^{2+} sensor protein. SOS^3 interacts with SOS2 and the SOS2–SOS3 complex is phosphorylated. The phosphorylation is due to the kinase activity of SOS2 and the complex gets confined to the plasma membrane. Certain antiporters also participate during this process. One such is the Na^+/H^+ antiporter (NHX) which gets activated by phosphorylation of SOS1 by SOS2 (Grattan and Grieve, 1998). Activation of Na^+/H^+ antiporter results into efflux of Na^+ from the cell. During this process, vacuolar compartmentalization by Na^+ occurs via NHXC $Na^{+/}$ H^+ exchanger. These activities are in response to stress.

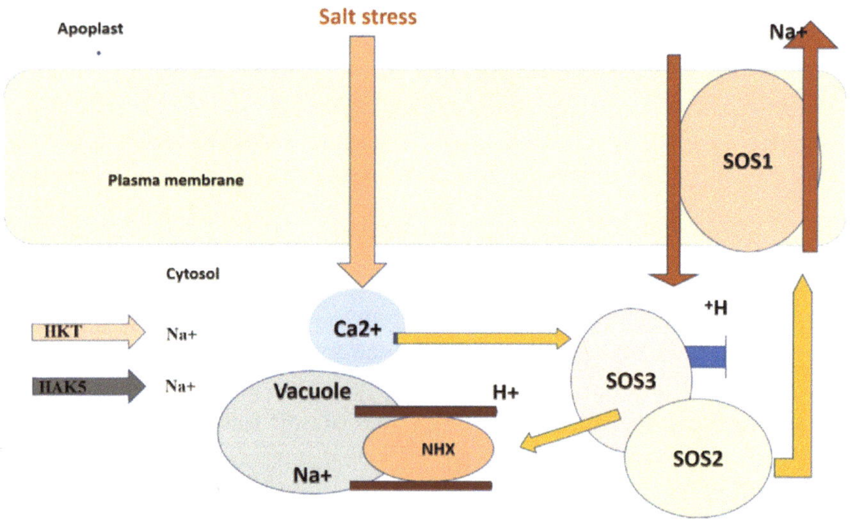

FIGURE 4.1 SOS pathway in halophytes.

4.4 MECHANISM OF NA INFLUX INTO CYTOSOL

Normally the $K^{+/}Na^+$ ratio across the plasma membrane is high and electrical membrane potential difference in plants remains negative. On the contrary, under salt stress, there is a passive transport of Na^+ from the soil to the cytosol as the increased Na^+ concentration in the soil establishes an electrochemical gradient. As mentioned earlier, there are transporters and channels facilitating the process; here NSCCs along with cyclic nucleotide-gated channels (CNGCs) and glutamate receptor (GLRs) are involved in Na^+ influx into roots (Agarwal et al., 2010). Similarly, HKTs and HAKs having high potassium affinity are also involved. Na^+ gets transported in a gradient manner from roots to the xylem and finally the shoots. Several channels and transporters such as voltage-insensitive nonselective cation channel (ViNSCC) play an important role during the root transport of Na^+ and hence have been classified into two types according to the cation they block for example Ca^{2+}, Mg^{2+}, Zn^{2+}, and so on. Quinine is the inhibitor of this ViNSCC. Certain relevant evidence shows that ViNSCC also participates in the influx of ions into the roots. In certain species of *Haemanthus* and *Clivia* fruits, the permeability of Na^+ in endosperm cells has been demonstrated. Also, epidermal cells of *Pisum sativum* and leaves and roots of *rye* cells. CNGCs and GLRs are the two subclasses of NSCC that participate in Na^+ influx. The presence of K^+ transporter having high affinity is also observed in bacteria, fungi, and plants (Ali and Yun, 2017). HKT's are also classified into two subclasses differing in the specificity for Na^+ and K^+ in context with motifs. Class-1 HKTs have SGGG (S = serine; G = glycine) pore-forming motif and is Na^+ specific while class II has GGGG pore-forming motif and are Na^+ and K^+ specific. Class-II is found in both monocots and dicots, and class I in only monocots (Abideen et al., 2014; Agarwal et al., 2010).

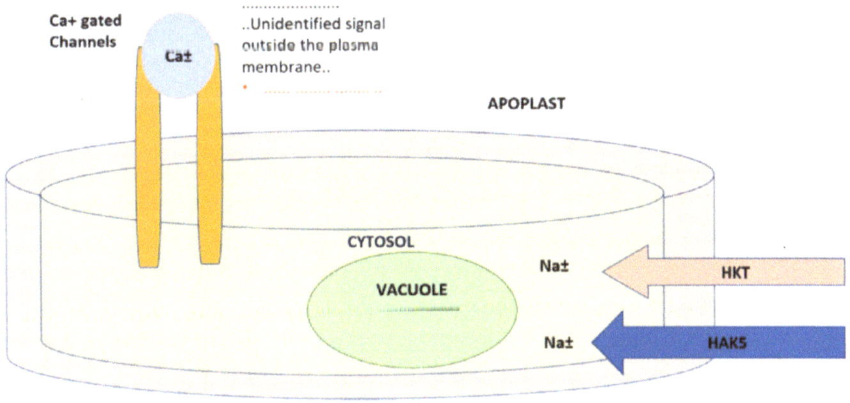

FIGURE 4.2 Sodium sensing and entry in halophytes.

4.4.1 *MECHANISM OF NA EFFLUX INTO VACUOLES*

Plants confine the compartmentalization of Na^+ to vacuoles to reduce Na^+ concentration in cytosol.

The process is facilitated by NHXs antiporters. It proves to be an important regulator. Overexpression of certain regulators identified in plants like the *Arabidopsis thaliana* NHX1 is found to improve salinity stress tolerance. This also has confined itself to the tonoplast of plants. Similarly, Brassica, cotton, maize, and some more transgenic plants tend to improve their salt tolerance after overexpression of the NHX regulator (Abideen et al., 2014). NHX2 also plays an important role in several processes such as Na compartmentalization, pH regulation in vacuoles, and potassium uptake into vacuoles. The efficiency of NHX increases when the concentration of NaCl is between 100 and 200 mM which enhances salt tolerance for these NHXs. These NHXs have been reported recently in *A. thaliana* to not only involved in Na compartmentalization but also regulation of pH and uptake of potassium/K+ into the vacuoles. The activity of induction of NHX1 and NHX2 is regulated by ABA. The role of NHX2 and NHX5 is yet to be determined apart from the activity of Na sequestration. The NHX plays an important role in salt tolerance when activated under salinity stress.

4.5 LONG-DISTANCE TRANSPORT OF SODIUM

There are two pathways in plants by which the water and food along with the nutrients and ions are translocated. The two processes or pathways are symplastic and the apoplastic. Sodium once travels to the epidermis is soon transported to the xylem via apopalastic or symplastic way. During this transport via the apoplastic pathway, some proportion of sodium remains confined to the root apoplast, and the remaining travels to the endodermis. The transport across the plasma membrane is vital for further radial transport. While the travel of solutes to the endodermis and exodermis there appears a barrier the Casparian strips and suberin lamellae which facilitates movement of selective ions in and out to the stele. At times the Na^+ may also cross the endodermis through apoplastic this is called the bypass flow. Bypass flow bypasses the symplastic region as the apoplastic barriers there have underdeveloped or the flow is lacking similar to young roots and root tips in lateral roots. The reduction of apoplastic pathway bypass can be possible by supplementation of silicon. The bypass flow is seen as significant in the rice the *Oryza sativa* and not in *Arabidopsis*. The other physical barrier to Na is reported in *Brassica* by the phi cells with a thick cell wall. The concentration of Na^+ is higher in apoplast in both glycophytes and halophytes as compare to cytoplasm as the enzyme present in apoplast is more salt tolerant.

4.6 REACTIVE OXYGEN SPECIES

Reactive oxygen species are molecules generated in response to salinity stress on redox homeostasis and cause oxidative stress. The oxidative stress proves detrimental to the cellular components. Reactive oxygen species are also one of the sensor molecules which responds to various signaling during stress conditions. These are produced in cytoplasm and subcellular compartments. Several studies have been carried out on halophytes and this led to determining the antioxidant defense system which play a vital role in the maintenance of plant growth in response to abiotic stresses. Hydrophytes possibly act on the halophyte antioxidant enzyme system particularly in two ways, first antioxidant system that nullifies ROS, second maintenance of decreased ROS production.

TABLE 4.1 Channels and Transporters in Halophytes.

Name	Role	Species
ViNSCC	Sodium permeability and influx into roots	*P. sativum*, *Triticum aestivum*
CNGC	Sodium uptake in the roots and unidirectional sodium flux	*A. thaliana*, *O. sativa*
GLR	Sodium permeability and influx into roots	*A. thaliana*
HKT	Influx of sodium in roots, retrieval from xylem	*A. thaliana*, *O. sativa*
HAK5	Sodium permeability and low-affinity influx into roots	*A. thaliana*, *Phragmites australis*
NHX	Sodium sequestration into vacuoles	*A. thaliana*, *T. aestivum*
SOS1	Sodium efflux from roots	*A. thaliana*, *Cynodon dactylon*

4.6.1 CALCIUM SIGNALING

In salinity, stress calcium plays an important role as a secondary messenger in the signaling pathway and also in abiotic stress. A sudden hike of cytosolic calcium generates several molecules and thus activates and regulates calcium levels. For the regulation of cytoplasm calcium level, the calcium transporters are involved including CNBC TPCs double-pore calcium channels, calcium ATP's vacuolar hydrogen calcium antiporter CAX (Ali and Yun, 2016). Also, the external signals function downstream targets that result in the regulation of varied pathways and salt-responsive genes.

4.7 SALT RESPONSIVE GENES

Under salinity, plants deal with osmotic stress that shows water deficiency, also ion specific stress is observed. The tolerance mechanism of halophytes responds in two ways; first, salt tolerance followed by the other technique, salt exclusion. The technique of salt exclusion includes secretion, shedding, and succulence. It was seen that the overexpression of salt tolerance genes under control conditions enhanced abiotic stress tolerance. *Salicornia brachyiata* and *Ricinus communis* were some of the plants on which the salt tolerance due to overexpression of genes was assessed (Abideen et al., 2014).

4.8 CONCLUSION

The chapter deals briefly with cellular signaling, the regulators, transporters, and various channels that facilitate the regulation in halophytes. It also briefly describes the physiological impacts on halophytes due to abiotic stress. Differential expression of various salt-responsive genes due to advanced techniques and studies have also been brief with certain examples of plants on which sustainable studies are carried out in halophytes. Calcium signaling, Na^+ influx and efflux from the cell and the transport in halophytes, and SOS signaling have been discussed in detail. The chapter toward the end leaves some gap to focus on the studies regarding up regulation and down-regulation of certain salt-responsive genes and to remain focus on improving salt tolerance genes for future studies and research.

KEYWORDS

- **salt stress**
- **protein kinases**
- **SOS2**
- **SOS3**
- **cGMP**
- **Na+ import**
- **K+ influx**

REFERENCES

Abideen, Z.; Koyro, H.-W.; Huchzermeyer, B.; Ahmed, M. Z.; Gul, B.; Khan, M. A. 2014. Moderate Salinity Stimulates Growth and Photosynthesis of Phragmites Karka by Water Relations and Tissue Specific Ion Regulation. *Environ. Exp. Bot. 105,* 70–76. https:/doi.org/10.1016/j.envexpbot.2014.04.009.

Abideen, Z.; Koyro, H.-W.; Huchzermeyer, B.; Ahmed, M. Z.; Gul, B.; Khan, M. A.

Agarwal, P. K.; Gupta, K.; Jha, B. Molecular Characterization of the Salicornia Brachiata, SbMAPKK Gene and Its Expression by Abiotic Stress. *Mol. Biol. Rep.* **2020,** *37* (2), 981–986.

Ali, A.; Cheol Park, H.; Aman, R.; Ali, Z.; Yun, D.-J. Role of HKT1 in Thellungiella2 2013.

Ali, A.; Yun, D. J. Differential Selection of Sodium and Potassium Ions by TTsHKT1, 2016.

Ali, A.; Yun, D. J. Salt Stress Tolerance; What Do We Learn From Halophytes? *J. Plant Biol.* **2017,** *60* (5), 431–439. https://doi.org/10.1007/s12374-017-0133-9. Springer New York.

Blumwald, E.; Aharon, G. S.; Apse, M. P. Sodium Transport in Plant Cells. *Biochim. Biophys. Acta* **2000,** *1465,* 140–151.

Bolarian, M.; Fernandez, F.; Cruz, V.; Cuartero, J. Salinity Tolerance in Four Wild Tomato Species using Vegetative Yield-salinity Response Curves. *J. Am. Soc. Hortic. Sci.* **1991,** *116,* 286–290.

Ghoulam, C.; Foursy, A.; Fares, K. Effects of Salt Stress on Growth, Inorganic Ions and Proline Accumulation in Relation to Osmotic Adjustment in Five Sugar Beet Cultivars. *Environ. Exp. Bot.* **2002,** *47,* 39–50.

Grattan, S. R.; Grieve, C. M. Salinity-mineral Nutrient Relations in Horticultural Crops. *Sci. Hort.* **1998,** *78,* 127–157.

Gupta, B.; Huang, B. Mechanism of Salinity Tolerance in Plants: Physiological, Biochemical, and Molecular Characterization. *Int. J. Genom.* **2014,** 701596.

Hasegawa, P. M.; Bressan, R. A.; Zhu, J. K.; Bohnert, H. J. Plant Cellular and Molecular Responses to High Salinity. *Annu. Rev. Plant Physiol. Plant. Mol. Biol.* **2000,** *51,* 463–499.

Maathuis, F. J. M. Sodium in Plants: Perception, Signalling, and Regulation of Sodium Fluxes. *J. Exp. Bot.* **2014**.

Moderate Salinity Stimulates Growth and Photosynthesis of Phragmites Karka by Water Relations and Tissue Specific Ion Regulation. *Environ. Exp. Bot.* **2014,** *105,* 70–76. https://doi.org/10.1016/j.envexpbot.2014.04.009.

Munns, R.; Tester, M. Mechanisms of Salinity Tolerance. *Annu. Rev. Plant Biol.* **2008,** *59,* 651–681.

Pardo, J. M.; Rubio, F. Na^+ and K^+ transporters in Plant Signaling. In *Transporters and Pumps in Plant Signaling*; Geisler, M., Venema, K., Eds.; Springer: Berlin/Heidelberg, Germany, 2011; Vol. 7, pp 65–98.

Tester, M.; Davenport, R. Na^+ Tolerance and Na^+ Transport in Higher Plants. *Ann. Bot.* **2003,** *91,* 503–527.

CHAPTER 5

Metabolomic, Proteomic, and Ionomic Adjustment and Adaptive Mechanism of Halophytes

RITU L. JAIN

Department of Botany, G.M. Momin Women's College Bhiwandi University of Mumbai, India

ABSTRACT

Due to salt stress, osmotic disbalance occurs in halophytes. It causes significant changes at the physiological level and molecular levels. Plant metabolites regulate the effect of salinity. Many metabolites have been reported to date such as proline, methyl proline, betaine hydroxy proline, sugar, sugar alcohol, tocopherols, and polyphenols accumulated under salt stress. Responses to salinity at the metabolomic level would enable a better understanding of the physiological mechanisms driving halophytes' salt tolerance. Studies of proteome have given information about some salt–tolerant genes and proteins in halophytes. Halophytes adopt various mechanisms to prevent accumulation of Na^+ in the cytoplasm which include Na^+ extrusion and/or intracellular compartmentalization of Na^+ along with recirculation of Na^+ out of the shoot and upregulation of Na^+/H+ antiporters in the plasma membrane).

5.1 INTRODUCTION

Halophytes, also known as salt-loving plants, are plants that can withstand salinity stress. Salinity do not harm much due to some genes and proteins

Physiology of Halophytes: Signaling, Omics, and Tolerance Mechanisms.
Nivas Manohar Desai & Ganesh Chandrakant Nikalje (Eds.)

(Askari et al., 2006; Yu et al., 2011), whereas glycophytes, also known as salinity-sensitive plants, are sensitive toward salinity. The salt–tolerant mechanisms of halophytes is dependent on the proteomic, metabolomic, and ionomic characteristics of these plants. Halophytes use a variety of mechanisms to survive in high salt environments, including (i) selective ion accumulation or exclusion (Mahajan and Tuteja, 2005), (ii) K^+ ion is selectively absorbed by roots, control of ion uptake, particularly, K^+ by roots and transport to leaves (K^+ is necessary for maintaining the osmotic balance, plays a role in the opening and closing of stomata, and is an essential co-factor for many enzymes, including the pyruvate (Shabala and Mackay, 2011) (iii) Production of antioxidant enzyme (Gagneul et al., 2007; Sanchez et al., 2008; Slama et al., 2015); and (iv) Formation of suitable solutes and osmo-protectants. By combining the data from halophytes and comparing it with that from glycophytes, it is possible to gain an overview of the mechanisms that allow halophytes to tolerate salt, paving the way for the development of salt–tolerant crop plants through genetic engineering and successful breeding techniques. Halophytes may possess a variety of effective adaptation mechanisms that allow them to endure and finish their life cycle under challenging circumstances. With the long-term goal of producing transgenic plants in commercial crops that have significantly lower salt tolerance efficacy, halophytes are therefore plants of great interest for metabolomic, proteomic, and economic investigations to unravel their salt tolerance processes. In the post-genomic age, proteomics and metabolomics are two crucial "-omic" approaches (Fernandez-Garcia et al., 2011). The study of big proteins in an organism that is encoded by its genome is known as proteomics (Zhang et al., 2013). A potent method for describing entire proteomes at the organelle, cell, organ, or tissue level is proteomics. Additionally, it can be used to compare proteomes under different stress circumstances, such as exposure to high salt (Witzel et al., 2009; Fernandez-Garcia et al., 2011). The metabolic profile of halophytes also alters in response to salinity for improved adaption in the saline environment. Many earlier studies have suggested the upregulation and down-regulation of specific metabolites in halophytes. For example, Brosché et al. (2005) reported a study on the salt–tolerant tree *Populus euphratica* that revealed changes in the metabolism of sugar and polyols as well as increased levels of the amino acids proline, valine, and alanine. Glycerol, glyceric acid, and myoinositol levels have increased while fructose and mannitol levels have somewhat reduced. Similar to this, Gong et al. (2005) found that *Thellungiella halophila* had constitutively greater levels of proline, citric acid, malic acid, and succinic acids than did *Arabidopsis thaliana*. These sugars included sucrose, glucose, and fructose. Additionally, in response to salt stress, *T. halophila* accumulated raffinose, myoinositol, and galactinol to a higher extent than *A. thaliana*, whereas the levels of malic, fumaric, aspartic, and phosphoric acid declined to a greater extent in the halophyte. Protein profiles of stressed plants have been widely studied using proteomics to identify the specific genes and proteins involved in salt tolerance and halophytes' ability to survive in salinity (Wang et al., 2014). Several people use proteomic technology.

To prevent Na+ accumulation in the cytoplasm, halophytes use a variety of mechanisms, including Na^+ extrusion and/or intracellular compartmentalization of Na^+, as well as recirculation of Na^+ out of the shoot and upregulation of Na^+/H^+ antiporters in the plasma membrane (Hamada et al., 2001; Shi et al., 2003). The responses of root cell membrane potential and net H^+, Na^+, and K^+ fluxes to NaCl-salinity have been studied in two halophytes, *Atriplex lentiformis* and *Chenopodium quinoa*, demonstrating that small depolarization of the root cell plasma membrane causes a fast response of plasma membrane. A specific enzyme H^+ ATPase restores plasma membrane potential and transports Na^+ out of the cell H^+ electrochemical gradient plays a significant role (Bose et al., 2014b). Kumari et al 2015 has given proteomic, metabolomic and ionomic perspective of salinity tolerance in halophytes.

5.2 METABOLOMIC RELATED TO HALOPHYTES

Studying and analyzing metabolites is known as metabolomics (Liu et al., 2011). It is the most widely used method in plant sciences because plants have the unique ability to synthesize a large number of molecules that do not exist in any other life form. The term "metabolome" refers to the complete set of small-molecule metabolites found in a biological sample (such as metabolic intermediates, hormones, and other signaling molecules, as well as secondary metabolites) (Brosché et al., 2005). A complete profile of metabolites of a particular sample is known as metabolome The term was coined in analogy with transcriptomics and proteomics; the metabolome, like the transcriptome and proteome, is dynamic, and variable with time. The biomolecules produced by plants naturally and under stress conditions are known as metabolites, bioactive compounds, or compatible solutes. Metabolites in plant-based metabolomics can be classified as "primary" or "secondary." Primary metabolites are involved in the regular activities of plants whereas secondary metabolites do some specific functions.

Salt stress causes severe osmotic disbalance, resulting in detrimental physiological and molecular changes in cellular components (Vinocur and Altman, 2005). Plants respond to such conditions by up and down regulating a wide range of metabolites, which protects them and prevents the negative effects of salinity. A wide variety of metabolites have been identified, including mono-, di-, oligo-, and polysaccharides such as glucose, fructose, sucrose, trehalose, raffinose, and fructans; amino acids such as proline, pipecolic acid; methylated proline-related compounds such as methyl-proline, proline betaine, and hydroxyproline betaine; sugar alcohols (polyols) such as sorbi (Rhodes and Hanson, 1993; Ashraf and Foolad, 2007; Fig. 5.1, Table 5.3). The most common halophyte metabolites are described in detail in the following sections.

5.2.1 ROLE OF AMINO ACID IN OSMOREGULATION

Amino acid production increased in response to salt stress, according to the Metabolome analysis (Brosché et al., 2005; Gagneul et al., 2007; Sanchez et al., 2008; Liu et al., 2011). During salt stress, amino acids play a significant role in osmoregulation (Slama et al., 2015). They also defend macromolecular subcellular structures from oxidative harm and lessen the impact of free radicals generated in response to salt stress (Hasegawa et al., 2000). Salt tolerance is increased by production of amino acid and organic acid responsible to form protective enzyme and free radical scavenging (Flowers and Colmer, 2015). According to Sonam Yadav et al. proline, serine, threonine, aspartic acid, glutamic acid, glutamine, and tryptophan are among the amino acids that accumulate differently in both *Suaeda* species under salt and drought stress when ambient CO^2 levels are present.

5.2.1.1 PROLINE

Low-molecular-weight chaperones such as proline function as Osmoprotectants. It works by decreasing the inhibition of certain enzyme activity by Na^+ ions. Thus, prevents the dissociation of enzyme complexes such as the oxygen-evolving complex (OEC) of photosystem II (PSII) by increasing the stability of enzymes (Hasegawa et al., 2000). Using two crucial enzymes, pyrroline-5-carboxylate synthetase (P5CS) and pyrroline-5-carboxylate reductase (P5CR), this stress-responsive amino acid is mainly produced from glutamate (Hu et al., 1992). Glutamate serves as the main precursor in osmotically stressed cells (Gupta and Huang, 2014). In *Distichlis spicata* cells exposure to 200-mM NaCl induced proline concentration greater than 230 mM (Ketchum et al., 1991).

5.2.1.2 BETAINE

Typically, betaine acts as an organic osmolyte. Betaine has reportedly been identified as a crucial secondary metabolite for *Suaeda salsa*'s defense against osmotic stress (Liu et al., 2011). Choline monooxygenase (CMO), a metabolite that can be detected in *S. salsa* converts choline into betaine aldehyde, and betaine aldehyde turns into betaine.

5.2.1.3 GLYCINE BETAINE

One of the quaternary ammonium compounds (QACs) that is most frequently seen in plants that have been exposed to salinity-induced dehydration is glycine betaine (GB) (Lokhande and Suprasanna, 2012). Numerous halophytic organisms have been well-documented to accumulate GB (>90 mol dry weight) (Rhodes and Hanson, 1993; Flowers and Colmer 2008). Numerous investigations have demonstrated that the membrane and photosynthetic machinery can be protected by relatively modest concentrations of GB (Fitzgerald et al., 2009). When plants are subjected to dehydration stress, GB is generated at higher concentrations in the plants (Lokhande and Suprasanna, 2012).

5.2.2 ROLE OF CARBOHYDRATES IN OSMOREGULATION

5.2.2.1 SUGARS

As a result of the accumulation of compatible solutes, which intensifies cellular osmolarity and causes an influx of water into the cell or at the very least reduces efflux from the cell, maintaining the turgor required for cell expansion, accumulation of sugars has been linked to a mechanism for tolerance to salinity and drought (Hare et al., 1998). The main functions of these carbohydrates in the reduction of stress are osmoprotection, carbon storage, and reactive oxygen species (ROS) scavenging. In comparison to fructose and glucose, sucrose has been found to be a molecule that plants use to store energy. NMR spectrum intensities from samples of S. salsa showed that sucrose was 5–10% of betaine (Li et al., 2011). According to Sanchez et al. (2008), *T. halophila* has greater concentrations of sucrose, fructose, and glucose in response to high salinity. Sucrose, glucose, and fructose all accumulated in response to salinity in *Juncus acutus* and *Juncus maritimus*, but sucrose accumulated in larger proportions than glucose and fructose (Gil et al., 2011).

5.2.2.2 SUGAR ALCOHOLS OR POLYOLS

Polyols compound levels are higher during environmental stress. Pinnitol and osmonitol showed the highest scavenging activity in comparison to myoinositol, mannitol, and sorbitol (Orthen et al., 1994).

5.2.3 METABOLITES RELATED TO DETOXIFICATION AND ANTIOXIDATION

Due to their stronger antioxidant capacity when compared to glycophytes, halophytes are better able to withstand high salinity levels (Flowers and Colmer, 2008; Kosová et al., 2013; Ozgur et al., 2013). To effectively detoxify ROS brought on by salinity stress, there are some nonenzymatic substances, and antioxidant enzymes are increased. enzymes such as superoxide dismutase (SOD), catalase (CAT), ascorbate peroxidase (APX), glutathione peroxidase (GPX), and glutathione reductase (GR) are played significant roles in salinity tolerance (Gupta and Huang, 2014). These nonenzymatic antioxidant chemical accumulations have also demonstrated antioxidative activity (Bose et al., 2013). Due to some nonenzymatic antioxidants, for example, proline, polyamines, and GB polyols their capacity to scavenge HO, numerous suitable solutes, including proline, polyamines, GB, polyols, etc., have been regarded as nonenzymatic antioxidants.

Proline has been demonstrated to carry out a variety of antioxidant tasks, including quenching H^2O^2 and HO, stabilizing enzymes related to ROS absorption protection of complex II of the mitochondrial electron transport chain, and prevention of PCD (Bose et al., 2013). Cuin and Shabala (2007) demonstrate that exogenous proline administration reduces hydroxyl radical-induced K^+ efflux from the roots of *Arabidopsis*.

Some molecules such as polyols, mannitol, Myo-inositol, and pinitol compounds are effective hydroxyl radical scavengers in addition to being used for osmotic regulation. Later studies revealed that a pinitol. If we compare the extract of *Mesembryanthemum crystallinum* and glycophyte *Lactuca sativa* for DPPH (1, 1-diphenyl-2-picrylhydrazl) scavenging extract of halophyte is more effective. In addition to directly scavenging, polyols can interact with proteins, enzymes, and membranes to communicate during times of stress and to safeguard cellular structures. Salinity also induced an antioxidative defense system in halophytes.

5.2.4 SALT TOLERANCE AND PHYTOHORMONES

Chemical messengers called phytohormones to serve crucial roles in controlling how plants react to stress even at very low concentrations. Numerous halophytic species have been found to alter their levels of phytohormones as well as enzymes involved in their biosynthesis like in response to high salt concentration. Abscisic acid (ABA) is the primary phytohormone whose expression is abundant. For a plant to survive under challenging environmental conditions like salt stress, ABA is a critical internal signal that is triggered. Plants exposed to salt have been found to have higher concentrations of the amino acid ABA. For the majority of the time, it is connected with the water potential of the leaf or soil, indicating that the rise in endogenous ABA caused by salinity is caused by a water shortage rather than a particular salt effect (Zhang et al., 2006). Salt stress causes an increase in ABA biosynthesis enzymes, which leads to an abundance of ABA production. *Thellungiella salsuginea* has been found to have a high level of NCED (Taji et al., 2004). Many early and delayed ABA-responsive genes (ABA-dependent transcription factors, such as Lea/Rab genes) have increased in expression in the halophyte. When exposed to salt stress.

5.2.4.1 TOCOPHEROLS

Tocopherols are vitamin E compounds. These molecules are lipid-soluble and found in three forms α-, β-, and γ-forms. It works as an antioxidant for $^1O^2$ and lipid peroxyl radicals. Plastid envelops form α tocopherol and it is stored in the chloroplast.

5.2.4.2 POLYPHENOLS

Polyphenols are antioxidants. It is water soluble. In halophytes, polyphenols accumulate and play a vital role in the scavenging of O^{2-}, OH^-, and O^2 (Ozgur et al., 2013).

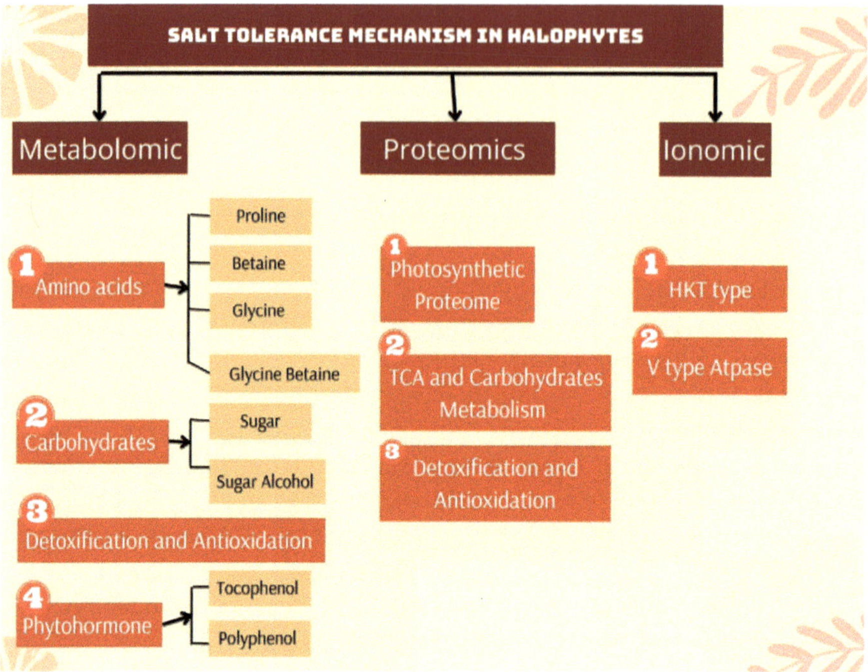

FIGURE 5.1 Salt-tolerant mechanism in halophytes.

5.3 PROTEOMIC RELATED TO HALOPHYTES

Protein profiles of stressed plants have been widely studied using proteomics (Table 5.1) to identify the specific genes and proteins involved in salt tolerance and halophytes' ability to survive in salinity (Wang et al., 2014). To increase salt tolerance, related genes have been cloned and transferred to glycophytes such as tobacco, rice, and *Arabidopsis* (Tang et al., 2011). Since the stomatal aperture is susceptible to salinity, stomatal conductance reduces in response to salinity. This drop in intracellular CO^2 in the leaves caused by the stomata closing reduces photosynthesis (Centritto et al., 2003; Redondo-Gómez et al., 2007). According to research, as salinity rose, *Puccinellia tenuiflora*'s photosynthetic rate, stomatal conductance, intercellular CO^2, and transpiration rate all declined but water use efficiency rose (Yu et al., 2011).

5.3.1 PHOTOSYSTEM PROTEOME

Upon salt stress, various proteins connected to photosynthesis showed variable expression, according to proteomic analyses. According to earlier research, OEE, which consists of OEE1, OEE2, and OEE3, plays an important role in a reaction such as oxidation of water in PSII. However, it is simple to separate these PSII complex subunits

In response to salt stress, *Bruguiera gymnorrhiza* has been shown to upregulate OEE2, which raises the possibility that it may be required to repair damage to the PSII complex and maintain the oxygen evolution reaction (Tada and Kashimura, 2009). Similar to this, OEE and PEP expression have been found to be elevated in some plants such as *Aeluropus lagopoides* (Sobhanian et al., 2010). *P. tenuiflora* is a good example of the downregulation of OEE expression. (Yu et al., 2011). A part of PSI's light-harvesting complex called CAB2 makes it easier to absorb light and transport excitation energy to reaction centers so that NADP can be reduced to NADPH (Wan and Liu., 2008; Yu et al., 2011; Wang et al., 2012). *A. lagopoides* (Table 5.1) on the other hand, has seen an increase in its accumulation (Sobhanian et al., 2010). Similar to how some PSII assembly/stabilization-related proteins, such as D1, have decreased in expression in *P. tenuiflora* (Yu et al., 2011), *A. lagopoides* had increased accumulation of D2 protein (Sobhanian et al., 2010). Sobhanian et al (2010), Cheng et al. (2015), Yamada et al. (1995) Tada and Kashimura (2009) studied major intrinsic protein (MIPs) OEE in *A. lagopoides*, *Tangut nitraria*, *M. crystallinum* (Ice plant) *B. gymnorrhiza.* Functional protein Rubisco LSU, RCA HSP60. PGAM, and CAB in various plants were studied by Wang et al. (2008b, 2014, 2009), Sengupta and Majumder (2009), Yu et al. (2011), Liska et al. (2004), and Debeza et al. (2012).

5.3.2 PROTEIN RELATED TO TCA CYCLE AND CARBOHYDRATE METABOLISM

Under salt stress, plants need a significant quantity of energy to grow and develop. According to Höper et al. (2006), this energy is mostly generated by the carbohydrate metabolic processes of glycolysis (EMP) and the TCA cycle. Upregulation of two enzymes PGAM and FRK are 600 and 450 mM salt stress, respectively (Wang et al., 2013). The most significant doorway in the regulation of sugar influx into EMP is FRK which forms fructose-1-phosphate in glycolysis (Hoper et al., 2006). Proteins such as AH, SD, and MD have also been discovered to be upregulated. The TCA cycle depends on the first two enzymes. As a result, the increase in FRK under NaCl, together with the other EMP-TCA enzymes, is the TCA cycle's important enzyme. As a result, the increase in FRK under NaCl, together with the other EMP-TCA enzymes, would aid in the breakdown of glucose for the production of energy to combat salt stress. These proteins have been found to be upregulated in *P. tenuiflora* (Yu et al., 2011) and *Kandelia candel* in response to high salt concentration (Zhao et al., 2013).

TABLE 5.1 Salinity Induced Proteomic in Different Halophytes Studied by Tada and Kashimura (2009), Wang et al. (2008a), Askari et al. (2006), Ueda et al. (2003) Pang et al. (2010), Sobhanian et al. (2010).

Plants	Concentration of salt (mM)	Proteomic techniques
B. gymnorrhiza	50–500	2-DE LC–MS/MS
Thinopyrum ponticum	50–200	2-DE MALDI-TOF/TOF
Suaeda aegyptiaca	150–600	2-DE LC/MS/MS
Aster tripolium	50–300	2-DE MALDI-TOF/TOF
T. halophila	50–150	2-DE MALDI-TOF/TOF nano-ESI-MS/MS
A. lagopoides	150–750	2-DE Edman sequencing nano-LC–MS/MS

5.3.3 PROTEINS INVOLVED IN DETOXIFICATION AND ANTIOXIDATION

Salinity affects plants in two different ways, the first is that the osmotic stress is generated by salinity, and the second effect of salinity is a disruption of nutrient balance (Munns and Tester, 2008). Due to the loss of coordination across several metabolic pathways, ROS generation is elevated in both situations (Asada, 2006). ROS creates a significant disturbance by directly activating NSCC channels, which results in extra K^+ efflux in addition to Na-induced membrane depolarization, in intracellular ionic balance (Demidchik et al., 2010). According to Shabala (2009) and Demidchik et al. (2010), this activation causes the cytosolic K content to drop, which is followed by the activation of caspase-like proteases and endonucleases. Due to the oxidative damage that ROS causes to proteins, lipids, and DNA, cell membrane integrity, enzyme activity, and the operation of the photosynthetic apparatus are all impacted (Yu et al., 2011). However, ROS might be viewed in plants as a signaling molecule that prompts an increase in antioxidant enzymes to prepare for salinity (Abogadallah, 2010; Jaspers and Kangasjärvi, 2010). Plants have developed a comprehensive antioxidant defense mechanism with both enzymatic and nonenzymatic components to deal with the excessive buildup of these chemicals. Due to the loss of coordination between several metabolic pathways for adjusting to high salt levels in both circumstances (Abogadallah, 2010; Jaspers and Kangasjärvi, 2010), ROS generation is enhanced. Some antioxidant enzymes regulate ROS (Yu et al., 2011; Bose et al., 2013; Wang et al., 2013). SOD is an enzyme that defends plants from free radicle damage, which can catalyze the conversion of O^2 to H^2O and O^2 during diverse stressors. Canola, *Salicornia europaea*, and *Solanum chilense* have all been shown to benefit from the accumulation of SOD in response to salt stress (Wang et al., 2009; Zhou et al., 2011). In a study on *K. candel*, SOD levels have risen in response to salinity (Wang et al., 2013). *T. nitraria* had similar results, according to Cheng et al. (2015); however, *P. tenuiflora* showed just a minor increase in activity at 50-mM NaCl but a sharp decline at 150-mM NaCl treatment. In earlier studies, salt stress was shown to increase DHAR protein expression. Stress-induced ascorbic acid synthesis may remove water by increasing the quantity of DHAR. DHAR is commonly mentioned as an enzyme that can shield plants from oxidative stress (see Fig. 5.2).

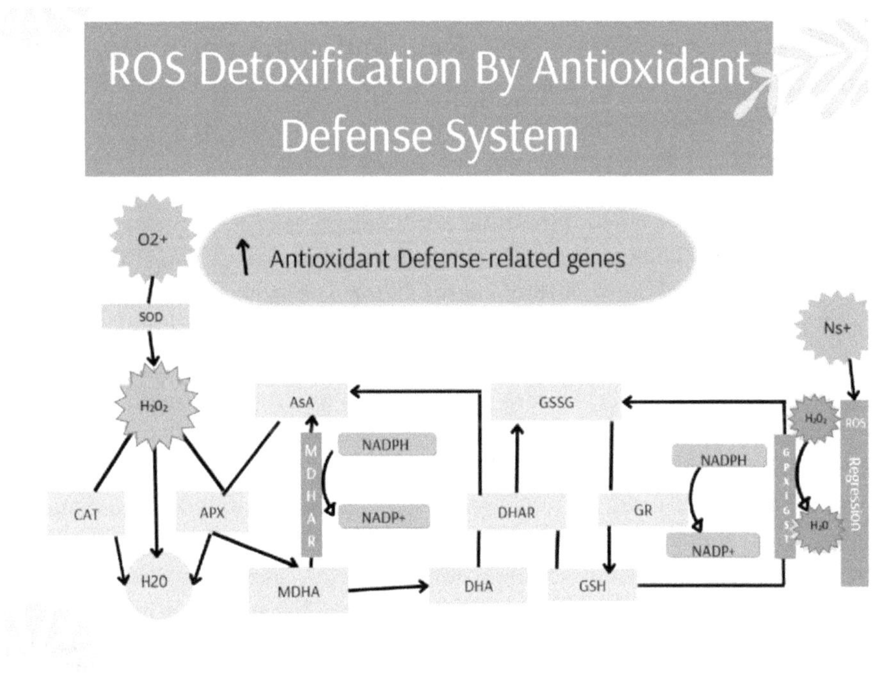

FIGURE 5.2 ROS detoxification by the antioxidant defense system.

5.4 STUDIES ON HALOPHYTES IONOMICS

Ionome is the term for the entire inorganic component of cellular and biological systems, and its study is known as ionomics (Satismruti et al., 2013). It includes the makeup of an organism's minerals, nutrients, and trace elements. Ionomics comprises the quantitative analysis of an organism's elemental content (Salt et al., 2008). The effects of physiological stimuli, developmental stage, genetic variations, and stress tolerance are also taken into account. Ionomics uses high-throughput genetic and bioinformatics methods along with elemental analysis technologies. It offers details on how an organism functions in response to various biotic and abiotic influences. Ionomic analysis offers a potent method for the functional investigation of the genes, gene networks, and physiological processes that indirectly affect the ionome (Baxter, 2010). Ionomics provides an easy way to analyze because it produces data with high efficiency and minimal cost. To get knowledge of all ionomics of the plant. profiling of plants, many analytical techniques are applied. The database used to store all of the ionic profiles of plants is called the Purdue Ionomics Information Management System (PiiMS) (Baxter et al., 2007). This database's data is useful for many genetic investigations on an organism's structure and function. To lessen the toxicity brought by excessive salt concentration, plants must get acclimated to saline soils through modifications in the absorption of mineral ions (Sanchez et al., 2008). A high K^+/Na^+ ratio is necessary for plants to operate normally at the cellular level. Lack of K^+ retention at increasing Na^+ concentrations results from Na-induced membrane depolarization, which causes K^+ efflux through depolarization-activated KOR channels (Shabala and Mackay, 2011; Bose et al., 2014b). Additionally, the buildup of ROS under salt stress may cause K^+ to leak via NSCCs (Bose et al., 2014a). Additionally, Na^+ can interfere with K^+-specific transporters, having toxic effects, and causing fatal alterations in protein structure (Zhu, 2003; Mahajan and Tuteja, 2005) In the saline environment, Na^+ transporter and cation channel help Na^+ to enter the cell cytosol (Sanchez et al., 2011). Halophytes adopt various mechanisms to prevent accumulation of Na^+ in the cytoplasm which include Na^+ extrusion and/or intracellular compartmentalization of Na^+ along with recirculation of Na^+ out of the shoot and upregulation of Na^+/H^+ antiporters in the plasma membrane (Hamada et al., 2001; Shi et al., 2003). The roots plasma membrane is depolarized due to salinity in *Atriplex* and *Chenopodium* and it causes a fast response of plasma membrane H^+ ATPase for restoration of plasma membrane potential and H^+ electrochemical gradient to drive transport of Na^+ out of the cell (Bose et al., 2014b). Halophytes develop several survival mechanisms, such as Na^+ extrusion and/or intracellular compartmentalization of Na^+, Na^+ recirculation out of the shoot, and upregulation of Na^+/H^+ antiporters in the plasma membrane, to prevent accumulation of Na+ in the cytoplasm (Hamada et al., 2001; Shi et al., 2003).

5.4.1 NA+ TRANSPORTER OF THE HKT TYPE

Na^+ and K+ are transported by proteins of the carrier type known as high-affinity potassium transporters. The Na+-specific transporters of the HKT family mediate Na^+ transport and Na^+-K^+ symport. Low cytosolic Na^+ concentrations are maintained in large part by HKT-type Na^+ transporters. HKT1 transporters work as a Na^+ and K^+ symporters as well as a picky Na^+ uniporter. The halophytes *M. crystallinum* and *S. salsa* have demonstrated the significance of HKT1 in the maintenance of K^+ (Su et al., 2003; Shao et al., 2014).

5.4.2 V-TYPE ATPASE FOR H+

To protect enzymes in the cytoplasm and chloroplasts that are sensitive to high salt concentrations, halophytes place an important level of significance on the compartmentation of Na^+ in the vacuole (Li et al., 2011). The activity of V-type H^+-ATPase and H^+-PPase controls the sequestration of Na^+ into the vacuole. Since they produce the proton gradient necessary for the influx of Na^+ into the vacuole, these phosphatases are essential for the action of Na^+/H^+ antiporters. A specific H^+-ATPase found on tonoplast is called V-ATPase. It is a head-stalk-base-arranged multi-heteromeric complex with at least 11 distinct subunits (Shabala et al., 2014).

5.5 CONCLUSION

Salinity has a massive effect on the metabolism and structure of plant cells. As they have a negative impact on cytosolic enzymes, photosynthesis, K^+ feeding, and general metabolism, Na^+ ions in soil are hazardous to plants. Studying how proteins, metabolites, and ions respond to salinity may help us better understand the physiological processes behind halophytes' ability to tolerate salt. The proteomic, metabolomic, and ionomic bases of salt tolerance in halophytes are discussed in this chapter. To reduce the toxic effects, efforts have been made to provide the regulation of different proteins and metabolites in halophytes Na^+ accumulating capacity. This chapter gives recent information that provides some strategies for improving salt tolerance as was previously mentioned, proteins are essential for halophytes' ability to withstand salt stress and reduce the negative effects of Na^+. Studies of quantitative modifications in the overall plant proteome or in particular subcellular proteomes such as the nuclear, mitochondrial, or plastid proteomes have mostly been the focus of studies on halophyte tolerance to salinity. These studies used a comparative proteomics approach that included 2-DE followed by MS analysis. Furthermore, with new advances in proteomic approaches, it appears that studying post-translational modifications (PTMs), particularly phosphoproteomics and redox proteomics, as well as protein interactions (interactomics), will be used to prepare all information of protein function which will undoubtedly aid us in understanding better comprehend the processes of salt tolerance acquisition in halophytes Similarly, under salt stress, halophytes can redirect ions and nutrients to other direction and induce adequate metabolic shift for survival in high salt concentrations. However, this is only a simplified way of expressing things; to gain a better understanding, we must consider metabolite and ion regulation as a web of interconnected reactions, Thus, the mechanism of salt tolerance can be understood by physiological and molecular studies of halophytes These studies have revealed that when exposed to salt stress, halophytes employ systematic mechanisms to develop salt tolerance, and the identification and characterization of the phenomenon of salt tolerance will guide us in improving crop plant salinity tolerance.

As a result, advances in omics approaches will improve our understanding of the strategy of halophyte salt tolerance, allowing us to identify the specific proteins, metabolites, and ions for maintaining tolerance. As a result, this information could be extremely useful for improving crop plants through genetic engineering. This review concludes that key proteins and metabolites used for genetically engineering crop plants include the downregulation of photosynthesis-related proteins and metabolites, as well as upregulation of energy metabolism and ion channels to compensate for Na^+ toxicity and efflux or compartmentalization as a result, the use of the delivered information

This review would provide a foundation to produce genetically engineered salt-tolerant crops and plants, which would be critical in solving the world's food and energy problems.

KEYWORDS

- **halophytes**
- **osmotic disbalance**
- **metabolomic**
- **proline**
- **salinity**
- **proteome**

REFERENCES

Abogadallah, G. M. Antioxidative Defense under Salt Stress. *Plant Signal.* **2010,** *5,* 369–374.

Agarie, S.; Kawaguchi, A.; Kodera, A.; Sunagawa, H.; Kojima, H.; Nose, A.; et al. Potential of the Common Ice Plant, *Mesembryanthemum crystallinum* as a New High-Functional Food as Evaluated by Polyol Accumulation. *Plant Prod. Sci.* **2009,** *12,* 37–46.

Askari, H.; Edqvist, J.; Hajheidari, M.; Kafi, M.; Salekdeh, G. H. Effects of Salinity Levels on Proteome of *Suaeda aegyptiaca* Leaves. *Proteomics* **2006,** *6,* 2542–2554.

Asada, K. Production and Scavenging of Reactive Oxygen Species in Chloroplasts and their Functions. *Plant Physiol.* **2006,** *141,* 391–396.

Ashraf, M.; Foolad, M. R. Roles of Glycine Betaine and Proline in Improving Plant Abiotic Stress Resistance. *Environ. Exp. Bot.* **2007,** *59,* 206–216.

Baxter, I. Ionomics: The Functional Genomics of Elements. *Brief. Funct. Genom.* **2010,** *9,* 149–156.

Baxter, I.; Ouzzani, M.; Orcun, S.; Kennedy, B.; Jandhyala, S. S.; Salt, D. E. Purdue Ionomics Information Management System (Piims): An Integrated Functional Genomics Platform. *Plant Physiol.* **2007,** *143,* 600–611.

Bose, J.; Rodrigo-Moreno, A.; Shabala, S. ROS Homeostasis in Halophytes in the Context of Salinity Stress Tolerance. *J. Exp. Bot.* **2013,** *65,* 1241–1257.

Bose, J.; Shabala, L.; Pottosin, I.; Zeng, F.; Velarde-Buendía, A. M.; Massart, A.; et al. Kinetics of Xylem Loading, Membrane Potential Maintenance, and Sensitivity of K+-Permeable Channels to Reactive Oxygen Species: Physiological Traitsth at Differentiate Salinity tolerance Between Pea and Barley. *Plant Cell Environ.* **2014a,** *37,* 589–600.

Bose, J.; Rodrigo-Moreno, A.; Lai, D.; Xie, Y.; Shen, W.; Shabala, S. Rapid Regulation of the Plasma Membrane H+-ATPase Activity is Essential to Salinity tolerance in Two Halophyte Species, *Atriplex lentiformis* and *Chenopodium quinoa. Ann Bot.* **2014b,** *115,* 481–494.

Brosché, M.; Vinocur, B.; Alatalo, E. R.; Lamminmäki, A.; Teichmann, T.; Ottow, E. A.; et al. Gene Expression and Metabolite Profiling of *Populus euphratica* Growing in the Negev Desert. *Genome Biol.* **2005,** *6,* R101.

Centritto, M.; Loreto, F.; Chartzoulakis, K. The Use of Low (CO^2) to Estimate Diffusional and Non-Diffusional Limitations of Photosynthetic Capacity of Salt-Stressed Olive Saplings. *Plant Cell Environ.* **2003,** *26,* 585–594.

Cheng, T.; Chen, J.; Zhang, J.; Shi, S.; Zhou, Y.; Lu, L.; et al. Physiological and Proteomic Analyses of Leaves from the Halophyte *Tangut nitraria* Reveals Diverse Response Pathways Critical for High Salinity tolerance. *Front. Plant Sci.* **2015,** *6,* 30.

Chinnusamy, V.; Jagendorf, A.; Zhu, J. K. Understanding and Improving Salt Tolerance in Plants. *Crop Sci.* **2005,** *45,* 437–448.

Chinnusamy, V.; Zhu, J.; Zhu, J. K. Salt Stress Signalling and Mechanisms of Plant Salt tolerance. *Genet. Eng.* **2006,** *27,* 141–177.

Cuin, T. A.; Shabala, S. Compatible Solutes Reduce ROS-Induced Potassium Efflux in *Arabidopsis* Roots. *Plant Cell Environ.* **2007,** *30,* 875–885.

Debeza, A.; Braunc, H.; Pichd, A.; Taamallia, W.; Koyroe, H.; Abdellya, C.; et al. Proteomic and Physiological Responses of the Halophyte Cakile Maritima to Moderate Salinity at the Germinative and Vegetative Stages. *J. Proteom.* **2012,** *75,* 5667–5694.

Fan, P.; Feng, J.; Jiang, P.; Chen, X.; Bao, H.; Nie, L.; et al. Coordination of Carbon Fixation and Nitrogen Metabolism in *Salicornia Europaea* Under Salinity: Comparative Proteomic Analysis on Chloroplast Proteins. *Proteomics* **2011,** *11,* 4346–4367.

Fernandez-Garcia, N.; Hernandez, M.; Casado-Vela, J.; Bru, R.; Elortza, F.; Hedden, P.; et al. Changes to the Proteome and Targeted Metabolites of Xylem Sap in *Brassica Oleracea* in Response to Salt Stress. *Plant Cell Environ.* **2011,** *34,* 821–836.

Flowers, T. J.; Colmer, T. D. Salinity Tolerance in Halophytes. *Tansley Rev. New Phytol.* **2008,** *179,* 945–963.

Flowers, T. J.; Colmer, T. D. Plant Salt Tolerance: Adaptations in Halophytes. *Ann. Bot.* **2015,** *115,* 327–331.

Gagneul, D.; Aïnouche, A.; Duhazé, C.; Lugan, R.; Larher, F. R.; Bouchereau, A. A Reassessment of the Function of the So-Called Compatible Solutes in the Halophytic Plumbaginaceae Limonium Latifolium. *Plant Physiol.* **2007,** *144,* 1598–1611.

Gil, R.; Lull, C.; Boscaiu, M.; Bautista, I.; Lidón, A.; Vicente, O. Soluble Carbohydrates as Osmolytes in Several Halophytes from a Mediterranean Salt Marsh. *Not. Bot. Hort. Agrobot. Cluj-Napoca* **2011,** *39,* 9–17.

Gong, Q.; Li, P.; Ma, S.; Indu Rupassara, S.; Bohnert, H. J. Salinity Stress Adaptation Competence in the Extremophile *Thellungiella halophila* in Comparison with Its Relative *Arabidopsis thaliana*. *Plant J.* **2005,** *44,* 826–839.

Gupta, B.; Huang, B. Mechanism of Salinity Tolerance in Plants: Physiological, Biochemical, and Molecular Characterization. *Int. J. Genom.* **2014,** 1–18.

Hamada, A.; Shono, M.; Xia, T.; Ohta, M.; Hayashi, Y.; Tanaka, A.; et al. Isolation and Characterization of a Na+/H+ Antiporter Gene from the Halophyte *Atriplex gmelini*. *Plant Mol. Biol.* **2001,** *46,* 35–42.

Hare, P. D.; Cress, W. A.; Staden, J. V. Dissecting the Roles of Osmolyte Accumulation during Stress. *Plant Cell Environ.* **1998,** *21,* 535–553.

Hasegawa, P. M.; Bressan, R. A.; Zhu, J. K.; Bohnert, H. J. Plant Cellular and Molecular Responses to High Salinity. *Annu. Rev. Plant Physiol. Mol. Biol.* **2000,** *51,* 463–499.

Höper, D.; Bernhardt, J.; Hecker, M. Salt Stress Adaptation of *Bacillus subtilis*: A Physiological Proteomics Approach. *Proteomics* **2006,** *6,* 1550–1562.

Hu, C. A. A.; Delauney, A. J.; Verma, D. P. S. A Bifunctional Enzyme (1-Pyrroline-5-Carboxylate Synthetase) Catalyses the first Two Steps in Proline Biosynthesis in Plants. *Proc. Natl. Acad. Sci. U. S. A.* **1992,** *89,* 9345–9358.

Ketchum, R. E. B.; Warren, R. C.; Klima, L. J.; Lopez-Gutierrez, F.; Nabors, M. W. The Mechanism and Regulation of Proline Accumulation in Suspension Cultures of the Halophytic Grass *Distichlis spicata* L. *J. Plant Physiol.* **1991,** *137,* 368–374.

Khan, M. S. Role of Sodium and Hydrogen (Na+/H+) Antiporters in Salt tolerance of Plants: Present and Future Challenges. *Afr. J. Biotechnol.* **2011,** *10,* 13693–13704.

Kosová, K.; Vítámvás, P.; Urban, M. O.; Prášil, I. T. Plant Proteome Responses to Salinity Stress–Comparison of Glycophytes and Halophytes. *Funct. Plant Biol.* **2013,** *40,* 775–786.

Kumari, A.; Das P.; Parida, A. K.; Agrawal, P. K. Proteomics, Metabolomics and Ionomics Perspectives of Slinity tolerance in Halophytes. *Front. Plant Sci.* **2015,** *6,* 537.

Li, W.; Zhang, C. Y.; Lu, Q. T.; Wen, X. G.; Lu, C. M. The Combined Effect of Salt Stress and He at Shock on Proteome Profiling in *Suaeda salsa. J. Plant Physiol.* **2011,** *168,* 1743–1752.

Liska, A. J.; Shevchenko, A.; Pick, U.; Katz, A. Enhanced Photosynthesis and Redox Energy Production Contribute to Salinity Tolerance in *Dunaliella* as Revealed by Homology-Based Proteomics. *Plant Physiol.* **2004,** *136,* 2806–2817.

Liu, X.; Yang, C.; Zhang, L.; Li, L.; Liu, S.; Yu, J.; et al. Metabolic Profiling of Cadmium-Induced Effects in One Pioneer Intertidal Halophyte *Suaeda salsa* by NMR-Based Metabolomics. *Ecotoxicol.* **2011,** *20,* 1422–1431.

Lokhande, V. H.; Suprasanna, P. Prospects of Halophytes in Understanding and Managing Abiotic Stress tolerance. In *Environmental Adaptations and Stress Tolerance of Plants in the Era of Climate Change*; 2012; pp 26–59.

Mahajan, S.; Tuteja, N. Cold, Salinity and Drought Stresses: An Overview. *Arch. Biochem. Biophys.* **2005,** *444,* 139–158.

Munns, R.; Tester, M. Mechanisms of Salinity Tolerance. *Annu. Rev. Plant Biol.* **2008,** *59,* 651–681.

Orthen, B.; Popp, M.; Smirnoff, N. Hydroxyl Radical Scavenging Properties of Cyclitols. *Proc. R. Soc. Edinb. Sec. B Biol. Sci.* **1994,** *102,* 269–272.

Ozgur, R.; Uzilday, B.; Sekmen, A. H.; Turkan, I. Reactive Oxygen Species Regulation and Antioxidant Defence in Halophytes. *Funct. Plant Biol.* **2013,** *40,* 832–847.

Redondo-Gómez, S.; Mateos-Naranjo, E.; Davy, A. J.; Fernandez Munoz, F.; Castellanos, E. M.; Luque, T.; et al. Growth and Photosynthetic Responses to Salinity of the Salt-Marsh Shrub *Atriplex portulacoides. Ann. Bot.* **2007,** *100,* 555–563.

Rhodes, D.; Hanson, A. D. Qu aternary Ammonium and Tertiary Sulfonium Compounds in Higher Plants. *Ann. Rev. Plant Physiol. Mol. Biol.* **1993,** *44,* 357–384.

Salt, D. E.; Baxter, I.; Lahner, B. Ionomics and the Study of the Plant Ionome. *Annu. Rev. Plant Biol.* **2008,** *59,* 709–733.

Sanchez, D. H.; Pieckenstain, F. L.; Escaray, F.; Erban, A.; Kraemer, U. T. E.; Udvardi, M. K. Comparative Ionomics and Metabolomics in Extremophile and Glycophytic Lotus Species under Salt Stress Challenge the Metabolic Pre-Adaptation Hypothesis. *Plant Cell Environ.* **2011,** *34,* 605–617.

Sanchez, D. H.; Siahpoosh, M. R.; Roessner, U.; Udvardi, M.; Kopka, J. Plant Metabolomics Reveals Conserved and Divergent Metabolic Responses to Salinity. *Physiol. Plant.* **2008,** *132,* 209–219.

Satismruti, K.; Senthil, N.; Vellaikumar, S.; Ranjani, R. V.; Raveendran, M. Plant Ionomics: a Platform for Identifying Novel Gene Regulating Plant Mineral Nutrition. *Am. J. Plant Sci.* **2013,** *4,* 1309–1315.

Saxena, S. C.; Kaur, H.; Verma, P.; Petla, B. P.; Andugula, V. R.; Majee, M. N. T.; Singh, S. G. Osmoprotectants: Potential for Crop Improvement under Adverse Conditions. In *Plant Acclimation to Environmental Stress*; Eds.; Springer: New York, NY, 2013; pp 197–232.

Sengupta, S.; Majumder, A. L. Insight into the Salt tolerance Factors of a Wild Halophytic Rice, *Porteresia Coarctata*: A Physiological and Proteomic Approach. *Planta* **2009,** *229,* 911–929.

Shabala, S. Salinity and Programmed Cell Death: Unravelling Mechanisms for Ion Specific Signalling. *J. Exp. Bot.* **2009,** *60,* 709–712.

Shabala, S. N.; Mackay, A. S. Ion Transport in Halophytes. *Adv. Bot. Res.* **2011,** *57,* 151–187.

Shao, Q.; Han, N.; Ding, T.; Zhou, F.; Wang, B. SsHKT1;1 is a Potassium Transporter of the C3 Halophyte *Suaeda salsa* that is Involved in Salt tolerance. *Funct. Plant Biol.* **2014,** *41,* 790–802.

Shi, H.; Lee, B. H.; Wu, S. J.; Zhu, J. K. Overexpression of a Plasma Membrane Na+/H+ Antiporter Gene Improves Salt tolerance in *Arabidopsis Thaliana. Nat. Biotech.* **2003,** *21,* 81–85.

Slama, I.; Abdelly, C.; Bouchereau, A.; Flowers, T.; Savoure, A. Diversity, Distribution and Roles of Osmoprotective Compounds Accumulated in Halophytes under Abiotic Stress. *Ann. Bot.* **2015,** *115,* 433–447.

Sobhanian, H.; Motamed, N.; Jazii, F. R.; Nakamura, T.; Komatsu, S. Salt Stress Induced Differential Proteome and Metabolome Response in the Shoots of *Aeluropus lagopoides* (Poaceae), A Halophyte C4 Plant. *J. Proteome Res.* **2010,** *9,* 2882–2897.

Sonar, B. A.; Desai, N.; Gaikwad, D. K.; Chavan, P. D. Assessment of Salinity-Induced Antioxidative Defense System in *Colubrina asiatica. Brongn J. Stress Physiol. Biochem.* **2011,** *7,* 193–200.

Stepien, P.; Johnson, G. N. Contrasting Responses of Photosynthesis to Salt Stress in the Glycophyte *Arabidopsis* and the Halophyte *Thellungiella*: Role of the Plastid Terminal Oxidase as An Alternative Electron Sink. *Plant Physiol.* **2009,** *149,* 1154–1165.

Tada, Y.; Kashimura, T. Proteomic Analysis of Salt-Responsive Proteins in the Mangrove Plant, *Bruguiera Gymnorhiza. Plant Cell Physiol.* **2009,** *50,* 439–446.

Tang, M.; Liu, X.; Deng, H.; Shen, S. Over-Expression of Jcdreb, a Putative AP2/EREBP Domain-Containing Transcription Factor Gene in Woody Biodiesel Plant *Jatropha Curcas*, Enhances Salt and Freezing tolerance in Transgenic *Arabidopsis thaliana. Plant Sci.* **2011,** *181,* 623–631.

Uzilday, B.; Ozgur, R.; Sekmen, A. H.; Yildiztugay, E.; Turkan, I. Changes in the Alternative Electron Sinks and Antioxidant Defense in Chloroplasts of the Extreme Halophyte *Eutrema* Parvulum (*Thellungiella parvula*) under Salinity. *Ann. Bot.* **2015,** *115,* 449–463.

Valluru, R.; Van Den Ende, W. Myo-Inositol and Beyond Emerging Networks under Stress. *Plant Sci.* **2011,** *181,* 387–400.

Vinocur, B.; Altman, A. Recent Advances in Engineering Plant tolerance to Abiotic Stress: Achievements and Limitations. *Curr. Opin. Biotechnol.* **2005,** *16,* 123–32.

Wan, X. Y.; Liu, J. Y. Comparative Proteomics Analysis Reveals an Intimate Protein Network Provoked by Hydrogen Peroxide Stress in Rice Seedling Leaves. *Mol. Cell Proteom.* **2008,** *7,* 1469–1488.

Wang, L.; Liu, X.; Liang, M.; Tan, F.; Liang, W.; Chen, Y.; et al. Proteomic Analysis of Salt-Responsive Proteins in the Leaves of Mangrove *Kandelia Candel* During Short-Term Stress. *PLoS One* **2014,** *9,* 83141.

Wang, X.; Fan, P.; Song, H.; Chen, X.; Li, X.; Li, Y. Comparative Proteomic Analysis of Differentially Expressed Proteins in Shoots of *Salicornia europaea* under Different Salinity. *J. Proteome Res.* **2013,** *8,* 3331–3345.

Wang, X. C.; Chang, L. L.; Wang, B. C.; Wang, D.; Li, P. H.; et al. Comparative Proteomics of *Thellungiella halophila* Leaves from Plants Subjected to Salinity Reveals the Importance of Chloroplastic Starch and Soluble Sugars in Halophyte Salt tolerance. *Mol. Cell Proteom.* **2013,** *12,* 2174–2195.

Witzel, K.; Weidner, A.; Surabhi, G. K.; Börner, A.; Mock, H. P. Salt Stress-Induced Alterations in the Root Proteome of Barley Genotypes with Contrasting Response Towards Salinity. *J. Exp. Bot.* **2009,** *60,* 3546–3557.

Yamada, S.; Katsuhara, M.; Kelly, W. B.; Michalowski, C. B.; Bohnert, H. J. A Family of Transcripts Encoding Water Channel Proteins: Tissue-Specific Expression in the Common Ice Plant. *Plant Cell* **1995,** *7,* 1129–1142.

Yu, J.; Chen, S.; Zhao, Q.; Wang, T.; Yang, C.; Diaz, C.; et al. Physiological and Proteomic Analysis of Salinity Tolerance in *Puccinellia tenuiflora. J. Proteome Res.* **2011,** *10,* 3852–3870.

Zhao, Q.; Zhang, H.; Wang, T.; Chen, S. X.; Dai, S. J. Proteomics-Based Investigation of Salt-Responsive Mechanisms in Plant Roots. *J. Proteom.* **2013,** *82,* 230–253.

Zhou, S.; Sauvé, R. J.; Liu, Z.; Reddy, S.; Bhatti, S.; Hucko, S. D.; et al. Identification of Salt-Induced Changes in Leaf and Root Proteomes of the Wild tomato, *Solanum chilense. J. Am. Soc. Hortic. Sci.* **2011,** *136,* 288–302.

Zhu, J. K. Regulation of Ion Homeostasis under Salt Stress. *Curr. Opin. Plant Biol.* **2003,** *6,* 441–445.

CHAPTER 6

Habitat, Growth Response, and Adaptation under Salinity in *Sonneratia* sp.

RAUL D[1], PAWAR GP[2], TELAVE AB[2], PATIL JR[1], and NIKALJE GC[1]

[1]*Post Graduate Research Centre, Department of Botany, Seva Sadan's R. K. Talreja College of Arts, Science and Commerce, University of Mumbai, Ulhasnagar, Maharashtra, India*

[2]*Post Graduate Research Centre, Department of Botany, AES's T. C. College of Arts, Science and Commerce (Autonomous), Baramati, Maharashtra, India*

ABSTRACT

Sonneratia sp. (Family: Lythraceae) is a pioneer tree mangrove species. It naturally grows in tropical and subtropical coastlines in intertidal zones. It colonizes on newly formed mudflats of moderate-to-high saline conditions. It consists of major six species, namely, *Sonneratia apetala* (Sa), *Sonneratia alba*, *Sonneratia caseolaris* (Sc), *Sonneratia griffithii*, *Sonneratia lanceolatae*, and *Sonneratia ovata*. These species are categorized into both salt excluders and/or salt accumulators. The accumulation of ionic and nonionic solutes in the cell generates higher osmotic pressure as compared to seawater. This difference in osmotic pressure helps *Sonneratia* to extract water from the sea via the ultrafiltration process. Being a mangrove plant, *Sonneratia* sp. requires a certain amount of salt (5–50% seawater for *S. alba*) for maximum growth. The anatomical study of Sc revealed the importance of periderm tissues in roots and tanniferous cells in pneumatophores for salt adaptation. At the biochemical level, *Sonneratia* relies on the increased content of soluble

Physiology of Halophytes: Signaling, Omics, and Tolerance Mechanisms.
Nivas Manohar Desai & Ganesh Chandrakant Nikalje (Eds.)

sugar and antioxidant enzyme activity for protection against high salinity. In addition to salinity, *Sonnenratia* can uptake and tolerate high amounts of metal ions (Cr, Pb, Cu, Cd, and Zn). In this chapter, the diversity, habitats, growth and adaptation to salinity, and phytoremediation ability of different species of *Sonneratia* are discussed.

6.1 INTRODUCTION

The term "mangroves" includes a group of halophytes that have evolved to survive in a high-salinity environment. They have many specialized characteristics including salt tolerance, presence of aerial roots, and high tannin content (Feng et al., 2020). They are found in tropical intertidal estuarine regions, sheltered coastlines, and creeks (Telave, 2015). Mangroves help in reducing erosion due to waves in coastal areas (Yang et al., 2015). The benefits of the mangrove ecosystem are well documented but many aspects of the relationships between interspecific differences in salt tolerance and species distributions are still unknown (Ball and Pidsley, 1995).

The genus *Sonneratia* is a major mangrove component. Members of this genus are solitary, having vestigial or apetalous flowers with numerous stamens. The height of pneumatophores is up to 1.5 m. The genus belongs to the family *Lythraceae* with five species in India, namely, *Sonneratia alba*, *Sonneratia apetala* (Sa), *Sonneratia caseolaris* (Sc), *Sonneratia griffithii*, *Sonneratia lanceolata*, and *Sonneratia ovata.* Among them, *S. alba* is widely distributed in estuaries and creeks of Maharashtra. It is found in the estuaries of Goa in the mesohaline zone with 18–30% salinity (Telave, 2015). It has tree habit, with a spreading, broad canopy. Usually, it grows up to 15–20 m in height but sometimes 30 m in height is also recorded. The tree trunk contains cream to brown-colored bark, having fine, longitudinal, and smooth fissures. The pneumatophores (conical roots that emerge vertically from the ground) are thick and blunt and can grow up to 1 m in height. The leaves are stalked, opposite, obovate, or elliptic to ovate shaped, leaf apex is rounded or broad, 5–11 cm long, and 4–8 cm wide. Leaf stalks are 6–15 mm long and have vestigial glands at the base. Fruits are berry type, fleshy, hard with flattened and round shaped, persistent calyx present at the base, 2–4.5 cm across, ripening green. It is mainly found in sandy beaches, and mudflats of mangrove forests on tropical coasts (Mariano et al., 2019). Sa is found in certain regions of Maharashtra, India. It is reported from Mumbai, Devgadh, and Achara estuaries (Telave, 2015). Sa is a common species on the river-facing intertidal mudflat. Sc is reported from the West coast of the Goa region. It is becoming a rare species in Maharashtra due to increased pollution and human encroachment. It is mainly found in the inner estuary and prefers less salinity (Telave, 2015; Naskar, 2015). Pneumatophores show differentiation among the species. Sa shows a smaller number of pneumatophores. The size of these pneumatophores is also small. Sc has larger pneumatophores. The color of the fruit also varies depending on the location. *S. griffithii* has a narrow distribution and is categorized as "critically endangered" (Polidoro et al., 2010). The Plant List (2013), categorized *S. griffithii* as an "unresolved" species. Because, although this species is locally common, there is a lack of collection and herbarium specimens (Ragavan et al., 2019). In India, *S. ovata* is reported from Havelock Island of South Andaman (Goutham-Bharathi et al., 2012). The salt tolerance also shows differentiation among species. *S. alba* shows optimal growth in 5–50% salinity. It indicates the capability to tolerate high salinity (Telave, 2015). Sc does not show maximum growth in high-saline regions (Loan et al., 2020). *S. lanceolata* is also found along the edges of freshwater ponds in the river floodplain. *S. lanceolata* shows maximum growth in 0–5% seawater and its growth is inhibited by an increase in salinity to 50% seawater. The relative growth rate of *S. lanceolata* was maximal in salinities ranging from 0 to 5% seawater (Ball and Pidsley, 1995).

The salt tolerance mechanism of a species plays a crucial role in the development of plants in saline regions. Understanding salt tolerance mechanisms enable to development of various methods to grow plants in saline conditions (Kawana and Sasamoto, 2008).

6.2 HABITAT

The mangrove species is mainly found in intertidal zones, estuaries, and creeks. The members of this genus are distributed in various regions of Maharashtra, Goa, Karnataka, Andaman, and Nicobar Island.

6.2.1 HABITAT OF S. ALBA AND S. LANCEOLATA

S. alba is native to low intertidal habitats with moderate-to-high rainfall and receives input from both freshwater and saltwater. It grows along the banks of rivers on mud and sand. It requires areas with low wave energy and islands with sheltered shores. The soil is soft, waterlogged, and compacted.

The compacted soil creates anaerobic conditions. The mangrove habitat harbors sulfur-oxidizing bacteria that grow on dead and decaying animals, and plants release hydrogen sulfide gas with the smell of rotten eggs. This smell is characteristic of mangrove habitats (Teo, 2013). *S. alba* is commonly encountered along the seaward margins of coastal wetlands having at least 1000 mm average annual rainfall, whereas the distribution of *S. lanceolata* is more restricted. It has been reported only from 10 river systems of Northern territory in areas. The average annual rainfall in this area is 1250–1500 mm. The *S. alba* and *S. lanceolata* are found in waterlogged, muddy banks at the lowest elevations colonized by mangroves but populations of the two species rarely overlap. *S. alba* is usually found near the mouth of tidal rivers, whereas *S. lanceolata* typically occurs further upstream (Ball and Pidsley, 1995). *S. alba* is often found in tropical and subtropical regions in brackish water. The *S. alba* is the only mangrove that grows toward the seaside because of its high salt tolerance (Niu et al., 2018).

FIGURE 6.1 Habitat structure of *S. alba*. (A) Habitat. (B) Flower. (C) Fruit. (D) Pneumatophores.

6.2.2 HABITAT OF S. CASEOLARIS

Sc is one of the pioneer mangrove species that colonizes riverbanks and mudflats in new habitats (Tatongjai et al., 2021). It grows in deep muddy soils along the brackish backwater of mangrove habitats and slow-moving water of tidal rivers. Sc is found toward the inner side of the estuary, which has low salinity, inundation takes place only during spring tide. It is recorded from the junction of saline rivers and urban sewage disposal channels and along the prawn cultivation fishery plants at the edges of saline water feed (Naskar, 2015).

6.2.3 HABITAT OF S. APETALA

Sa is one of the ecologically important tree mangrove species in the Indo-Malayan region of mangrove forests. In Sundarban forests, it forms mono-specific strands. It is a well-known pioneer species in succession and is found in association with *Avicennia officinalis*. It prefers moderate-to-high salinity for growth (Rahman et al., 2020). It naturally grows along the tropical and subtropical coasts. It occurs in upstream estuarine zones of low to mid-intertidal regions. In the ecological succession process, it can act as pioneer species and form monophyletic strands. In general, it tends to colonize newly formed mudflats with moderate-to-high saline conditions (Nasrin et al., 2017).

FIGURE 6.2 Habitat structure of Sa. (A) Habitat. (B) Flower. (C) Fruit. (D) Pneumatophores.

6.2.4 HABITAT OF S. OVATA

S. ovata is not a common species. In the whole Asia, it is considered a rare mangrove species. It grows as a terrestrial fringe along the coastal region. However, it does not grow on hard substrates such as coral reefs. As per the present record, this mangrove is found in the islands of Andaman and Nicobar, which is a new record for the mangrove flora of India. This is convincing proof, that *S. ovata* has floristic affinities with other Southeast Asian countries (Goutham-Bharathi et al., 2012).

6.2.5 HABITAT OF S. GRIFFITHII

S. griffithii is a rare species observed in India. The record of the distribution of this species in India is inaccurate. Hence, there is a need for critical studies on the distribution of *S. griffithii* on an urgent basis. It is categorized under threatened species of mangrove at the global level. The population of this mangrove is declining. Therefore, studies on its conservation, accurate identification, and occurrence are important (Ragavan et al., 2019).

6.3 EFFECT OF SALINITY ON THE GROWTH

Salinity is one of the abiotic factors responsible for the growth, establishment, and zonation of mangroves. Most of the mangrove species prefer less or moderate salinity levels (~25 ppt). However, there is a significant difference between the salt requirement and tolerance level among the mangroves. Although mangroves are known for their salt tolerance, high salinity causes the lower osmotic potential of sediments which disturbs the physiological disturbance in mangroves and leads to altered water relations and stunted growth of mangroves (Chowdhury, 2015). The salt responses of different *Sonneratia* sp. are species specific.

6.3.1 EFFECT OF SALINITY ON GROWTH OF S. ALBA AND S. LANCEOLATA

S. alba grows in salinities ranging from 0 to 100% seawater, with maximal biomass developing in 5–50% seawater. However, growth in 0% and 100% seawater is extremely poor and it is doubtful whether these plants would survive. In contrast, *S. lanceolata* grows best in low salinity (0–5% seawater) while the decrease in growth starts with an increment in salinity to 50% seawater. Only 50% of plants can survive in the 50% seawater treatment at the time of harvest. Both species attain maximum growth at 5% salinity. *S. lanceolata* produces roughly twice the biomass, leaf area, and height of *S. alba.* Thus, both species show maximal growth in relatively low salinities but differ in maximum growth rates and in the range of salinities over which maximal growth rates are sustained. Salinity differentially affects canopy development in *S. alba* and *S. lanceolata*. In *S. alba*, the lowest number of leaf pairs are developed in plants grown in 0% seawater, and the highest number of leaf pairs are produced in plants grown in 5 and 25% seawater. In contrast, the average number of leaf pairs per plant in *S. lanceolata* declines with an increase in salinity from 0 to 50% seawater. In 5% seawater, both species show a similar number of leaf pairs but the total leaf area averaged 10 and 250 cm^2 in *S. alba* and *S. lanceolata*, respectively, because of differences in the areas of individual leaves. Hence, salinity affects both leaf initiation and leaf area expansion (Loan et al., 2020).

6.3.2 EFFECT OF SALINITY ON GROWTH OF S. CASEOLARIS

The Sc is sensitive to high salinity. It fails to attain sufficient growth under high salinity. The relative water content and succulence of matured Sc leaves are not greater in the highly saline area. This may be the reason for the development of succulence during the transition of leaves from maturity to senescence. In addition, it shows a reduction in chlorophyll b content in high salinity (Loan et al., 2020). The different types of Sc roots show the presence of periderm in the young stage due to exposure to saline water. Generally, the periderm tissues develop at the second growth stage, however, salt exposure leads to the early development of periderm tissues, as a part of the salt adaptation mechanism. This early development helps in the infiltration of salt from seawater and soil. In both fresh and saltwater conditions, it shows the presence of mucilaginous cells in the adaxial palisade surface which protects the plant from wilting by storing the water in these cells. This character is common among all the species of *Sonneratia.* The presence of tanniferous cells is a common feature of most mangrove species. However, under salt exposure plants accumulate more tannins as compared to freshwater. These cells are localized in pneumatophores and palisade parenchyma under salt exposure and absent in freshwater conditions. The tannins can chelate the excessive iron ions to retard the oxidation process. In Sc, tissues accumulate more tannins under salt conditions than in freshwater which proves its role in salt adaptations (Tatongjai et al., 2021). Also, the thickness of the leaf blade of Sc is associated with its salinity tolerance mechanism. Under salt conditions, the length of spongy parenchyma and palisade cells gets reduced which leads to a decrease in the thickness of mesophyll cells. This causes a reduction of intercellular spaces and structural alterations in mesophyll cells. Ultimately, it reduces the thickness of the leaf blade as compared to freshwater conditions (Tatongjai et al., 2021).

FIGURE 6.3 (A) Fruits of *S. alba.* (B) Nest of weaver ant (*Oecophylla smaragdina*) on *S. alba.* (C) Pneumaophores of a mature tree of *S. apetala.* (D) Leaves of *S. alba* showing salt glands.

6.3.3 *EFFECT OF SALINITY ON S. APETALA GROWTH*

Sa can tolerate strong salinity but grow well in moderate saline areas with newly accreted land. It can tolerate soil salinity of up to 3–8 ppt and water salinity of up to 44 ppt (Nasrin et al., 2017). The Sundarbans forests are divided into three distinct zones depending on salinity levels. The first zone consists of low salinity (14 ppt salinity), the second with medium or moderate salinity (14–25 ppt salinity), and the third zone, possesses high salinity (>25 ppt salinity). Sa prefers the first zone, that is, the low salinity zone for its maximum growth. In this zone, it grows better as compared to medium and high salinity zones. The higher salinity imposes a negative impact on the radial growth of Sa. High salt disturbs the physiology of this tree mangrove by affecting the decline in photosynthetic rate if salinity exceeds 15 ppt under controlled conditions. The low input of freshwater in estuaries might be the reason for increasing salinity in mangrove habitats which subjects mangroves to unfavorable conditions. The freshwater inputs from river discharge and monsoon precipitation are key factors in the growth of Sa (Rahman et al., 2020).

TABLE 6.1 Growth of *Sonneratia* sp. in Different Salinity (Chowdhury, 2015; Nguyen et al., 2020; Rahman et al., 2020; Tatongjai et al., 2021).

Sonneratia **species**	**Growth in different salinity (%)**				
	0	**0–25**	**25–50**	**50–100**	**100**
S. alba	Extremely poor	Optimum	Average	Poor	Extremely poor
S. lanceolata	Optimum	Optimum	Poor	Extremely poor	Zero
Sc	Optimum	Optimum	Poor	Extremely poor	Zero
Sa	Optimum	Optimum	Poor	Poor	Extremely poor

6.4 MORPHOLOGICAL AND ANATOMICAL ADAPTATIONS TO SALT STRESS

Unlike other species of mangroves, *Sonneratia* does not show the presence of salt glands to eliminate salt. Sc shows the presence of multiple root systems to eliminate salt. Sc and *S. alba* possess four different types of roots such as anchor, cable, feeding, and pneumatophores. Interestingly, in Sc the anatomical characteristics of the petiole, secondary growth of the stem, feeding root, and anchor root are similar in both saltwater and freshwater. The anatomy of Sc with reference to salt adaptation is as follows:

a. **Multiple root system:** It consists of three types of roots.

   1. **Cable root:** The periderms are composed of three layers namely phelloderm, phellogen, and phellem form periderm. The cortex shows the presence of aerenchyma. The cortex is made up of two

types of cells having round and arm-shaped structures. It also shows the presence of druse crystals and sclereids. Endodermis is made up of rectangular cells. Casparian strips are absent. The stele is composed of pericycles, rings of the xylem and phloem, and pith.

2. **Pneumatophores:** The pneumatophores are consisting of a periderm, cortex, and stele. The periderm is the outermost layer. The parenchymatous tissues in the cortex are of two distinct types, namely, aerenchyma and round-shaped parenchyma. The "round-shaped parenchyma contains tannins in mangroves." In mangroves, the round parenchyma shows the presence of tannin. The cortex shows the presence of druse crystals and sclereids. The stellar region consists of a pericycle, xylem phloem rings, and parenchymatous pith.

3. **Feeding and anchor roots:** Both feeding and anchor roots are similar in structure. It is made up of an outer layer, that is periderm. Round elliptical parenchyma cells are also present. Druse crystals are present in the cortex. The Casparian strips are present on endodermis and radial vascular bundles. The pith is made up of only parenchymatous cells and is the root's innermost structure.

b. **Stem:** During the primary growth stage, in the stem, the cortex shows the presence of tenniferous cells. During secondary development, the periderm, cortex region, vascular tissue, and parenchymatous pith are present in the stem. The cortex shows the presence of sclereids, druse crystals, and tenniferous cells. The phloem consists of gelatinous fibers, and druse crystals. Pith is composed of parenchyma and sclereids.

c. **Wood:** It is porous in structure. Vessels show the presence of simple perforations. The comparison between the anatomy of plant samples collected from saltwater and freshwater then salt-water samples shows a higher density of vessels than that of freshwater samples. The diameter of vessels is larger in saltwater samples than that in freshwater samples. The comparison between the anatomy of different parts of Sc revealed that the cortex and stele of cable roots accumulated a high amount of salinity, that is, 583.70 ppm and wood accumulated a low amount of salinity, that is, 64.75 ppm. This implies that the NaCl concentration is different in different parts of the plant (Purnobasuki and Suzuki, 2004; Tatongjai et al., 2021). The flooding causes anaerobic conditions in sediments, to survive under these conditions, mangroves develop high porosity for survival. In *S. alba,* high porosity facilitates the supply of oxygen from aerial parts to the underground roots. The porosity of pneumatophores is 60.7% while cable roots possess 50.1% which is the maximum as compared to porosity in anchor roots (41.1%) and feeding roots (38.9%). The lenticels present on the pneumatophores above ground provide oxygen. Pneumatophores supply oxygen to the feeding and anchor roots (Purnobasuki and Suzuki, 2004).

6.4.1 PHYSIOLOGICAL AND BIOCHEMICAL ADAPTATIONS

Halophytes can be classified on the basis of the different mechanisms of adaptation to salty conditions into (1) salt-excluding, (2) salt excreting, and (3) salt-accumulating plants (Taia, 2020). *S. alba*, Sa, and Sc show a salt accumulation mechanism to tolerate high salinity. Plants absorb the maximum amount of water to dilute the salt content and get rid of the toxic effects of salt ions. The mechanism of salt exclusion includes restriction of salt uptake by ultrafiltration at the root level in a high salinity environment. In saline conditions, pneumatophores provide access to atmospheric oxygen. It shows the presence of air cavities that help in the oxygenation of root tissues (Teo, 2013).

TABLE 6.2 Adaptation Mechanisms used by different *Sonneratia* Species (Teo, 2013; Hasanuzzaman et al., 2014; Taia, 2020)

Salt adaptation strategy	Plant response and examples
Salt exclusion	Minimizing salt stress using the ultrafiltration mechanism of the root system; e.g., Sc.
Salt accumulation	Cells and tissues accumulate high concentrations of salt to maintain water flux; e.g., Sa, Sc.
Pneumatophores	Vertical roots present above the soil allow access to atmospheric oxygen in anaerobic soil conditions; e.g., Sa, Sc.

At a biochemical level, the net photosynthesis rate and stomatal conductance of Sc and Sa under low salinity were stable, that is, adapted to low salinity. At high salinity, both parameters were adversely affected. The transpiration rate decreased with increments in salt treatment. In Sc, the transpiration rate decreased sharply even in low salinity. The stomatal conductance follows the same trend as that of the transpiration rate. Thus, decreased stomatal conductance helps to slow down the transpiration rate and subsequent water loss (Yan and Guizhu, 2007). The total soluble sugar of Sa and Sc increased significantly at high salinity and increased slightly at low salinity. The addition of salt in the substratum causes increased osmotic potential which restricts water absorption by roots and imposes physiological drought on plants. In this situation, the balance of ions in the vacuole is controlled by low molecular weight compatible solutes such as soluble sugars. They act as a solvent in biochemical reactions and protect water circulation and cell structure. In both Sa and Sc, increased total soluble sugars enhance permeability and maintenance of water metabolism balance (Yan and Guizhu, 2007).

The salt stress causes increased membrane peroxidation in all parts of plants. Under low salinity, it is maintained in Sa and Sc but increment in salt concentration rapidly increased lipid peroxidation. However, there was no significant induction of superoxide dismutase (SOD) activity. The anti-oxidant enzymes such as SOD activity were decreased in Sc and Sa under salinity in leaves. This may be because these plants rely on other antioxidant enzymes such as catalase, peroxidase, ascorbate peroxidase, etc. The roots of this mangrove are insensitive to salinity as there are no significant changes in malondialdehyde (MDA) content and SOD activity. Mangrove roots possess a unique adaptation mechanism. They are submerged in seawater perennially with low oxygen supply. It leads to the development of a special kind of root metabolism which may control changes in membrane peroxidation of roots even under high salinity (Yan and Guizhu, 2007).

6.5 APPLICATIONS IN PHYTOREMEDIATION

Due to increasing industrialization and human encroachment, the mangrove ecosystems are degrading at high rates. Industrial wastes, agricultural run-offs, and urban waste are dumped into the seashore. It is increasing the heavy/toxic metal content of these sites. Among *Sonneratia* sp., Sa has been studied for its phytoremediation ability in Malaysian mangrove sites. These results indicated that Sa have the capacity to uptake some heavy metals (Cr, Pb, Cu, Cd, and Zn) via roots and store them in leaves without any sign of injury. It can take up a high amount of Cu and Pb as compared to other plants without toxicity (Nazli and Hashim, 2010). It can accumulate a high amount of Pb (35.5 μg/g), followed by Cu (26.8 μg/g), Cr (9.5 μg/g), Zn (5.9 μg/g), and Cd (μg/g) (Nazli and Hashim, 2010). The minimal damage is due to metal stress in Sa reveals the high phytoremediation ability of this species. The phytostabilization of the mangrove ecosystem can be achieved with *Sonneratia* sp.

6.6 CONCLUSION AND FUTURE PROSPECTS

Sonneratia sp. is one of the pioneer mangrove species. The morphological and anatomical modification enables it to survive in saline conditions. It shows a wide range of habitats including Sundarbans and intertidal regions. Among the six species *S. ovata* and *S. griffithii* need to be studied in detail to understand their growth in different salinities. *S. alba* shows optimum growth in moderate and high saline regions. Sa shows optimum growth in low and moderate salinity. Hence these two species are widely distributed in coastal regions. Sl and Sa show optimum growth in low salinity. The tolerance to salinity is because of the different physiological and biochemical adaptations of these mangroves. The phytoremediation ability of these species needs to be further exploited for the stabilization of mangrove ecosystems.

ACKNOWLEDGMENT

Author GCN is thankful to Science and Engineering Research Board (SERB), India for the fellowship under the SERB-SIRE program (Grant code: SIR/2022/000391).

KEYWORDS

- **growth**
- **biochemical response**
- **ion homeostasis**
- **salt adaptation**
- **phytoremediation**
- **Sonneratia sp.**

REFERENCES

Ball, M.; Pidsley, S. Growth Responses to Salinity in Relation to Distribution of Two Mangrove Species, *Sonneratia alba* and *S. lanceolata*, in Northern Australia. *Funct. Ecol.* **1995,** *9* (1), 77.

Chowdhury, M. Effect of Salinity on the Above-Ground Biomass and Growth of Dominant Mangrove Plants in Indian Sundarbans. *Ann. Biol. Res.* **2015,** *6* (5), 20–25.

Deshmukh, S.; Balaji, V. Conservation of Mangrove Forest Genetic Resources; M. S. Swaminathan Research Foundation, **1994**; pp 425–426.

Feng, X.; Xu, S.; Li, J.; Yang, Y.; Chen, Q.; Lyu, H.; et al. Molecular Adaptation to Salinity Fluctuation in Tropical Intertidal Environments of a Mangrove Tree Sonneratia alba. *BMC Plant Biol.* **2020,** *20* (1).

Goutham-Bharathi, M. P.; Kaliyamoorthy, M.; Dam Roy, S.; Krishnan, P.; George, G.; Murugan, C. *Sonneratia ovata* (Sonneratiaceae) – A New Distributional Record for India from Andaman and Nicobar Islands. *Taiwania* **2012,** *57*, 406–409. 10.6165/tai.2012.57.406

Hasanuzzaman, M.; Nahar, K.; Alam, M.; Bhowmik, P.; Hossain, M.; Rahman, M.; et al. Potential Use of Halophytes to Remediate Saline Soils. *Biomed. Res. Int.* **2014,** *2014*, 1–12.

Kader, A.; Sinha, S.; Ghosh, P. Impact of Various Factors on in vitro Investigation of Rare Mangrove Species *Sonneratia apetala* Buch.-Ham. (Myrtales: Lythraceae) and Suaedamaritima (L.) Dumort (Caryophyllales: Amaranthaceae). *Braz. J. Biol. Sci.* **2016,** *3* (5), 181.

Kawana, Y.; Sasamoto, H. Stimulation Effects of Salts on Growth in Suspension Culture of a Mangrove Plant, *Sonneratia alba*, Compared with Another Mangrove, Bruguierasexangula and Non-mangrove Tobacco BY-2 cells. *Plant Biotechnol.* **2008,** *25* (2), 151–155.

Lacerda, L.; Carvalho, C.; Tanizaki, K.; Ovalle, A.; Rezende, C. The Biogeochemistry and Trace Metals Distribution of Mangrove Rhizospheres. *Biotropica* **1993,** *25* (3), 252.

Loan, N.; Tan, D.; Huyen, T.; Quang, N.; Hue, L.; Nga, P.; et al. Comparison of Several Secondary Metabolite and Elemental Ion Contents of Leaves from Kandelia obovata and *Sonneratia caseolaris* Forests Located in the Red River Delta. *Acad. J. Biol.* **2020,** *42* (4). https://doi.org/10.15625/2615-9023/v42n4.15068

Mariano, H.; Dagoc, F. L.; Dagoc, S.; Espra, A. S.; Amparado, R. Mangrove Diversity, Taxonomic Classification, and Morphological Characteristics of Natural and Reforested Mangrove Forests in Selected Municipalities of Zamboanga Del Sur, Mindanao Island, Philippines. *J. Biodivers. Environ. Sci.* **2019,** *15*, 86–89.

Naskar, S. Leaf Epicuticular and Pollen Ultrastructural Comparisons of *Sonneratia apetala* Buch.-Ham. and *S. caseolaris* (L.) Engler (Sonneratiaceaea). *Mod. Phytomorphol.* **2015,** *7*, 47–53. 10.5281/zenodo.160360.

Nasrin, S.; Hossain, M.; Alam, M. In *A Monograph on Sonneratia apetala Buch.- Ham*, **2017**; pp 4–10.

Niu, D.; Li, S.; Yi, S.; Shen, L.; Huang, M.; Zhou, R.; et al. Gene Cloning, Protein Expression and Functional Analysis of a Type 3 Metallothionein with Bioaccumulation Potential from *Sonneratia alba*. *Polish J. Environ. Stud.* **2018,** *27* (5), 2203–2212.

Polidoro, B. A.; et al. The Loss of Species: Mangrove Extinction Risk and Geographic Areas of Global Concern. *PLoS One* **2010,** *5* (4), e10095.

Purnobasuki, H.; Suzuki, M. Aerenchyma Formation and Porosity in Root of a Mangrove Plant, *Sonneratia alba* (Lythraceae). *J. Plant Res.* **2004,** *117* (6), 465–472.

Qiu, Y.; Qiu, H. In *Comparison of Metals Levels in Two Mangrove Species (Rhizophora stylosa and Sonneratia hainanensis) from Hainan Island, South China*, IOP Conference Series: Earth And Environmental Science, **2017**; vol *52*, p 012050.

Ragavan, P.; Dubey, S.; Panda, M.; Deyroy, M.; Ravichandran, K.; Trivedi, R.; et al. Critical Notes on the Identity and Distribution of *Sonneratia griffithii* Kurz (Lythraceae) in India – A Critically Endangered Mangrove Species. *Nordic J. Bot.* **2019,** *37* (1). https://doi.org/10.1111/njb.02119

Rahman, M.; Sass-Klaassen, U.; Zuidema, P.; Chowdhury, M.; Beeckman, H. Salinity Drives Growth Dynamics of the Mangrove Tree *Sonneratia apetala* Buch. -Ham. in the Sundarbans, Bangladesh. *Dendrochronologia* **2020,** *62*, 125711.

Rumanta, M. The Potential of Rhizophora mucronata and Sonneratia caseolaris for Phytoremediation of Lead Pollution in Muara Angke, North Jakarta, Indonesia. *Biodivers. J. Biol. Divers.* **2019,** *20* (8).

Sasamoto, H.; Iwashina, T.; Suzuki, S.; Azumi, Y.; Fujii, Y. Evaluation of an Anthocyanin, Cyanidin 3,5-di-O-glucoside, as an Allelochemical in Red Callus of a Mangrove *Sonneratia ovata*, Using Protoplast Co-Culture Bioassay Method with Digital Image Analysis. *J. Plant Stud.* **2018,** *7* (2), 1.

Shete, A.; Gunale, V.; Pandit, G. Bioaccumulation of Zn and Pb in Avicennia marina (Forsk.) Vierh. and *Sonneratia apetala*; Buch. Ham. from Urban Areas of Mumbai (Bombay), India. *J. Appl. Sci. Environ. Manag.* **2010,** *11* (3).

Srikanth, S.; Lum, S.; Chen, Z. Mangrove Root: Adaptations and Ecological Importance. *Trees* **2015,** *30* (2), 451–465.

Taia, W. Wide Scene on Halophytes. *Int. J. Contemp. Res. Rev.* **2020,** *11* (01), 20219–20232.

Tatongjai, S.; Kraichak, E.; Kermanee, P. Comparative Anatomy and Salt Management of *Sonneratia caseolaris* L. Engl. (Lythraceae) Grown in Saltwater and Freshwater. *PeerJ* **2021,** *9*, e10962.

Telave, A. Ecophysiological Studies on Sonneratia L. from the Coast of Maharashtra, India. *Indian J. Geo-Marine Sci.* **2015,** *44*, 1239–1244.

Teo, J. *Sonneratia alba* - Mangrove Apple, **2013**; p 9.

Tsuchiya, S.; Ogita, S.; Kawana, Y.; Oyanagi, T.; Hasegawa, A.; Sasamoto, H. Relation between Amino Acids Profiles and Recalcitrancy of Cell Growth or Salt Tolerance in Tissue and Protoplast Cultures of Three Mangrove Species, *Avicennia alba*; Bruguierasexangula and *Sonneratia alba. Am. J. Plant Sci.* **2013,** *04* (07), 1366–1374.

Wang, Y.; Dai, A.; Tang, T. Weak Effect of Gypsy Retrotransposon Bursts on *Sonneratia alba* Salt Stress Gene Expression. *Front. Plant Sci.* **2022,** *12*.

Yamamoto, R.; Kawana, Y.; Minagawa, R.; Sasamoto, H. Effects of Sea Salts on Induction of Cell Proliferation in Liquid Cultures of Mangrove Plants, *Sonneratia caseolaris* and *S. alba. Am. J. Plant Sci.* **2011,** *02* (01), 35–42.

Yang, Y.; Guo, W.; Shen, X.; Li, J.; Yang, S.; Chen, S.; et al. Identification and Characterization of Evolutionarily Conserved Alternative Splicing Events in a Mangrove Genus Sonneratia. *Sci. Rep.* **2018,** *8* (1).

Yang, E.; Yi, S.; Bai, F.; Niu, D.; Zhong, J.; Wu, Q.; et al. Cloning, Characterization and Expression Pattern Analysis of a Cytosolic Copper/Zinc Superoxide Dismutase (SaCSD1) in a Highly Salt Tolerant Mangrove (*Sonneratia alba*). *Int. J. Mol. Sci.* **2015,** *17* (1), 4.

CHAPTER 7

Physiology and Potential Applications of Sand Dune Halophytes

GANESH PAWAR and AJIT TELAVE

Post-Graduate Research Centre, Department of Botany, Tuljaram Chaturchand College of Arts, Science and Commerce, Baramati, District Pune, Maharashtra, India (Autonomous)

ABSTRACT

The coastal ecosystem provides ecosystem services such as livelihoods through fisheries, tourism, ports, and other industries for the population settled in coastal areas. These areas have higher average population densities and are also economically important. The global population continued to grow and human activities drastically increased in coastal regions of the world. Besides human settlement, coastal areas are severely affected by natural disasters, tropical cyclones, high tides, hurricanes tsunamis, and coastal erosion. The intensity of natural calamities due to rapid sea level rise is intensively increased in recent periods, especially in tropical coastal areas. Natural disaster in coastal areas causes high economic and human life loss which retards the development of affected areas. To fight this concern, many engineering solutions are implemented worldwide but high cost and less effectiveness requires new strategies for coastal protection. The coastal forests such as mangroves, sand dune halophytes, and associated plants can be used for conservation, restoration, and management of the coastal environment. The well-established sand dune halophytes act as a bioshield which block wave action with their rhizomes or stem to several meters of circumference. Sand dune halophytes can trap sediment by a root, which forms a sand dune structure and absorbs the energy of tidal actions. This

Physiology of Halophytes: Signaling, Omics, and Tolerance Mechanisms.
Nivas Manohar Desai & Ganesh Chandrakant Nikalje (Eds.)

chapter deals with the importance of sand dune halophytes against natural disasters, their salt-tolerant mechanism and the conservation of fragile coastal ecosystems.

7.1 INTRODUCTION

Halophytes are the group of plants that possess high salt tolerance and have ability to grow in substrates that are rich in high salinity. They are found worldwide in different habitats such as sandy beaches, deserts, mangrove swamps, and coastal areas (Kapler, 2019). In world flora, halophytes are represented by ~1% monocot and dicot plants. On the demand and resistance to salts (NaCl), halophyte plants are classified into two types—facultative and obligate halophytes. Facultative halophytes require some freshwater supply for growth; in opposite, obligate halophytes require some salt for growth and development (Kumari et al., 2015). Based on water availability and habitats, halophytes are further classified into two types—hydrohalophytes and xerohalophytes. Hydrohalophytes are found in aquatic habitats where wet soil is present, and xerohalphytes are grown in saline and dry soil such as desert areas (Kumari et al., 2015; Etesami and Beattie, 2018).

Coastal sand dune (CSD) ecosystems are found in between terrestrial and marine ecosystems which show characteristic vegetation known as "psammophytes or strand plant community" (Saye and Pye, 2007). In coastal regions, sand dunes play an important role, that is, protect coastal barriers against natural disasters. CSD consists mainly of shrubs and grasses. Because of their ecological role and bioactive potential, coastal halophytes grab the attention of researchers and government agencies. CSDs are a fragile and dynamic area of sand formed by three main components, that is, availability of sand, suitable wind velocity, and vegetation for accumulation of sand (Rodrigues, 2011). CSD plants show some external and internal adaptations in their structure to survive in harsh climatic and geographical conditions such as deficiency of nutrients, high temperature, lack of moisture, and shifting sand. Changes in external and cellular structure are also found in halophytes (Fig. 7.1).

A chain of series of metabolic, molecular, and physiological processes is involved in plants for tolerance to salinity stress. Tolerance to salinity stress is a complex process, both genetically and physiologically (Shao et al., 2016). Coastal halophytes consist of creepers and grasses that bind

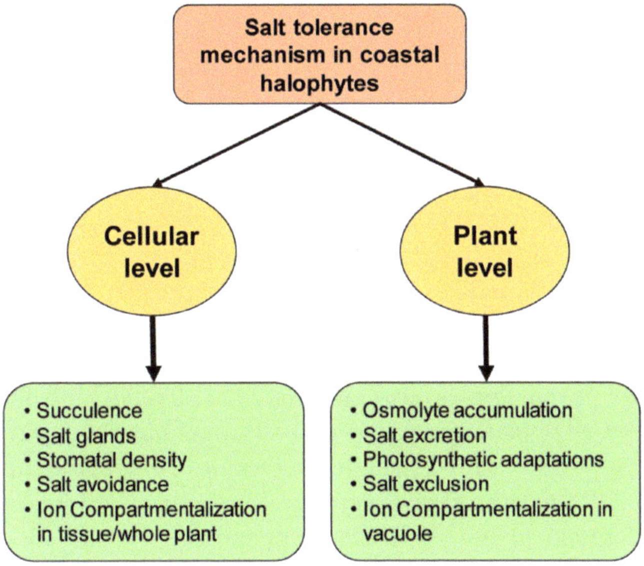

FIGURE 7.1 Cellular and plant level adaptations in halophytes.

unstable sand using their network of rhizomes and act as a sand binder, protecting coastal habitats (Naikoo et al., 2020). The result of high intensity of waves, wind velocity, direction, and beach grasses establishes a fore dune which may be two or more meters in height. This formation of foredune and other dune types depends extensively on the rhizome system of vegetation (Short and Hesp, 1982). The CSD flora provides many direct uses and indirect ecosystem services to society, which is important for the well-being of coastal populations (Kapler, 2019). In view of such a great bioactive potential, in recent decades, CSD flora and dependent fauna have continuously decreased due to over-exploitation and anthropogenic activities in coastal areas (Vecchio, 2015) thus such valuable ecosystems need to be protected from degradation and conservation of CSD native diversity and their ecological functions.

7.2 DISTRIBUTION OF COASTAL SAND DUNE HALOPHYTES

7.2.1 WORLD

CSDs are part of unique coastal ecosystems which is present in between terrestrial and marine ecosystem environments. Coastal dunes are composed of shifting mobile beaches and stable sand dunes. Different

CSD types are found above the high tide level of sandy beaches and occur on seas, oceans, and estuaries of coastal shorelines. They occur from tropics and temperate zones, the Arctic to the Equator, and all climatic zones, that is, worldwide distribution. Due to worldwide distribution, CSDs are found in all climate zones, coastal dunes show very diverse flora covering different ecological habitats from tropical to polar latitudes, and from tropical rain forests to deserts (Hesp, 1991; Martinez and Vanquez, 2009). The angiosperm population in CSD area is very diverse (Table 7.1). Some areas are rich in dune flora. In, The Netherlands, 700 plant species are found in the coastal environment. The sand dune flora of Turkey is also the richest in the coastal Mediterranean regions. West Indies, Cuba, and Mexico in the tropics also show greater diversity of coastal dune flora. Many CSD plant species are found throughout the world, but some of those show restricted distribution. *Ammophila breviligulata*, *Cakile maritima*, *Salsola kali*, *Pinus*, *Leymus arenarius*, *Honkenya peploides*, *Ammophila arenaria* spp., *Quercus* spp., and *Fagus* spp., are frequently found in temperate regions. In the Mediterranean and subtropical regions of the world, *Scaevola plumier*, *Ipomoea stolonifera*, *Juniperus phoenicea*, and *Juncus* spp. are found. Some CSD species found in pantropical regions such as *Sesuvium portulacastrum*, *Ipomoea pescaprae*, *Canavalia rosea*, *Sporobolus virginicus*, *S. plumier*, *Acacia* spp., *Uniola paniculata*, and *Tournefortia gnaphalodes*, etc. In view of the endemic plant species distribution in CSD areas, New Zealand (40), West Indies (200), Cuba (>30), Mexico (18), Brazil (8), France (9), Portugal (10), and Norway (1) species are found (Martinez and Vanquez, 2009).

TABLE 7.1 Worldwide Distribution of Coastal Sand Dune Halophytes.

S. No.	Region	Angiosperms		
		Family	Genera	Total
1	Europe and the Polar Region	131	715	846
2	Asia, Africa, America, and Oceania	219	1414	1633

Source: Modified from (Martinez and Vanquez, 2009).

The families such as Poaceae, Fabaceae, Asteraceae, and Brassicaceae are dominant in polar regions and Europe while Asteraceae, Poaceae, Fabaceae Rubiaceae, and Chenopodiaceae are dominant families found in Africa, Asia, and American CSD areas (Martinez and Vanquez, 2009).

7.2.2 INDIA

The Indian coastline is about 7500 km long bounded by the Arabian Sea on the West Coast and Bay of Bengal to the East Coast. Coastal regions show various important ecosystems such as coral reefs, sand dunes, mangrove swamps, wetlands, and seagrass beds (Mukhopadhyay et al., 2014). The West Coast of India shows estuaries, sand dunes, mangrove swamps, and cliffs, while the East Coast shows the formation of a large delta plain (Ahmad, 1972). These coastal habitats are the nursery grounds for many important fishes, crustaceans, and medicinal flora. A lot of work regarding the physical, restoration, and geological aspects of the CSD ecosystem is being carried out in India (Mascarenhas and Jayakumar, 2008). Besides this, the floristic and taxonomic study of CSD vegetation is poorly understood. A lot of gaps exist in the ecology, diversity, and importance of coastal dune vegetation (Desai et al., 2003; Jagtap et al., 2003; Sridhar, 2009). In India, CSD ecosystems are categorized as "ecologically sensitive" (MoEF, 1991). Due to the tropical region, India receives southern winds which are responsible for high wind velocity and high wave action on the West Coast in the monsoon season (September–December). Because of this reason, the West Coast shows more sand dune structures (Fig. 7.2) and vegetation diversity as compared to the East Coast (Table 7.2).

FIGURE 7.2 Coastal sand dune types. (A and B) Embryo dunes (Baby dunes), (C) Incipient Dune (Foredunes), (D) Secondary dunes (Mature dunes). (According to Short and Hesp, 1982).

TABLE 7.2 Distribution of Coastal Sand Dune Halophytes from the Coastal States of India.

S. No.	States	Genera	Species
	West Coast		
1	Karnataka	113	145
2	Goa	91	106
3	Lakshadweep	60	67
4	Gujarat	57	66
5	Maharashtra	40	46
6	Kerala	34	37
	Total West Coast	176	267
	East Coast		
7	Tamil Nadu	79	89
8	Orissa	62	73
9	Andhra Pradesh	51	57
10	West Bengal	40	43
11	Andaman and Nicobar	23	29
	Total East Coast	120	163
	Total West Coast + Total East Coast	**208**	**338**

Source: Desai, 1995; Kulkarni et al., 1997; Arun et al., 1999; Beena et al., 2000; NIO, 2000; Jaiswal and Rodrigues, 2001; Banerjee et al., 2002; Environmental Information System (ENVIS) Newsletter, 2002; Desai et al., 2003; Jagtap et al., 2003; Ingole, 2005; KSCSTE, 2005; Arun and Sridhar, 2006; Kumaran et al., 2008; Pattanaik et al., 2008; Rodrigues et al., 2011.

The total 338 sand dune halophyte species belonging to 69 families were observed in Indian coastline. Mostly, coastal halophytes shows herbaceous habit (62%), followed by shrubs (21%), trees, climbers, creepers, and trees (17%) of the total vegetation. The diversity of sand dune halophytes on the West Coast (267) is greater as compared to the East Coast (163). The maximum number of species were recorded in Karnataka (145) from the West Coast and Tamil Nadu (89) from the East Coast. *I. pescaprae* (L.) R. Br., *Spinifex littoreus* (Burm.f.) Merr., *S. portulacastrum* (L.) L., *Suaeda maritime* (L.) Dumort., *Cyperus arenarius* Retz., *Calotropis gigantea* (L.) R. Br., *Vitex trifolia* L., *Casuarina equisetifolia* Forst, *Clerodendrum inerme* (L.) Gaertn. and *Cocos nucifera* L. are the commonly found coastal halophytic species (Fig. 7.3). The species such as *Anacardium occidentale* L., *Calophyllum inophyllum* L., and *Calotropis gigantea* (L.) R. Br. which have medicinal potential were also found in sand dune areas. Fabaceae is the dominant family followed by Poaceae, Asteraceae, and Cyperaceae.

The CSD halophytes of the West Coast of India are useful for green manure, mulch, cover crops, pasture, fodder, oil, and medicinal values (Sridhar and Bhagya, 2007). The floral diversity studies of coastal dune halophytes identified 42 species belonging to 38 genera, 26 families from Tamilnadu, (Muthukumar and Samual, 2011); 36 species belonging to 35 genera, 25 families from Puducherry, India (Poyyamoli et al., 2012); 39 species, belonging to 33 genera, 28 families Southeast India (Senthilkumar and Silambarasan, 2015) and a total of 27 coastal halophyte species belonging to 14 families and 23 genera were reported from the Gulf of Kutch, Gujarat, India (Salvi et al., 2017).

FIGURE 7.3 Coastal sand dune halophytes from coastal Maharashtra region. (A) *S. portulacastrum* (L.) L, (B) *I. pescaprae* (L.) R. Br, (C) *S. littoreus* (Burm.f.) Merr, (D) *V. trifolia* L. (E) *Crotolaria retusa* L., (F) *Crotolaria verrucosa* L.

7.3 SALT RESPONSES IN SAND DUNE HALOPHYTES

Coastal areas show a variety of vegetation types which includes mangroves, psammophytes, and sand dune plant communities constituted by grasses, shrubs, herbs, and creeping vines (Medina, 2016). Halophytes in coastal areas have to fight against a number of environmental and climatological stress factors for growth and development. The environmental factors include tidal fluctuations, salt spray, salinity, nutrient deficiency, and soil dryness (Brown et al., 2006). The separate or combined effects of these environmental factors leads to low species diversity (Rozema et al., 1985). Along with environmental stresses, coastal halophytes also face various abiotic stresses such as toxic pollutants, heavy metals, soil acidification, excessive use of fertilizers, and biotic stress such as pathogen attacks (Flowers and Colmer, 2008). When biotic and abiotic stress affects halophytes, they show a tolerance mechanism against combined stress (Hamed et al., 2013). Halophytic plants develop several genetically regulated strategies to survive in these harsh environmental conditions (Ezawa and Tada, 2009; Nikalje, 2019).

The stress tolerance mechanism in halophytes is regulated by osmotic adjustment (Flowers and Colmer, 2008) and various osmoregulatory reactions such as changes in leaf morphological characteristics including leaf succulence, higher leaf tissue density (Denaxa et al., 2012), and stomatal resistance to transpiration (Maricle et al., 2007). The variation in leaf water potential is mainly dependent on soil matric potentials (Veste and Breckle, 1995), especially in sand dunes where water penetrates in the sand by lateral movement at the monsoon season (Yair, 1990). In such harsh saline conditions, optimum osmotic adjustment is required for sufficient water uptake and growth. The osmotic adjustment mechanism is achieved by the synthesis of osmotically active solutes and ion uptake or both (Cheeseman, 1988). In salt marshes, even if no rain for a few years, removing excess salts from the root zone is carried out with the help of tidal flushing which helps plant growth and survival (Noor and Khan, 1995). However, very little information is available on plant distribution and physiological responses which affects plant diversity and abundance in the coastal zone (Rasool et al., 2019).

In the coastal region, during the summer season, salt concentration of the soil was higher as compared to the autumn and winter seasons. Many species of chenopodiaceae family accumulate a large amount of sodium ions in the summer season (Ramos et al., 2004) however, higher concentration of sodium accumulation can cause direct harmful effects on the plant growth (Niu et al., 1995). Besides this, halophytes need sodium uptake for osmoregulation process which starts from seed germination, important to seedlings for maintaining water uptake, turgor pressure, and growth (Koyro and Eisa, 2008). Halophytic plants control high Na^+ concentrations in two ways—either ion exclusion by root or excretion of sodium ions through salt glands of leaves. In the opposite way, some halophytes store or compartmentalize salts in their tissues or vacuoles for osmotic adjustment. Insufficient osmotic adjustment leads to growth reduction (Barhoumi et al., 2007). In coastal areas, many plant species such as *S. portulacastrum, Zygophyllum xanthoxylum, I. pescaprae, Bruguiera cylindrica*, and *Atriplex halimus* show positive effect of salinity under water scarcity and drought conditions (Slama et al., 2008; Atreya et al., 2009; Ma et al., 2012). In the combined stress situation of water scarcity and salinity, plants increased stomatal conductance, net photosynthesis rate, and higher leaf Na^+ concentration, which leads to triggered photosynthesis rate and perform osmotic adjustment for water storage (Martinez et al., 2005). In view of halophyte germination, NaCl has been the center of the majority of studies. The study suggests that Na^+ ion concentration may be the most suitable ionic correlation with glycophyte germination (Zhang et al., 2018) also Na^+ may be influential on seed germination of halophytes. For example, K^+ ions are more toxic than Na^+ in the germination of *Zygophyllum propinquum* (Manzoor et al., 2017). Temperature, light, and water potential are major environmental factors that can regulate the time and duration of seed germination, which controls the emergence and survival of seedlings (El- Keblawy et al., 2016).

Halophytes also synthesize protective compounds in response to ion toxicity and show the ability to tolerate heavy metal stress (Rastgoo and Alemzadeh, 2011). Halophytes show accumulation of proline in cell cytosol which is commonly associated with vacuolar Na^+ sequestration (Lutts and Lefevre, 2015). The osmotic function protecting the quaternary structure of protein and stabilizing cell membrane through phospholipid (Rastgoo and Alemzadeh, 2011). In *Aeluropus littoralis*, the proline amount increased of heavy metals such as Co, Cd, Ag, and Pb (Lutts and Lefevre, 2015). In *Mesembryanthemum crystallinum*, a facultative halophyte shows Cd exposure caused accumulation of total soluble sugars and proline (Lefevre et al., 2009).

7.4 APPLICATIONS OF SAND DUNE HALOPHYTES

CSD halophytes are part of a coastal ecosystem which are highly threatened by the current decade by mismanagement including construction, mineral extraction, and replacement of the coastal vegetation (FAO, 2015). The main reason for the mismanagement of coastal ecosystems is the lack of awareness about their values and defining norms for their appropriate use in coastal area habitats. The awareness among human society will lead to take better decisions for the appropriate use and protection of the ecosystem (Daily et al., 2009). Coastal ecosystems work as "coastal bioshield" against natural disasters, which protects the coastal population (Leyva et al., 2009). The direct and indirect uses of coastal halophytes are explained to people by using the concept of "environmental services (ESs)". Ecosystem services are the direct and indirect benefits of the ecosystem to the community. The enlisting and classification of various ESs provided by coastal ecosystems are categorized (Table 7.3) into four main functions (De Groot et al., 2012). These are (1) supporting function (e.g., photosynthesis and nutrient cycles), (2) provisioning function (firewood, food, and fibers), (3) regulating functions (climate regulation, regulation of diseases, pollination), and (4) cultural functions (touristic, recreational, spiritual, and esthetical).

7.4.1 MEDICINAL PROPERTIES

In recent years, isolation of important medicinal compounds from, seagrass, seaweeds, mangroves, and coastal halophytic plants acquired more attention. Coastal halophytes show biologically and chemically unique properties and great scope for the discovery of new pharmaceutical compounds (Boopathy and Kathiresan, 2010).

Recent studies proved that coastal halophytes contain highly bioactive compounds with useful medicinal properties (Table 7.4) such as anticancer, antimalarial, antiviral, antimicrobial, antifungal, antiplasmodial, antioxidant, antiparasitic, and anti-inflammatory properties due to presence of secondary metabolites and other phytochemical compounds such as alkaloids, steroids, polyphenols, glycosides, saponins, tannin, quinones, terpenoids, and polysaccharides (Devi et al., 2009; Kannan et al., 2010; Prakash et al., 2014; Prakash et al., 2016). So, the addition of coastal halophytes to daily diet may prevent serious diseases such as cardiac disorders, cancer, etc. (Kumari et al., 2015). The oil obtained from *Cynometra ramiflora* is used to cure skin diseases (Abeysinghe, 2010). The other important use of coastal halophytes is *Pandanus* spp. in scabies, heart diseases, and leprosy; *Salvadora persica*, in rheumatism, cough, tooth-ache, and piles; *Tamarix articulata* in diarrhea, ulcers, sore throat, and liver disorders (Dager, 2003). For a long time, halophytes have been used as medicine by villagers and coastal tribals enough documentation is not available to prove this. Little information is available regarding the use of halophytes as medicine in India.

TABLE 7.3 List and Importance Value of Ecosystem Services Provided by Coastal Dune Halophytes (Modified, *Everard et al., 2010).

Ecosystem function		Importance value*
A. Regulating functions		
1	Natural hazard	3
2	Water regulation	3
3	Soil fertility	2
4	Climate regulation	2
5	Biological control	1
6	Air quality regulation	1
7	Water flow	1
8	Waste	1
9	Prevention of erosion	1
10	Pollination	1
11	Erosion regulation	0
B. Provisioning functions		
12	Freshwater	3
13	Medical resources	3
14	Mineral extraction	2
15	Military use	2
16	Food	1
17	Genetic resources	1
18	Ornamental	0
C. Cultural functions		
19	Aesthetic value	2
20	Recreation	2
21	Spiritual experience	2
22	Inspiration for culture	1
23	Cognitive development	1
D. Supporting functions		
24	Refuge	3
25	Protection of genebank	1

7.4.2 PROTECTIVE ROLE AGAINST NATURAL DISASTERS

Coastal environments are vulnerable to the effects of different natural disasters, due to the expansion of urban areas, encroachment, and human activities such as tourism, agriculture, industries, fisheries, trades, and transportation in coastal regions (Kantamaneni et al., 2019). Globally, 40% of the human

TABLE 7.4 Medicinal uses of Coastal Sand Dune Halophytes.

No.	Species	Family	Uses
1	*Aeluropus* spp.	Poaceae	Pain killer and wound healing.
2	*A. occidentale* L.	Anacardiaceae	Snakebites, antipyretic, and antidiarrheal treatment.
3	*Barringtonia acutangala* (L.) Gaertn.	Lecythidaceae	Astringent anthelmintic. Leaf juice for diarrhea applications.
4	*Caesalpinia cristata* Prowazek	Fabaceae	Herbal tea making.
5	*C. inophyllum* L.	Calophyllaceae	Seed oil for rheumatism and scabies.
6	*C. gigantean* (L.) R. Br	Apocynaceae	Diaphoretic, mild stimulant. Root bark powder in dysentery.
7	*C. equisetifolia* Forst.	Casuarinaceae	The bark is an astringent and tonic. Useful in treating diarrhea.
8	*C. inerme* (L.) Gaertn	Lamiaceae	Root extract is used in jaundice and kidney infection.
9	*C. nucifera* L.	Arecaceae	Water use for urinary disorders, fever, and cholera. Fruit shows, laxative, stomachic diuretic, sedative properties.
10	*Cyperus rotundus* L.	Cyperaceae	Root extract oil is used for eye pain and redness. Treatment of digestive system disorders and fever.
11	*Cyperus arenarius* Retz.	Cyperaceae	Treatment of fever.
12	*Dactyloctenium aegyptium* (L.) Willd.	Poaceae	Bitter tonic, anthelminthic. Externally used for ulcers and wounds. Intestinal and urinary diseases.
13	*I. pescaprae* (L.) R. Br.	Convolvulaceae	Laxative, anti-diarrheal, analgesic. anti-inflammatory. Used externally in rheumatism. The oil of leaves is antagonistic. Leaf extract is used for the treatment of inflammations by poisonous jellyfish.
14	*Launaea procumbens* (Roxb.) Ramayya and Rajgopal	Asteraceae	Treatment of rheumatism.
15	*Lantana camara* L.	Verbanaceae	Cuts and wound healing, diabetes treatment.
16	*Oldenlandia biflora* L.	Rubiaceae	Root paste used in snake bite. Leaf paste for glandular swellings.
17	*Opuntia* spp.	Cactaceae	Fruit is used for cough.
18	*S. portulacastrum* (L.) L.	Aizoaceae	Indigestion, diarrhea, and dysentery.
19	*Sida cordifolia* L.	Malvaceae	Very useful in joint pain and arthritis. Oil is used to reduce inflammation and pain.
20	*Terminalia catappa* L.	Combretaceae	Antimicrobial.

TABLE 7.4 *(Continued)*

No.	Species	Family	Uses
21	*Thespesia populnea* (L.) Soland. ex Correa	Malvaceae	Roots extract is used for cholera and dysentery.
22	*Tridax procumbens* L.	Asteraceae	Wound healing, anticoagulant, and antifungal. Relief from diarrhea and dysentery. Ayurvedic medicines for liver disorders.
23	*Vernonia cinerea* (L.) Less.	Asteraceae	Used to cure skin diseases and urinary disorders.
24	*Vitex negundo* L.	Lamiaceae	Leaves and roots are used in ringworm and skin diseases, liver disorders, rheumatic pain, gout, and backache.
25	*Waltheria indica* L.	Malvaceae	Flowers and roots are used in making Ayurvedic Kadhas.
26	*Abutilon indicum* (L.) Sweet.	Malvaceae	Treatment of piles.
27	*Achyranthes aspera* L.	Amaranthaceae	Leaf extract is used for respiratory problems.
28	*Aristolochia bracteolate* Lam.	Aristolochiaceae	Snake-bites and treatment of skin diseases.
29	*Boerhaevia diffusa* L. nom. Cons.	Nyctaginaceae	Intestinal trouble, indigestion.
30	*Catharanthus roseus* (L.) G.Don	Apocynaceae	Anticancer properties.
31	*Euphorbia hirta* L.	Euphorbiaceae	Leaf and latex extracts are used to cure skin diseases.
32	*Leucas aspera* (Willd.) Link	Lamiaceae	Treatment of brain migraine.
33	*Melia azedarach* L.	Meliaceae	Leaf juice is antihelminthic.
34	*Pandanus odorattissmus* Linn.	Pandanaceae	Root extract used to treat jaundice
35	*Solanum nigrum* L.	Solanaceae	Treatment of mouth ulcer.

Source: Aparna and Raja Sekhar, 2015; Prakash et al., 2016; Nikalje et al., 2018; Dhaarani et al., 2018; Vineeth et al., 2020.

population lives within 50 km of the coastal zone (UNEP, 2012) and 25% and more people of the total human population of India settled within 50 km of the coastal stretch (Vivek et al., 2014). Risks of natural disasters in coastal areas in India are rapidly increasing; the major threats are sea level rise, floods, and cyclone-induced high tides in coastal areas which are the reasons causing tsunamis (Ramesh et al., 2015).

In view of this, coastal vegetation, such as sand dune vegetation and mangroves, plays an important role in decreasing the energy distribution of high tides such as tsunamis. The artificial tsunami model experiment suggest that 90% of tsunami effects are reduced with the help of a 100-m wide vegetation belt. The post-tsunami survey observed that, damage to *Casuarina* and coconut plantations is less in the Cuddalore district, Tamil Nadu on the East Coast of India (Danielsen et al., 2005). Kathiresan and Rajendran (2005) carried out a field survey in 18 coastal villagses on southeastern coast of India and observed the loss of properties and the number of casualties at study sites. They found that mangrove vegetated areas were less damaged by tsunami waves and recommended that the mangrove plantation program is important for future coastal area protection. Tanaka et al. (2007) studied the coastal vegetation along the coasts of Thailand and Sri Lanka after the Indian Ocean tsunami in 2004. After investigation, they identified two distinctive features that are useful in reducing the effect of tsunami disasters. First is vegetation that shows complex root systems, such as mangroves, that is, *Rhizophora* species and *Pandanus* spp., is more effective as compared to other coastal trees against wave action. Second is multiple layers of vegetation in horizontal and vertical directions trapped debris, which were useful in reducing the damage caused by flooding. According to Laso Bayas et al. (2011), coastal vegetation reduced the magnitude of damage and casualties by the tsunami on the West Coast of Aceh, Indonesia. Many factors such as shore bathymetry, distance from the coast, plant density, vegetation characteristics, and topography are useful and their statistical analyses confirmed that the coastal vegetation played an important role in causing less damage.

7.4.3 HABITAT OF IMPORTANT FLORA AND FAUNA

CSDs have a rich flora and fauna which includes a number of endangered and rare species. For example, in Wales, 91 species (9%) of sand dune flora are rare or endangered, and about 8% of plant species are endemic or near-endemic to coastal Europe (Rhind and Jones, 1999). The dune slacks are highly important for the habitat of rare plants. *Liparis loeselii* and *Gentianella uliginosa* are two rare plants found in the dune slacks of the UK. Dune slacks are also the habitat of many rare mosses and liverworts. Petalwort *Petalophyllum ralfsii* is a liverwort in Wales. *Bryum calophyllum*, a small moss colonizing on open sand in dune slacks, is thought to be extinct in England but found at one locality of Wales and other three sites in Scotland (Plantlife International, 2003). There are 502 taxa recorded in Welsh sand dunes of which 34 fungal species were included in the endangered, vulnerable, and rare categories (Rotheroe, 1993). CSDs also provide habitats for small populations of rare species. For example, in Canada, *Camissonia contorta*, an evening primrose is restricted to sandy flats on dunes (COSEWIC, 2006).

Sand dunes also provide suitable habitat or nesting sites for sea turtles such as Loggerhead turtles *Caretta caretta* and Olive ridley turtle *Lepidochelys olivacea*. The sea turtles return to the same coastline sandy beaches in every breeding season and lay 80–100 eggs in a nest which is situated above high water levels (Kelaskar et al., 2016). Inland sand dunes also support a wide range of avifaunal species such as nesting gull *Larus ridibundus*, duck *Somateria mollissim*, short-eared owl *Asio flammeus*, Merlin *falco columbarius*, and *Calidris alpine*. Sand dunes are also the habitat of particular mammal species. They provide a suitable habitat to *Oryctolagus cuniculus* (European rabbit) hence predatory species such as *Vulpes vulpes* (fox) and *Mustela ermine* (Stoat) are also found in sand dunes. *Sciurus vulgaris* (Red squirrel) is found at sand dunes of the Sefton coast. Sandy Coast is also rich in reptile diversity. The *Lacerta agilis*, sand lizard is present all over the European coastal areas but rare on the UK coast. The *Liolaemus multimaculatus*, sand dune lizard is another endemic species found in CSD areas of Argentina. Sand dunes provide habitat and food for snakes. For example, in the USA, rattlesnakes *Crotalus admanteus*, coachwhip snakes *Masticophis flagellum*, green snakes *Opheodrys aestivus carinatus*, and crowned snake *Tantilla relicta pamlica* are found in sand dunes (Doody, 2013).

7.4.4 RECREATIONAL AND AESTHETIC VALUES

Sand dunes and sandy beaches support different types of recreational activities, such as photography, fishing, hiking, horseback riding, canoeing, and camping. The sandy beach provides a suitable place for sunbathing, water sports, surfing, playing beach games, swimming, and access to the sea. Natural wave actions and associated beach vegetation habitats help to create scenes of great esthetic scenery which are found in many photographs and paintings. Globally, sandy beaches are preferred places for resting and relaxation. The other activities, such as walking on sand, and watching birds are also favorite activities. Many photographers, painters, and poets are motivated by the serene beauty of sandy beaches. Coastal areas tourism generate great economic benefits to the coastal population (Doody, 2013).

7.5 CONCLUSION

The CSDs have important nature conservation values. The beach or foredune is an important component of sand dunes present in a coastal location. It supports a rich variety of plants and animals, which are dependent on the dynamic sand dune ecosystem. These values change with the intensity of anthropogenic pressures on sand dunes and the changes or loss of CSD and its associated flora and fauna is leading to the loss of nature conservation values. Hence, the protection and management of this CSD ecosystem is the need of the time. In addition, the salt tolerant, CSD halophytes can be utilized for the reclamation and revegetation of degraded coastal areas.

KEYWORDS

- **sand dune halophytes**
- **natural disasters**
- **ecosystem**
- **bioshield**
- **coastal ecosystem**

REFERENCES

Abeysinghe, P. D. Antibacterial Activity of Some Medicinal Mangroves Against Antibiotic Resistant Pathogenic Bacteria. *Indian J. Pharm. Sci.* **2010,** *72*, 167–172. DOI: 10.4103/0250-474X.65019

Ahmad, E. In *Coastal Geomorphology of India*; Orient Longman, 1972; p 222.

Aparna, S.; Rajasekhar Studies on the Coastal Sand Dune Phytoresources at Visakhapatnam, Bay of Bengal, India. *Asian J. Plant Sci. Res.* **2015,** *5* (6), 69–76.

Arun, A. B.; Beena, K. R.; Raviraja, N. S.; Sridhar, K. R. Coastal Sand Dune: A Neglected Ecosystem. *Curr. Sci.* **1999,** *77*, 19–21.

Arun, A. B.; Sridhar, K. R. Arbuscular Mycorrhizal Association with CSD Legumes of South West Coast of India. *Mycorrhiza News* **2006, vol** *18*, pp 13–17.

Atreya, A.; Vartak, V.; Bhargava, S. Salt Priming Improves Tolerance to Desiccation Stress and to Extreme Salt Stress in Bruguiera cylindrica. *Int. J. Integr. Biol.* **2009,** *6*, 68–73.

Baird, A. H.; Kerr, A. M. Landscape Analysis and Tsunami Damage in Aceh: Comment on Iverson and Prasad. *Landsc. Ecol.* **2008,** *23* (1), 3–5. DOI: 10.1007/ s10980-007-9152-0.

Banerjee, L. K.; Rao, T. A.; Sastry, A. R. K.; Ghosh, D. Diversity of Coastal Plant Communities in India. ENVIS- EM CBTAP Publication, Botanical Survey of India, Ministry of Environment and Forests, Kolkata, 2002.

Barhoumi, Z.; Djebali, W.; Smaoui, A.; Chaïbi, W.; Abdelly, C. Contribution of NaCl Excretion to Salt Resistance of *Aeluropus littoralis* (Willd) Parl. *J. Plant Physiol.* **2007,** *164*, 842–850.

Beena, K. R.; Raviraja, N. S.; Arun, A. B.; Sridhar, K. R. Diversity of Arbuscular Mycorrhizal Fungi on the Coastal Sand Dunes of the West Coast of India. Curr. Sci. **2000,** *79*, 1459–1466.

Brown, C. E.; Pezeshki, S. R.; DeLaune, R. D. The Effects of Salinity and Soil Drying on Nutrient Uptake and Growth of Spartina alterniflora in a Simulated Tidal System. *Environ. Exp. Bot.* **2006,** *58*, 140–148.

Cheesman, J. M. Mechanisms of Salinity Tolerance in Plants. *Plant Physiol.* **1988,** *87*, 745–755.

, V. The Formation of Sand Dunes. *Ann. Sci. Nat.* **1897,** *6*, 16–22.

COSEWIC. COSEWIC Assessment and Status Report on the Contorted-Pod Evening Primrose. Committee on the Status of Endangered wildlife in Canada, 2006; p 21. http://dsp-psd.pwgsc.gc.ca/Collection/CW69-14-472-2006E.

Dagar, J. C. Biodiversity of Indian Saline Habitats and Management and Utilization of High Salinity Tolerant Plants with Industrial Application for Rehabilitation of Saline Areas. In *Desertification in the Third Millennium*; Alsharhan, A. S., Wood, W., Goudie, A. S., Fowler, A., Abdellatif, E. M., Eds.; Swets and Zeitlinger Publishers: Lisse, 2003; pp 151–172.DOI: 10.1201/ NOE9058095718.ch17

Daily, G. C.; Polasky, S.; Goldstein, J.; et al. Ecosystem Services in Decision Making: Time to Deliver. *Front. Ecol. Environ.* **2009,** *7* (1), 21–28

Danielsen, F.; Sorensen, M. K.; Olwig, M. F.; Selvam, V.; Parish, F.; Burgess, N. D.; Topp-Jorgensen, E.; et al. Coastal Vegetation and the Asian Tsunami - Response. *Science* **2006,** *311*, 37–38. DOI: 10.1126/science.311.5757.37.

De Groot, R.; Brander, L.; Van Der Ploeg, S.; et al. Global Estimates of the Value of Ecosystem Sand their Services in Monetary Units. Ecosyst. Serv. **2012,** *1*, 50–61.

Denaxa, N. K.; Roussos, P. A.; Damvakaris, T.; Stournaras, V. Comparative Effects of Exogenous Glycine Betaine, Kaolin Clay Particles and Ambiol on Photosynthesis, Leaf Sclerophylly Indexes and Heat Load of Olive cv. Chondrolia Chalkidikis Under Drought. *Sci. Hortic.* (2012, *137*, 87–94.

Desai, V. V.; Komarpant, D. S.; Jagtap, T. G. Distribution and Diversity of Marine Flora in Coral Reef Ecosystems of Kadmat Island in Lakshadweep Archipelago, Arabian Sea, India. *Atoll Res. Bull.* **2003,** *506*, 1–23.

Dhaarani, V.; Sarvalingam, A.; Rajendran, A. Medicinal uses of Psammophytic Plants in Tranquebar Regions of Tamil Nadu, India. *J. Herbs SpicesMed. Plants* (2018. DOI: 10.1080/10496475.2018.1471765

Desai, K. N. The Structure and Function of Sand Dune Vegetation Along the Coast of Goa; Ph.D. Thesis, Goa University, Goa, India, 1995.

Doody, J. Sand Dune Conservation, Management and Restoration, 2013. DOI: 10.1007/978-94-007-4731-9.

Drius, M.; Malavasi, M.; Acosta, A. T.; Ricotta, C.; Carranza, M. L. Boundary-Based Analysis for the Assessment of Coastal Dune Landscape Integrity Over Time. *Appl. Geo.* **2013,** *45*, 41–48.

Devi, K.; Thirumaran, G.; Arumugam, R.; Anantharaman, P. Antibacterial Activity of Selected Medicinal Plants from Parangipettai Coastal Regions; Southeast Coast of India. *World Appl. Sci. J.* **2009,** *7*, 1212–1215.

EEA. Article 17 Technical Report 2001–2006. European Environment Agency, Brussels, Belgium, 2008.

Elliott, M.; Cutts, N. D.; Trono, A. A Typology of Marine and Estuarine Hazards and Risks as Vectors of Change: A Review for Vulnerable Coasts and their Management. *Ocean Coast. Manag.* **2014,** *93*, 88–99.

Environmental Information System (ENVIS) Newsletter. Botanical Survey of India; Ministry of Environment and Forest, Kolkata, 2002, vol 8.

Etesami, H.; Beattie, G. A. Mining Halophytes for Plant Growth-Promoting Halotolerant Bacteria to Enhance the Salinity Tolerance of Non-Halophytic Crops. *Front. Microbiol.* **2018,** *9*, 148. DOI: 10.3389/fmicb.2018.00148.

Everard, M.; Jones, L.; Watts, B. Have we Neglected the Societal Importance of Sand Dunes? An Ecosystem Services Perspective. *Aquat. Conserv.: Mar. Freshw. Ecosyst.* **2010,** *20*, 476–487.

Ezawa, S.; Tada, Y. Identification of Salt Tolerance Genes from the Mangrove Plant Bruguiera gymnorhiza using Agrobacterium Functional Screening. *Plant Sci.* **2009,** *176* (2), 272–278.

FAO. Fisheries Ecosystem Approach. Fundamental concepts and their Application in Small-Scale Fisheries in Latin America, 2015, p 94.

Few, R. Flood Hazards, Vulnerability and Risk Reduction. In *Flood Hazards and Health*; Routledge: London, UK, 2013; pp 20–39.

Flowers, T.; Colmer, T. Salinity Tolerance in Halophytes. *New Phytol.* **2008,** *179*, 945–963.

Gari, S. R.; Newton, A.; Icely, J. D. A Review of the Application and Evolution of the DPSIR Framework with an Emphasis on Coastal Social-Ecological Systems. *Ocean Coast. Manag.* **2015,** *103*, 63–77.

Goldman, R. L.; Tallis, H. A Critical Analysis of Ecosystem Services as a Tool in Conservation Projects. *Ann. N. Y. Acad. Sci.* **2009,** *1162* (1), 63–78.

Hamed, K.; Ellouzi, H.; Talbi, O.; Hessini, K.; Slama, I.; Ghnaya, T.; Bosch, S.; Savoure, A.; Abdelly, C. Physiological Response of Halophytes to Multiple Stresses. *Funct. Plant Biol.* **2013,** *40*, 883–896.

Heslenfeld, P.; Jungerius, P.; Klijn, J. European Coastal Dunes: Ecological Values, Threats, Opportunities and Policy Development. In *Coastal Dunes, Ecology and Conservation*; Martinez, M. L., Psuty, N., Eds.; Springer: Berlin, Heidelberg, New York, 2004; vol *171*, pp 335–351.

Hesp, P. A. Ecological Processes and Plant Adaptations on Coastal Dunes. *J. Arid Environ.* **1991,** *21*, 165–191.

Ingole, B. S. Indian Ocean Coasts, Coastal Ecology. In *Encyclopedia of Coastal Science*; Schwartz, M. L., Eds.; Springer: Dordrecht, 2005; pp 546–554.

Jaiswal, V.; Rodrigues, B. F. Occurrence and Distribution of Arbuscular Mycorrhizal Fungi in CSD Vegetation of Goa. *Curr. Sci.* **2001,** *80*, 826–827.

Jagtap, T. G.; Desai, K.; Rodrigues, R. Coastal Vegetation: Patterns in a Tourist Region. In*Coastal Tourism, Environment, and Sustainable Local Development*; Noronha, L., Lourenco, N., Lobo-Ferreira, J. P., Lleopart, A., Feoli, E., Sawkar, K., Chachadi, A., Eds.; Tata Energy Research Institute: New Delhi, 2003; pp 131–150.

Kapler, A. Habitats of Halophytes. In *Halophytes & Climate Change: Adaptive Mechanisms and Potential Uses*; Hassanuzzaman, M.; Shabala, S., Fujita, M. CAB International, 2019; pp 19–37.

Kannan, R. R.; Arumugam, R.; Anantharaman, P. In Vitro Antioxidant Activities of Ethanol Extract from Enhalus acoroides (L.F.) Royle. *Asian Pac. J. Trop. Med.* **2010,** *3*, 898–901.

Kantamaneni, K.; Gallagher, A.; Du, X. Assessing and Mapping Regional Coastal Vulnerability for Port Environments and Coastal Cities. *J. Coast. Conserv.* **2019,** *23*, 59–70.

Karavas, N.; Georghiou, K.; Arianoutsou, M.; Dimopoulos, D. Vegetation and Sand Characteristics Influencing Nesting Activity of *Caretta caretta* on Sekania Beach. *Biol. Conserv.* **2005,** *121*, 177–188.

Kathiresan, K.; Rajendran, N. Coastal Mangrove Forests Mitigated Tsunami. *Estuar. Coast. Shelf Sci.* **2005,** *65* (3), 601–606. DOI: 10.1016/j.ecss.2005.06.022.

Kelaskar, A.; Upadhay, M.; Korgaonkar, S. Lesser Known Natural Threats to Protected Olive Ridley Nests in Coastal Maharashtra. Indian Ocean Turtle Newsletter No. 24, 2016, pp 6–7.

Khan, M. A.; Ungar, I. A.; Showalter, A. M. Effects of Salinity on Growth, Water Relations and Ions Accumulation of the Subtropical Perennial Halophyte, Atriplex griffithii var. stockii. *Ann. Bot.* **2000,** *85*, 225–232.

Koyro, H. W.; Eisa, S. S. Effect of Salinity on Composition, Viability and Germination of Seeds of Chenopodium quinoa Willd. *Plant Soil* **2008,** *302* (1), 79–90.

Kulkarni, S. S.; Raviraja, N. S.; Sridhar, K. R. Arbuscular Mycorrhizal Fungi of Tropical Sand Dunes of West Coast of India. *J. Coast. Res.* **1997,** *13*, 931–936.

Kumari, A.; Das, P.; Parida, A. K.; Agarwal, P. K. Proteomics, Metabolomics and Ionomics Perspectives of Salinity Tolerance in Halophytes. *Front. Plant Sci.* **2015,** *6*, 537. DOI: 10.3389/fpls.2015.00537

Kunte, P. D.; Wagle, B. G.; Sugimori, Y. A Review and Re-assessment of Sediment Transport Along the Goa Coast India. *J. Coast. Res.* **2002,** *18*, 612–621.

Kumaran, K. N.; Limaye, R. B.; Nair, K. M.; Padmalal, D. Palaecological and Paleoclimate Potential of Subsurface Palynological Data from the Data Quaternary Sediments of South Kerala Sedimentary Basin, South West India. *Curr. Sci.* **2008,** *95*, 515–526.

KSCSTE. State of Environment Report for Kerala. Kerala State Council for Science, Technology and Environment, Kerala, 2005.

Laso Bayas, J. C.; Marohn, C.; Dercon, G.; Dewi, S.; Piepho, H. P.; Joshi, L.; Van Noordwijk, M.; Cadisch, C. Influence of Coastal Vegetation on the 2004 Tsunami Wave Impact in West Aceh. *Proc. Natl.. Acad. Sci.* **2011,** *108* (46), 18612–18617. DOI: 10.1073/pnas.1013516108.

Leyva, C.; Espejel, I.; Pombo, A.; et al. Protective Natural Area, A Concept for the Management of Spaces with Native Vegetation in Arid-Coastal Cities: The Case of the City of Ensenada, BC. In Strategy for the Urban Management of Native Vegetation Spaces with Multifunctional Purposes: Case Study Population Center of Ensenada, B.C.; Leyva, C., Eds., 2009; p 140.

Lutts, S.; Lefevre, I. How Can We Take Advantage of Halophyte Properties to Cope with Heavy Metal Toxicity in Salt-Affected Areas? *Ann. Bot.* **2015,** *115*, 509–528.

Ma, Q.; Yue, L. J.; Zhang, J. L.; Wu, G. Q.; Bao, A. K.; Wang, S. M. Sodium Chloride Improves Photosynthesis and Water Status in the Succulent Xerophyte *Zygophyllum xanthoxylum*. *Tree Physiol.* **2012,** *32*, 4–13.

Malcolm, C. V.; Lindley, V. A.; O'Leary, J. W.; Runciman, H.V.; Barrett-Lennard, E. G. Halophyte and Glycophyte Salt Tolerance at Germination and the Establishment of Halophyte Shrubs in Saline Environments. *Plant Soil* **2003,** *253* (1), 171–185.

Maricle, B. R.; Cobos, D. R.; Campbell, C. S. Biophysical and Morphological Leaf Adaptations to Drought and Salinity in Salt Marsh Grasses. *Environ. Exp. Bot.* **2007,** *60*, 458–467.

Manzoor, S.; Hameed, A.; Khan, M. A.; Gul, B. Seed Germination Ecology of a Medicinal Halophyte *Zygophyllum propinquum*: Responses to Abiotic Factors. *Flora* **2017,** *233*, 163–170.

Martinez, J. P.; Kinet, J. M.; Bajji, M.; Lutts, S. NaCl Alleviates Polyethylene Glycol-Induced Water Stress in the Halophyte Species *Atriplex halimus* L. In *J. Exp. Bot.* **2005,** *56*, 2421–2431.

Mascarenhas, A.; Jayakumar, S. An Environmental Perspective of the Post-Tsunami Scenario Along the Coast of Tamil Nadu, India: Role of Sand Dunes and Forests. *J. Environ. Manag.* **2008,** *89*, 24–34.

Medina, E. Physiological Ecology of Psammophytic and Halophytic Plant Species from Coastal Plains in Northern South America. In *Sabkha Ecosystems Volume V: The Americas, Tasks for Vegetation Science*; Khan, et al., Eds., 2016; vol *48*, pp 29–56. DOI: 10.1007/978-3-319-27093-7_3.

Moreno-Casasola, P. A Case Study of Conservation and Management of a Tropical Sand Dune System: la Mancha-El Llano. In *Coastal Dunes, Ecology and Conservation*; Martinez, M. L., Psuty, N., Eds., Springer: Berlin, Heidelberg, New York, 2004; pp 319–334.

Mukhopadhyay, R.; Karisiddaiah, S. M. The Indian Coastline: Processes and Landforms, 2014. DOI. 10.1007/978-94-017-8029-2_8.

Munns, R.; Tester, M. Mechanisms of Salinity Tolerance. *Ann. Rev. Plant Biol.* **2008,** *59*, 651–681.

Muthukumar, K.; Samual, S. Coastal Sand Dune Flora in the Thoothukudi District, Tamilnadu, Southern India. *J. Threat. Taxa* **2011,** *3* (11), 2211–2216.

MoEF. Declaration of Coastal Stretches as Coastal Regulation Zone (CRZ). Notification, S.O. no. 114 (E), 19 February. Ministry of Environment and Forests, New Delhi, 1991; p 14.

Martinez, M. L.; Vazquez, G. Coastal Sand Dunes And Barrier Islands. In *Coastal Zone and Estuaries*-I; Isla, F. I., Iribarne, O., Eds.; UNESCO-EOLSS, 2009; pp.540. ISBN: 978-1-84826-466-3.

Naidoo, G.; Rughunanan, R. Salt Tolerance in the Succulent, Coastal Halophyte, Sarcocornia natalensis. *J. Exp. Bot.* **1990,** *41* (225), 497–502.

Naikoo, M.; Kafeel, U.; Naushin, F.; Kha, F. A. Halophytes in India and Their Role in Phytoremediation. In *Handbook of Halophytes*; Grigore, M. N., Ed., 2020. https://doi.org/10.1007/978-3-030-17854-3_115-1.

Niu, X.; Bressan, R. A.; Hasegawa, P. M.; Pardo, J. M. Ion Homeostasis in NaCl Stress Environments. *Plant Physiol.* **1995,** *109*, 735–742.

Nikalje, G. C.; Yadav, K.; Penna, S. Halophyte Responses and Tolerance to Abiotic Stresses. In *Ecophysiology, Abiotic Stress Responses and Utilization of Halophytes*; Hasanuzzaman, M., et al., Eds., 2019; pp 1–23. DOI: org/10.1007/978-981-13-3762-8_1.

, G. C.; Srivastava, A. K.; Pandey, G. K.; Suprasanna, P. Halophytes in Biosaline Agriculture: Mechanism, Utilization and Value Added Products. *Land Degrad. Dev.* **2018,** *29* (4), 1081–1095. DOI: 10.1002/ldr.2819

NIO. Development of GIS Based Information System for Selected Marine Critical Habitat: Data Collection around Kadmat Island (Lakshadweep), Arabian sea (Report no. NIO/TR-4/2000). National Institute of Oceanography, Goa, India, 2000.

Padmavathy, K.; Poyyamoli, G.; Balachandran, N. Puducherry. *Check List* **2010,** *6* (2), 198.

Plantlife International. Looking After Rare Mosses and Liverworts in Coastal Dune Slacks. Backfrom the Brink Management Series, 2003. Source: http://www.plantlife.org.uk/uploads/documents.

Pattanaik, C.; Reddy, C. S.; Dhal, N. K. Phytomedical Study of Coastal Sand Dune Species of Orissa. *Indian J. Tradit. Knowl.* **2008,** *7*, 263 268.

Prakash, S.; Ramasubbunarayan, C.; Ramkumar, K. E.; Palavesami, A.; Immanuel, G. In Vitro—Scientific Evaluation on Antimicrobial, Antioxidant, Cytotoxic Properties and Phytochemical Constituents of Traditional Coastal Medicinal Plants. *Biomed. Pharmacother.* **2016,** *83*, 648–657.

Prakash, S.; Ravikumar, S.; Reddy, K. V. R.; Kannapiran, E. Spermicidal Activity of Indian Seaweed: An in vitro Study. *Andrologia* **2014,** *46*, 408–416.

Poyyamoli, G.; Padmavathy, P.; Balachandran, N. Coastal Sand Dunes—Vegetation Structure, Diversity and Disturbance in Nallavadu Village, Puducherry, India. In *Coastal Environment: Focus on Asian Coastal Regions*; Subramanian, V., Eds., 2012; pp 204–218. DOI: 10.1007/978-90-481-3002-3_14

Rani, N. S.; Satyanarayana, A.; Bhaskaran, P. K. Coastal Vulnerability Assessment Studies Over India: A Review. *Nat. Hazards* **2015,** *77*, 405–428.

Ramesh, R.; Chen, Z.; Cummins, V.; Day, J.; D'Elia, J.; Dennison, B.; Forbes, D. L.; Glaeser, B.; Glaser, M.; Glavovic, B.; et al. Land–Ocean Interactions in the Coastal Zone: Past, Present & Future. *Anthropocene* **2015,** *12*, 85–98.

Ramos, J.; Lopez, M. J.; Benlloch, M. Effect of NaCl and KCl Salts on the Growth and Solute Accumulation of the Halophyte Atriplex Nummularia. *Plant Soil* **2004,** *259* (1–2), 163–168.

Rastgoo, L.; Alemzadeh, A. Biochemical Responses of Gouan (*Aeluropus littoralis*) to Heavy Metals Stress. *Aust. J. Crop Sci.* **2011,** *5*, 375–383.

Rasool, S.; Siddiqui, H.; Hameed, A.; Hussain, T.; Aziz, I.; Khan, M.; Gulzar, S. Temporal Variations in Water and Ion Relations in Coastal Halophytes. In *Sabkha Ecosystems, Tasks for Vegetation Science VI*; B. Gul, B., et al., Eds., 2019; pp 447–458. DOI: org/10.1007/978-3-030-04417-6_27.

Rhind, P. M.; Jones, P. S. The Floristics and Conservation Status of Sand-Dune Communities in Wales. *J. Coast. Conserv.* **1999,** *5* (1), 31–42.

Rotheroe, M. The Larger Fungi of Welsh Sand Dunes; Countryside Council for Wales, 1993; p 44.

Rozema, J.; Bijwaard, P.; Prast, G.; Broekman, R. Ecophysiological Adaptations of Coastal Halophytes from Foredunes and Salt Marshes. *Vegetation* **1985,** *62*, 499–521. DOI: 10.1007/BF00044777.

Rodrigues, R. S.; Mascarenhas, A.; Jagtap, T. G. An Evaluation of Flora from Coastal Sand Dunes of India: Rationale for Conservation and Management. *Ocean Coast. Manag.* **2011,** *5*, 181–188.

Salvi, H.; Das, L.; Brahmbhatt, B.; Vaghela, N.; Kamboj, R. D. Diversity of Halophytes in Gulf of Kachchh, Gujarat. *Int. J. Life Sci. Sci. Res.* **2017,** *3* (3), 995–1002. DOI: 10.21276/ijlssr.2017.3.3.6

Saye, S. E.; Pye, K. Implications of Sea Level Rise for Coastal Dune Habitat Conservation in Wales, UK. *J. Coast. Conserv.* **2007,** *11*, 31–52.

Seal, C. E.; Barwell, L. J.; Flowers, T. J.; Wade, E. M.; Pritchard, H. W. Seed Germination Niche of the Halophyte Suaeda Maritima to Combined Salinity and Temperature is Characterized by a Halothermal Time Model. Environ. Exp. Bot. **2018,** *155*, 177–184.

Short, A. D.; Hesp, P. A. Wave, Beach and Dune Interactions in Southeastern Australia. *Mar. Geol.* **1982,** *48* (3-4), 259–284.

Shao, H.; Xua, C.; Tang, X.; Wang, H. Salinity Tolerance Mechanism of Economic Halophytes from Physiological to Molecular Hierarchy for Improving Food Quality. *Curr. Genomics* **2016,** *17*, 207–214

Silambarasan, K.; Senthilkumar, P. Distribution and Diversity of Coastal Sand Dunes (CSD) of Marakkanam Coastal Belts, Southeast Coast of India. *Species* **2015,** *12* (35), 101–105.

Slama, I.; Ghnaya, T.; Savoure, A.; Abdelly, C. Combined Effects of Long-Term Salinity and Soil Drying on Growth, Water Eelations, Nutrient Status and Proline Accumulation of *Sesuvium portulacastrum*. *C R Biol.* **2008,** *331*, 442–451.

Sridhar, K. R.; Bhagya, B. Coastal Sand Dune Vegetation : A Potential Source of Food, Fodder & Pharmaceuticals. *Livest. Res. Rural Dev.* **2007,** *19*, 84.

Sridhar, K. R. Bioresources of Coastal Sand Dunes- are They Neglected? In *Coastal Environments: Problems and Perspectives*; Jayappa, K. S., Narayana, A. C., Eds.; International Publishing House: New Delhi, 2009; pp 53–76.

Touchette, B. W. Seagrass Salinity Interactions: Physiological Mechanisms used by Submersed Marine Angiosperms for a Life at Sea. *J. Exp. Mar. Biol. Ecol.* **2007,** *350*, 194–215.

Tanaka, N.; Sasaki, Y.; Mowjood, M. I. M.; Jinadasa, K. B. S. N.; Homchuen, S. Coastal Vegetation Structures and Their Functions in Tsunami Protection: Experience of the Recent Indian Ocean Tsunami. *Landsc. Ecol. Eng.* **2007,** *3* (1), 33–45. DOI: 10.1007/s11355-006-0013-9.

UNEP. Global Environment Outlook 3: Past, Present and Future Perspectives; UNEP/Earthscan: Nairobi, Kenaya, 2012.

Vecchio, D. S.; Prisco, I.; Acosta, A. T. R.; Stanisci, A. Changes in Plant Species Composition of Coastal Dune Habitats Over a 20-Year Period. *AoB Plants* **2015,** *7*, plv018. DOI: 10.1093/aobpla/plv018.

Verson, L. R.;Prasad, A. M. Modeling Tsunami Damage in Aceh: A Reply. *Landsc. Ecol.* **2008,** *23* (1), 7–10. DOI:10.1007/s10980-007-9180-9.

Veste, M.; Breckle, S. W. Xerohalophytes in a Sandy Desert Ecosystem. In *Biology of Salt Tolerant Plants*; University of Karachi: Karachi City, 1995; pp 161–165.

Vivek, G.; Mahendra, R. S.; Mohanty, P. C.; Kumar, S.; Nanda, S. Coastal Vulnerability Assessment for North East Coast of Andhra Pradesh, India. *Int. J. Remote Sens. Geoscie.* **2016,** *5*, 1–7.

Vineeth, T. V.; Kumar, S.; Shukla, M.; Chinchmalatpure, A.; Chandersharma, P. Ecological and Economic Potential of Major Halophytes and Salt Tolerant Vegetation in India, 2020. DOI: 10.5772/intechopen.93841.

Wijayasinghe, M. M.; Jayasuriya, K. G.; Gunatilleke, V. S.; Gunatilleke, I. N.; Walck J. L. Effect of Salinity on Seed Germination of Five Mangroves from Sri Lanka: Use of Hydrotime Modelling for Mangrove Germination. *Seed Sci. Res.* **2019,** *29*, 55–63.

Yair, A. Runoff Generation in a Sandy Area-the Nizzana Sands, Western Negev, Israel. *Earth Surf. Process. Landf.* **1990,** *15* (7), 597–609.

Zhang, H.; Tian, Y.; Guan, B.; Zhou, D.; Sun, Z.; Baskin C. C. The Best Salt Solution Parameter to Describe Seed/Seedling Responses to Saline and Sodic Salts. *Plant Soil* **2018,** *426*, 313–325.

CHAPTER 8

Physiological Responses of Halophytes to Salinity: A Review

DARSHANA PATIL[1], PRIYANKA GUPTA[2], and AVINASH PATIL[2]

[1]*Department of Botany, Smt. C.H.M. College, Ulhasnagar, Maharashtra, India*

[2]*Department of Biotechnology, B. K. Birla College of Arts, Science & Commerce (Autonomous), Kalyan (West), Maharashtra, India*

ABSTRACT

One of the most significant abiotic stresses is soil salinity, which affects plant growth and worldwide productivity. This problem has become considerably worse due to high surface evaporation, weathering of native rocks, low precipitation, irrigation with saline water, the use of excessive fertilizers, and inappropriate agricultural practices. Halophytes, a group of plants, are salt-tolerant or salt-resistant plants and remarkably capable of completing their life cycle in saline soils (wetland, desert, and temperate zones). During evolution, they adapted different morphological and anatomical strategies to survive in a highly saline environment. In addition, the strategy behind salt tolerance involves certain physiological and/or biochemical changes that enable the plants to maintain protoplasm viability during the accumulation of salt ions within the cells. Halophytes are fascinating plants, and studying their physiology will improve our understanding of the detailed mechanisms adapted by them to cope with salt stress. This study will also be helpful in finding practical uses for them, from revegetation to forage for livestock and developing genetically engineered salt-tolerant plants. Some of them could even be domesticated to generate new crops for food, feed, fiber,

Physiology of Halophytes: Signaling, Omics, and Tolerance Mechanisms.
Nivas Manohar Desai & Ganesh Chandrakant Nikalje (Eds.)

biofuel production, or other industrial uses. It is possible to use halophytes in future agriculture to provide salt-tolerant genes for prospective plants. In this chapter, an effort has been made to present a comprehensive account of halophytes and their physiological adaptations to saline environments.

8.1 INTRODUCTION

Salinization is the gradual accumulation of salt in soil that is exacerbated by climate change and inappropriate anthropogenic activities. It has emerged as a serious environmental problem all over the world. As estimated, 400 million hectares of land are affected by salinity due to excessive irrigation practices, and this area is growing daily (Aslam and Safdar, 2011). This may threaten the sustainability of agriculture and food security in the future. A large amount of water-soluble salts get accumulated in the soil during this process, affecting crop productivity, microbial communities, and agricultural economics (FAO, 2008). Eventually, salinization transforms fertile land into barren land. Therefore, the process is detrimental to environmental health as it destroys all the vegetation and all other organisms that live in the soil (Flowers and Colmer, 2008; Grigore et al., 2004).

Soil salinity is widely known to affect plant growth and productivity throughout the world and has become a major topic of research (Zhu, 2001). In addition, it imposes negative effects on the germination, growth, and reproduction of plants. Salinity is also associated with detrimental effects on physiological processes, including respiration, transpiration, photosynthesis, nutrient balance, membrane properties, enzymatic activity, and metabolism, as well as cellular homeostasis and hormonal regulation. This results in the production of reactive oxygen species (ROS), and during severe stress, it is associated with the death of the plant (Gupta and Huang, 2014; Fahad et al., 2015; Hasanuzzaman et al., 2014). There is an assumption that no toxic substance limits plant growth and productivity more than salt on a global scale (Xiong and Zhu, 2002).

Salinization is an ongoing process that is costly and labor-intensive to remediate. The world's population is growing tremendously, so there is a great need for food supplies. Despite the fact that crops cannot be grown on salt-affected land, nature gave us a unique class of plants, the halophytes (Aslam and Safdar, 2011). Such halophyte species can also be used to decipher the mechanisms of salt and metal tolerance, as well as to isolate stress-responsive genes that can be used to improve the tolerance of susceptible glycophytes (Nikalje and Penna, 2018). A novel concept of "Biosaline Agriculture" has been emerging for the past few decades. In this, different halophytes (salt-tolerant plants) are cultivated using saline/brackish water irrigation as a substitute for conventional crop plants (Nikalje et al., 2019).

Elevated levels of salt in the soil adversely affect the growth of the majority of plants on salt-affected land (Glenn et al., 1999). Such plants are often called nonsalt-tolerating plants, nonhalophytes, or glycophytes (Xiong and Zhu, 2002). In some plants, salt tolerance is facilitated by different mechanisms present in them, and they can grow in saline conditions. Such plants are called salt-tolerant plants or halophytes (Flowers et al., 1986). Flowers and Colmer (2008) define halophytes as plants which can thrive and reproduce in environments with salt concentrations greater than 200 mM of NaCl (~20 dS m^{-1}). Apart from being able to survive in salinity-prone soils (wetland, desert, arid, semiarid, and temperate zones), halophytes can also provide fodder, biofuels, medicinal benefits, and can desalinate salt-affected areas (phytodesalination) (González, 2019).

As explained by O'leary and Glenn (1994), a halophyte does not belong to any particular taxon or to any particular geographical or climatic area. We must therefore know that halophytes occur worldwide, whether in the desert or near the sea. It refers to any plant that can grow and reproduce in an environment that is subjected to high salinity for the majority of the time. It literally means "salt plant."

Only 1–2% of the world's flora is salt-tolerant, but it represents a diverse range of plant forms, including dicots and monocots (Kumari et al., 2015). There are around 50% higher plant families that include halophytes, and they have a polyphyletic origin (Glenn et al., 1999). Flowers and Colmer (2015) pointed out that salt tolerance occurs in relatively few species, suggesting that this trait has either been acquired and lost over time or that salt-tolerant species have gone extinct. This statement implies that we should pay great attention to this group and to the variations they experience on an internal and external level.

8.2 CLASSIFICATION OF HALOPHYTES

8.2.1 BASED ON THE DIFFERENT MECHANISMS OF SALT ADAPTATION

Walter (1961) classified halophytes into three types according to their mechanisms of adaptation to salty conditions:

8.2.1.1 SALT-EXCLUDING

The root system of these plants possesses an ultrafiltration mechanism to exclude excessive salts. Apoplastic barriers at the roots increase bypass flow resistance, which effectively excludes salts from the roots and, as a result, lowers the buildup of harmful ions through the transpiration stream in the aboveground shoots. This characteristic enables them to become the dominant species in mangrove vegetation. Examples include *Ceriops candolleana, Bruguiera gymnorrhiza, Rhizophora mucronata,* and *Kandelia candel.*

8.2.1.2 SALT-EXCRETING

These plants expel excess salts without damaging their cells through some specialized structures like salt glands, which are present in *Tamarix, Limonium, Avicennia, Spartina,* and *Frankenia* species or from epidermal bladders present on the leaves of *Chenopodium* and *Atriplex* species.

8.2.1.3 SALT-ACCUMULATING

The cells and tissues of these plants accumulate high concentrations of salts that are compartmentalized into vacuoles and used for osmotic adjustment. Examples include *Sonneratia apetala, Sonneratia alba, Sonneratia acida, Sesuvium portulacastrum, Suaeda nudiflora, Excoecaria agallocha, Salvadora persica, Limnitzera racemosa,* and *Pentatropis sianshoides.*

8.2.2 CLASSIFICATION ON THE BASIS OF RESPONSE TO THE INTERNAL SALT CONTENT OF THE PLANT

Steiner (1934) classified halophytes based on their response to internal salt levels. In his opinion, salt marsh halophytes can be categorized into two types: salt-accumulating and salt-regulating. Later, Ashraf et al. (2006) also classified halophytes as excluders or includers based on their internal salt content.

8.2.3 CLASSIFICATION ON THE BASIS OF MORPHOLOGY

According to their morphology, halophytes are often classified as excretives and succulents.

8.2.3.1 EXCRETIVES

A halophyte that excretes excess salt is known as an excretive. There may be salt crystals visible on the leaf surface of such halophytes; the glandular cells help to eradicate some excess salt from the plant (Marschner, 1995). According to the structural similarities, there are four categories of these salt glands: salt bladders, bicellular salt glands, multicellular salt glands, and unicellular vacuolated secretory hairs (Dassanayake and Larkin, 2017).

8.2.3.2 SUCCULENTS

These halophytes have leaves with a salt bladder on them. A succulent stores a large amount of water within its body, dilutes salts, and thus reduces salt toxicity (Weber, 2008). Halophytes in deserts are almost exclusively found in this group.

8.2.4 CLASSIFICATION BASED UPON TOLERANCE AND SALT DEMAND OF HALOPHYTES

Depending on their requirements and tolerance for sodium salts, Sabovljevic and Sabovljevic (2007) classified halophytes as facultative or obligate. In addition, ecophysiological characteristics can also be used to identify obligates, facultatives, and habitat-indifferent halophytes (Cushman, 2001).

8.2.4.1 OBLIGATE HALOPHYTES

These halophytes are also known as true halophytes as they need salt for their development and thrive in an area where there is at least 0.5–1% sodium chloride present (Ungar, 1978). Due to the need for salt in their diet to activate or deactivate a number of salt-sensitive enzymes, they are unable to develop at lower salinities. Increasing salt levels in the media optimizes the growth of obligatory halophytes. Members of the family Amaranthaceae (formerly Chenopodiaceae) like *Arthrocnemum macrostachyum*, *Sarcocornia fruticosa*, *Sesuvium portulacastrum* L., and *Salicornia brachiata* belong to this category (Cushman, 2001).

8.2.4.2 FACULTATIVE HALOPHYTES

Despite their ability to grow under saline stress, they can also grow well when exposed to nonsaline or low saline conditions. Several dicotyledons, like *S. portulacastrum*, *Aster tripolium*, *Glaux maritima*, *Chenopodium quinoa*, and *Plantagomaritima*, and some Cyperaceae, Poaceae, and Juncaceae species, belong to this group (Sabovljevic and Sabovljevic, 2007).

8.2.4.3 HABITAT-INDIFFERENT HALOPHYTES

Some plants are not native to their environment, but they are capable of adapting to the salty soil in nature. Although these plants prefer salt-free soil, they can also live in salty soil and compete with salt-sensitive species (Cushman, 2001). It is possible for such types of plants to differ in their genetic makeup from those that live on salt-free soil and those that live on salt-enriched soil, such as *Festucarubra*, *Agrostis stolonifera*, and *Juncusbufonius* (Sabovljevic and Sabovljevic, 2007).

8.2.5 CLASSIFICATION OF HALOPHYTES ON THE BASIS OF HABITAT

According to their geographical distribution or habitat, halophytes (Youssef, 2009) are of two types:

8.2.5.1 HYDROHALOPHYTES

These plants can grow in wet conditions or in aquatic habitats. This includes salt marsh species along coastlines and most of the mangroves (Youssef, 2009).

8.2.5.2 XEROHALOPHYTES

These plants can grow in soil that is saline but contains less water due to evaporation (i.e., arid and semiarid inlands that have unpredictable rainfall). Succulents and the majority of desert plants fall under this category. Some examples are *Atriplex halimus* and *A. canescens* (Youssef, 2009).

8.2.5.3 EXTREMOPHILES

These plants are capable of germinating, growing, and reproducing in high-salinity conditions ($\geq$500 mM NaCl) (Ungar, 1978).

Extreme halophytes (irreversible and reversible) and mesohalophytes were proposed as a new classification of halophytes by Grigore and Toma (2010) based on integrated anatomical studies with ecological considerations (salinity).

Extreme halophytes are halophytes that have adapted well to saline environments and only thrive there. They have the strongest anatomical adaptations, in correlation with the salinity factor. Furthermore, two types of extreme halophytes were described, that is, irreversible and reversible.

It was concluded that Chenopodiaceae (which is now included in the Amaranthaceae family) succulent species such as *Petrosimonia oppositifolia, P. triandra, Salicornia europaea, Halimione verrucifera, and Suaeda maritima* are irreversible extreme halophytes because they are well-adapted to high salinity environments. Halophytes, for example, *Atriplex tatarica, Atriplex littoralis, Atriplex prostrata, Bassia hirsuta, Camphorosma annua,* and *C. monspeliaca,* are reversible halophytes as they display adaptations that are a little bit less developed.

Mesohalophytes are halophytes which exhibit intermediate anatomical adaptations between extreme halophytes and glycophytes. Examples include *Aster tripolium* subsp. *pannonicus, Lactuca saligna, Scorzonera cana, Irishalophila, Plantago schwarzenbergiana, Lepidium cartilagineum* subsp. *crassifolium, Lepidium latifolium, Lepidium perfoliatum, Trifolium fragiferum,* and *Spergularia media.* They are able to grow in various conditions, apart from salinized ones.

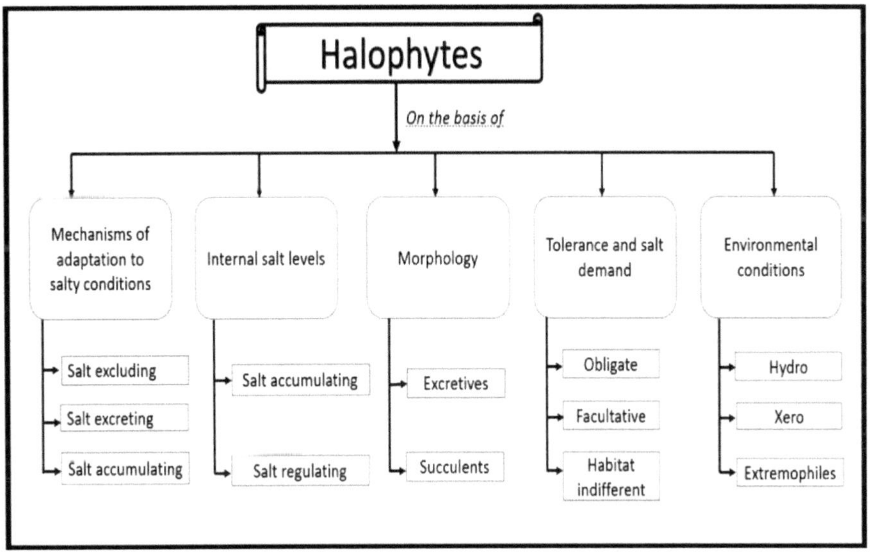

FIGURE 8.1 Classification of halophytes.

It has been concluded that salts are not essential for the development of halophytic species, but the availability of water and nutrients is also a key limiting factor for growth in natural saline habitats (Grigore et al., 2012). They also hypothesized that halophytes prefer saline habitats to avoid competition with glycophytes.

The detrimental effects of salinity stress are observed at the cellular, organ, and whole plant levels at the osmotic phase (early/short-term response) and at the ionic phase (late/long-term response). The presence of high salinity negatively impacts plant processes such as osmotic and ionic equilibrium, protein synthesis, photosynthesis, energy metabolism, and lipid metabolism. In order to adapt to salt stress, plants have evolved physiological and biochemical mechanisms orchestrated by several biochemical pathways, including ion homeostasis, osmolyte synthesis, ROS scavenging, and hormonal balance (Muchate et al., 2016). This chapter focused on the physiological adaptations of halophytes that enable these plants to live in challenging environments that most species cannot.

8.3 MECHANISMS OF SALT TOLERANCE AT PHYSIOLOGICAL AND BIOCHEMICAL LEVEL

The evolution of halophytes (ephemerals, shrubs, and trees) has resulted in a variety of mechanisms for regulating growth and development, ensuring their survival in salty environments (inland, coastal, salt marshes, dunes, and deserts) (Gul et al., 2013; Flowers and Colmer, 2015; Flowers and Colmer, 2008). Anatomical and morphological adaptations of halophytes include succulence (dilution of ion concentration), salt glands, or salt bladders (for selective accumulation or exclusion of ions) in the plant. Plants, which survive in soils with high salt concentrations, develop various mechanisms of salt tolerance at the physiological and biochemical levels. Principal mechanisms of salt tolerance include but are not limited to: (1) ion uptake and compartmentalization; (2) biosynthesis of osmoprotectants and compatible solutes; (3) activation of antioxidant enzymes and synthesis of antioxidant compounds; (4) synthesis of polyamines; (5) generation of nitric oxide (NO); and (6) plant hormone induction and modulation (Gupta and Huang, 2014)

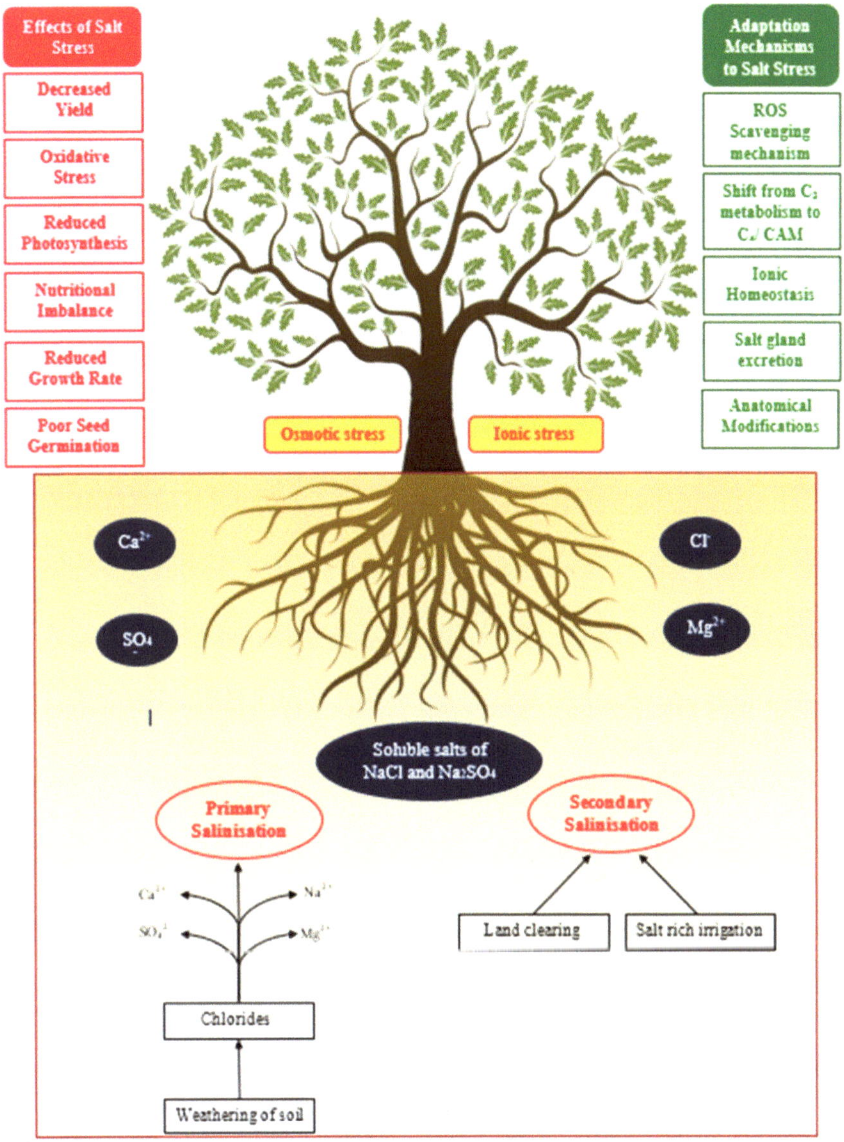

FIGURE 8.2 Salt stress: effects and adaptive mechanisms.

8.3.1 ION UPTAKE AND COMPARTMENTALIZATION

Ion uptake and compartmentalization are crucial processes for maintaining ion homeostasis in plants not only during normal growth but also during salt stress (Niu et al., 1995; Serrano, 1999). Neither glycophytes nor halophytes are able to tolerate high salt concentrations in their cytoplasm. As a result, the excess salt is either sequestered in the vacuole or in older tissues that eventually die, which is helpful for the plant to protect it from salinity stress (Reddy, 1992; Zhu, 2003). NaCl is the major salt found in soil; therefore, the key areas of investigation are the transport mechanisms of Na^+ ions and their compartmentalization. Both halophytes and nonhalophytes (glycophytes) compartmentalize Na^+ into vacuoles for ion regulation. However, halophytes are smart enough to not only sequester excess sodium into vacuoles but also use it as a cheap osmolyte to maintain low water potentials under salt stress. Thus, a hallmark of halophytes is their high efficiency of sequestering Na^+ in vacuoles (Xu et al., 2016). Through the Na^+/H^+ antiporter, Na^+ enters the cytoplasm and is transported to the vacuole. The vacuolar membrane contains two different types of H^+ pumps: the vacuolar-type H^+-ATPase (V-ATPase) and the vacuolar pyrophosphatase (V-PPase) (Dietz et al., 2001; Oliveira Otoch et al., 2001; Wang et al., 2001). The most prominent H^+ pump found inside the plant cell is the V-ATPase. During nonstress conditions, it facilitates vesicle fusion, energizing secondary transport and maintaining solute homeostasis. In stressed conditions, the V-ATPase is critical to the plant's survival. An experiment conducted by Oliveira Otoch et al. (2001) demonstrated that VATPase pump activity increases under salinity stress in *Vigna unguiculata*, but V-PPase activity is inhibited, whereas in the case of the halophyte *Suaeda salsa*, V-ATPase activity is upregulated and V-PPase plays a minor role. There is increasing evidence that the Salt Overly Sensitive (SOS) stress signaling pathway plays a role in salt tolerance and ion homeostasis. Tada et al. (2014) found that the accumulation of Na^+ and Cl^- was increased quickly in shoots and roots of *Sporobolus virginicus* due to salinity stress.

Most plants have developed methods to maintain a low level of ion concentration in their cytoplasm. Ion uptake and transport are regulated by membranes and their associated components during times of stress to maintain ion concentration in the cytosol (Sairam and Tyagi, 2004). The process of transport is carried out by a wide variety of channel proteins, carrier proteins, symporters, and antiporters. NHX1, Salt Overly Sensitive 1 (SOS1), and the high-affinity K^+ transporter (HKT) are the foremost transporters involved in Na^+ accumulation in plants (Wang, 2020). Vacuolar NHX1 is a membrane protein that is crucial for the segregation of Na^+ into vacuoles and the exchange of Na^+ for H^+ across the vacuolar membrane (Hasegawa, 2013). HKT1 and SOS1 are located on the plasma membrane of xylem parenchyma cells in the roots and have opposite functions in regulating Na^+ influx and efflux, respectively, across the plasma membranes (Wang et al., 2020). A plant's survival in saline environments depends on maintaining Na^+/K^+ homeostasis. Under salt stress, Ma et al. (2012) have found that NADPH oxidases AtrbohD and AtrbohF regulate Na^+/K^+ homeostasis in an ROS-dependent manner in *Arabidopsis*.

8.3.2 BIOSYNTHESIS OF OSMOPROTECTANTS AND COMPATIBLE SOLUTES

A strong osmotic gradient across the vacuolar membrane is created when salts are collected from the cytoplasm and enter the vacuole. This gradient is balanced by the biosynthesis of compatible solutes in the cytoplasm and a process known as osmotic adjustment (Wyn Jones and Gorham, 1983; Waisel, 1972). It is widely agreed that osmotic adjustment is a crucial adaptation of plants to salinity since it maintains the turgor and volume of cells.

Compatible solutes are also known as compatible osmolytes. They are a class of chemically varied organic compounds that are uncharged, polar, and soluble in nature and have no adverse effects on the cellular metabolism at high concentrations. This includes proline (Hoque et al., 2007; Nounjan et al., 2012; Ashraf and Foolad, 2007), glycine betaine (Hoque et al., 2007; Ashraf and Foolad, 2007), sugar (Kerepesi and Galiba, 2000; Bohnert et al., 1995), and polyols (Bhattacharya and Kundu, 2020). There is a variation in the amount and type of organic osmolytes synthesized and accumulated by different species of plants. For example, it must be noted that beta-alanine betaine accumulates in a limited number of Plumbaginaceae members (Hanson et al., 1994), while amino acid proline accumulates in several taxonomically diverse collections of plants (Saxena et al., 2013). Cells maintain compatible solute levels either through irreversible synthesis or through a combination of synthesis and degradation. It has been thoroughly studied that genes and the biochemical pathways are involved in these processes. They function by protecting the cell structure and by maintaining osmotic balance inside the cell by continuously influxing water as their accumulation is proportionate to the external osmolarity (Hasegawa et al., 2000).

When exposed to salinity stress, cysteine, arginine, and methionine, which account for about 55% of total free amino acids, decrease in concentration, while proline concentration rises (El-Shintinawy and El-Shourbagy, 2001). It is well-known that proline accumulation is an effective way of alleviating salinity stress (Matysik et al., 2002; Ben Ahmed et al., 2010). The intracellular proline accumulated under salinity stress not only provides tolerance toward stress but also serves as a nitrogen source during stress recovery. A proline molecule can be synthesized from glutamate or ornithine. The primary precursor in osmotically stressed cells is glutamate. There are two major enzymes in the biosynthetic pathway: pyrroline carboxylic acid reductase and pyrroline carboxylic acid synthetase. Plants use both of these regulatory steps to overproduce proline. A study by Matysik et al. (2002) demonstrated that it functions as an O^2 quencher, thereby demonstrating its antioxidant properties. A study by Ben Ahmed et al. (2010) observed that proline supplements improved salt tolerance in olive (*Oleaeuropaea*) by improving antioxidant enzyme activities, photosynthetic activity, growth, and plant water status in saline environments. Proline increases the activity of antioxidant defense enzymes in *Nicotiana tabacum*, which improves salt tolerance (Hoque et al., 2008). A study by Benjamin et al. (2019) also showed that the level of proline increases with increasing concentrations of NaCl in *Sesuvium portulacastrum* and *Suaeda maritima*. Tada et al. (2014) observed an increase in proline concentration in *Sporobolus virginicus* proportionately with an increase in the concentration of NaCl in the culture solution. Poddar et al. (2019) found an increase in the level of free proline by 20% in *P. triandra* leaves in the presence of higher soil salinity.

Glycine betaine (GB) is an amphoteric quaternary ammonium compound that is found in many microorganisms, animals, and higher plants and is an electrically neutral compound at a wide range of pH. In addition to being highly soluble in water, it contains a nonpolar moiety that is composed of three methyl groups. As a result of its distinct structural characteristics, it combines with both hydrophobic and hydrophilic domains of macromolecules, including enzymes and proteins. When stress occurs, glycine betaine, which is a nontoxic cellular osmolyte, contributes to the reduction of stress by raising the osmolarity of the cell. GB also defends the cell by adjusting osmotic pressure (Gadallah, 1999), stabilizing proteins (Makela et al., 2000), preventing stress damage to the photosynthetic system (mad-Um and Kirdmanee, 2010), and reducing ROS (Ashraf and Foolad, 2007; Saxena et al., 2013). A wide variety of plants belonging to different taxonomic groups accumulate GB. In general, crops do not accumulate GB when stressed, but halophytic Poaceae and Chenopodiaceae species do (Flowers et al., 1998). Cells synthesize glycine betaine from either choline or glycine. A two-step process that uses two or more enzymes is required to synthesize glycine betaine from choline. As choline undergoes its first oxidation, it turns into betaine aldehyde, which is again oxidized in the next step into glycine betaine. The first step of the conversion is catalyzed by the enzyme choline monooxygenase (CMO) in higher plants, while the second step is catalyzed by betaine aldehyde dehydrogenase (BADH). The phospho ethanolamine methyl transferase (PEAMT) gene is related to the synthesis of betaine in *Suaeda Glauca* (Jin et al., 2015). The CMO gene expression was increased in *Suaeda salsa* (Wu et al., 2012) and *Suaeda aegyptiaca* (Askari et al., 2006). The BADH gene, which is involved in the production of betaine to maintain the osmotic balance and improve salt tolerance, was stimulated to express in *Suaeda liaotungensis* (Li et al., 2003) and *S. corniculate* (Wang et al., 2016) seedlings under salt conditions. Thus, during salt stress, the genes CMO, BADH, and PEAMT involved in the production of osmolytes are increased. These could act as positive regulators in response to NaCl, which helps *Suaeda* cells to tolerate salt stress by checking the osmotic balance of cells.

GB is also synthesized from glycine in some plants, mainly halophytes. It is synthesized by three successive N-methylation reactions catalyzed by two S-adenosyl methionine-dependent methyltransferases that is GSMT (glycine sarcosine N-methyl transferase) and SDMT (sarcosine dimethylglycine N-methyl transferase). The roles of these two enzymes overlap as GSMT catalyzes both the first and second steps while SDMT catalyzes the second and third steps (Kimura et al., 2010). Pretreating seedlings with glycine betaine largely prevented these damages. The foliar application of glycine betaine to stressed plants resulted in pigment stabilization and growth and photosynthetic rate increases (mad-Um and Kirdmanee, 2010; Sulpice et al., 2003). Benjamin et al. (2019) found that glycine betaine and polyols were accumulated in *Salicornia maritima* and *Suaedabrachiata* with increasing concentrations of salt. Firstly, choline monooxygenase (CMO) converts choline into betaine aldehyde, which is subsequently catalyzed by betaine aldehyde dehydrogenase (BADH) to synthesize betaine (Wang et al., 2016).

A polyol is a compound that contains multiple hydroxyl functional groups for organic reactions. Polyols such as sugar alcohols function as compatible solutes, low-molecular-weight chaperones, and ROS scavengers (Reddy et al., 2021). In general, they can be categorized into two main types: acyclic (e.g., mannitol) and cyclic (e.g., pinitol). When plants are stressed, NADPH-dependent mannose-6-phosphate reductase induces mannitol synthesis. As protectors or stabilizers, these compatible solutes safeguard enzymes or membrane structures at risk of dehydration or damage from ions.

In response to salinity stress, pinitol is accumulated within plant cells. The biosynthetic pathway involves two major steps: the methylation of myoinositol and the subsequent epimerization of an intermediate compound, ononitol, to form pinitol. Pinitol is synthesized by the enzyme inositol methyl transferase, encoded by the imt gene. Plants with imt genes show similar results to those with mltd genes. Hence, pinitol also contributes to stress reduction. Myoinositol plays a significant role in the supply of Na^+ in the shoot of *M.crystallinum* (Sengupta et al., 2008), and it is the substrate for IMT in the production of pinitol. In *M. crystallinum* and *P. coarctata*, IMT has been characterized (Sengupta et al., 2008; Sengupta and Majumder, 2010). The accumulation of polyols, such as straight-chain polyols like sorbitol and mannitol or cyclic polyols like myoinositol and its methylated derivatives, is associated with tolerance to salt and drought based on polyol distribution in various species, including microorganisms, animals, and plants. Cockburn and Paul (1989) found that crassulacean acid metabolism (CAM) in *Mesembryanthemum crystallinum L.* plants was accompanied by pinitol accumulation during irrigation with 400 mol m^{-3} NaCl.

Carbohydrates, for example, sugars (e.g., glucose, fructose, fructans, and trehalose) and starch, accumulate under salt stress. These carbohydrates play a key role in stress mitigation through osmoprotection, carbon storage, and scavenging reactive oxygen species. Various plants belonging to diverse species have been observed to have increased levels of reducing sugars (i.e., fructans and sucrose) in their cells when exposed to salt stress. Trehalose accumulation protects organisms against a variety of physical and chemical challenges, including salt stress, in addition to serving as a carbohydrate reserve. In physiological responses, they serve as osmoprotectants (Parvaiz and Satyawati, 2008). According to Sanchez et al. (2008), in response to high salinity, *T. halophila* showed increased concentrations of glucose, fructose, and sucrose. Similarly, glucose, fructose, and sucrose all accumulated in response to salinity in *Juncus acutus* and *Juncus maritimus*, although sucrose accumulated in larger proportions than glucose and fructose (Gil et al., 2011). In the same context, it was noted that *Bruguiera parviflora* leaves had decreased starch content and increased reducing and nonreducing sugar content (Muchate et al., 2016). Significantly higher contents of glycine betaine, proline, and total soluble sugars were observed in all the salt-treated pot-grown *Sesuvium portulacastrum* halophyte plants as compared to the control (Parida et al., 2004).

Aquaporin is another type of osmolyte involved in the intracellular compartmentalization of water. In halophytes, these pore-forming proteins conduct water molecules. It suggests that the gating of water channels may affect water movement within compartments. Under salt stress, these aquaporins are thought to contribute to salt tolerance by maintaining the turgor and osmotic equilibrium of plant cells (Maurel, 1997). Vaziriyeganeh et al. (2022) demonstrated that the ability of the halophytic grass *Puccinellia nuttalliana* to tolerate salt is facilitated by the critical involvement of Na in aquaporin-mediated root water transport.

8.3.3 SYNTHESIS AND ACTIVATION OF ANTIOXIDANT ENZYMES AND COMPOUNDS

In chloroplasts and mitochondria, stress from both abiotic and biotic sources can lead to electron transport chain overflow, dysregulation, or even disruption. In these conditions, molecular oxygen (O^2) serves as an electron acceptor, causing ROS to accumulate. Singlet oxygen (1 O^2), the hydroxyl radical (OH^-), the superoxide radical (O^{-2}), and hydrogen peroxide (H^2O^2) are all strong oxidizing compounds that are detrimental to cell integrity (Groß et al., 2013). During salinity stress, antioxidant enzymes and nonenzymatic compounds are critical for detoxifying ROS. It has been shown that salt tolerance is positively related to antioxidant enzyme activity, including superoxide dismutase (SOD), catalase (CAT), glutathione peroxidase (GPX), glutathione reductase (GR), and ascorbate peroxidase (APX), as well as to the accumulation of nonenzymatic antioxidant compounds (Ahmad et al., 2010). Uzilday et al. (2015) investigated the ability of halophytes to handle ROS. Halophytes are famous for their ability to endure and quench toxic reactive oxygen species (ROS), for they possess a powerful antioxidant system (Kong et al., 2017). Researchers have detected Mn-SOD and several isoforms of CuZn-SOD and Fe-SOD in *S. salsa* leaf extracts and also cloned several genes related to oxidative stress tolerance in *S. salsa*, including Ss. sAPX, SsCAT2, SsPrxQ, SsGST, SsCAT1, SsAPX, and SsTypA1 (Li et al., 2012; Pang et al., 2011). According to Gill et al. (2013) and Tuteja et al. (2013), helicase proteins such as OsSUV3 dual helicase and DESD-box helicase are involved in the salt tolerance of plants through improvements in photosynthesis and antioxidant machinery.

It has been well-documented that plants under salt stress accumulate anthocyanin, which is a flavonoid. The anthocyanin-impaired response-1 (air1) mutant, isolated by Van Oosten et al. (2013), is incapable of accumulating anthocyanins under salt stress. Salt stress causes anthocyanin production to be impaired in the air1 mutant, but not other stresses such as high light, low phosphorus, high temperature, or drought. Anthocyanin biosynthesis in response to salt stress is not affected by the air1 mutation but rather by its regulation. AIR1 is a key discovery in dissecting the link between abiotic stress and the buildup of antioxidant compounds such as flavonoids and anthocyanins.

One of the most important antioxidants found within the cell is ascorbate. APX activity and s-nitrosylated APX content increased in pea plants under saline (150 mM NaCl) stress, along with an increase in H^2O^2, NO, and S-nitrosothiol (SNO) that can justify introducing APX activities. Based on proteomic analysis, APX has been identified as a potential target of PTMs mediated by NO-derived compounds (Begara-Morales et al., 2014).

Glutathione is another antioxidant that reduces stress by reacting with free radicals, such as superoxide, hydroxyl radicals, and hydrogen peroxide. In addition, it can contribute to the regeneration of ascorbate through the ascorbate–glutathione cycle (Foyer et al., 1997). During externally applied salinity stress in *Allium cepa*, glutathione maintained plasma membrane permeability and cell viability (Salama and Al-Mutawa, 2009). Poddar et al. (2019) found that the activity of ROS processing enzymes such as catalase (CAT), superoxide dismutase (SOD), guaiacol peroxidase (GPX), ascorbate peroxidase (APX), and glutathione transferase (GST) was significantly increased in *Petrosimonia triandra,* grown at higher salinity, in comparison to those living at lower soil salinity.

Ghanem et al. (2021) investigated the responses of three obligatory halophytes, *Sarcocornia fruticosa*, *Salicornia europaea*, and *Arthrocnemum macrostachyum*, to salt stress (0, 100, 200, 400, and 600 mM NaCl) throughout their vegetative development. They suggested that the biosynthesis of antioxidant compounds (carotenoids, phenolics, and flavonoids) and the activation of antioxidant enzymes are responsible for salt tolerance.

As Roy et al. (2014) suggested in their review, plants have three main traits that help them adapt to salinity stress: tissue tolerance, ion exclusion, and salinity tolerance. Apparently, antioxidants play an important role in tissue and salinity tolerance.

8.3.4 SYNTHESIS OF POLYAMINES

Polyamines (PAs) are ubiquitous, polycationic aliphatic molecules of small size and low molecular weight found throughout plants. Polyamines have a wide range of functions in the normal growth and development of plants, including regulating cell proliferation, differentiation and morphogenesis, somatic embryogenesis, breaking dormancy of tubers and germination of seed, the development of fruits and flowers, and senescence (Gupta et al., 2013). Moreover, it is also important for the abiotic stress tolerance of plants, including salinity (Gupta et al., 2013; Groppa and Benavides, 2008). There are three major polyamines found in plant systems: tetraamine spermine (SPM), triamine spermidine (SPD), and diamine putrescine (PUT) (Kuznetsov and Shevyakova, 2007; Hussain et al., 2011).

The endogenous polyamine level of plants improved when subjected to salt stress. Naidu (2003) reported that some halophytes are able to cope with highly saline soils by producing proline analogues, which occur in Australian soils. Melanleuca species, such as *Melanleuca bracteata*, build up the analogue of proline, that is,, 4-hydroxy-N-methyl proline (MHP). The ability of such proline analogues to regulate, compartmentalize, and increase production outlay increases plants' ability to survive under salinity stress (Bohnert and Shen, 1998).

Polyamines are catabolized by oxidative means through amine oxidases, including FAD-binding polyamine oxidases and copper-binding diamine oxidases. These enzymes regulate the intracellular polyamine level and play a key role in stress tolerance (Takahashi and Kakehi, 2010). Cellular polyamine levels change in response to stress but not in a manner that is consistent with their role in reducing stress. Exogenous application of polyamines prior to or during stress has thus been investigated to determine whether polyamines protect cells from stress-induced damage (Tisi et al., 2008). In response to stress, exogenous polyamine increased endogenous polyamine levels; polyamines have been linked to maintaining membrane integrity, the regulation of gene expression for osmotically active solute synthesis, the reduction of reactive oxygen species (ROS), and controlling the accumulation of Na^+ and Cl^- ions in different organs (Wang et al., 2007; Afzal et al., 2009; Yiu et al., 2009). Plants deficient in ADC1 and ADC2 are hypersensitive to stress (Koji et al., 2006). Salinity stress increases the expression of ADC and SPMS in Arabidopsis. Polyamine biosynthesis mutants appear to be sensitive to salinity (Yamaguchi et al., 2006). Salt tolerance is enhanced in rice, tobacco, and Arabidopsis by overproducing PUT, SPD, and SPM (Roy and Wu, 2002). Salt stress controls polyamine synthesis and catabolism by functioning as a biological signal in hormonal pathways modulating abscisic acid (ABA) in response to stress (Shevyakova et al., 2013). Shevyakova et al. (2006) were the first to investigate the free polyamine content in *Mesembryanthemum crystallinum* L. under severe salinity conditions.

8.3.5 GENERATION OF NITRIC OXIDE

Nitric oxide (NO) is a small, volatile, and gaseous compound that regulates different growth and developmental processes in plants, such as root growth, respiration, the closing of stomata, flowering, cell death, and seed germination. It also functions as a stress-signaling molecule (Besson-Bard et al., 2008; Zhao et al., 2009). Several redox-regulated genes are triggered by NO, either directly or indirectly. As NO reacts with lipid radicals, it prevents lipid oxidation, scavenging superoxide radicals and forming peroxynitrite that can be neutralized by other cellular processes. Additionally, it activates antioxidant enzymes (GPX, APX, CAT, SOD, and GR) (Bajgu et al., 2017). There has been evidence that exogenous NO can mitigate stress (Sung and Hong, 2010; Hossain et al., 2010; Xiong et al., 2010), but the effects are dependent on the concentration of NO. Sodium nitroprusside (SNP), an exogenous NO donor, stimulated seed germination and root growth in *Lupinus luteus* seedlings exposed to salt stress (Kopyra and Gwozdz, 2003).

It has been demonstrated that NO has a positive effect on salinity tolerance and stress mitigation as a result of its antioxidant properties and modulation of ROS detoxification (Mishra et al., 2011). Exogenous NO application improved plant growth under salinity stress by increasing antioxidant enzyme activities like GPX, APX, CAT, SOD, and GR (Zhao et al., 2004) and suppressing malondialdehyde (MDA) production or lipid peroxidation (Nalousi et al., 2012). The effects of NO on tolerance to salt are also associated with its control of cell membrane H^+-ATPase and Na^+/K^+ ratio (Zhang et al., 2006).

Hasanuzzaman et al. (2021) demonstrated that nitric oxide controls plant physiology, antioxidant defense, ion homeostasis, and growth to give the mangrove species *Kandelia obovata* salt tolerance.

8.3.6 PLANT HORMONE INDUCTION AND MODULATION

Phytohormones are chemical messengers that serve crucial roles in controlling how plants react to stress, even at very low concentrations. Numerous halophytic species have been found to respond to high salt concentrations by changing the levels of phytohormones like auxin, indole acetic acid (IAA), gibberellin (GA), zeatin (ZR), ethylene (ET), jasmonic acid (JA), brassinolide (BR), and ABA, as well as enzymes related to their biosyntheses like DWARF3 (GA biosynthesis), NCED

(ABA biosynthesis), SAMS (ET biosynthesis), lipoxygenase (LOX) (JA biosynthesis), and allene oxide cyclase (AOC). After being exposed to saline treatment, *S. salsa* showed upregulation of the auxin, ET, and JA signaling transduction pathways, which are crucial for the gene control of ion transport and antioxidation. Moreover, in *S. salsa* exposed to NaCl, genes encoding GID1, NCED, AHP, TCH4, CYP735A, CYP85A, PIF4, SnRK2, and ABF were increased. These genes are connected to the production of IAA, GA, ZR, BR, and ABA as well as to plant hormone signal transduction (Guo et al., 2020).

The chief phytohormone with a high expression level is ABA (Zhang et al., 2006). As an important phytohormone, ABA alleviates the effects of stress conditions on plants. Historically, it has been recognized as a hormone that is upregulated by soil water deficits around the roots. As a result of osmotic stress and water deficits caused by salt stress, shoots and roots produce more ABA (Cabot et al., 2009; Cramer and Quarrie, 2002). ABA accumulation can lessen the negative effects of salt on assimilate translocation, growth, and photosynthesis (Jeschke et al., 1997). The accumulation of K^+, Ca^{2+}, and compatible solutes, including proline and sugars, in the vacuoles of roots, which counteract with the absorption of Na^+ and Cl^- ions, is thought to be the cause of the positive association between ABA accumulation and salt tolerance (Chen et al., 2001; Gurmani et al., 2011).

Plants also respond to abiotic stress with the assistance of some other compounds with hormonal properties, like brassinosteroids (BRs) and salicylic acid (SA) (Fragnire et al., 2011; Clause et al., 1998). It has been observed that the endogenous level of SA increases and salicylic acid, ethylene, and polyamine metabolisms are involved in salt resistance in the wild-related halophyte *Solanum chilense* under salinity stress (Gharbi et al., 2016). Recently, Jayakannan et al. (2013) demonstrated that SA enhances the ability of Arabidopsis to tolerate salinity by restoring membrane potential and inhibiting salt-induced K^+ loss through a guard cell outward rectifying $K^{(+)}$ (GORK) channel.

In mung bean cultivars, Nazar et al. (2011) have argued that by improving nitrogen and sulfur absorption and antioxidant metabolism differently, SA prevents photosynthesis from declining under salt stress. BR may also mitigate the negative effects of salinity (Vikram et al., 2022). It has been shown that BR stimulates antioxidant enzyme activity (SOD, POX, APX, and GPX) as well as the accumulation of nonenzymatic antioxidants (tocopherol, ascorbate, and reduced glutathione) (El-Mashad and Mohamed, 2012). BRs and SAs are ubiquitous in the plant kingdom, helping plants grow and develop in many different ways. They are also known to improve plants' resistance to stress. A review and discussion of the current knowledge and applications of BRs and SA for the reduction of salt stress in plants was conducted by Ashraf et al. (2010). Furthermore, BRs and SA have also been discussed in terms of how they can regulate physiological and biochemical processes that result in an improvement in salt tolerance in plants through exogenous application.

8.4 CONCLUSIONS AND FUTURE PERSPECTIVES

India has abundant halophytic, mangrove, and other saline vegetation, which has enormous economic and environmental potential. Tangible evidence suggests that the climate is changing and land is being degraded because of salinization, so it is imperative that one can cultivate as much salt-affected land as possible to sustain our burgeoning population. Considering this, this chapter reviewed the adaptive physiological strategies used by halophytes to survive in a diverse array of habitats.

The response of plants to salinity involves a complex set of molecular, cellular, physiological, metabolic, and whole-plant responses. A comprehensive study of cellular, metabolic, and physiological processes has demonstrated that mechanisms and strategies controlling ion-uptake, transport, balance, hormone metabolism, antioxidant metabolism, osmotic regulation, and stress signaling are key mechanisms in plant adaptations to salinity stress. In order to determine key pathways or processes controlling salt tolerance, there needs to be the integration of genomic, transcriptomic, proteomic, and metabolomic studies. Although significant progress has been made in understanding plant stress responses, there are still large gaps in our understanding of transmembrane ion transport, sensors and receptors in signal transmission, and metabolites in energy supply. Molecular interactions between cells and within cells should be studied in the future as they contribute to the salinity stress response. For several halophytes, the metabolic, biophysical, and biochemical mechanisms underlying these processes are relatively well-understood. In this way, physiology and biotechnology can be coupled to identify the key genes underlying these processes and introduce them into nontolerant plants. It will enable glycophytic species to be grown on arid and saline land, providing food in some of the world's poorest regions.

KEYWORDS

- **soil salinity**
- **salt stress**
- **halophytes**
- **physiological adaptations**
- **salt tolerance, osmoprotectants**

REFERENCES

Ashraf, M.; Akram, N. A.; Arteca, R. N.; Foolad, M. R. The Physiological, Biochemical and Molecular Roles of Brassino Steroids and Salicylic Acid in Plant Processes and Salt Tolerance. *Crit. Rev. Plant Sci.* **2010,** *29* (3), 162–190.

Afzal, I.; Munir, F.; Ayub, C. M.; Basra, S. M. A.; Hameed, A.; Nawaz, A. Changes in Antioxidant Enzymes, Germination Capacity and Vigour of Tomato Seeds in Response of Priming with Polyamines. *Seed Sci. Technol.* **2009,** *37* (3), 765–770.

Ahmad, P.; Jaleel, C. A.; Salem, M. A.; Nabi, G.; Sharma, S. Roles of Enzymatic and Nonenzymatic Antioxidants in Plants During Abiotic Stress. *Crit. Rev. Biotechnol.* **2010,** *30* (3), 161–175.

Ashraf M.; Foolad, M. R. Roles of Glycine Betaine and Proline in Improving Plant Abiotic Stress Resistance. *Environ. Exp. Bot.* **2007,** *59* (2), 206–216.

Ashraf, M.; Hameed, M.; Arshad, M.; Ashraf, Y.; Akhtar K. Salt Tolerance of Some Potential Forage Grasses from Cholistan Desert of Pakistan. In In *Ecophysiol High Salinity Tolerance Plants*; Khan, M. A., Weber, D.J., Eds.; Springer: Netherlands, 2006; pp 31–54.

Askari, H.; Edqvist, J.; Hajheidari, M.; Kafi, M.; Salekdeh, G. H. Effects of Salinity Levels on Proteome of *Suaeda aegyptiaca* Leaves. *Proteomics* **2006,** *6*, 2542–2554.

Aslam, R.; Safdar, W. A Critical Review on Halophytes: Salt Tolerant Plants. *J. Med. Plants Res.* **2011,** *5* (33), 7108–7118. 10.5897/JMPRX11.009.

Bajgu, A. Nitric Oxide: Role in Plants Under Abiotic Stress. In *Physiological Mechanisms and Adaptation Strategies in Plants Under Changing Environment*; Springer, 2014; pp 137–159.

Begara-Morales, J. C.; Sanchez-Calvo, B.; Chaki, M.; et al. Dual Regulation of Cytosolic Ascorbate Peroxidase (APX) by Tyrosine Nitration and S-nitrosylation. *J. Exp. Bot.* **2014,** *65* (2), 527–538.

Ben Ahmed, C.; Ben Rouina, B.; Sensoy, S.; Boukhriss, M.; Ben Abdullah, F. Exogenous Proline Effects on Photosynthetic Performance and Antioxidant Defense System of Young Olive Tree. *J. Agric. Food Chem.* **2010,** *58* (7), 4216–4222.

Benjamin, J. J.; Lucini, L.; Jothiramshekar, S.; Parida, A. Metabolomic Insights into the Mechanisms Underlying Tolerance to Salinity in Different Halophytes. *Plant Physiol. Biochem.* **2019,** *135*, 528–545.

Besson-Bard, A.; Pugin, A.; Wendehenne, D. New Insights into Nitric Oxide Signaling in Plants. *Ann. Rev. Plant Biol.* **2008,** *59*, 21–39.

Bhattacharya, S.; Kundu, A. Sugars and Sugar Polyols in Overcoming Environmental Stresses. In *Protective Chemical Agents in the Amelioration of Plant Abiotic Stress*, 2020. 10.1002/9781119552154.ch4.

Bohnert, H. J.; Nelson, D. E.; Jensen, R. G. Adaptations to Environmental Stresses. *Plant Cell* **1995,** *7* (7), 1099–1111.

Bohnert, H. J.; Shen, B. Transformation and Compatible Solutes. *J. Hortic. Sci. Biotechnol.* **1998,** *191*, 41–46.

Cabot, C.; Sibole, J. V.; Barcelo, J. Poschenrieder, C. Abscisic Acid Decreases Leaf Na+ Exclusion in Salt-Treated *Phaseolus vulgaris* L. *J. Plant Growth Regul.* **2009,** *28* (2), 187–192.

Chen, S.; Li, J.; Wang, S.; Huttermann, A.; Altman, A. Salt Nutrient Uptake and Transport, and ABA of *Populus euphratica*; A Hybrid in Response to Increasing Soil NaCl. *Trees Struct. Funct.* **2001,** *15* (3), 186–194.

Clause, S. D.; Sasse, J. M. Brassino Steroids: Essential Regulators of Plant Growth and Development. *Ann. Rev. Plant Biol.* **1998,** *49*, 427–451.

Cockburn, W.; Paul, M. J. Pinitol, A Compatible Solute in *Mesembryanthemum crystallinum* L.?, *J. Exp. Bot.* **1989,** *40* (10), 1093–1098. https://doi.org/10.1093/jxb/40.10.1093.

Cramer, G. R.; Quarrie, S. A. Abscisic Acid is Correlated with the Leaf Growth Inhibition of Four Genotypes of Maize Differing in their Response to Salinity. *Funct. Plant Biol.* **2002,** *29* (1), 111–115.

Cushman, J. C. Osmoregulation in Plants: Implications for Agiculture. *Am. Zool.* **2001,** *414*, 758–769.

Dassanayake M.; Larkin M. D. Making Plants Break a Sweat: The Structure, Function, and Evolution of Plant Salt Glands. *Front. Plant Sci.* **2017,** *8*, 406.

Dietz, K. J.; Tavakoli, N.; Kluge, C.; et al. Significance of the Vtype ATPase for the Adaptation to Stressful Growth Conditions and its Regulation on the Molecular and Biochemical Level. *J. Exp. Bot.* **2001,** *52* (363), 1969–1980.

El-Mashad, A. A. A.; Mohamed, H. I. Brassinolide Alleviates Salt Stress and Increases Antioxidant Activity of Cowpea Plants (*Vigna sinensis*). *Protoplasma* **2012,** *249* (3), 625–635.

El-Shintinawy, F. El-Shourbagy, M. N. Alleviation of Changes in Protein Metabolism in NaCl-Stressed Wheat Seedlings by Thiamine. *Biol. Plant.* **2001,** *44* (4), 541–545.

Fahad, S.; Hussain, S.; Matloob, A.; Khan, F. A.; Khaliq, A.; Saud, S. et al. Phytohormones and Plant Responses to Salinity Stress: A Review. *Plant Growth Regul.* **2015,** *75*, 391–404. DOI: 10.1007/s10725-014-0013-y.

FAO. Land and Plant Nutrition Management Service; Rome: FAO, 2008.

Flowers, T. J.; Colmer, T. D. Salinity Tolerance in Halophytes. *New Phytol.* **2008,** *179*, 945–963. DOI: 10.1111/j.1469-8137.2008. 02531.x

Flowers, T. J.; Colmer, T. D. Plant Salt Tolerance: Adaptations in Halophytes. *Ann. Bot.* **2015,** *115*, 327–331. DOI: 10.1093/aob/mcu267.

Flowers, T. J.; Hajibagheri, M. A.; Clipson, N. J. W. Halophytes. *Q. Rev. Biol.* **1986,** *6*, 313–337.

Foyer, C. H.; Lopez-Delgado, H.; Dat, J. F.; Scott, I. M. Hydrogen Peroxide- and Glutathione-Associated Mechanisms of Acclimatory Stress Tolerance and Signalling. *Physiologia Plantarum* **1997,** *100* (2), 241–254.

Fragnire, C.; Serrano, M.; Abou-Mansour, E.; Metraux, J. P.; L'Haridon, F. Salicylic Acid and its Location in Response to Biotic and Abiotic Stress. *FEBS Lett.* **2011,** *585* (12), 1847–1852.

Gadallah, M. A. A. Effects of Proline and Glycine Betaine on *Vicia faba* Responses to Salt Stress. *Biol. Plant.* **1999,** *42* (2), 249–257.

Gharbi, E.; Martínez, J. P.; Benahmed, H.; Fauconnier, M. L.; Lutts, S.; Quinet, M. Salicylic Acid Differently Impacts Ethylene and Polyamine Synthesis in the Glycophyte Solanum lycopersicum and the Wild-Related Halophyte Solanum chilense Exposed to Mild Salt Stress. *Physiol. Plant.* **2016,** *158*, 152–167. https://doi.org/10.1111/ppl.12458.

Ghanem, A.; Mohamed, E.; Kasem, A.; El-Ghamery, A. A. Differential Salt Tolerance Strategies in Three Halophytes from the Same Ecological Habitat: Augmentation of Antioxidant Enzymes and Compounds. *Plants* (Basel, Switzerland) **2021,** *10* (6), 1100. https://doi.org/10.3390/plants10061100

Gil, R.; Lull, C.; Boscaiu, M.; Bautista, I.; Lidón, A.; Vicente, O. Soluble Carbohydrates as Osmolytes in Several Halophytes from a Mediterranean Salt Marsh. *Not. Bot. Hort. Agrobot. Cluj-Napoca* **2011,** *39*, 9–17.

Gill, S. S.; Tajrishi, M.; Madan, M.; Tuteja, N. A DESDbox Helicase Functions in Salinity Stress Tolerance by Improving Photosynthesis and Antioxidant Machinery in Rice (*Oryza sativa* L. cv. PB1). *Plant Mol. Biol.* **2013,** *82* (1-2), 1–22.

Glenn, E.; Brown, J.; Blumwald, E. Salt Tolerance and Crop Potential of Halophytes. Critical Reviews in Plant Sciences – *Crit. Rev. Plant Sci.* **1999,** *18*, 227–255. 10.1080/07352689991309207.

González, M. In *Adaptation of Halophytes to Different Habitats. Seed Dormancy and Germination*; Jimenez-Lopez, J., Ed.; IntechOpen, 2019. DOI: 10.5772/intechopen.87056.

Grigore, M. N.; Ivanescu, L.; Toma, C. In *Halophytes: An Integrative Anatomical Study*; Springer: New York, NY, 2014.

Grigore, M. N.; Toma, C. A Proposal for a New Halophytes Classification, Based on Integrative Anatomy Observations. *Oltenia J. Stud. Nat. Sci.* **2010,** *26*, 45–50.

Grigore, M. N.; Villanueva, M.; Boscaiu, M.; Vicente, O. Do Halophytes Really Require Salts for their Growth and Development? An Experimental Approach. *Not. Sci. Biol.* **2012,** *4*, 23–29. DOI: 10.15835/nsb427606.

Groß, F.; Durner, J.; Gaupels, F. Nitric Oxide, Antioxidants and Prooxidants in Plant Defence Responses. *Front. Plant Sci.* **2013,** *4*, 419.

Groppa, M. D.; Benavides, M. P. Polyamines and Abiotic Stress: Recent Advances. *Amino Acids* **2008,** *34* (1), 35–45.

Gul, B.; Ansari, R.; Flowers, T. J.; Khan, M. A. Germination Strategies of Halophyte Seed Under Salinity. *Environ. Exp. Bot.* **2013,** *92*, 4–18. DOI: 10.1016/ envexpbot.2012.11.006.

Guo, J.; Lu, C.; Zhao, F.; Gao, S.; Wang, B. Improved Reproductive Growth of Euhalophyte *Suaeda salsa* Under Salinity is Correlated with Altered Phytohormone Biosynthesis and Signal Transduction. *Funct. Plant Biol.* **2020,** *47*, 170–183.

Gupta, K.; Dey, A.; Gupta, B. Plant Polyamines in Abiotic Stress Responses. *Acta Physiol. Plant.* **2013,** *35* (7), 2015–2036.

Gupta, K.; Dey, A.; Gupta, B. Polyamines and their Role in Plant Osmotic Stress Tolerance. In *Climate Change and Plant Abiotic Stress Tolerance*; Tuteja, N.; Gill, S. S., Eds.; Wiley-VCH: Weinheim, Germany, 2013; pp 1053– 1072.

Gupta, B.; Huang, B. Mechanism of Salinity Tolerance in Plants: Physiological, Biochemical, and Molecular Characterization. *Int. J. Genomics* **2014,** *2014*, 701596. DOI: 10.1155/2014/701596.

Hanson, A. D.; Rathinasabapathi, B.; Rivoal, J.; Burnet, M.; Dillon, M. O.; Gage, D. A. Osmoprotective Compounds in the Plumbaginaceae: A Natural Experiment in Metabolic Engineering of Stress Tolerance. *Proc. Natl.. Acad. Sci. U. S. A.* **1994,** *91* (1), 306–310.

Hasanuzzaman, M.; Inafuku, M.; Nahar, K.; Fujita, M.; Oku, H. Nitric Oxide Regulates Plant Growth, Physiology, Antioxidant Defense, and Ion Homeostasis to Confer Salt Tolerance in the Mangrove Species, *Kandelia obovata. Antioxidants* **2021,** *10*, 611. https:// doi. org/10.3390/antiox10040611.

Hasanuzzaman, M.; Nahar, K.; Alam, M. M.; Bhowmik, P. C; Hossain, M A.; Rahman, M. M.; Prasad, M. N.; Ozturk, M.; Fujita, M. Potential use of Halophytes to Remediate Saline Soils. *Biomed. Res. Int.* **2014,** *2014*, 589341. DOI: 10.1155/2014/589341. Epub 2014 Jul 6. PMID: 25110683; PMCID: PMC4109415.

Hasegawa, P. M. Sodium Na+ Homeostasis and Salt Tolerance of Plants. *Environ. Exp. Bot.* **2013,** *92*, 19–31.

Hasegawa, P. M.; Bressan, R. A.; Zhu, J. K.; Bohnert, H. J. Plant Cellular and Molecular Responses to High Salinity. *Ann. Rev. Plant Biol.* **2000,** *51*, 463–499.

Hoque, M. A.; Banu, M. N. A.; Nakamura, Y.; Shimoishi, Y.; Murata, Y. Proline and Glycine Betaine Enhance Antioxidant Defence and Methylglyoxal Detoxification Systems and Reduce NaCl-Induced Damage in Cultured Tobacco Cells. *J. Plant Physiol.* **2008,** *165* (8), 813–824.

Hoque, M. A.; Banu, M. N. A.; Okuma, E.; et al. Exogenous Proline and Glycine Betaine Increase NaCl-Induced Ascorbate Glutathione Cycle Enzyme Activities, and Proline Improves Salt Tolerance more than Glycine Betaine in Tobacco Bright Yellow-2 Suspension-Cultured Cells. *J. Plant Physiol.* **2007,** *164* (11), 1457–1468.

Hossain, K. K.; Itoh, R. D.; Yoshimura, G.; et al. Effects of Nitric Oxide Scavengers on Thermos Inhibition of Seed Germination in *Arabidopsis thaliana. Russ. J. Plant Physiol.* **2010,** *57* (2), 222–232.

Hussain, S. S.; Ali, M.; Ahmad, M.; Siddique, K. H. M. Polyamines: Natural and Engineered Abiotic and Biotic Stress Tolerance in Plants. *Biotechnol. Adv.* **2011,** *29* (3), 300–311.

Jayakannan, M.; Bose, J.; Babourina, O.; et al. Salicylic Acid Improves Salinity Tolerance in *Arabidopsis* by Restoring Membrane Potential and Preventing Salt-Induced K+ Loss via a GORK Channel. *J. Exp. Bot.* **2013,** *64* (8), 2255–2268.

Jeschke, W. D.; Peuke, A. D.; Pate, J. S.; Hartung, W. Transport, Synthesis and Catabolism of Abscisic Acid (ABA) in Intact Plants of Castor Bean (*Ricinus communis* L.) under Phosphate Deficiency and Moderate Salinity. *J. Exp. Bot.* **1997,** *48* (314), 1737–1747.

Jin, H.; Dong, D.; Yang, C.; Yuan, F.; Yu, X.; Fu, X.; Zhu, D. Cloning and Analysis of PEAMT Gene in *Suaeda Glauca. Chin. Agric. Sci. Bull.* **2015,** *31*, 178–183.

Kimura,Y.; Kawasaki, S.; Yoshimoto, H.; Takegawa, K. Glycine Betaine Biosynthesized from Glycine Provides an Osmolyte for Cell Growth and Spore Germination during Osmotic Stress in *Myxococcus xanthus. J Bacteriol.* **2010,** *192* (5), 1467–1470. DOI: 10.1128/ JB.01118-09. Epub 2009 Dec 18. PMID: 20023011; PMCID: PMC2820835.

Kerepesi, I.; Galiba, G. Osmotic and Salt Stress-Induced Alteration in Soluble Carbohydrate Content in Wheat Seedlings. *Crop Sci.* **2000,** *40* (2), 482–487.
Koji, Y.; Yoshihiro, T.; Thomas, B.; Akihiko, I.; Atsushi, M.; Taku, T.; Anthony, M.; Tomonobu, K. The Polyamine Spermine Protects against High Salt Stress in *Arabidopsis thaliana. FEBS Lett.* **2006,** *580*, 30, 6783–6788. ISSN 0014-5793, https://doi.org/10.1016/j.febslet.2006.10.078.
Kong, L. A.; Xie, Y.; Hu, L.; Si, J. S.; Wang, Z. S. Excessive Nitrogen Application Dampens Antioxidant Capacity and Grain Filling in Wheat as Revealed by Metabolic and Physiological Analyses. *Sci Rep.* **2017,** *7*, 43363.
Kopyra, M.; Gwozdz, E. A. Nitric Oxide Stimulates Seed Germination and Counteracts the Inhibitory Effect of Heavy Metals and Salinity on Root Growth of *Lupinus luteus*. *Plant Physiol. Biochem.* **2003,** *41* (11-12), 1011–1017.
Kumari, A.; Das, P.; Parida, A. K.; Agarwal, P. K. Proteomics, Metabolomics, and Ionomics Perspectives of Salinity Tolerance in Halophytes. *Front. Plant Sci.* **2015,** *6*, 537. DOI: 10.3389/ fpls.2015.00537.
Kuznetsov, V. V.; Shevyakova, N. I. Polyamines and Stress Tolerance of Plants. *Plant Stress* **2007,** *1*, 50–71.
Li, Q.; Liu, D.; Gao, X.; Su, Q.; Jia, A. Cloning of cDNA Encoding Choline Monooxygenase from *Suaeda liaotungensis*and Salt Tolerance of Transgenic Tobacco. *J. Integr. Plant Biol.* **2003,** *45*, 242–247.
Li, K.; Pang, C. H.; Ding, F.; Sui, N.; Feng, Z. T.; Wang, B. S. Overexpression of *Suaeda salsa* Stroma Ascorbate Peroxidase in Arabidopsis Chloroplasts Enhances Salt Tolerance of Plants. *S. Afr. J. Bot.* **2012,** *78*, 235–245. DOI: 10.1016/j.sajb.2011.09.006.
Ma, L.; Zhang, H.; Sun, L.; et al. NADPH Oxidase AtrbohD and AtrbohF Function in ROS-Dependent Regulation of Na+/K+ Homeostasis in *Arabidopsis* under Salt Stress. *J. Exp. Bot.* **2012,** *63* (1), 305–317.
mad-Um, S.; Kirdmanee, C. Effect of Glycine Betaine on Proline, Water Use, and Photosynthetic Efficiencies, and Growth of Rice Seedlings under Salt Stress. *Turk. J. Agric. For.* **2010,** *34* (6), 517–527.
Makela, P.; Karkkainen, J.; Somersalo, S. Effect of Glycine Betaine on Chloroplast Ultrastructure, Chlorophyll and Protein Content, and RuBPCO Activities in Tomato grown Under Drought or Salinity. *Biol. Plant.* **2000,** *43* (3), 471– 475.
Marschner, H. In *Mineral Nutrition of Plants*, 2nd Ed.; Academic Press: Boston, 1995.
Matysik, J.; Alia, A.; Bhalu, B.; Mohanty, P. Molecular Mechanisms of Quenching of Reactive Oxygen Species by Proline Under Stress in Plants. *Curr. Sci.* **2002,** *82* (5), 525– 532.
Maurel, P. Aquaporins and Water Permeability of Plant Membranes. *Ann. Rev. Plant Physiol. Plant Mol. Biol.* **1997,** *48*, 399–429.
Mishra, S.; Jha, A. B.; Dubey, R. S. Arsenite Treatment Induces Oxidative Stress, Upregulates Antioxidant System, and Causes Phytochelatin Synthesis in Rice Seedlings. *Protoplasma* **2011,** *248* (3), 565–577.
Muchate, N. S.; Nikalje, G. C.; Rajurkar, N. S.; Suprasanna, P.; Nikam, T. D. Plant Salt Stress: Adaptive Responses, Tolerance Mechanism and Bioengineering for Salt Tolerance. *Bot. Rev.* **2016,** *82* (4), 371–406. http://www.jstor.org/stable/45212019.
Muchate, N. S.; Nikalje, G. C.; Rajurkar, N. S.; Suprasanna, P.; Nikam, T. D. Physiological Responses of the Halophyte *Sesuvium portulacastrum* to Salt Stress and their Relevance for Saline Soil Bio-Reclamation. *Flora* **2016,** *224*, 96–105. DOI: 10.1016/j.flora.2016.07.00.

Naidu, B. P. Production of Betaine from Australian *Melaleuca* spp for Use in Agriculture to Reduce Plant Stress. *Austr. J. Exp. Agric.* **2003,** *43*, 1163–1170.

Nalousi, A. M.; Ahmadiyan, S.; Hatamzadeh, A.; Ghasemnezhad, M. Protective Role of Exogenous Nitric Oxide Against Oxidative Stress Induced by Salt Stress in Bell-Pepper (*Capsicum annum* L.). *Am. Eurasian J. Agric. Environ. Sci.* **2012,** *12* (8), 1085–1090.

Nazar, R.; Iqbal, N.; Syeed, S.; Khan, N. A. Salicylic Acid Alleviates Decreases in Photosynthesis under Salt Stress by Enhancing Nitrogen and Sulfur Assimilation and Antioxidant Metabolism Differentially in Two Mung Bean Cultivars. *J. Plant Physiol.* **2011,** *168* (8), 807–815.

Nikalje, G. C.; Bhaskar, S. D.; Yadav, K.; Penna, S. Halophytes: Prospective Plants for Future. In *Ecophysiology, Abiotic Stress Responses and Utilization of Halophytes*, 2019; pp 221–234. DOI: 10.1007/978-981-13-3762-8_10.

Nikalje, G.; Penna, S. Coping With Metal Toxicity – Cues From Halophytes. *Front. Plant Sci.* **2018,** *9*, 777. 10.3389/fpls.2018.00777.

Niu X.; Bressan, R. A.; Hasegawa, P. M.; Pardo, J. M. Ion Homeostasis in NaCl Stress Environments. *Plant Physiol.* **1995,** *109* (3), 735–742.

Nounjan, N.; Nghia, P. T., Theerakulpisut, P. Exogenous Proline and Trehalose Promote Recovery of Rice Seedlings from Salt-Stress and Differentially Modulate Antioxidant Enzymes and Expression of Related Genes. *J. Plant Physiol.* **2012,** *169* (6), 596–604.

O'leary, J. W.; Glenn, E. P. *Global Distribution and Potential for Halophytes.* In *Halophytes as a Resource for Livestock and for Rehabilitation of Degraded Lands. Tasks for Vegetation Science*; Squires, V. R., Ayoub, A. T., Eds.; Springer: Dordrecht, 1994; vol *32*. https://doi.org/10.1007/978-94-011-0818-8_2.

Oliveira Otoch, M. D. L.; Menezes Sobreira, A. C.; Farias De Aragao, M. E.; Orellano, E. G.; Guia Silva Lima, M. D.; Fernandes De Melo, D. Salt modulation of vacuolar H+-ATPase and H+-Pyrophosphatase Activities in *Vigna unguiculata. J. Plant Physiol.* **2001,** *158* (5), 545–551.

Pang, C. H.; Li, K.; Wang, B. S. Overexpression of SsCHLAPXs Confers Protection against Oxidative Stress Induced by High Light in Transgenic *Arabidopsis thaliana. Physiol. Plant* **2011,** *143* (4), 355–366. DOI: 10.1111/j.1399-3054.2011.01515.x

Parida, A. K.; Das, A. B.; Mohanty, P. Investigations on the Antioxidative Defence Responses to NaCl Stress in a Mangrove, *Bruguiera parviflora*: Differential Regulations of Isoforms of Some Antioxidative Enzymes. *Plant Growth Regul.* **2004,** *42* (3), 213–226.

Parvaiz, A.; Satyawati, S. Salt Stress and Phyto-Biochemical Responses of Plants-A Review. *Plant Soil Environ.* **2008,** *54* (3), 89.

Podar, D.; Macalik, K.; Réti, K. O.; Martonos, I.; Török, E.; Carpa, R.; Weindorf, D. C.; Csiszár, J.; Székely, G. Morphological, Physiological and Biochemical Aspects of Salt Tolerance of Halophyte *Petrosimonia triandra* Grown in Natural Habitat. *Physiol. Mol. Biol. Plants* **2019,** *25* (6), 1335–1347. https://doi.org/10.1007/s12298-019-00697-x.

Reddy, R.; Puli, C.; Pandita, D.; Wani, S. Sugar Alcohols and Osmotic Stress Adaptation in Plants. In *Compatible Solutes Engineering for Crop Plants Facing Climate Change*, 2021; pp 189–203. 10.1007/978-3-030-80674-3_8.

Reddy, M. P.; Sanish, S.; Iyengar, E. R. R. Photosynthetic Studies and Compartmentation of Ions in Different Tissues of *Salicornia brachiata Roxb.* under Saline Conditions. *Photosynthetica* **1992,** *26*, 173–179.

Roy, S. J.; Negrao, S.; Tester, M. Salt Resistant Crop Plants. *Curr. Opin. Biotechnol.* **2014,** *26*, 115–124.

Roy, M.; Wu, R. Overexpression of S-adenosylmethionine Decarboxylase Gene in Rice Increases Polyamine Level and Enhances Sodium Chloride-Stress Tolerance. *Plant Sci.* **2002,** *163* (5), 987–992.

, M.; Sabovljevic, A. Contribution to the Coastal Bryophytes of the Northern Mediterranean: Are there Halophytes among Bryophytes? *Phytol. Balcanica* **2007,** *13*, 131–135.

Sairam, R. K.; Tyagi, A. Physiology and Molecular Biology of Salinity Stress Tolerance in Plants. *Curr. Sci.* **2004,** *86* (3), 407–421.

Salama, K.; Al-Mutawa, M. M. Glutathione-Triggered Mitigation in Salt-induced Alterations in Plasmalemma of Onion Epidermal Cells. *Int. J. Agric. Biol.* **2009,** *11*, 639–642.

Sanchez, D. H.; Siahpoosh, M. R.; Roessner, U.; Udvardi, M.; Kopka, J. Plant Metabolomics Reveals Conserved and Divergent Metabolic Responses to Salinity. *Physiol. Plant* **2008,** *132*, 209–219. DOI: 10.1111/j.1399-3054.2007.00993.x

Saxena, S. C.; Kaur, H.; Verma, P.; et al. Osmoprotectants: Potential for Crop Improvement under Adverse Conditions. In *Plant Acclimation to Environmental Stress*; Springer: New York, NY, USA, 2013; pp 197–232.

Sengupta, S.; Majumder, A. L. Porteresia coarctata (Roxb.) Tateoka, A Wild Rice: A Potential Model for Studying Salt-Stress Biology in Rice. *Plant Cell Environ.* **2010,** *33*, 526–542. DOI: 10.1111/j.1365-3040.2009.02054.x.

Sengupta, S.; Patra, B.; Ray, S.; Majumder, A. L. Inositol Methyl Tranferase from a Halophytic Wild Rice, Porteresia Coarctata Roxb. (Tateoka): Regulation of Pinitol Synthesis Under Abiotic Stress. *Plant Cell Environ.* **2008,** *31*, 1442–1459. DOI: 10.1111/j.1365-3040.2008.01850.x.

Serrano, R.; Mulet, J. M.; Rios, G. A Glimpse of the Mechanisms of Ion Homeostasis during Salt Stress. *J. Exp. Bot.* **1999,** *50*, 1023–1036.

Shevyakova, N. I.; Musatenko, L. I.; Stetsenko, L. A.; et al. Effects of Abscisic Acid on the Contents of Polyamines and Proline in Common Bean Plants under Salt Stress. *Russ. J. Plant Physiol.* **2013,** *60*, 200–211.

Shevyakova, N. I.; Shorina, M. V.; Rakitin, V. Y.; Kuznetsov, V. Stress-Dependent Accumulation of Spermidine and Spermine in the Halophyte *Mesembryanthemum crystallinum* Under Salinity Conditions. *Russ. J. Plant Physiol.* **2006,** *53*, 739–745. DOI: 10.1134/S1021443706060021.

Steiner, M. To the Ecology of the Salt March of the Nordostlichen United Countries of Nordamerika. *Jahrb. Know. Offered.* **1934,** 81, 94.

Sulpice, R.; Tsukaya, H.; Nonaka, H.; Mustardy, L.; Chen, T. H; Murata, N. Enhanced Formation of Flowers in Salt-Stressed Arabidopsis after Genetic Engineering of the Synthesis of Glycine Betaine. *Plant J.* **2003,** *36*, 165–176.

Sung, C. H.; Hong, J. K. Sodium Nitroprusside Mediates Seedling Development and Attenuation of Oxidative Stresses in Chinese Cabbage. *Plant Biotechnol. Rep.* **2010,** *4* (4), 243–251.

Tada, Y.; Shiho, K.; Takamitsu, K. Growth and Physiological Adaptation of Whole Plants and Cultured Cells from a Halophyte Turf Grass under Salt Stress. *AoB Plants* **2014,** *6*, plu041. https://doi.org/10.1093/aobpla/plu041.

Takahashi, T.; Kakehi, J. I. Polyamines: Ubiquitous Polycations with Unique Roles in Growth and Stress Responses. *Ann. Bot.* **2010,** *105* (1), 1–6.

Tisi, A.; Angelini, R.; Cona, A. Wound Healing in Plants: Cooperation of Copper Amine Oxidase and Flavin-Containing Polyamine Oxidase. *Plant Signal. Behav.* **2008,** *3* (3), 204–206.

Tuteja, N.; Sahoo, R. K.; Garg, B; Tuteja, R. OsSUV3 Dual Helicase Functions in Salinity Stress Tolerance by Maintaining Photosynthesis and Antioxidant Machinery in Rice (*Oryza sativa* L. cv. IR64). *Plant J.* **2013,** *76* (1),115–127.

Ungar, I. A. Halophyte Seed Germination. *Bot. Rev.* **1978,** *44*, 233–364.

Uzilday, B.; Ozgur, R.; Sekmen, A. H.; Yildiztugay, E.; Turkan, I. Changes in the Alternative Electron Sinks and Antioxidant Defence in Chloroplasts of the Extreme Halophyte *Eutremaparvulum* (*Thellungiella parvula*) Under Salinity. *Ann. Bot.* **2015,** *115*, 449–463.

Vaziriyeganeh, M.; Carvajal, M.; Du, N.; Zwiazek, J. J. Salinity Tolerance of Halophytic Grass *Puccinellia nuttalliana* Is Associated with Enhancement of Aquaporin-Mediated Water Transport by Sodium. *Int. J. Mol. Sci.* **2022,** *23* (10), 5732. https://doi.org/10.3390/ijms23105732

Van Oosten, M. J.; Sharkhuu, A.; Batelli, G.; Bressan, R. A.; Maggio, A. The Arabidopsis thaliana Mutant Air1 Implicates SOS3 in the Regulation of Anthocyanins under Salt Stress. *Plant Mol. Biol.* **2013,** *83*, 405–415.

Vikram, P.; Sharma, J.; Sharma, A. Role of Brassinosteroids in Plants Responses to Salinity Stress: A Review. *JANS* **2022,** *14*, 582–599.

Waisel, Y. Biology of Halophytes, Academic Press: New York, 1972.

Walter, H. Salinity Problems in the Acid Zones. The Adaptations of Plants to Saline Soils. *Arid Zone Res.* **1961,** *14*, 65–68.

Wang, W.; Liu, Y.; Duan, H.; Yin, X.; Cui, Y.; Chai, W.; Song, X.; Flowers, T.; Wang, S. SsHKT1; 1 is coordinated with SsSOS1 and SsNHX1 to regulate Na^+ homeostasis in Suaeda salsa under saline conditions. *Plant Soil* **2020,** *449*, 117–131.

Wang, B.; Luttge, U.; Ratajczak, R. Effects of Salt Treatment and Osmotic Stress on V-ATPase and V-PPase in Leaves of the Halophyte *Suaeda salsa*. *J. Exp. Bot.* **2001,** *52* (365), 2355–2365.

Wang, X.; Shi, G.; Xu, Q.; Hu, J. Exogenous Polyamines Enhance Copper Tolerance of *Nymphoides peltatum*. *J. Plant Physiol.* **2007,** *164* (8), 1062–1070.

Wang, F.; Wang, M.; Guo, C.; Wang, N.; Li, X.; Chen, H.; Dong, Y.; Chen, X.; Wang, Z.; Li, H. Cloning and Characterization of a Novel Betaine Aldehyde Dehydrogenase Gene from *Suaeda corniculata*. *Genet. Mol. Res.* **2016,** *15*, gmr15027848.

Weber, D. J. Adaptive Mechanisms of Halophytes in Desert Regions. In *Salinity and Water Stress*. Springer Netherlands: Neteherlands, 2008; vol 44, pp 179–185.

Wu, H.; Liu, X.; You, L.; Zhang, L.; Zhou, D.; Feng, J.; Zhao, J.; Yu, J. Effects of Salinity on Metabolic Profiles, Gene Expressions, and Antioxidant Enzymes in Halophyte *Suaeda salsa*. *J. Plant Growth Regul.* **2012,** *31*, 332–341.

Wyn Jones, R. G.; Gorham, J. Osmoregulation. In *Physiological Plant Ecology*; In *Encyclopaedia of Plant Physiology*; Lange, O. L., Nobel, P. S., Osmond, C. B., Ziegler, H., Eds.; Springer-Verlag: Berlin, Germany, 1983; vol 12C, p 35–38.

Xiong, J.; Fu, G.; Tao, L.; Zhu, C. Roles of Nitric Oxide in Alleviating Heavy Metal Toxicity in Plants. *Arch. Biochem. Biophys.* **2010,** *497* (1-2), 13–20.

Xiong, L.; Zhu, J. Molecular and Genetic Aspects of Plant Responses to Osmotic Stress. *Plant Cell Environ.* **2002,** *25*, 131–139. 10.1046/j.1365-3040.2002.00782.x.

Xu, C.; Tang, X.; Shao, H.; Wang, H. Salinity Tolerance Mechanism of Economic Halophytes from Physiological to Molecular Hierarchy for Improving Food Quality. *Curr Genomics* **2016,** *17* (3), 207–214. DOI: 10.2174/1389202917666160202215548. PMID: 27252587; PMCID: PMC4869007.

Yamaguchi, K.; Takahashi, Y.; Berberich, T.; et al. The Polyamine spermine Protects against High Salt Stress in *Arabidopsis thaliana*. *FEBS Lett.* **2006,** *580* (30), 6783–6788.

Yiu, J. C.; Juang, L. D.; Fang, D. Y. T.; Liu, C. W.; Wu, S. J. Exogenous Putrescine Reduces Flooding-Induced Oxidative Damage by Increasing the Antioxidant Properties of Welsh onion. *Sci. Hortic.* **2009,** *120* (3), 06–314.

Youssef, A. M. Salt Tolerance Mechanisms in Some Halophytes from Saudi Arabia and Egypt. *Res. J. Agric. Biol. Sci.* **2009,** *5*, 191–206.

Zhang, J.; Jia, W.; Yang, J.; Ismail, A. M. Role of ABA in Integrating Plant Responses to Drought and Salt Stresses. *Field Crop Res.* **2006,** *97*, 111–119. DOI: 10.1016/j.fcr.2005.08.018.

Zhang, Y.; Wang, L.; Liu, Y.; Zhang, Q.; Wei, Q.; Zhang, W. Nitric Oxide Enhances Salt Tolerance in Maize Seedlings Through Increasing Activities of Proton-Pump and Na+/H+ Antiport in the Tonoplast. *Planta* **2006,** *224* (3), 545–555.

Zhao, M. G.; Chen, L.; Zhang, L. L.; Zhang, W. H. Nitric Reductase-Dependent Nitric Oxide Production is Involved in Cold Acclimation and Freezing Tolerance in *Arabidopsis*. *Plant Physiol.* **2009,** *151* (2), 755–767.

Zhao, L.; Zhang, F.; Guo, J.; Yang, Y.; Li, B.; Zhang, L. Nitric Oxide Functions as a Signal in Salt Resistance in the Calluses from Two Ecotypes of Reed. *Plant Physiol.* **2004,** *134* (2), 849–857.

Zhu, J. K. Plant Salt Tolerance. *Trends Plant Sci.* **2001,** *6*, 66–71. http://dx.doi.org/10.1016/S1360-1385(00)01838-0

Zhu J. K. Regulation of Ion Homeostasis under Salt Stress. *Curr. Opin. Plant Biol.* **2003,** *6* (5), 441–445.

CHAPTER 9

Adaptive Pathways of Halophytes and the Prospects of Increasing Plant Salinity Tolerance

CHIRAG U. NARAYANKAR[1], SUDARSHAN T. CHARAPALE[2], UMESH R. PAWAR[3], MANASI S. PATIL[2], and D. K. GAIKWAD[4]

[1]*Department of Botany, Rajarshi Chhatrapati Shahu College, Kolhapur, India*

[2]*Department of Botany, Sadguru Gadage Maharaj College, Karad, India*

[3]*Department of Botany, Shri Pancham Khemraj Mahavidyalaya, Sawantwadi, India*

[4]*Department of Botany, Shivaji University Kolhapur, India*

ABSTRACT

As a result of inappropriate human activity and climate change, soil salinization has become a severe environmental issue that threatens future food security and sustainable agriculture. The majority of crop cultivars produced in salty soils still show a loss in production, prompting the search for alternatives. Despite significant progress being made in the development of crop types by incorporating salt tolerance-associated features, this problem still exists. Halophytes are recognized to possess tremendous potential for rehabilitating salt-contaminated soils to sustain plant development in saline soils by using various tactics, including phytoremediation. This is due to their inherent salt-tolerance qualities. Additionally, the recent identification and characterization of genes related to salt tolerance that encode signaling elements from halophytes, which are naturally grown in high salinity, has enabled the development of transgenic crops with improved salt tolerance.

Physiology of Halophytes: Signaling, Omics, and Tolerance Mechanisms.
Nivas Manohar Desai & Ganesh Chandrakant Nikalje (Eds.)

The goal of this study is to offer a thorough update on the detrimental effects of salinity on soils and plants, including changes to the physicochemical characteristics of soils and adjustments to physiological and biochemical processes as well as ion imbalances in plants. We also go through the physiological and biochemical coping mechanisms that halophytes use to develop and endure in salinity-affected environments. Additionally, we demonstrate the phytoremediation process mediated by halophytes in salinity-affected locations as well as their potential effects on soil characteristics. In order to keep agriculture in salinity-prone locations, we also comprehensively address the potential for boosting salt tolerance in crop plants by importing potential genes related to antiporters, ion transporters, and antioxidants, including resistance proteins from halophytes. This is crucial since it is found on current research on the methods halophytes employ to tolerate salt.

9.1 INTRODUCTION

A loss of USD 27.5 billion per year is caused by soil degradation brought on by salinity, which affects 3600 million hectares (Mha) of the world's 5200 Mha of agricultural land (Zhang and Shi, 2013; Qadir et al., 2014). Moreover, according to reports, the world's population will reach 9.7 billion in 2050, necessitating a 70% increase in food production, further placing pressure on the world's decreasing agricultural land (Calone et al., 2021). Many drought-prone regions of the world still use salinated groundwater for irrigation in order to feed the world's expanding population (Minhas et al., 2019). As shown in Punjab in Pakistan and Fubei in China (Wang et al., 2007), inappropriate agricultural management techniques, such as incorrect drainage, aggravate soil salinity and render the land unusable for cultivation. Large masses of concentrating seawater also accumulate on agricultural land as a consequence of climate-related factors such as sea level rise, tidal changes, saltwater droplets that may drift, and even catastrophic events like tsunamis (Omuto et al., 2020). Thus, 1.5 Mha of land is at risk from salinity each year, which raises the possibility that 50% of cultivable lands may disappear more by the middle of the 21st century (Rahman et al., 2016; Karakas et al., 2021).

About 1130 Mha of land is vulnerable to the invasion of various salinity levels, as per the Harmonized World Soil Database (HWSD). The bulk of salinity-affected soils is composed of saline, sodic, and saline-sodic soils, making up 60, 26, and 14% of these, respectively (Wicke et al., 2011) (Fig. 9.1A). The additional components of all salinity-affected lands are 736 (65%), 227 (20%), 53 (5%), as well as 114 (10%) Mha of slightly, fairly, significantly, excessively salinity-affected soils, etc., (Wicke et al., 2011)

(Fig. 9.1B). The size of the global population and the extent of soils impacted by salinity vary by region. With 176, 169, 161, and 126 Mha of area, the Middle East, Australia, North Africa, and the former Soviet Union are the most salinity-affected locations (Wicke et al., 2011) (Fig. 9.1C).

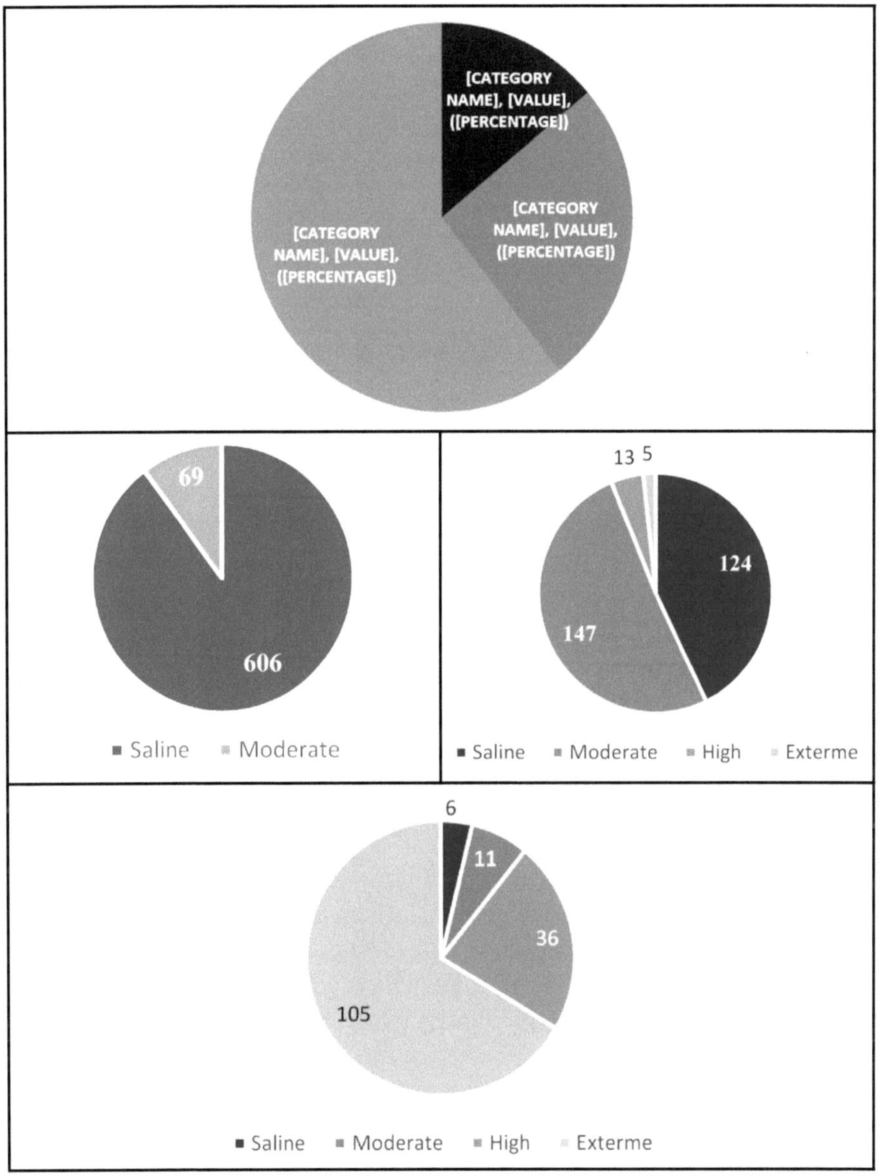

(Continued)

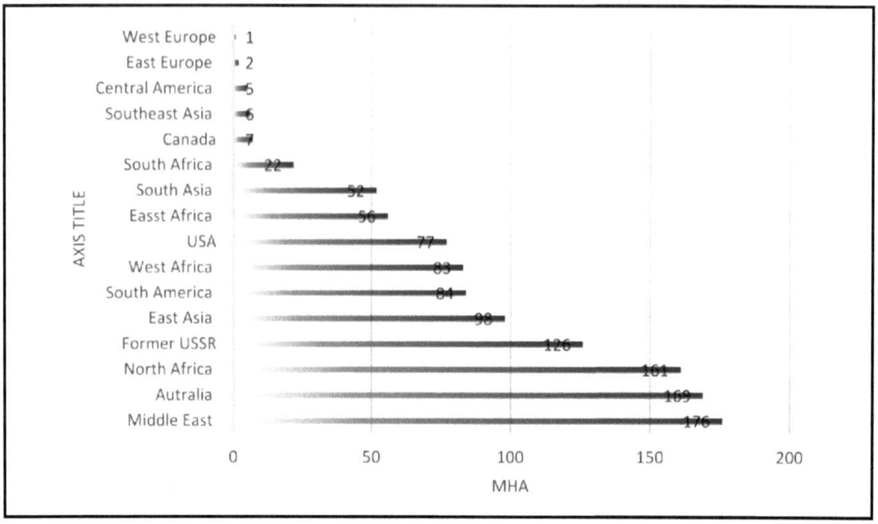

FIGURE 9.1 Shows a global overview of soil salinization. (A) Soil types that are impacted by salinity. (B) Types and degrees of salinity-affected soil severity. (C) Soils influenced by salinity in various places around the globe *. The total may not accurately reflect the salinity-affected lands' actual data as a result of rounding. Million hectares, or Mha.

Growing in salinity-affected areas results in plants accumulating high levels of toxic particles that lead to a variety of physiological abnormalities, including ionic imbalance, poor gas exchange, damage to underwater homeostasis, modifications in tiers of metabolic pathways, as well as oxidative impairment toward the cell functions brought on by reactive oxygen species (ROS) (Rahman et al., 2019; Zhao et al., 2020).

Trees use a variety of basic processes to combat the negative effects of salt stress, such as (1) toxic ion (Na^+ and Cl) prohibitions or with their compactment into pores or old cells; (2) providing these accumulations; (3) synthesis of the many stress adaptation-related endogenous metabolic pathways; and (4) the formulation and initiation of antioxidative enzymes (such as catalase (CAT) and superoxide dismutase) (Rahman et al., 2019; Zaid and Wani, 2019; Mostofa et al., 2021; Patel et al., 2020).

A wide range of genetic variants, including regional landraces, for salinity tolerance in conventional agriculture, including local landraces, remain elusive. Sadly, traditional crops cannot use their salt tolerance potential to live in salinity-affected soils.

It is interesting to note that some plants, such as halophytes, comprise the adaptation of intrinsic mechanisms related to selective ion uptake, improved antioxidant capabilities, salt glands and salt bladder development, ion homeostasis, generation of defensive metabolites, and Na^+ compartmentalization, allowing only about 2% of the world's flora to survive and thrive under extreme salinity (Patel et al., 2020; Flowers et al., 2008). The growth of halophytes on salt-contaminated soils has been proposed as an alternative to affordable desalination and restoration of severely degraded salty and sodic soils through the procedure known as phytoremediation (Litalien and Zeeb, 2020) because of their capacity to tolerate high salinity. It is important to note that several species' gene expression patterns in response to salt stress have been extensively studied using molecular studies, such as transcriptome sequencing (Liu et al., 2018). To be sure of sustainable agriculture in salinity-affected areas, it is necessary to investigate the methods timely by which halophyte species tolerate salt with the ultimate aim of transferring their characteristics and/or genes to conventional crops.

Considering the aforementioned information, the current review concentrates on the most recent information regarding the detrimental effects of salinity on both soils and plants, as well as the inherent halophytes' physiologic, metabolic, and genetic pathways used to adapt to salinity. To improve agricultural production from salt-contaminated soils, the mechanistic features of desalinization and the phytoremediation potential of halophytes will also be discussed and evaluated. The ability of halophytes to improve salt tolerance in conventional crops will be highlighted by considering the genes involved with salt tolerance-related features of halophytes. This will provide a thorough update on transcriptome investigations in halophytes.

9.2 SALINITY'S ADVERSE EFFECTS ON THE PROPERTIES OF SOIL

Significant changes in soil quality are brought on by salinity. For instance, excessive salt content in soils may cause surface crusting, soil compaction, poor soil structure, high bulk density, and clay dispersion (Rengasamy, 2016). Additionally, water penetration can be substantially hampered by salinity-induced low hydraulic conductivity, which can negatively affect plant development and production (Yuan et al., 2018). The microbial populations and their enzymatic activity in soil are negatively impacted by salt buildup (Luo et al., 2018). Soil microbiomes and related enzyme activities are thought of as early indicators for determining the level of soil deterioration and modifications to soil quality due to their heightened sensitivity to environmental perturbations (Guangming et al., 2017). The mineralization of soil, the breakdown of soil organic matter (OM), and the stability of soil aggregates are all processes that depend heavily on microbial activity (Zhang et al. 2019; Tian et al., 1792). Osmotic imbalance brought on by salinity can cause water to seep from microbial cells, which then causes plasmolysis and the microorganism's demise (Yan et al., 2015). Alternatively, microbes may survive in settings with low osmotic potential by collecting osmolytes, which requires a lot of energy (Rath and Rousk, 2015). CATs, reactive phosphatases, ureases, dehydrogenases, and glucosidases are only some of those soil enzymes that are produced by plant root exudates, beneficial microorganisms, and other plant and animal residues. These enzymes are essential for the regulation of all OM decomposition processes as well as the soil's nutrient cycle (Xie et al., 2017). Only a few studies were undertaken to evaluate the potential of these enzymes in modifying soil quality in coastal settings, despite their significant effect on soil quality change. Due to the complexity of soils, complete knowledge of how salt affects soil quality is currently lacking (Rath and Rousk, 2015). Because of this, future studies should concentrate on elucidating these mechanisms in a spatiotemporal manner.

9.3 ADVERSE IMPACTS OF SALINITY ON PLANTS: MECHANISMS

There are various ways in which increased soil salt might harm plant performance. By lowering the soil water's osmotic pressure and interfering with plant seeds' ability to absorb water, salinity-induced water scarcity adversely impacts seed germination (Uçarlı, 2020; El-Keblawy et al., 2021). Any problem with seed imbibition can change protein synthesis and metabolism, upset hormonal equilibrium, slow down the use of seed reserves, and frequently affect the intrinsic internal structure of tissues, including organs (Miransari and Smith, 2014; Mwando et al., 2020). The primary adverse effect of soil salinity, which varies depending on salinity levels, length of susceptibility to salt, and the types of plants, is a reduction in the development and biomass of plants following germination (Uçarlı, 2020). Within a few seconds of exposure to salt, plant cells can begin to lose water, leading to cell shrinkage (Munns and Tester, 2008). Over the course of hours, cells return to their initial volume; however, cell elongation rates decrease (Munns, 2002). Cell division rates also significantly decrease over the course of days, which inhibits plant development (Arif et al., 2020). Salinity can also result in obvious damage to the oldest leaves and the loss of younger and older leaves, which may cause the plants to die before the seeds have a chance to mature (Munns and Tester, 2008; Munns, 2002). However, as roots are immediately exposed to salty soils and serve as an important pathway for mineral translocation to aerial portions of the plant, they are primarily responsible for controlling the lower tree rate of growth amid saltwater (Arif et al., 2020). As was already noted, salt stress reduces the ground solution's water propensity and interferes with roots' ability to transport water, which eventually reduces the amount of water and minerals that reach the plant's body (Basu et al., 2020). High levels of Na^+ and Cl^- in soil solutions prevent macronutrients like Ca^{2+} and Mg^{2+} from being absorbed and accumulated, which disrupts signal transduction, membrane stability, chlorophyll production, and stomatal opening (Arif et al., 2020; Geilfus, 2018). Additionally, salinity's detrimental effects on photosynthesis are closely linked to the oxidation of photosynthetic pigments, although the underlying details are still not fully understood. Plants exposed to salt stress have a reduction in stomatal density, conductance, photosynthetic activity, and carbon uptake; however, there is a rise in mesophyll resistance, which lowers the efficiency of photosystems (PS)-I and PS-ability II to absorb light (Gururani et al., 2015). A crucial enzyme involved in assimilating carbon dioxide (CO^2) throughout photosynthesis's overnight period, ribulose-1,5-bisphosphate carboxylase/oxygenase (RuBisCO), is also inhibited by salt stress (Miteva et al., 1992).

Additionally, salt causes some cellular organelles to produce ROS such as the hydroxyl radical (OH), superoxide (O^{2-}), hydrogen peroxide (H^2O^2), and singlet oxygen ($1O^2$) (Zaid and Wani, 2019; Mostofa et al., 2021). Increased amounts of ROS can harm a plant's capacity to grow and develop by reducing protein synthesis, destroying cell membranes, inducing genomic instability, and harming the photosynthetic system (Mostofa et al., 2021; Czarnocka and Karpi ´nski, 2018). Despite being the most prevalent anion in saline soils, Cl^- impacts on plant performance and its absorption and transport pathways inside plants have received less research than those of Na^+(Geilfus, 2018). As a crucial micronutrient, Cl^- is thought to play a part in the cytoplasmic activation of enzymes, oxygen evolution during photosynthesis, stability of membrane potential, and upkeep of the pH and turgor pressure of plant cell walls (Geilfus, 2018). However, if overly accumulated, Cl^- destabilizes the normal activities of the cells even more than Na^+ does (Geilfus, 2018). Crop productivity is reduced by all of the salinity-induced deleterious impacts on plants outlined above, although the severity of the damage varies depending on the kind of plant.

9.4 THE CATEGORIZATION OF HALOPHYTES

The plants known as halophytes are able to survive and complete their entire lifespans in highly salinized settings without experiencing significant detrimental effects on their growth or progress (Flowers and Colmer, 2008). Halophytes may be roughly divided into two groups: facultative halophytes and obligatory halophytes depending on how well they can tolerate various types of salty soils. Obligate halophytes may maintain a sufficient rate of growth and development in highly salinized environments with salinities that are comparable to those of seawater (Flowers and Colmer, 2015). Conversely, facultative halophytes may survive in surroundings free of salt but are typically found in low-saline settings (Parida and Jha, 2010). Furthermore, Von Sengbusch (2003) distinguished between habitat-independent and habitat-dependent halophytes using ecophysiological traits. Although they often prefer to flourish in soils devoid of salt, habitat-independent halophytes may tolerate saline environments and are not necessarily native to salty areas (Von Sengbusch, 2003). Although halophytes were characterized by scientists by taking into account several factors, such as the integration of anatomical findings, salinity intensity, and processes of salt tolerance (Grigore and Toma, 2010; Breckle, 2002), the precise categorization of halophytes is still unclear. Even more significantly, a database was established by Santos et al. (2015) that compiles information on plant species that can endure salt concentrations of 80 mM or more. Hydrohalophytes, Chasmophytes, Phreatophytes, Psammophiles, Xerophytes, Weedy Halophytes, and Xero Halophytes are the seven classifications that the authors used to categorize the halophytes. Phreatophytes are deep-rooted plants that obtain their water from a salt-free or saline deeper subterranean source; psammophiles tend to typically find xerophytes being evolved to extremely dry climates in sandy soils; weedy halophyte taxa are primarily invading and colonizing highly disturbed sites or regions; and chasmophytes are found on clifftops, rocky and sandy seashores. Furthermore, Breckle et al. (1995) divided halophytes into three basic categories: recretohalophytes, euhalophytes, and salt-exclusion halophytes based on the physiological foundation of tolerance toward salt as well as the buildup and movement of ions. Euhalophytes may absorb salts into their succulent leaves or stems, but recretohalophytes have special salt-secreting organs such as salt glands and salt bladders. Contrarily, salt excluders can stop the absorption of salt ions from soils or discarded leaves that contain dangerous quantities of salt (Chen et al., 2018; Lu et al., 2021). With these intricate classifications, it is critical to understand how much salt they can tolerate and what internal defenses they have in place to counteract the impacts of salinity.

9.5 HALOPHYTIC PHYSIOLOGICAL MECHANISMS OF SOIL SALINITY ADAPTATION

A thorough understanding of halophytic plants' processes of salt tolerance and their study for breeding objectives are both essential. Halophytic plants use a variety of morphophysiological techniques, such as salt excretion via specialized organs, salt exclusion such as salt glands or bladders, as well as the diluting of salt ions through succulence, to tolerate and continue to develop healthily in a saline environment. In light of these morphophysiological systems' roles in halophyte resistance to salt stress, we shall evaluate them.

9.5.1 SALT EXCLUSION

It has been claimed that halophytic plants, notably hydrohalophytes and certain phreatophytes, use ultrafiltration processes caused by their root systems to keep out excess salts. Establishing root-level apoplastic shields, in particular, increases bypass flow resistance, which effectively excludes roots' salts and, as a consequence, reduces the buildup of harmful ions through travel and transpiration streams somewhere above shoots (Hussain et al., 2021) (Fig. 9.2). The Casparian band (CB) and the suberin lamella (SL), two separate cell barrier layers that make up the root endodermis and exodermis, each have highly specific functions (Fig. 9.2). All apoplastic transport is compelled into the carefully controlled symplastic system by a buildup of paracellular lignin within the CB cell walls of the endodermis and exodermis (Lee et al., 2013; Naseer et al., 2012). Additionally, SL, which is predominantly composed of long-chain fatty acids, penetrates the whole exodermis and endodermis cell walls and offers a hydrophobic barrier that helps with the control of ion and water intake (Krishnamurthy et al., 2011; Zhang et al., 2021). The direct transfer of water and ions into endodermal protoplasts first from the extracellular medium is assumed to be prevented by SL (Cui et al., 2021; Wang et al., 2020). CT transport of water and ions into the endodermal protoplasts from the apoplast (Cui et al., 2021; Wang et al., 2020). Recent research has shown that SL serves as an apoplastic barrier where lateral root emergence occurs and CBs have not yet formed (Wang et al., 2020). A few studies on halophytes have revealed that *S. maritime* (Hajibagheri et al., 1985), *A. officinalis* (Krishnamurthy et al., 2014), *S. salsa (*Liu et al., 2020), and *A. marina* (Cheng et al., 2020) all require apoplastic barriers for salt adaptation. More study is necessary to determine the involvement of SL in plants during salinity and the ionic barrier because there is less proof that SL contributes to plant salt stress than there is for CBs.

Importantly, the cortical cells' salt overly sensitive 1 (SOS1) protein is crucial for keeping Na^+ out of roots (Fig. 9.2). Liu et al. (2018) recently discovered that *S. salsa* could survive in an environment with high levels of salt because the roots of intertidal S. salsa populations showed stronger expression of S. salsa SOS1 (SsSOS1) than those of inland S. salsa populations. Similar to this, higher Kareliniacaspia SOS1 (KcSOS1) gene expression was noted in the roots of the recretohalophyte *K. caspia* (Guo et al., 2020). In-depth research is still needed to fully understand the processes underpinning root system-induced ultrafiltration in salty environments. Numerous studies have suggested the salt exclusion mechanism as a treatment to lessen hazardous amounts of Na^+ and Cl^- buildup in plant bodies (Roy et al., 2014). However, a compelling case might be made against this process since it could increase soil salinity by limiting root salt intake.

9.5.2 SALT EXCRETION

Through a special epidermal structure known as salt glands, halophytes, especially Recret halophytes, effectively emit noticeably more salt ions over onto the surface of the leaf in contend with the maintenance of cellular ion homeostasis, making them superior to Recret halophytes in that they effectively emit noticeably more salt ions over onto the surface of the leaf and contend (Lu et al., 2021) (Fig. 9.3). The Limonium genus has many species that are excellent examples of recretohalophytes, which increase salt tolerance by enhancing salt production through the salt glands (Leng et al., 2018; Li et al., 2020; Mi et al., 2021). These species include some species of *Laccaria.* By increasing the number of salt glands and the rate of salt production per salt gland, species of the recretohalophyte genus Tamarix are salt-tolerant (Wei et al., 2020). However, many halophyte species' salt glands exhibit a variety of structural properties, and salt glands are composed of various numbers of cells (Lu et al., 2021). Nevertheless, they share the following traits: (1) thicker cuticles that surround the salt glands; (2) many plasmodesmata between cells; (3) numerous highly developed mitochondria; and (4) numerous tiny vesicles in the cytoplasm (Lu et al., 2021; Yuan et al., 2016). Dicotyledonous recretohalophytes, such as L. bicolor, have multicellular salt glands with anywhere between 4 and 40 cells (Leng et al., 2018), while the Poaceae family is more likely to have bicellular salt glands (Céccoli et al., 2015). Depending mostly on the environment, salt glands release a variety of cations (Na^+, K^+, Ca^{2+}, Mg^{2+}, Fe^{2+}, Mn^{2+}, and Zn^{2+}) and anions (Cl^-, Br^-, I^-, SO^{42-}, PO^{43-}, and NO^{3-}) (Lu et al., 2021). Although numerous potential mechanisms have been proposed based on recent investigations, the precise processes via which salt is secreted from salt glands are still unknown. Although Yuan et al. (2016) and Lu et al. (2021) have proposed numerous potential mechanisms based on recent investigations, the precise processes via which salt is secreted from salt glands are yet unknown. However, it was also shown that salt glands may pose harmful hazards to plants by leading to leaf dehydration because the glands create granules of salt on the exterior of the leaf (Shabala et al., 2014). Synthesis of suitable solutes from inception can reduce this issue, but it uses a lot of energy between 30 and 109 ATP molecules to create a single suitable solute molecule, which reduces crop production (Shabala et al., 2014).

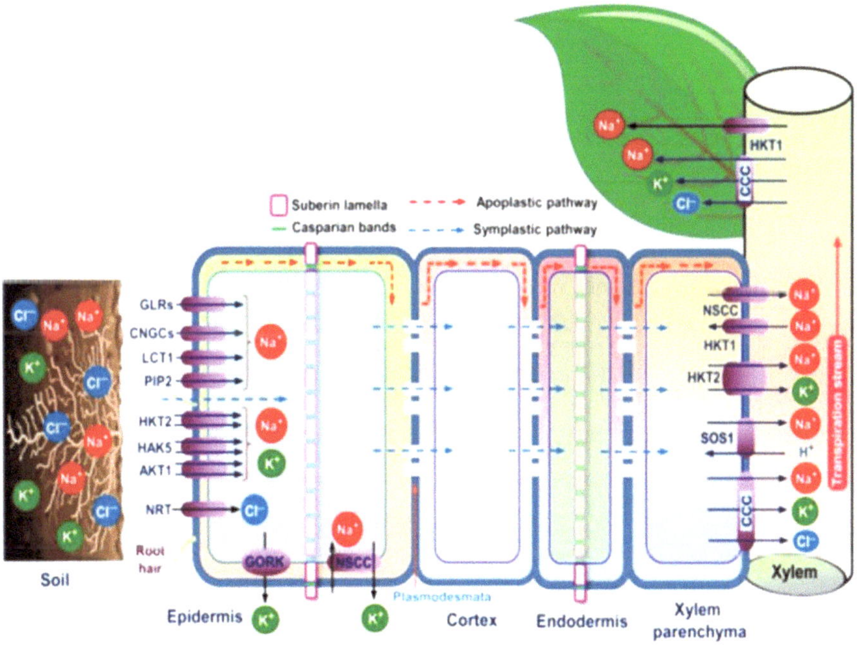

FIGURE 9.2 Potential pathways for salt absorption and buildup in a halophyte plant in its optimal state.

More than half of the halophytes include epidermal bladder cells (EBCs), commonly referred to as salt bladders (Shabala et al., 2014; Kiani-Pouya et al., 2017). The apical cell of the EBC develops a globular form, normally having a 500 nL cell capacity as well as a 1 mm wall thickness., and the EBC has a modified nonglandular trichome structure (Shabala et al., 2014; Kiani-Pouya et al., 2017; Barkla et al., 2018) (Fig. 9.3). In order to sequester excessive salts, the width of EBCs frequently expands by up to 1000 times more than the capacity of conventional leaf cell vacuoles (Shabala et al., 2014).

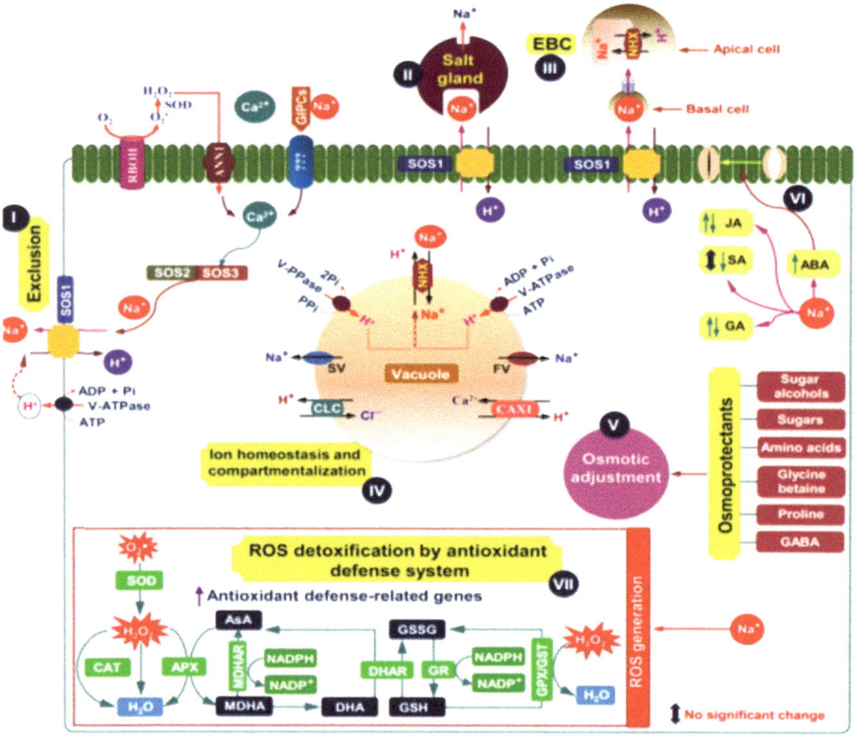

FIGURE 9.3 Potential pathways for a perfect halophyte plant to tolerate salt.

Fundamentally, EBCs have long been thought to be essential for preserving low Na^+ concentrations in leaves, particularly in young leaves (Bonales-Alatorre et al., 2013). Younger leaves primarily rely on intracellular salt sequestration using EBCs since the mesophyll cells in those leaves have tiny and undeveloped vacuoles (Bonales-Alatorre et al., 2013). EBCs may store other metabolic substances, such as ROS-scavenging metabolites and organic osmoprotectants, in addition to acting as a potential hotspot for the storage of too much Na^+ and Cl (Otterbach et al., 2021). In Chenopodium quinoa, it was shown that there was a positive link between the existence and/or a number of EBCs and salt tolerance (Barkla et al., 2018; Otterbach et al., 2021). Using transcriptome analysis, certain candidate genes for EBC formation have recently been found in the species of *M. crystallinum*, *A. centralasiatica*, and *C. quinoa* (Roeurn et al., 2017; Böhm et al., 2018; Yao et al., 2019). More importantly, Imamura et al. (2020) laid the groundwork for understanding underlying halophytes' EBC-associated stress response pathway by identifying a gene called reduced epidermal bladder cells (REBCs), which is implicated in the establishment of EBC in *C. quinoa*.

9.5.3 SUCCULENCE

To sustain cell turgor pressure in salty circumstances, halophytes also use pathways for ion homeostasis as well as succulence in the accumulation of osmoprotectants (Lokhande and Suprasanna, 2012; Nikalje et al., 2017). Halosucculents, or succulent plants, are known for their thick leaves and stems, larger mesophyll cells, a smaller intercellular gap, greater increased turgor tension, as well as tissue moisture content (Lokhande and Suprasanna, 2012). More stretched-out mitochondria are seen in succulent leaves, which are thought to supply the extra energy needed for salt compartmentalization and sequestration (Nikalje et al., 2017). Salsoladrummondii, Achrassapota, and Sarcocorniafruticosa are just a few examples of the many halophytes that have been found to have effective succulent mechanisms for conferring salt tolerance through the accumulation of higher photosynthetic ability despite the presence of increased salts within leaf and stems (García-Caparrós et al., 2017; Elnaggar et al., 2020; Rahman et al., 2018).

9.6 BIOCHEMICAL MECHANISMS RELATED TO HALOPHYTE SALINITY ADAPTATION

Halophytes and other plants utilize a variety of biochemical methods to respond to the negative effects of salt. For instance, among the crucial biochemical strategies utilized by plants to safeguard their cellular components from the harmful effects of excessive salt are ion homeostasis, osmotic adjustment, and ROS detoxification (Himabindu et al., 2016) (Fig. 9.2). Here, we provide a summary of the three biochemical processes involved in halophyte responses and salt stress tolerance.

9.6.1 ION HOMEOSTASIS

The tissues of halophytic plants actively accommodate Na^+, Cl^-, and K^+ through a variety of channels and transporters. It is widely known that the apoplastic and symplastic pathways serve as the primary routes by which Na^+ is taken up by roots from soils (Arif et al., 2020). Transporters and channels are required for the symplastic route to move salt ions. In general, high-affinity K^+ transporters (like HKT2 and HAK5), nonselective NSCCs, CNGCs, GLRs, and aquaporins (PIP) isoforms remain primarily in charge of root-Na^+ uptake (Fig. 9.2). LCT1, besides Arabidopsis K^+ transporter 1 (AKT1), can also facilitate the inflow of Na^+ (Fig. 9.2) (Arif et al., 2020). Nitrate transporters, Cl/H^+ cotransporters, and H^+/Cl symporters all play a role in the inward migration of Cl^- in root cells (NRTs). Slow anion channels (SLAC channels), among others, participate in the passive transfer of Cl^- at high salinity (Wu and Li, 2019). In the xylem and shoot tissues, cation chloride cotransporters (CCCs) aid in Cl^- inflow (Li et al., 2017). The tonoplast's chloride channels (CLCs) are in charge of sequestering Cl (Tran et al., 2019) (Fig. 9.3). K^+ is competitively taken up by HKT2, HAK5, as well as AKT1 about Na^+ (Zhao et al., 2020).

Halophytes primarily utilize two techniques to stop the buildup of extra salt in their cytosol: (1) preventing salt from entering the roots and (2) sequestering salt ions in vacuoles (Himabindu et al., 2016) (Fig. 9.3). These systems, however, are not mutually exclusive, and a specific plant may use several tactics in different situations. For instance, in the cytosol, homeostasis is indeed a special approach that plants use to efficiently compartmentalize salt ions and limit their accumulation, which is crucial for plant survival under salinity-induced hostile circumstances. For the roots to exclude excess Na^+, for Na^+ to be loaded into the xylem, and for excess Na^+ to be sequestered into the vacuoles utilizing Na^+/H^+ exchangers (NHXs), the SOS system is essential (Zhao et al., 2020; Saddhe et al., 2020) (Fig. 9.3). SOS3/SOS3-like calcium-binding proteins (SCaBPs), SOS2 (a serine/threonine-protein kinase), and SOS1 (a plasma membrane Na^+/H^+ antiporter) make up the majority of the SOS system (Zhao et al., 2020; Himabindu et al., 2016) (Fig. 9.3). It is significant to note that SOS4 (pyridoxal kinase) and SOS5 have been most recently identified; SOS4 is involved in B6 vitamin, which is crucial for maintaining Na^+/K^+ homeostasis in plants, and SOS5 is a hypothesized protein that acts as a cell surface adhesive necessary for regular cell expansion (Saddhe et al., 2020). Typically, SOS3 transmits salt-triggered Ca^{2+} signals to SOS2 kinase, leading to the SOS3–SOS2 protein complex to assemble, which phosphorylates SOS1 and causes Na^+ resistance

(Zhao et al., 2020; Nikalje et al., 2017) (Fig. 9.3). The obligatory halophyte Cakile maritime (Arbelet-Bonnin et al., 2018), the showy halophyte *L. maritime* (Ben Hsouna et al., 2020), as well as the halophyte *K. caspia* (Guo et al., 2020) have all shown increased expression of the SOS1 gene and its relationship to an improvement in salt tolerance. Similar findings exist for the phreatophyte Pongamia pinnata and the obligatory halophyte C. maritima species' high salinity tolerance and concurrent manifestation of SOS1, SOS2, and SOS3 (Arbelet-Bonnin et al., 2018; Marriboina et al., 2017).

Additionally, NHXs aid in regulating vacuolar Na^+ inflow and preserving ideal intracellular Na^+/K^+ equilibrium (Himabindu et al., 2016) (Fig. 9.3). For example, in saline circumstances, the hardwood halophyte *N.sibirica*, which is perennial, may maintain an ideal K^+/Na^+ balance by increasing the expression of NsVHA, NsVP1, and NsNHX1 somewhere at the tissue and cell stages and increasing the activities of vacuolar H^+ -ATPase and H^+ -PPase (Tang et al., 2020). In the meantime, *N. sibirica* enhanced the transcriptional production of two-pore K^+ (NsTPK), facilitating K^+ efflux from vacuoles and mediating cytosolic transport K^+ balance (Tang et al., 2020). Additionally, Tran et al. (2019) found that M. crystallinum revealed that salt increased the expression of genes encoding tonoplast antiporters H^+/Cl^- (i.e., McCLC1) and Na^+/H^+ McNHX1 exchangers involving McVmac1 V-ATPase subunits c for such sequester of Cl^- and Na^+ into the vacuoles to confer salt.

9.6.2 OSMOTIC ADJUSTMENT

Two methods are used by plants to achieve osmotic adjustment: (1) enhanced absorption for maintenance of ionic species and (2) de novo production of organic osmolytes organellar volume and turgor pressure within developing plant cells (Zhao et al., 2020). In order to provide osmotic modification in cells in an osmotic area affected by salt, it is fundamentally important for Na^+ and Cl^- to be in vacuoles, as well as K^+ and suitable organic solutes like proline and carbohydrates to be in the cytoplasm (Fig. 9.3). Guard cell outward rectifying K^+ (GORK) channels and/or ROS-activated NSCCs are two key kinds of channels that are often involved in salinity-induced K^+ efflux from the cell (Zhao et al., 2020) (Fig. 9.2). Although a thorough study into the molecular underpinnings regulating osmotic equilibrium in halophytes that is regulated by GORK channels remains difficult, GORK channels play a critical role in stress-induced K^+ loss from the cytosol. However, Kiani-Pouya et al. (2020) recently found that roots' CqGORK transcript expression remained relatively stable (constant) or increased in salt-sensitive accessions while decreasing in salt-tolerant accessions, suggesting that GORK plays a role in K^+ retention for improved osmotic adjustment in the cells of salt-sensitive C. quina. Additionally, it has been noted that Eutremaparvula and Hordeum brevisubulatum exhibited increased K^+ uptake as a result of the ion selectivity strategy of HAKs/HKTs, which enhanced K^+/Na^+ balance(Zhang, 2020; Ali et al., 2018). Future research should focus on understanding how K^+ transporters' transportation of K^+ between intracellular and intercellular membranes affects how the osmotic environment in salt-exposed halophytic plants is adjusted.

9.6.3 ROS DETOXIFICATION

For the purpose of protecting themselves from oxidative stress, which is really brought on by an excessive buildup of ROS under high salinity, plants rely on effective antioxidant defense systems (Castro et al., 2021). Plants have a strong antioxidant defense system that includes enzymatic anti-oxidants like SOD, CAT, APX, GR, MDHAR, DHAR, and GPX, as well as tocopherols and nonenzymatic radicals such as tocopherols, AsA, GSH, and other flavonoids to GST (Czarnocka et al., 2018) (Fig. 9.3). Appropriate antioxidant pathway activity during salt stress allows different halophytes to maintain ROS homeostasis, which contributes to their increased salt tolerance (Altuntas and Terzi, 2021; Yıldız and Terzi, 2021; Ben Hsouna et al., 2020; Wang et al., 2020; Panda et al., 2019; Bose et al., 2013).

For instance, in the H. salicornicum plant, increased levels of GSH, SOD activity (DHA), and GSSG ratios are crucial for scavenging ROS, which helps the plant endure oxidative stress and improves its ability to tolerate salt (Panda et al., 2019).

Halophytic plants alter the levels of endogenous phytohormones in response to salt stress in order to control a number of defensive mechanisms, including antioxidant defense (Wiszniewska et al., 2021; Llanes et al., 2018). In-depth reviews of the precise functions that phytohormones play in halophyte responses to abiotic stressors may be found elsewhere (Llanes et al., 2018). Phytohormone accumulation in halophytes varies according to the kinds of halophytes that are subjected to salt stress. For instance, during salt stress, S. persica jasmonic acid (JA) levels rose while a fall in Aster tripolium and H. salicornicum levels was recorded (Fig. 9.3). Salicylic acid levels in *S. persica* (Kumari et al., 2018) and H. californium (Panda et al., 2020) did not change considerably; however, they dramatically decreased in A. tripolium (Wiszniewska et al., 2021) (Fig. 9.3). Gibberellic acid (GA) levels significantly increased in S. salsa (Guo et al., 2020) but they were reduced in H. salicornicum (Panda et al., 2020) (Fig. 9.3). The amount of JA in plants is related to the quantity of ROS and the rate of ROS-induced lipid peroxidation (Reinbothe et al., 2009), which is created by linolenic acid oxidation (Gupta et al., 2020), In fact, during salt stress, a reduced concentration of JA was favorably linked with a lower amount of ROS (Panda et al., 2020). However, Lycium humile, S. persica, H. salicornicum, and A. tripolium dramatically raised their levels of abscisic acid (ABA), a crucial stomatal closure enzyme that lowers transpiration, in response to salt stress (Panda et al., 2020; Wiszniewska et al., 2021; Palchetti et al., 2021; Kumari et al., 2018). According to Panda et al. (2020), increased ABA accumulation assisted in lowering oxidative stress in H. salicornicum that had been exposed to salt. In addition to managing plant growth and stress tolerance, the production of ROS and subsequent redox processing are essential components of hormone regulation (Xia et al., 2015). Future research should thus concentrate on the discovery of proteins that might monitor ROS changes and the spatial–temporal control of ROS formation. The interaction between several hormone signaling mechanisms in halophytes and the associated phenotype of stress response that is hormone-modulated may be better understood with the use of such information.

9.7 PHYTOREMEDIATION OF SALINITY-AFFECTED SOILS THROUGH HALOPHYTES

Leaching and the addition of chemicals and organic amendments are often used as techniques to restore salt-affected regions (Srivastava, 2020). Leaching makes it easier for salt ions to travel from the shallow deeper horizon in the distance. Moreover, its use is constrained in salt-contaminated locations due to the absence of water during leaching and its impacts on reduced soil stability, total organic carbon, as well as total nitrogen levels, topsoil fertility, and microbial activity (Jesus et al., 2015). The use of calcium carbonate ($CaCO^3$) or other chemical alternatives as part of a chemical method requires an ion exchange process; however, due to their poor solubility, poor quality, and high cost, chemical approaches to the reclamation of salty soils are currently limited (Jesus et al., 2015; Qadir et al., 2008). It is interesting to note that using the right amounts of each, not any more than 50 T of organic amendment, can effectively boost the solubility of $CaCO^3$ while also enhancing the soil's physical, chemical, and physical–chemical characteristics, which support microbial activity and plant development. The danger of releasing numerous substances, such as toxic metals and unclassified organic waste, however, limits the advantages (Leogrande and Vitti, 2018; Ouni et al., 2013). The farmers have been compelled by all of these limitations to search for cost-efficient, environmentally beneficial, and visually appealing alternatives. A viable alternative to existing remediation strategies is phytoremediation, also known as environmental restoration or vegetal biotransformation. This method includes planting certain salt-tolerant plants to eliminate soil ions that contain salt and recover soils impacted by salinity (Luo et al., 2018; Jesus et al., 2015; Wang et al., 2020). Its acceptability is growing steadily as a result of several advantages. In contrast to gypsum, phytoremediation (1) removes excess salts more uniformly, even from greater depths; (2) improves plant access to nutrients in the soil, which increases agricultural yield; and (3) is environmentally friendly because it sequesters neither loads of different organics nor below-ground organics that contain carbon (Santos et al., 2015). Even so, knowledge of mechanistic insights into the use of halophytes in the phytoremediation of salty soils remains elusive. The whole and holistic modes of action of plants in the rehabilitation of salty soil, as well as their advantageous impacts on soil parameters, are difficult to envision due to complex interactions. To the best of our knowledge, Figure 9.4 depicts an underlying phytoremediation process of halophytes that is well-known and acknowledged. Due to its ability to absorb salt, S. salsa has lately been utilized to restore salty land in northwest China that varies between 3.75 and 3.91 T ha^{-1} $year^{-1}$(Wang et al., 2020). It is significant to remember that with a population density of 15 plants/m^2, S. salsa may hypothetically remove 3.0–3.8 T Na^+ from the environment (Liang and Shi, 2020). The aboveground biomass of S. salsa was also reported by Liang and Shi (2020). Salsa plants gathered around 27 mg of salt per g of dry weight. The shoots of Tecticorniapergranulata were shown by Shaygan et al. (2018) to be able to acquire 98 g of salt Kg^{-1} dry weight.

To reduce the danger of car accidents and injuries, sodium chloride is frequently used to reduce water's core temperature on Northern American soil roadways (Mann et al., 2020). The identical study (Mann et al., 2020) revealed that if the salt application was stopped, it would take A. patula six years to completely remove all of the salt from a region that had 1540 g of Cl^- per m^2. However, given that salt is always used in the winter, A. patula might be an option to handle salt management quantity built up over time in the groundwater and soil beside the roadway.

Second, according to some research, halophyte planting might lower salt ion buildup in soils by enhancing the physical characteristics of the soil. In a nutshell, halophytic plants with deep roots may be used as tillage tools. This process, commonly known as biological drilling, has been shown to increase soil porosity and hydraulic properties, which enhance Na^+ leaching and can substitute other cations in its place while also increasing nutrient-use efficiency (Arevalo-Gardini et al., 2015; Silva et al., 2016). It also has the ability to reduce soil bulk density and stabilize soil quality (Fig. 9.4). As a result, as compared to a conventional monocropping system, intercropping S. salsa with cotton (Gossypium hirsutum) dramatically reduced soil salinity and bulk density while boosting soil porosity, soil organic carbon, root development, total aboveground biomass, and cotton production (Liang and Shi, 2020). Additionally, A. nummularia demonstrated improved hydraulic conductivity and a decrease in bulk density, which improved macropores and increased water penetration rates under a salinity level of 40.86 dS m^{-1}(Silva et al., 2016). Halophyte planting decreased bulk density while increasing soil porosity and hydraulic conductivity according to Ashraf et al. (2010), and this subsequently enhanced salt ion loss from salty, alkaline soil.

Tertiary, since they can raise the CO^2 concentration in the soil environment, root respiration and the breakdown of organic matter are crucial in lowering the Na^+ level in saline-sodic soils (Qadir et al., 2005). In essence, soil water that is made available by soil moisture and irrigation water reacts with CO^2 to create carbonic acid (H^2CO^3) (Fig. 9.4). In halophyte roots, H^2CO^3 dissociation and N^2-fixation result in the release of a proton (H^+), which causes calcite ($CaCO^3$) to dissolve and release Ca^{2+}, H^2O, and CO^2 (Qadir et al., 2005) (Fig. 9.4). The removal of Na^+ from the cation exchange sites of the soil colloid may be made easier by the released Ca^{2+} in the soil (Qadir et al., 2005). The exchanged Na^+ content is subsequently removed from soils by two different processes: (1) absorption through the roots of halophytic plants and (2) leaching caused by irrigation being flooded before planting, which eventually aids in the decrease of salinity as well as sodicity of the soil (Fig. 9.4). However, in noncalcareous soils, increasing CO^2 causes the pH to decrease, which improves $CaCO^3$ dissolution and increases the amount of Ca^{2+} that is available for ion exchange with Na^+ (Qadir et al., 2005). Two facultative halophytes were sown by S. verrucosum, and Bacopa monnieri improved the soil between 54.71% and 57.23% porosity while lowering the electrical conductivity (EC) between 11.13 dS m^{-1} and 7.97 dS m^{-1} and lowering the pH range between 7.84 and 7.42 over the course of 240 days. The combined planting overall phytodesalination ability among these 2 halophytes was 1.21 $T.Na^+$ ha^{-1}, which made the soil more conducive to the growth of Zea mays and increased yielding between 7.2 Tha^{-1} and 8.5 Tha^{-1} (Lastiri-Hernández et al., 2019). Salt marsh halophyte S. monoica has been demonstrated to reduce soil EC values between 4.75 dSm^{-1} and 2.10 dSm^{-1} and pH values between 8.40 and 6.73 (Malik and Ravindran, 2020).

Halophytes can be used to desalinate water, but this approach is not always reliable because it is susceptible to environmental variations like temperature and rainfall, cultivation methods like irrigation and nonirrigation, and the possibility of salt particles being recycled through plant parts (Shabala, 2013). Prior to their use in the desalinization of salt-affected areas, thorough evaluations of the potential of various halophytes by examining their mode of action are thus required. Further study will be needed to confirm the effectiveness of the plant-uptake-mediated phytoremediation process, and if it is shown to be substantial, it will provide a new pathway for repairing salinity-affected regions, particularly noncalcareous soils and soils under nonleaching conditions.

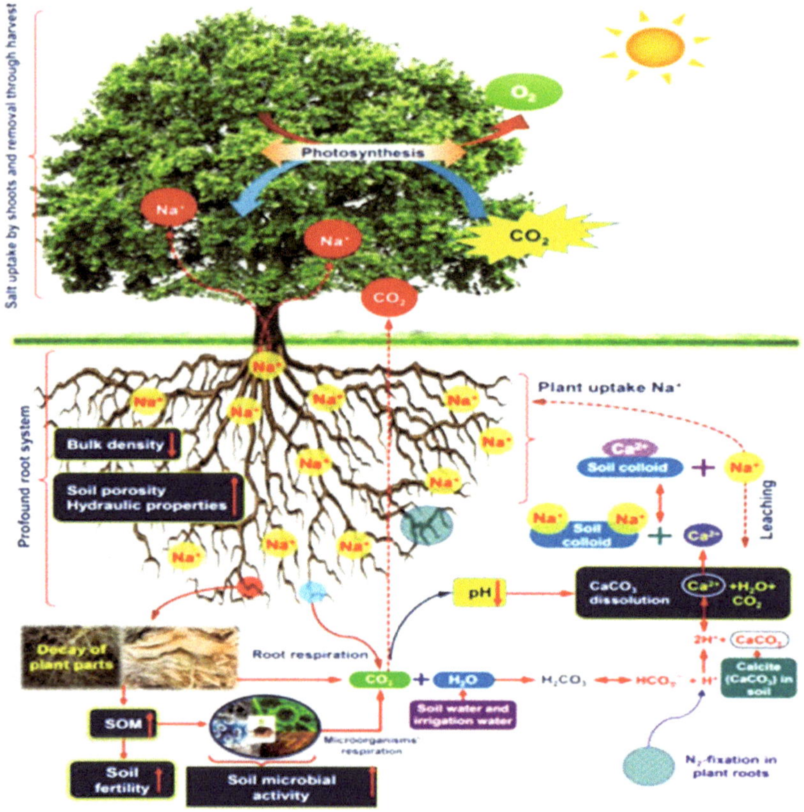

FIGURE 9.4 Halophyte plant potential for phytoremediation and the advantageous impacts it has on the soil's characteristics in places impacted by salt.

9.8 POSSIBILITY OF USING HALOPHYTES AS GENETIC RESOURCES TO CREATE CROPS WITH IMPROVED SALT TOLERANCE

Halophytes can withstand ambient salinity better than glycophytes, which may be a result of good gene regulation that activates their salt tolerance mechanisms (Himabindu et al., 2016; Barros et al., 2021). Increasingly, high-throughput transcriptome sequencing is being utilized to analyze gene expression in order to comprehend how plants react to abiotic stressors (Ye et al., 2020). To identify the underlying molecular reactions to severe salt stress, Guo et al. (2021) recently performed a comparative transcriptome study of the model halophyte Puccinelliatenuiflora and numerous glycophytic Gramineae plants. They discovered that enhanced development of P. tenuiflora in high salinity conditions may be related to K^+ absorption, sucrose formation, and the pentose phosphate, including flavonoid biosynthetic pathway gene relatives, all of which showed increased expression. Aliakbari et al. (2020) found 1595 differentially expressed genes (DEGs) in Salicornia persica between control and salt-treated plant shoots. These DEGs included transcription factor, transporter, and protein kinase-encoding genes. Improved energy homeostasis and the production of primary metabolites may be crucial for S. persica's ability to tolerate salt according to functional annotation studies. Using gene network analysis, it was discovered that Ca^+, ABA, and Na^+ are essential for developing salt tolerance; compartmentalization is essential for S. persica; and 327 expression level genes as well as 399 enhanced genes were found by Han et al. (2020). S. rigida plants react to various soil salinity regimes. The downregulated genes were thought to be involved in the transport of organic substances, nitrogen compounds, and intracellular proteins according to a gene ontology (GO) enrichment study. On the other hand, the elevated genes were related to cell wall biogenesis, including extracellular matrix organization, cell wall synthesis, and cell wall assembly (Han et al., 2020).

Additionally, under salt stress, RNA-seq discovered that *A. central-asiatica* leaves have 9144 DEGs, including 5325 dysregulated and 3819 activated genes (Yao et al., 2019). Salt stress significantly elevated the expression of genes linked to ion transport, the *ABA-dependent signaling system, ROS scavenging*, and the HB-7 and MYB78 transcriptional regulatory mechanisms according to GO enrichment analysis (Yao et al., 2019). Gene-generating proteins that operate as transporters, cellular membrane proteins, antioxidant, transcriptional, mitogen-activated protein kinases (MAPKs), and photosynthesis enzymatic enzymes, including enzymes for the synthesis of compatible solutes by salinity, were found to be upregulated in the halophyte U. setulosa by de novo transcriptome profiling (Mann et al., 2021). In reaction to 300 mM NaCl, Wang et al. (2020) found a large number of DEGs; precisely, 3800 genes were negatively regulated and 4903 genes were increased in Z. macrostachya. By modifying ion homeostasis and plant hormonal signaling pathways through salt secretion, osmoregulatory substance accumulation, and the action of an antioxidant to avoid oxidative injury-effective ROS-scavenging systems, such changes in gene expression in a large number of genes may help improve plant growth performance. Halophytes have the genetic capacity to tolerate salt; therefore, they might be a crucial genetic resource for creating transgenic plants that are more salt-tolerant. In reality, by introducing genes, several transgenic plants have been created to increase salt tolerance from halophytic plants. The majority of these genes produce antiporter proteins essential for ion-trap plants' functional stability and signal transduction (Xiang and Man, 2018; Liu et al., 2017). A transgenic poplar plant, for instance, that ectopically expressed the *N. sibirica*vacuolar Na^+/H^+ antiporter gene (NsNHX1) showed a more notable improvement in salt tolerance through improved Na^+ compartmentalization, more effective photosynthesis, increased activity of antioxidant enzymes, and improved osmotic adjustment (Geng et al., 2020). By increasing K^+ uptake capacity, K^+/Na^+ homeostasis, and the enhanced activity of antioxidant enzymes toward scavenging free radicals like SOD, POD, and CAT, ectopic overexpression of a high-affinity K^+ cotton's salt tolerance was increased by the *S. bigelovii* transporter-encoding gene SbHKT1 in transgenic plants (Guo et al., 2020). In contrast, the *Z.xanthoxylum* nitrate transporter 1-encoding gene ZxNRT1.5 was ectopically expressed in transgenic Arabidopsis, which improved NO^{3-} uptake and root-to-shoot NO^3 translocation. This, in turn, increased the transgenic plant's ability to synthesize amino acids, which aided in a salt-induced osmotic adjustment. The thorough molecular processes behind ZxNRT1.5's role in controlling Na^+, K^+, and Cl transport must be uncovered by more research. Additionally, transgenic plants exposed to salt stress exhibited altered transcription of genes encoding important proline metabolism-related enzymes and K^+ transportation when the NtCIPK11 gene from *N. tangutorum* was overexpressed in Arabidopsis (Lu et al., 2020; Lu et al., 2021). These changes included improved seed germination and longer roots. Several unique genes from S. brachiata are being expressed ectopically in transgenic tobacco plants. Brachiata species like S. protein associated with stress in brachiata (SbSRP), S. 1 (SbSI-1), S. brachiata salt inducible, S. myeloblastomabrachiata 15 (SbMYB15),

SbDREB2A, a protein that binds to a S. brachiata dehydration-responsive element (SbRPC5L) from the brachiata, and AlRab7 (AlRabring7) from the Aeluropuslagopoides both showed greater levels of suitable solutes and phytohormones (such as indole acetic acid) (for instance, proline), enhanced gas exchange capabilities, showed less membrane damage, greater antioxidant capability, lower harmful ion accumulations, and eventually better biomass and transgenic plants' responses to salt stress in terms of growth (Udawat et al., 2017; Agarwal et al., 2021). Additionally, transgenic plants' ability to tolerate salt was improved by the ectopic expression of L. maritima vacuolar H^+-ATPase subunit E1 (LmVHA-E1), Tamarixhispida dehydration-responsive element-binding (ThDREB), PutNHX1 or PutAPX from P. tenuiflora, or SeNHX1 from S. europaea in Arabidopsis plants (Liu et al., 2017; Dabbous et al., 2017; Yang et al., 2017). Additionally, transgenic tobacco that carried the Iris lactea tonoplast Na^+/H^+ antiporter gene IlNHX demonstrated an increase in the compartmentalization of Na^+ into vacuoles by improved vacuolar V-ATPase activity, which supported the preservation of K^+/Na^+ balance (Guo et al., 2020). Halophytes with inherent and effective capacities for salt tolerance can serve as models for enhancing plant and/or crop responses to salt stress. Furthermore, it is crucial to emphasize the molecular distinctions between thes;e two plant families because crops are mostly glycophytes. These discoveries may hold the key to directing and optimizing glycophyte transformation utilizing halophytic candidate genes, opening up a new field of investigation into how to get crops ready for high soil salinity. Despite its significant implications, there are currently little known facts about the discovery, metabolic processes, and isolation of key genes from halophytes that are connected to salt tolerance. In order to sustainably enhance crop output for attaining global food security in the context of climate change and defense, unique salt tolerance-associated genes from halophytes must be identified and characterized, and their future use in genetic engineering is an important field of plant study.

9.9 CONCLUSIONS AND FUTURE PERSPECTIVES

Crop production is hampered by salt-induced land degradation, endangering global food security and a sustainable food supply. The need to grow plant species that can survive ambient salt is pressing due to climate-driven changes and anthropogenic activities that have exacerbated salinity issues. Halophytes are a broad collection of plants that exhibit exceptional cell- and whole-plant adaptabilities to high salinities. As a result, they have a great deal of potential to overcome this cultivation difficulty and be effectively included in future agricultural systems. Through the adoption of inherent salt tolerance mechanisms, a number of halophytes are able to develop naturally in ambient salinity. Although halophytic products have only just begun to be commercialized, these plants have the potential to be powerful economic substitutes for traditional crops. The growth of halophytic plants in salinity-affected locations might also lessen soil salinization. So, the most effective and practical method to restore salinity-affected fields and promote crop production in salt-prone locations may be this phytoremediation technology. Despite having enormous potential for managing soil salinity, phytoremediation is constrained by the lengthy procedure involved and the need to screen out appropriate species. Because they can create more biomass for eliminating salt, halophytic plants with numerous salt tolerance mechanisms should be the focus of future studies on most salts possible from polluted soils. Furthermore, certain fundamental physiological, biochemical, and molecular pathways remain unknown despite significant scientific progress on halophytes (during the past 100 years). Therefore, much research is needed to improve salt tolerance in halophytes. It is imperative to understand the underlying processes that conserve dirt. Recent genetic and genomic research has also shown that salt tolerance is a complex feature that depends on entire plant responses as opposed to the activity of a single gene or protein. Therefore, in order to promote sustainable agriculture in salinity-affected areas and lessen the effects of global climate change, it is essential to explore the salt tolerance capability of diverse halophytes and to produce transgenic plants with better remediation properties.

KEYWORDS

- **halophytes**
- **coastal areas**
- **transgenic plants**
- **salt tolerance mechanisms**
- **soil salinity**
- **phytoremediation**

REFERENCES

Agarwal, P.; Baraiya, B. M.; Joshi, P. S.; Patel, M.; Parida, A. K.; Agarwal, P. K. AlRab7 from Aeluropus lagopoides ameliorates Ion Toxicity in Transgenic Tobacco by Regulating Hormone Signaling and Reactive Oxygen Species Homeostasis. *Physiol. Plant.* **2021**.

Alencar, N. L. M.; Gadelha, C. G.; Gallão, M. I.; Dolder, M. A. H.; Prisco, J. T.; Gomes-Filho, E. Ultrastructural and Biochemical Changes Induced by Salt Stress in Jatropha curcas Seeds during Germination and Seedling Development. *Funct. Plant Biol.* **2015,** *42*, 865–874.

Ali, A.; Khan, I. U.; Jan, M.; Khan, H. A.; Hussain, S.; Nisar, M.; Chung, W. S.; Yun, D. J. The High-Affinity Potassium Transporter EpHKT1;2 from the Extremophile Eutrema Parvula Mediates Salt Tolerance. *Front. Plant Sci.* **2018,** *9*, 1108.

Aliakbari, M.; Razi, H.; Alemzadeh, A.; Tavakol, E. RNA-seq Transcriptome Profiling of the Halophyte Salicornia Persica in Response to Salinity. *J. Plant Growth Regul.* **2020,** *40*, 707–721.

Altunta¸s, C.; Terzi, R. Concomitant Accumulations of Ions, Osmoprotectants and Antioxidant System-Related Substances Provide Salt Tolerance Capability to Succulent Extreme-Halophyte Scorzonera hieraciifolia. *Authorea Prepr.* **2021**.

Arbelet-Bonnin, D.; Ben Hamed-Laouti, I.; Laurenti, P.; Abdelly, C.; Ben Hamed, K.; Bouteau, F. Cellular Mechanisms to Survive Salt in the Halophyte Cakile maritima. *Plant Sci.* **2018,** *272*, 173–178.

Arbelet-Bonnin, D.; Ben-Hamed-Louati, I.; Laurenti, P.; Abdelly, C.; Ben-Hamed, K.; Bouteau, F. Cakile Maritima, A Promising Model for Halophyte Studies and a Putative Cash Crop for Saline Agriculture. *Adv. Agron.* **2019,** *155*, 45–78.

Arbelet-Bonnin, D.; Blasselle, C.; Palm, E. R.; Redwan, M.; Ponnaiah, M.; Laurenti, P.; Meimoun, P.; Gilard, F.; Gakière, B.; Mancuso, S.; et al. Metabolism Regulation during Salt Exposure in the Halophyte Cakile maritima. *Environ. Exp. Bot.* **2020,** *177*, 104075.

Arevalo-Gardini, E.; Canto, M.; Alegre, J.; Loli, O.; Julca, A.; Baligar, V. Changes in Soil Physical and Chemical Properties in Long Term Improved Natural and Traditional Agroforestry Management Systems of Cacao Genotypes in Peruvian Amazon. *PLoS One* **2015,** *10*, e0132147.

Arif, Y.; Singh, P.; Siddiqui, H.; Bajguz, A.; Hayat, S. Salinity Induced Physiological and Biochemical Changes in Plants: An Omic Approach Towards Salt Stress Tolerance. *Plant Physiol. Biochem.* **2020,** *156*, 64–77.

Ashraf, M. Y.; Mahmood, K.; Akhter, J.; Hussain, F.; Arshad, M. Phytoremediation of Saline Soils for Sustainable Agricultural Productivity. In *Plant Adaptation and Phytoremediation*; Ashraf, M., Ozturk, M., Ahmad, M. S. A., Eds.; Springer: Dordrecht, The Netherlands, 2010; pp 335–355.

Barkla, B. J.; Rhodes, T.; Tran, K. N. T.; Wijesinghege, C.; Larkin, J. C.; Dassanayake, M. Making Epidermal Bladder Cells Bigger: Developmental- and Salinity-Induced Endopolyploidy in a Model Halophyte. *Plant Physiol.* **2018,** *177*, 615–632.

Barros, N. L. F.; Marques, D. N.; Tadaiesky, L. B. A.; de Souza, C. R. B. Halophytes and Other Molecular Strategies for the Generation of Salt-Tolerant Crops. *Plant Physiol. Biochem.* **2021,** *162*, 581–591.

Basu, S.; Kumar, A.; Benazir, I.; Kumar, G. Reassessing the Role of Ion Homeostasis for Improving Salinity Tolerance in Crop Plants. *Physiol. Plant.* **2020,** *171*, 502–519.

Ben Hsouna, A.; Ghneim-Herrera, T.; Ben Romdhane, W.; Dabbous, A.; Ben Saad, R.; Brini, F.; Abdelly, C.; Ben Hamed, K. Early Effects of Salt Stress on the Physiological

and Oxidative Status of the Halophyte Lobularia maritima. *Funct. Plant Biol.* **2020,** *47*, 912–924.
Böhm, J.; Messerer, M.; Müller, H. M.; Scholz-Starke, J.; Gradogna, A.; Scherzer, S.; Maierhofer, T.; Bazihizina, N.; Zhang, H.; Stigloher, C.; et al. Understanding the Molecular Basis of Salt Sequestration in Epidermal Bladder Cells of Chenopodium quinoa. *Curr. Biol.* **2018,** *28*, 3075.e7–3085.e7.
Bonales-Alatorre, E.; Shabala, S.; Chen, Z. H.; Pottosin, I. Reduced Tonoplast Fast-Activating and Slow-Activating Channel Activity is Essential for Conferring Salinity Tolerance in a Facultative Halophyte, Quinoa. *Plant Physiol.* **2013,** *162*, 940–952.
Bose, J.; Rodrigo-Moreno, A.; Shabala, S. ROS Homeostasis in Halophytes in the Context of Salinity Stress Tolerance. *J. Exp. Bot.* **2013,** *65*, 1241–1257.
Breckle, S. W. Salinity, Halophytes and Salt Affected Natural Ecosystems. In *Salinity: Environment-Plants-Molecules*; Springer: Dordrecht, The Netherlands, 2002; pp 53–77.
Calone, R.; Bregaglio, S.; Sanoubar, R.; Noli, E.; Lambertini, C.; Barbanti, L. Physiological Adaptation to Water Salinity in Six Wild Halophytes Suitable for Mediterranean Agriculture. *Plants* **2021,** *10*, 309.
Castro, B.; Citterico, M.; Kimura, S.; Stevens, D. M.; Wrzaczek, M.; Coaker, G. Stress-Induced Reactive Oxygen Species Compartmentalization, Perception and Signalling. *Nat. Plants* **2021,** *7*, 403–412.
Céccoli, G.; Ramos, J.; Pilatti, V.; Dellaferrera, I.; Tivano, J.C.; Taleisnik, E.; Vegetti, A. C. Salt Glands in the Poaceae Family and their Relationship to Salinity Tolerance. *Bot. Rev.* **2015,** *81*, 162–178.
Chen, M.; Yang, Z.; Liu, J.; Zhu, T.; Wei, X.; Fan, H.; Wang, B. Adaptation Mechanism of Salt Excluders under Saline Conditions and its Applications. *Int. J. Mol. Sci.* **2018,** *19*, 3668.
Cheng, H.; Inyang, A.; Li, C. D.; Fei, J.; Zhou, Y. W.; Wang, Y. S. Salt Tolerance and Exclusion in the Mangrove Plant Avicennia Marina in Relation to Root Apoplastic Barriers. *Ecotoxicology* **2020,** *29*, 676–683.
Cui, B.; Liu, R.; Flowers, T. J.; Song, J. Casparian Bands and Suberin Lamellae: Key Targets for Breeding Salt Tolerant Crops? *Environ. Exp. Bot.* **2021,** *191*, 104600.
Czarnocka, W.; Karpi ´nski, S. Friend or Foe? Reactive Oxygen Species Production, Scavenging and Signaling in Plant Response to Environmental Stresses. *Free Radic. Biol. Med.* **2018,** *122*, 4–20.
Dabbous, A.; Ben Saad, R.; Brini, F.; Farhat-Khemekhem, A.; Zorrig, W.; Abdely, C.; Ben Hamed, K. Over-Expression of a Subunit E1 of a Vacuolar H+ -ATPase Gene (Lm VHA-E1) Cloned from the Halophyte Lobularia maritima Improves the Tolerance of Arabidopsis thaliana to Salt and Osmotic Stresses. *Environ. Exp. Bot.* **2017,** *137*, 128–141.
El-Keblawy, A.; Aljasmi, M.; Gairola, S.; Mosa, K. A.; Hameed, A. Provenance Determines Salinity Tolerance and Germination Requirements of the Multipurpose Tree Prosopis juliflora Seeds. *Arid Land Res. Manag.* **2021,** 1–16.
Elnaggar, A.; Mosa, K. A.; El-Keblawy, A.; Tammam, A.; El-Naggar, M. Physiological and Biochemical Insights for Salt Stress Tolerance in the Habitat-Indifferent Halophyte Salsola drummondii during the Vegetative Stage. *Botany* **2020,** *98*, 1–17.
Flowers, T. J.; Colmer, T. Salinity Tolerance in Halophytes. *New Phytol.* **2008,** *179*, 945–963.
Flowers, T. J.; Colmer, T. D. Plant Salt Tolerance: Adaptations in Halophytes. *Ann. Bot.* **2015,** *115*, 327–331.

García-Caparrós, P.; Llanderal, A.; Pestana, M.; Correia, P. J.; Lao, M. T. Nutritional and Physiological Responses of the Dicotyledonous Halophyte Sarcocornia fruticosa to Salinity. *Aust. J. Bot.* **2017,** *65*, 573–581.

Geilfus, C. M. Chloride: From Nutrient to Toxicant. *Plant Cell Physiol.* **2018,** *59*, 877–886.

Geilfus, C. M. Chloride in Soil: From Nutrient to Soil Pollutant. *Environ. Exp. Bot.* **2018,** *157*, 299–309.

Geng, X.; Chen, S.; Yilan, E.; Zhang, W.; Mao, H.; Qiqige, A.; Wang, Y.; Qi, Z.; Lin, X. Overexpression of a Tonoplast Na+/H+ Antiporter from the Halophytic Shrub Nitraria sibirica Improved Salt Tolerance and Root Development in Transgenic Poplar. *Tree Genet. Genomes* **2020,** *16*, 1–14.

Grigore, M. N.; Toma, C. A Proposal for a New Halophytes Classification, based on Integrative Anatomy Observations. *Muz. Olten. Craiova Stud. Com. ,St. Nat.* **2010,** *26*, 45–50.

Grigore, M. N.; Toma, C.; Boscaiu, M. Dealing with Halophytes: An Old Problem, the Same Continuous Exciting challenge. Analele Stiintifice ale Universitatii" Alexandru Ioan Cuza" din Iasi. *Biol. Veg.* **2010,** *56*, 21–32.

Guangming, L.; Xuechen, Z.; Xiuping, W.; Hongbo, S.; Jingsong, Y.; Xiangping, W. Soil Enzymes as Indicators of Saline Soil Fertility under Various Soil Amendments. *Agric. Ecosyst. Environ.* **2017,** *237*, 274–279.

Guo, J.; Lu, C.; Zhao, F.; Gao, S.; Wang, B. Improved Reproductive Growth of Euhalophyte Suaeda Salsa under Salinity is Correlated with Altered Phytohormone Biosynthesis and Signal Transduction. *Funct. Plant Biol.* **2020,** *47*, 170–183.

Guo, Q.; Meng, S.; Tao, S.; Feng, J.; Fan, X.; Xu, P.; Xu, Z.; Shen, X. Overexpression of a Samphire High-Affinity Potassium Transporter Gene SbHKT1 Enhances Salt Tolerance in Transgenic Cotton. *Acta Physiol. Plant.* **2020,** *42*, 36.

Guo, Q.; Tian, X.; Mao, P.; Meng, L. Overexpression of Iris lactea tonoplast Na+/H+ Antiporter Gene IlNHX Confers Improved Salt Tolerance in Tobacco. *Biol. Plant.* **2020,** *64*, 50–57.

Guo, R.; Zhao, L.; Zhang, K.; Lu, H.; Bhanbhro, N.; Yang, C. Comparative Genomics and Transcriptomics of the Extreme Halophyte Puccinellia tenuiflora Provides Insights into Salinity Tolerance Differentiation between Halophytes and Glycophytes. *Front. Plant Sci.* **2021,** *12*, 767.

Gupta, A.; Bhardwaj, M.; Tran, L. S. P. Jasmonic Acid at the Crossroads of Plant Immunity and Pseudomonas syringae virulence. *Int. J. Mol. Sci.* **2020,** *21*, 7482.

Gupta, K.; Jha, B.; Agarwal, P. K. A Dehydration-Responsive Element Binding (DREB) Transcription Factor from the Succulent Halophyte Salicornia brachiata Enhances Abiotic Stress Tolerance in Transgenic Tobacco. *Mar. Biotechnol.* **2014,** *16*, 657–673.

Gururani, M.; Venkatesh, J.; Tran, L. S. P. Regulation of Photosynthesis during Abiotic Stress-Induced Photoinhibition. *Mol. Plant* **2015,** *8*, 1304–1320.

Hajibagheri, M. A.; Yeo, A. R.; Flowers, T. J. Salt Tolerance in Suaeda Maritima (L.) Dum. Fine Structure and Ion Concentrations in the Apical Region of Roots. *New Phytol.* **1985,** *99*, 331–343.

Hamdani, F.; Derridj, A.; Rogers, H. J. Multiple Mechanisms Mediate Growth and Survival in Young Seedlings of Two Populations of the Halophyte Atriplex halimus (L.) Subjected to Long Single-Step Salinity Treatments. *Funct. Plant Biol.* **2017,** *44*, 761–773.

Han, Z. J.; Sun, Y.; Zhang, M.; Zhai, J. T. Transcriptomic Profile Analysis of the Halophyte Suaeda rigida Response and Tolerance under NaCl Stress. *Sci. Rep.* **2020,** *10*, 1–10.

Himabindu, Y.; Chakradhar, T.; Reddy, M. C.; Kanygin, A.; Redding, K. E.; Chandrasekhar, T. Salt-Tolerant Genes from Halophytes are Potential Key Players of Salt Tolerance in Glycophytes. *Environ. Exp. Bot.* **2016,** *124*, 39–63.

Hussain, S.; Khalid, M. F.; Sohail, M.; Anjum, M. A.; Ejaz, S.; Nafees, M.; Zakir, I.; Ahmad, M.; Ali, S.; Ahmad, S. Role of Transporters in Accumulating Salt Ions by Halophytes. *Approaches Remediat. Inorg. Pollut.* **2021,** 11–40.

Imamura, T.; Yasui, Y.; Koga, H.; Takagi, H.; Abe, A.; Nishizawa, K.; Mizuno, N.; Ohki, S.; Mizukoshi, H.; Mori, M. A Novel WD40-Repeat Protein Involved in Formation of Epidermal Bladder Cells in the Halophyte quinoa. *Commun. Biol.* **2020,** *3*, 1–14.

Jesus, J.; Danko, A. S.; Fiuza, A.; Borges, M. T. Phytoremediation of Salt-Affected Soils: A Review of Processes, Applicability, and the Impact of Climate Change. *Environ. Sci. Pollut. Res.* **2015,** *22*, 6511–6525.

Karakas, S.; Bolat, I.; Dikilitas, M. The Use of Halophytic Companion Plant (Portulaca oleracea L.) on Some Growth, Fruit, and Biochemical Parameters of Strawberry Plants under Salt Stress. *Horticulturae* **2021,** *7*, 63.

Kiani-Pouya, A.; Rasouli, F.; Shabala, L.; Tahir, A. T.; Zhou, M.; Shabala, S. Understanding the Role of Root-Related Traits in Salinity Tolerance of quinoa Accessions with Contrasting Epidermal Bladder Cell Patterning. *Planta* **2020,** *251*, 1–11.

Kiani-Pouya, A.; Roessner, U.; Jayasinghe, N. S.; Lutz, A.; Rupasinghe, T.; Bazihizina, N.; Bohm, J.; Alharbi, S.; Hedrich, R.; Shabala, S. Epidermal Bladder Cells Confer Salinity Stress Tolerance in the Halophyte Quinoa and Atriplex Species. *Plant Cell Environ.* **2017,** *40*, 1900–1915.

Krishnamurthy, P.; Jyothi-Prakash, P. A.; Qin, L.; He, J.; Lin, Q.; Loh, C. S.; Kumar, P. P. Role of Root Hydrophobic Barriers in Salt Exclusion of a Mangrove Plant Avicennia Officinalis. *Plant Cell Environ.* **2014,** *37*, 1656–1671.

Krishnamurthy, P.; Ranathunge, K.; Nayak, S.; Schreiber, L.; Mathew, M. K. Root Apoplastic Barriers Block Na+ Transport to Shoots in Rice (Oryza sativa L.). *J. Exp. Bot.* **2011,** *62*, 4215–4228.

Kumari, A.; Jha, B. Engineering of a Novel Gene from a Halophyte: Potential for Agriculture in Degraded Coastal Saline Soil. *Land Degrad. Dev.* **2018,** *30*, 595–607.

Kumari, A.; Parida, A. Metabolomics and Network Analysis Reveal the Potential Metabolites and Biological Pathways Involved in Salinity Tolerance of the Halophyte Salvadora persica. *Environ. Exp. Bot.* **2018,** *148*, 85–99.

Kumari, J.; Udawat, P.; Dubey, A.K.; Haque, I.; Rathore, M. S.; Jha, B. Overexpression of SbSI-1, A Nuclear Protein from Salicornia brachiata Confers Drought and Salt Stress Tolerance and Maintains Photosynthetic Efficiency in Transgenic Tobacco. *Front. Plant Sci.* **2017,** *8*, 1215.

Lakhdar, A.; Scelza, R.; Scotti, R.; A Rao, M.; Jedidi, N.; Gianfreda, L.; Abdelly, C. The Effect of Compost and Sewage Sludge on Soil Biologic Activities in Salt Affected Soil. *Rev. Cienc. Suelo Nutr. Veg.* **2010,** *10*, 40–47.

Lastiri-Hernández, M. A.; Álvarez-Bernal, D.; Ochoa-Estrada, S.; Contreras-Ramos, S. M. Potential of Bacopa monnieri (L.) Wettst and Sesuvium verrucosum Raf. as an Agronomic Management Alternative to Recover the Productivity of Saline Soils. *Int. J. Phytoremediat.* **2019,** *22*, 343–352.

Lee, Y.; Rubio, M. C.; Alassimone, J.; Geldner, N. A Mechanism for Localized Lignin Deposition in the Endodermis. *Cell* **2013,** *153*, 402–412.

Leng, B.; Yuan, F.; Dong, X.; Wang, J.; Wang, B. Distribution Pattern and Salt Excretion Rate of Salt Glands in Two Recretohalophyte Species of Limonium (Plumbaginaceae). *S. Afr. J. Bot.* **2018,** *115*, 74–80.

Leogrande, R.; Vitti, C. Use of Organic Amendments to Reclaim Saline and Sodic Soils: A Review. *Arid Land Res. Manag.* **2018,** *33*, 1–21.

Li, B.; Tester, M.; Gilliham, M. Chloride on the Move. *Trends Plant Sci.* **2017,** *22*, 236–248.

Li, J.; Yuan, F.; Liu, Y.; Zhang, M.; Liu, Y.; Zhao, Y.; Wang, B.; Chen, M. Exogenous Melatonin Enhances Salt Secretion from Salt Glands by Upregulating the Expression of Ion Transporter and Vesicle Transport Genes in Limonium Bicolor. *BMC Plant Biol.* **2020,** *20*, 1–11.

Liang, J.; Shi, W. Cotton/Halophytes Intercropping Decreases Salt Accumulation and Improves Soil Physicochemical Properties and Crop Productivity in Saline-Alkali Soils under Mulched Drip Irrigation: A Three-Year Field Experiment. *Field Crop. Res.* **2020,** *262*, 108027.

Litalien, A.; Zeeb, B. Curing the Earth: A Review of Anthropogenic Soil Salinization and Plant-Based Strategies for Sustainable Mitigation. *Sci. Total Environ.* **2020,** *698*, 134235.

Liu, X.; Cai, S.; Wang, G.; Wang, F.; Dong, F.; Mak, M.; Holford, P.; Ji, J.; Salih, A.; Zhou, M.; et al. Halophytic NHXs Confer Salt Tolerance by Altering Cytosolic and Vacuolar K+ and Na+ in Arabidopsis Root Cell. *Plant Growth Regul.* **2017,** *82*, 333–351.

Liu, Q.; Liu, R.; Ma, Y.; Song, J. Physiological and Molecular Evidence for Na+ and Cl− Key Systemic Regulator of Salt Secretion and K+/Na+ Homeostasis in the Recretohalophyte Karelinia caspia. *Environ. Exp. Bot.* **2020,** *177*, 104098.

Liu, A.; Xiao, Z.; Li, M. W.; Wong, F. L.; Yung, W. S.; Ku, Y. S.; Wang, Q.; Wang, X.; Xie, M.; Yim, A. K. Y.; et al. Transcriptomic Reprogramming in Soybean Seedlings under Salt Stress. *Plant Cell Environ.* **2018,** *42*, 98–114.

Liu, Q.; Yang, Y.; Zhang, X.; Liu, R.; Song, J. Chloride Allocation in the Euhalophyte Suaeda Salsa from Different Habitats in Field and Controlled Saline Conditions. *Aquat. Bot.* **2020,** *167*, 103292.

Llanes, A.; Reginato, M.; Devinar, G.; Luna, V. What is known about Phytohormones in Halophytes? A Review. *Biologia* **2018,** *73*, 727–742.

Lokhande, V. H.; Suprasanna, P. Prospects of Halophytes in Understanding and Managing Abiotic Stress Tolerance. In *Environmental Adaptations and Stress Tolerance of Plants in the Era of Climate Change*; Springer: Berlin, Germany, 2012; pp 29–56.

Lu, L.; Chen, X.; Wang, P.; Lu, Y.; Zhang, J.; Yang, X.; Cheng, T.; Shi, J.; Chen, J. CIPK11. A Calcineurin B-like Protein-Interacting Protein Kinase from Nitraria tangutorum, Confers Tolerance to Salt and Drought in Arabidopsis. *BMC Plant Biol.* **2021,** *21*, 1–16.

Lu, L.; Chen, X.; Zhu, L.; Li, M.; Zhang, J.; Yang, X.; Wang, P.; Lu, Y.; Cheng, T.; Shi, J.; et al. NtCIPK9: A Calcineurin B-Like Protein-InteractingProtein Kinase from the Halophyte Nitraria tangutorum, Enhances Arabidopsis Salt Tolerance. *Front. Plant Sci.* **2020,** *11*.

Lu, C.; Yuan, F.; Guo, J.; Han, G.; Wang, C.; Chen, M.; Wang, B. Current Understanding of Role of Vesicular Transport in Salt Secretion by Salt Glands in Recretohalophytes. *Int. J. Mol. Sci.* **2021,** *22*, 2203.

Luo, S.; Tian, L.; Chang, C.; Wang, S.; Zhang, J.; Zhou, X.; Li, X.; Tran, L. S. P.; Tian, C. Grass and Maize Vegetation Systems Restore Saline-Sodic Soils in the Songnen Plain of Northeast China. *Land Degrad. Dev.* **2018,** *29*, 1107–1119.

Ma, Q.; Hu, J.; Zhou, X. R.; Yuan, H. J.; Kumar, T.; Luan, S.; Wang, S. M. ZxAKT1 is Essential for K+ Uptake and K+/Na+ Homeostasis in the Succulent Xerophyte Zygophyllum xanthoxylum. *Plant J.* **2017,** *90*, 48–60.

Makni, H.; Ayed, L.; Ben Khedher, M.; Bakhrouf, A. Evaluation of the Maturity of Organic Waste Composts. *Waste Manag. Res.* **2009,** *28*, 489–495.

Malik, Z.; Ravindran, K. Bioaccumulation of Salts and Heavy Metals by Suaeda monoica: A Salt Marsh Halophyte from Paper Mill Effluent Contaminated Soil. *Int. J. Sci. Technol. Res.* **2020,** *9*, 7248–7254.

Mann, A.; Kumar, N.; Kumar, A.; Lata, C.; Kumar, A.; Meena, B. L.; Mishra, D.; Grover, M.; Gaba, S.; Parameswaran, C.; et al. De Novo Transcriptomic Profiling of Differentially Expressed Genes in Grass Halophyte Urochondra setulosa under High Salinity. *Sci. Rep.* **2021,** *11*, 1–14.

Mann, E.; Rutter, A.; Zeeb, B. Evaluating the Efficacy of Atriplex spp. in the Phytoextraction of Road Salt (NaCl) from Contaminated Soil. *Environ. Pollut.* **2020,** *265*, 114963.

Marriboina, S.; Sengupta, D.; Kumar, S.; Reddy, A. R. Physiological and Molecular Insights into the High Salinity Tolerance of Pongamia pinnata (L.) pierre, A Potential Biofuel Tree Species. *Plant Sci.* **2017,** *258*, 102–111.

Mi, P.; Yuan, F.; Guo, J.; Han, G.; Wang, B. Salt Glands Play a Pivotal Role in the Salt Resistance of Four Recretohalophyte Limonium Mill. Species. *Plant Biol.* **2021.**

Minhas, P.; Qadir, M.; Yadav, R. Groundwater Irrigation Induced Soil Sodification and Response Options. *Agric. Water Manag.* **2019,** *215*, 74–85.

Miransari, M.; Smith, D. Plant Hormones and Seed Germination. *Environ. Exp. Bot.* **2014,** *99*, 110–121.

Miteva, T.; Zhelev, N. Z.; Popova, L. Effect of Salinity on the Synthesis of Ribulose-1,5-bisphosphate Carboxylase/Oxygenase in Barley Leaves. *J. Plant Physiol.* **1992,** *140*, 46–51.

Mohammadi, F.; Kavousi, H. R.; Mansouri, M. Effects of Salt Stress on Physio-Biochemical Characters and Gene Expressions in Halophyte Grass Leptochloa fusca (L.) Kunth. *Acta Physiol. Plant.* **2019,** *41*, 1–10.

Mostofa, M. G.; Rahman, M.; Ansary, M. U.; Keya, S. S.; Abdelrahman, M.; Miah, G.; Tran, L. S. P. Silicon in Mitigation of Abiotic Stress-Induced Oxidative Damage in Plants. *Crit. Rev. Biotechnol.* **2021,** *41*, 918–934.

Muchate, N. S.; Nikalje, G.; Rajurkar, N. S.; Suprasanna, P.; Nikam, T. D. Physiological Responses of the Halophyte Sesuvium portulacastrum to Salt Stress and their Relevance for Aaline Soil Bio-Reclamation. *Flora* **2016,** *224*, 96–105.

Munns, R. Comparative Physiology of Salt and Water Stress. *Plant Cell Environ.* **2002,** *25*, 239–250.

Munns, R.; Passioura, J. B.; Colmer, T. D.; Byrt, C. S. Osmotic Adjustment and Energy Limitations to Plant Growth in Saline Soil. *New Phytol.* **2019,** *225*, 1091–1096.

Munns, R.; Tester, M. Mechanisms of Salinity Tolerance. *Annu. Rev. Plant Biol.* **2008,** *59*, 651–681.

Mwando, E.; Han, Y.; Angessa, T. T.; Zhou, G.; Hill, C. B.; Zhang, X. Q.; Li, C. Genome-Wide Association Study of Salinity Tolerance during Germination in Barley (Hordeum vulgare L.). *Front. Plant Sci.* **2020,** *11*, 118.

Naseer, S.; Lee, Y.; Lapierre, C.; Franke, R.; Nawrath, C.; Geldner, N. Casparian Strip Diffusion Barrier in Arabidopsis is made of a Lignin Polymer without Suberin. *Proc. Natl. Acad. Sci. USA* **2012,** *109*, 10101–10106.

Nikalje, G. C.; Srivastava, A. K.; Pandey, G. K.; Suprasanna, P. Halophytes in Biosaline Agriculture: Mechanism, Utilization, and Value Addition. *Land Degrad. Dev.* **2017,** *29*, 1081–1095.

Omuto, C. T. V.; Vargas, R. R.; El Mobarak, A. M.; Mohamed, N.; Viatkin, K.; Yigini, Y. In *Mapping of Salt-Affected Soils: Technical Manual*; FAO: Rome, Italy, 2020.

Ouni, Y.; Lakhdar, A.; Scelza, R.; Scotti, R.; Abdelly, C.; Barhoumi, Z.; Rao, M. A. Effects of Two Composts and Two Grasses on Microbial Biomass and Biological Activity in a Salt-Affected Soil. *Ecol. Eng.* **2013,** *60*, 363–369.

Otterbach, S. L.; Khoury, H.; Rupasinghe, T.; Mendis, H.; Kwan, K. H.; Lui, V.; Natera, S. H. A.; Klaiber, I.; Allen, N. M.; Jarvis, D. E.; et al. Characterization of Epidermal Bladder Cells in Chenopodium quinoa. *Plant Cell Environ.* **2021**.

Palchetti, M. V.; Reginato, M.; Llanes, A.; Hornbacher, J.; Papenbrock, J.; Barboza, G. E.; Luna, V.; Cantero, J. J. New Insights into the Salt Tolerance of the Extreme Halophytic Species Lycium humile (Lycieae, Solanaceae). *Plant Physiol. Biochem.* **2021,** *163*, 166–177.

Panda, A.; Rangani, J.; Parida, A. K. Cross Talk between ROS Homeostasis and Antioxidative Machinery Contributes to Salt Tolerance of the Xero-Halophyte Haloxylon salicornicum. *Environ. Exp. Bot.* **2019,** *166*, 103799.

Panda, A.; Rangani, J.; Parida, A. K. Unraveling Salt Responsive Metabolites and Metabolic Pathways using Non-Targeted Metabolomics Approach and Elucidation of Salt Tolerance Mechanisms in the Xero-Halophyte Haloxylon salicornicum. *Plant Physiol. Biochem.* **2020,** *158*, 284–296.

Parida, A. K.; Jha, B. Salt Tolerance Mechanisms in Mangroves: A Review. *Trees* **2010,** *24*, 199–217.

Patel, M. K.; Kumar, M.; Li, W.; Luo, Y.; Burritt, D. J.; Alkan, N.; Tran, L. S. P. Enhancing Salt Tolerance of Plants: From Metabolic Reprogramming to Exogenous Chemical Treatments and Molecular Approaches. *Cells* **2020,** *9*, 2492.

Qadir, M.; Noble, A.; Oster, J. D.; Schubert, S.; Ghafoor, A. Driving Forces for Sodium Removal during Phytoremediation of Calcareous Sodic and Saline-Sodic Soils: A Review. *Soil Use Manag.* **2005,** *21*, 173–180.

Qadir, M.; Quillérou, E.; Nangia, V.; Murtaza, G.; Singh, M.; Thomas, R. J.; Drechsel, P.; Noble, A. D. Economics of Salt-Induced Land Degradation and Restoration. *Nat. Resour. Forum* **2014,** *38*, 282–295.

Qadir, M.; Tubeileh, A.; Akhtar, J.; Larbi, A.; Minhas, P. S.; Khan, M. A. Productivity Enhancement of Salt-Affected Environments through Crop Diversification. *Land Degrad. Dev.* **2008,** *19*, 429–453.

Rahman, M.; Haque, A.; Nihad, S. A. I.; Howlader, R. A.; Akand, M. H. Morpho Physiological Response of Acacia auriculiformis as Influenced by Seawater Induced Salinity Stress. *For. Syst.* **2016,** *25*, e 071.

Rahman, M.; Mostofa, M. G.; Rahman, A.; Islam, R.; Keya, S. S.; Das, A. K.; Miah, G.; Kawser, A. Q. M. R.; Ahsan, S. M.; Hashem, A.; et al. Acetic Acid: A Cost-Effective Agent for Mitigation of Seawater-Induced Salt Toxicity in Mung Bean. *Sci. Rep.* **2019,** *9*, 1–15.

Rahman, M.; Mostofa, M. G.; Rahman, A.; Miah, G.; Saha, S. R.; Karim, M. A.; Keya, S. S.; Akter, M.; Islam, M.; Tran, L. S. P. Insight into Salt Tolerance Mechanisms of the Halophyte Achras sapota: An Important Fruit Tree for Agriculture in Coastal Areas. *Protoplasma* **2018,** *256*, 181 191.

Rath, K. M.; Rousk, J. Salt Effects on the Soil Microbial Decomposer Community and their Role in Organic Carbon Cycling: A Review. *Soil Biol. Biochem.* **2015,** *81*, 108–123.

Reinbothe, C.; Springer, A.; Samól, I.; Reinbothe, S. Plant Oxylipins: Role of Jasmonic Acid during Programmed Cell Death, Defence and Leaf Senescence. *FEBS J.* **2009,** *276*, 4666–4681.

Rengasamy, P. Soil Chemistry Factors Confounding Crop Salinity Tolerance—A Review. *Agronomy* **2016,** *6*, 53.

Roeurn, S.; Hoshino, N.; Soejima, K. T.; Inoue, Y.; Cushman, J. C.; Agarie, S. MYB and HD-ZIP IV Homologs Related to Trichome Formation are Involved in Epidermal Bladder Cell Development in the Halophyte Mesembryanthemum crystallinum L. *Plant Prod. Sci.* **2017,** *20*, 72–82.

Roy, S. J.; Negrão, S.; Tester, M. Salt Resistant Crop Plants. *Curr. Opin. Biotechnol.* **2014,** *26*, 115–124.

Saddhe, A. A.; Jamdade, R.; Gairola, S. Recent Advances on Cellular Signaling Paradigm and Salt Stress Responsive Genes in Halophytes. In *Handbook of Halophytes: From Molecules to Ecosystems Towards Biosaline Agriculture*, 2020; pp 1–26.

, J.; Al-Azzawi, M.; Aronson, J.; Flowers, T. J. eHALOPH a Database of Salt-Tolerant Plants: Helping put Halophytes to Work. *Plant Cell Physiol.* **2015,** *57*, e10.

48. Breckle, S. How do Halophytes Overcome Salinity. *Biol. Salt Toler. Plants* **1995,** *23*, 199–203.

Shabala, S. Learning from Halophytes: Physiological Basis and Strategies to Improve Abiotic Stress Tolerance in Crops. *Ann. Bot.* **2013,** *112*, 1209–1221.

Shabala, S.; Bose, J.; Hedrich, R. Salt Bladders: Do they Matter? *Trends Plant Sci.* **2014,** *19*, 687–691.

Shaygan, M.; Mulligan, D.; Baumgartl, T. The Potential of Three Halophytes (Tecticornia pergranulata, Sclerolaena longicuspis, and Frankenia serpyllifolia) for the Rehabilitation of Brine-Affected Soils. *Land Degrad. Dev.* **2018,** *29*, 2002–2014.

Shukla, P. S.; Gupta, K.; Agarwal, P.; Jha, B.; Agarwal, P. K. Overexpression of a Novel SbMYB15 from Salicornia brachiata Confers Salinity and Dehydration Tolerance by Reduced Oxidative Damage and Improved Photosynthesis in Transgenic Tobacco. *Planta* **2015,** *242*, 1291–1308.

Silva, Y. J. A. B.; Freire, M. G.; Lopes, E. A. P.; Santos, M. A. Atriplex nummularia Lindl. as Alternative for Improving Salt-Affected Soils Conditions in Semiarid Environments: A Field Experiment. *Chil. J. Agric. Res.* **2016,** *76*, 343–348.

Slama, I.; Abdelly, C.; Bouchereau, A.; Flowers, T.; Savouré, A. Diversity, Distribution and Roles of Osmoprotective Compounds Accumulated in Halophytes under Abiotic Stress. *Ann. Bot.* **2015,** *115*, 433–447.

Srivastava, N. Reclamation of Saline and Sodic Soil through Phytoremediation. In *Environmental Concerns and Sustainable Development: Volume 2: Biodiversity, Soil and Waste Management*; Shukla, V., Kumar, N., Eds.; Springer: Singapore, 2020; pp 279–306.

Tang, X.; Zhang, H.; Shabala, S.; Li, H.; Yang, X.; Zhang, H. Tissue Tolerance Mechanisms Conferring Salinity Tolerance in a Halophytic Perennial Species Nitraria sibirica Pall. *Tree Physiol.* **2020**.

Tian, L.; Lin, X.; Tian, J.; Ji, L.; Chen, Y.; Tran, L. S. P.; Tian, C. Research Advances of Beneficial Microbiota Associated with Crop Plants. *Int. J. Mol. Sci.* **2020,** *21*, 1792.

Tran, D. Q.; Konishi, A.; Cushman, J. C.; Morokuma, M.; Toyota, M.; Agarie, S. Ion Accumulation and Expression of Ion Homeostasisrelated Genes Associated with Halophilism, NaCl-Promoted Growth in a Halophyte Mesembryanthemum Crystallinum L. *Plant Prod. Sci.* **2019,** *23*, 91–102.

Uçarlı, C. Effects of Salinity on Seed Germination and Early Seedling Stage. In *Abiotic Stress in Plants*; IntechOpen: London, UK, 2020.

Udawat, P.; Jha, R. K.; Mishra, A.; Jha, B. Overexpression of a Plasma Membrane-Localized SbSRP-Like Protein Enhances Salinity and Osmotic Stress Tolerance in Transgenic Tobacco. *Front. Plant Sci.* **2017,** *8*, 582.

Von Sengbusch, P. Halophytes Botanik Online; University of Hamburg: Hamburg, Germany, 2003.

Wang, Y. Y.; Hsu, P. K.; Tsay, Y. F. Uptake, Allocation and Signaling of Nitrate. *Trends Plant Sci.* **2012,** *17*, 458–467.

Wang, H.; Tang, X.; Wang, H.; Shao, H. B. Proline Accumulation and Metabolism-Related Genes Expression Profiles in Kosteletzkya Virginica Seedlings under Salt Stress. *Front. Plant Sci.* **2015,** *6*, 792.

Wang, P.; Wang, C. M.; Gao, L.; Cui, Y. N.; Yang, H. L.; De Silva, N. D. G.; Ma, Q.; Bao, A. K.; Flowers, T. J.; Rowland, O.; et al. Aliphatic Suberin Confers Salt Tolerance to Arabidopsis by Limiting Na+ Influx, K+ Efflux and Water Backflow. *Plant Soil* **2020,** *448*, 603–620.

Wang, L.; Wang, X.; Jiang, L.; Zhang, K.; Tanveer, M.; Tian, C.; Zhao, Z. Reclamation of Saline Soil by Planting Annual Euhalophyte Suaeda salsa with Drip Irrigation: A Three-Year Field Experiment in Arid Northwestern China. *Ecol. Eng.* **2020,** *159*, 106090.

Wang, R.; Wang, X.; Liu, K.; Zhang, X. J.; Zhang, L. Y.; Fan, S. J. Transcriptome Analysis of Halophyte Zoysia Macrostachya in Response to Salinity Stress. *Plants* **2020,** *9*, 458.

Wang, Y.; Xiao, D.; Li, Y.; Li, X. Soil Salinity Evolution and its Relationship with Dynamics of Groundwater in the Oasis of Inland River Basins: Case Study from the Fubei Region of Xinjiang Province, China. *Environ. Monit. Assess.* **2007,** *140*, 291–302.

Wei, X.; Yan, X.; Yang, Z.; Han, G.; Wang, L.; Yuan, F.; Wang, B. Salt Glands of Recretohalophyte Tamarix under Salinity: Their Evolution and Adaptation. *Ecol. Evol.* **2020,** *10*, 9384–9395.

Wicke, B.; Smeets, E.; Dornburg, V.; Vashev, B.; Gaiser, T.; Turkenburg, W.; Faaij, A. The Global Technical and Economic Potential of Bioenergy from Salt-Affected Soils. *Energy Environ. Sci.* **2011,** *4*, 2669–2681.

Wiszniewska, A.; Ko´zmi ´nska, A.; Hanus-Fajerska, E.; Dziurka, M.; Dziurka, K. Insight into Mechanisms of Multiple Stresses Tolerance in a Halophyte Aster tripolium Subjected to Salinity and Heavy Metal Stress. *Ecotoxicol. Environ. Saf.* **2019,** *180*, 12–22.

Wiszniewska, A.; Ko ´zmi ´nska, A.; Hanus-Fajerska, E.; Dziurka, K.; Dziurka, M. Insight into Phytohormonal Modulation of Defense Mechanisms to Salt Excess in a Halophyte and a Glycophyte from Asteraceae Family. *Plant Soil* **2021,** *463*, 55–76.

Wu, H.; Li, Z. The Importance of Cl− Exclusion and Vacuolar Cl− Sequestration: Revisiting the Role of Cl− Transport in Plant Salt Tolerance. *Front. Plant Sci.* **2019,** *10*, 1418.

Xia, X. J.; Zhou, Y. H.; Shi, K.; Zhou, J.; Foyer, C. H.; Yu, J. Q. Interplay between Reactive Oxygen Species and Hormones in the Control of Plant Development and Stress Tolerance. *J. Exp. Bot.* **2015,** *66*, 2839–2856.

Xiang, D.; Man, L. EhEm1, A Novel Em-Like Protein from Eutrema Halophilum, Confers Tolerance to Salt and Drought Stresses in Rice. *Mol. Breed.* **2018,** *38*, 17.

Xie, X.; Pu, L.; Wang, Q.; Zhu, M.; Xu, Y.; Zhang, M. Response of Soil Physicochemical Properties and Enzyme Activities to Long-Term Reclamation of Coastal Saline Soil, Eastern China. *Sci. Total Environ.* **2017,** *607–608*, 1419–1427.

Yan, N.; Marschner, P.; Cao, W.; Zuo, C.; Qin, W. Influence of Salinity and Water Content on Soil Microorganisms. *Int. Soil Water Conserv. Res.* **2015,** *3*, 316–323.

Yang, G.; Yu, L.; Zhang, K.; Zhao, Y.; Guo, Y.; Gao, C. A ThDREB Gene from Tamarix hispida Improved the Salt and Drought Tolerance of Transgenic Tobacco and T. hispida. *Plant Physiol. Biochem.* **2017,** *113*, 187–197.

Yao, Y.; Zhang, X.; Wang, N.; Cui, Y.; Zhang, L.; Fan, S. Transcriptome Analysis of Salt Stress Response in Halophyte Atriplex centralasiatica Leaves. *Acta Physiol. Plant.* **2019,** *42*, 3.

Yazdanpanah, N.; Mahmoodabadi, M. Effect of Some Amendments on Leachate Properties of a Calcareous Saline- Sodic Soil: A Laboratory Experiment. *Int. Agrophys.* **2010,** *25*, 14738.

Ye, W.; Wang, T.; Wei, W.; Lou, S.; Lan, F.; Zhu, S.; Li, Q.; Ji, G.; Lin, C.; Wu, X.; et al. The Full-Length Transcriptome of Spartina alterniflora Reveals the Complexity of High Salt Tolerance in Monocotyledonous Halophyte. *Plant Cell Physiol.* **2020,** *61*, 882–896.

Yıldız, M.; Terzi, H. Comparative Analysis of Salt-Induced Changes in the Root Physiology and Proteome of the Xero-halophyte Salsola crassa. *Braz. J. Bot.* **2021,** *44*, 33–42.

Yuan, J. Z.; Cui, Y. N.; Li, X. T.; Wang, F. Z.; He, Z. H.; Li, X. Y.; Bao, A. K.; Wang, S.; Ma, Q. ZxNPF7.3/NRT1.5 from the Xerophyte Zygophyllum xanthoxylum Modulates Salt and Drought Tolerance by Regulating NO3-, Na+ and K+ Transport. *Environ. Exp. Bot.* **2020,** *177*, 104123.

Yuan, C.; Feng, S.; Wang, J.; Huo, Z.; Ji, Q. Effects of Irrigation Water Salinity on Soil Salt Content Distribution, Soil Physical Properties and Water use Efficiency of Maize for Seed Production in Arid Northwest China. *Int. J. Agric. Biol. Eng.* **2018,** *11*, 137–145.

Yuan, F.; Leng, B.; Wang, B. Progress in Studying Salt Secretion from the Salt Glands in Recretohalophytes: How do Plants Secrete Salt? *Front. Plant Sci.* **2016,** *7*, 977.

Zaid, A.; Wani, S. H. Reactive Oxygen Species Generation, Scavenging and Signaling in Plant Defense Responses. In *Bioactive Molecules in Plant Defense*; Springer Science and Business Media LLC: Berlin, Germany, 2019; pp 111–132.

Zhang, W. W.; Chong, W.; Rui, X.; Wang, L. J. Effects of Salinity on the Soil Microbial Community and Soil Fertility. *J. Integr. Agric.* **2019,** *18*, 1360–1368.

Zhang, S.; Quartararo, A.; Betz, O. K.; Madahhosseini, S.; Heringer, A. S.; Le, T.; Shao, Y.; Caruso, T.; Ferguson, L.; Jernstedt, J. Root Vacuolar Sequestration and Suberization are Prominent Responses of Pistacia spp. Rootstocks During Salinity Stress. *Plant Direct* **2021,** *5*, e00315.

Zhang, J.; Shi, H. Physiological and Molecular Mechanisms of Plant Salt Tolerance. *Photosynth. Res.* **2013,** *115*, 1–22.

Zhang, W. D.; Wang, P.; Bao, Z.; Ma, Q.; Duan, L. J.; Bao, A. K.; Zhang, J. L.; Wang, S. M. SOS1, HKT1;5, and NHX1 Synergistically Modulate Na+ Homeostasis in the Halophytic Grass Puccinellia tenuiflora. *Front. Plant Sci.* **2017,** *8*.

Zhang, H.; Xiao, W.; Yu, W.; Jiang, Y.; Li, R. Halophytic Hordeum brevisubulatum HbHAK1 Facilitates Potassium Retention and Contributes to Salt Tolerance. *Int. J. Mol. Sci.* **2020,** *21*, 5292.

Zhou, G.; Xia, Y.; Ma, B. L.; Feng, C.; Qin, P. Culture of Seashore Mallow under Different Salinity Levels using Plastic Nutrient-Rich Matrices and Transplantation. *Agron. J.* **2010,** *102*, 395–402.

Zhao, C.; Zhang, H.; Song, C.; Zhu, J. K.; Shabala, S. Mechanisms of Plant Responses and Adaptation to Soil Salinity. *Innovation* **2020,** *1*.

CHAPTER 10

Halophilic Cyanobacteria and Microalgae: Role in Environmental Clean-Up and Value-Added Products

MADHULIKA GUPTA

Department of Microbiology, Lady Brabourne College, P1/2 Suhrawardy Avenue, Kolkata, West Bengal, India

ABSTRACT

An increase in the salinity of soil and water poses a negative impact on environmental well-being and affects the agricultural and economic sectors as well. The main causes for the increase in salinization are widespread agriculture, mining, and lacking proper approaches for water management. This results in the increase in the salt ion content in various aquatic and terrestrial ecosystems which affects the biotic communities accompanied by a decline in their diversity and replacement of native species. Salinization deteriorates the quality of groundwater and surface water, decreasing soil fertility, a major element leading to desertification resulting in land degradation. The conventional methods to deal with the problem include salt leaching, better agricultural practices, soil modification approaches, salt-redirecting mechanisms, reverse and forward osmosis, electrodialysis, multiple effect distillation, and many more. Such processes are costly, demanding high energy expenditure, rigorous labor, and degrade the ecosystem when used haphazardly. Halophilic cyanobacteria and microalgae, the inhabitants of deep-sea basins, salt lakes, salt flats, and evaporation ponds can acclimatize in such hypersaline environments. The unique features of halophilic cyanobacteria and microalgae act as a natural solution to the problems that inhibit environmental renewal and restoration of the aquatic and terrestrial ecosystem.

Physiology of Halophytes: Signaling, Omics, and Tolerance Mechanisms.
Nivas Manohar Desai & Ganesh Chandrakant Nikalje (Eds.)

This chapter aims to provide an overview of the causes and effects of soil and water salinization, the study of halophilic cyanobacteria and microalgae, their classification, distribution, features, mode of adaptation, role in saline soil restoration, desalination of water, wastewater treatment, production of secondary metabolites and biofuels. Applications of halophilic cyanobacteria and microalgae in environmental clean-up indicate sustainability and growth of the ecosystem from various negative impacts.

10.1 INTRODUCTION

Salinity can be defined as the presence of soluble salt in all types of soil including fertile soils and water bodies (Srivastava et al., 2019). The concentration of salts in the soil reveals the type of soil that is, normal, saline, sodic, or saline-sodic. Similarly, the presence of salts in water decides whether the water is potable or of inferior quality (saline or brackish) (El-Swaify, 2000). Salinity can be classified into the following types-primary and secondary. Primary salinity is caused by natural activities such as weathering of parental rock materials, infiltration of ocean water into coastal regions and rivers, evapotranspiration, and natural calamities such as cyclones, outflow from the sea in low-lying areas, discharge of excess salts in arid zones while secondary salinity involves man-made activities such as inferior quality irrigation, application of fertilizers, deforestation, waterlogging, and improper drainage (Said-Al and Omer, 2011; Patel et al., 2011; Chaitanya et al., 2014; Srivastava et al., 2019) The largest bodies of saline water are exemplified by the oceans but hypersaline environments indicate a salt concentration which exceeds far more than that of seawater and are derived from the evaporation of sea-water (Oren, 2002; Kerkar, 2004). These hypersaline environments can be categorized into thalassohaline and athalassohaline environments. The thalassohaline environment resembles the proportion of salt concentration in seawater (3.5%), with the dominance of Na^+ and Cl^-, and has a salinity of 35%, for example, the "Great Salt Lake." In an athalassohaline environment, salinity ranges from 29 to 44% with the dominance of divalent cations exceeding monovalent ions, Na^+ and K^+ with low pH (5.8–6.0), for example, "the Dead Sea" (Oren, 2002; Kerkar, 2004; Das Sarma, 2012; Satyanarayana, 2012; Edbieb et al., 2016). It poses serious concern when an excessive accumulation of water-soluble salts, such as sodium, potassium, magnesium, calcium as cations and sulfate, bicarbonate, and chloride as anions, occurs leading to the salinization of soil and water (El-Swaify, 2000).

Occurrence of salinization by natural processes of weathering of rocks or mishandled irrigation due to soil drainage obstruction, the intervention of seawater into groundwater, and other anthropogenic activities (Srivastava et al., 2019). The detrimental effects of salinization include loss of soil fertility, inferior quality potable water, scarcity of water, modification in the composition of natural sources of water such as ponds, rivers, groundwater, breakdown of agriculture and fishing industries leading to loss of productivity, health problems, loss of biodiversity, all indicating environmental as well as economic losses (Williams, 2001a, 2001b; Briggs and Taws, 2003; Butcher et al., 2016; Harmon and Daigh, 2017; Majeed and Muhammad, 2019; Zorb et al., 2019; Rocha et al., 2020). A number of conventional approaches to combat the problem of salinization include salt leaching, better agricultural practices, soil modification approaches, salt redirecting mechanisms, reverse and forward osmosis, electrodialysis, multiple effect distillation, and many more (Qadir et al., 2000; Lattemann and Hopner, 2008; Vecchio et al., 2018; Rocha et al., 2020; Australian Government, 2020; Wichern et al., 2020; Minhas et al., 2020; Gautam and Kapoor, 2022). The drawbacks of such approaches are expenditure of high energy, severe environmental effects on flora and fauna, high cost, involvement of extensive labor, and degrading of the ecosystem when used haphazardly. The diverse variety of microorganisms inhabits in these hypersaline environments. Among them, halophilic cyanobacteria and microalgae with their unique features adapt to such extreme environments and play a major role as well as an alternative approach in environmental renewal, restoration, and clean-up. (Qadir et al., 2000; Lattemann and Hopner, 2008; Shankar and Evelin, 2019; Leogrande and Vitti, 2019; Rocha et al., 2020; Gautam and Kapoor, 2022) An overview of the causes and effects of soil and water salinization, the study of halophilic cyanobacteria and microalgae, their classification, distribution, features, mode of adaptation, role in saline soil restoration, desalination of water, wastewater treatment, production of secondary metabolites and biofuels have been discussed in this review.

10.2 HALOPHILES

Microorganisms that can withstand, flourish, and reproduce in intense saline environments are designated as halophiles. They maintain a balance between the intercellular or intracellular regions of the cell and the low water activity and high osmotic pressure present on the outer side of the cell.

The classification of halophiles is based on the amount of NaCl required by them for their growth (Kushner, 1993).

i. Mild halophiles: Optimum growth requirement 0.2–0.5 M (~1.2–2.9%) NaCl; they can be facultative halophiles.

ii. Moderate halophiles: Optimum growth requirement 0.5–2.5 M (~2.9–13.6%) NaCl; often considered as true halophiles.

iii. Borderline extreme halophiles: Optimum growth requirement 1.5–4.0 M (~8.8–23.4%) NaCl; mostly obligate halophiles.

iv. Extreme halophiles: Optimum growth requirement 2.5–5.2M (~13.6–30.4%) NaCl; modified and obligate halophiles.

v. Nonhalophiles: Growth requirement <0.2 M (<1.2%) NaCl; one of their nutrients (Kushner, 1993; Ali et al., 2016).

Most halophilic microorganisms are prokaryotic while few are eukaryotic. Examples of extensively studied halophiles include *Halobacterium* sp., *Aphanothece halophytica*, *Dunaliella salina*, and *Hortaea werneckii* (Al-Abri, 2011; Ali et al., 2016).

10.3 CYANOBACTERIA AND MICROALGAE AS HALOPHILES

Halophilic cyanobacteria and microalgae, the inhabitants of deep-sea basins, salterns, salt lakes, oceans, salt flats, evaporation ponds, crystallization ponds can acclimatize in such hypersaline environments. *A. halophytica*, an extreme cyanobacterial halophile forms microbial mats with a top brown layer in several hypersaline environments, and flourishes at NaCl concentrations of 2–5M. It has been reported from several hypersaline environments such as Solar Lake, the Dead Sea, the Great Salt Lake, solar ponds, etc. *Dactylococcopsis salina*, a cyanobacterium plankton described from the Great Salt Lake. A green layer of mats is formed by a variety of filamentous cyanobacteria, which are moderate halophiles, have optimum growth at 1–2.5M NaCl, have the ability to fix nitrogen (Kerkar, 2004; Burns et al., 2004; Ali et al., 2016). *Chroococcus*, *Spirulina platensis*, *Dunaliella*, *Synechococcus*, *Chloroflexus*, *Anabaenopsis arnoldii*, etc., are the halophilic microalgae reported from various saline lakes (Bhatt et al., 2016). Photosynthetic organisms *Dunaliella*, and cyanobacteria *Phormidium*, *Komvophoron*, and *Geitlerinema* are found at salinities higher than 30% (Kirkwood et al., 2007; Sommer et al., 2020). Polyextremophilic behavior is noted among halophilic species such as *Chlamydomonas nivalis* and *Cyanidium caldarum* (Ali et al., 2016). *Halospirulina tapeticola*, a tightly coiled filamentous cyanobacterium in benthic mats in saltern evaporation ponds at a salt concentration (~20%) is a true euryhaline species, which grows from 3–20% (Nubel et al., 2000b). *Halomicronema excentricum*, a filamentous cyanobacterium with salt range (3.2–12%) has been reported from the microbial mat of saltern evaporation ponds (Abed et al., 2002). *Chroococcus minutus, Aphanothece microscopica, Synechococcus aeruginosus, Synechocystis salina, Dermocarpa leibleiniae, Arthrospira platensis, Spirulina subsalsa, Oscillatoria salina, Phormidium valderianum, Lyngbya martensiana, Symploca hydnoides, Microcoleus chthonoplastes, Plectonema terebrans* reported from the hypersaline environment of Salt pans of southeastern coast of India (Oren, 2015).

Some of the selected features of halophilic microalgae and cyanobacteria are demonstrated under hypersaline environments as follows (Oren, 2015):

i. A variation in salinity was noted in halophilic cyanobacteria such as *Lyngbya aestuarii, Entophysalis* sp., *Chroococcus* sp., *C. chthonoplastes* forming mats during low tide from 6–20% resulting in color changes of the overlying water. The cyanobacterial mat showed an orange–reddish color when the salinity was above 15%. This phenomenon was designated as halotaxis, a relocation which was driven by salinity (Abed et al., 2008; Oren, 2015).
ii. Halophilic cyanobacteria such as *Nodularia, Spumigena,* and *Scytonema* possesses heterocysts at enhanced salt concentrations. Nitrogen fixation was performed by mats containing *Coleofasciculus* (Paerl et al., 2000; de Wit et al., 2005; Roney et al., 2009; Oren, 2015).
iii. Halophilic cyanobacteria show the property of fermentation of stored food material such as glycogen during night time which was formed during the day by photosynthesis to form hydrogen gas and other fermentation products. For example, mats of *Coleofasciculus* and *Lyngbya.*
iv. A shift from oxygenic to anoxygenic photosynthesis was observed by cyanobacteria and microalgae living in hypersaline environments. Sulfide supplied the electrons for CO^2 fixation. It was observed in *Oscillatoria limnetica.* The electrons were directly transferred to photosystem I (PSI), and photosystem II (PSII) was not involved in the process. *Aphanothece* and *Phormidium* also exemplify anoxygenic photosynthesis in hypersaline environments (Cohen et al., 1975a; Ionescu et al., 2007; Oren, 2015).

10.3.1 MODE OF ADAPTATION

Microorganisms subjected to hypersaline conditions often lose water and undergo shrinkage resulting in the damage of structure and function of the cell. To avoid such fate of cells, halophiles have developed two approaches:

i. **High Salt-in Strategy:** An adaptive process in which intercellular build-up of inorganic ions occurs to maintain the balance of salt concentration in their environment. Instead of employing organic osmotic solutes halophiles use chloride pumps which accumulate Cl^- ions into the cytoplasm from the external environment. Arginine and Lysine accelerate the uptake and release of Cl^- ions which are located at the terminal part of the channel (Muller and Oren, 2003; Edbeib et al., 2016; Devi and Rasool, 2017). Extreme halophiles gather K^+ ions to sustain the osmotic balance compared to that of the external environment (Oren, 2002c; Edbeib et al., 2016).

ii. **Low Salt-in-Strategy:** The second approach to maintain osmotic balance is that halophiles accumulate low molecular weight, osmo-regulatory compounds called compatible solutes or osmolytes such as amino acids, alcohols, betaine compounds, ectoine compounds, sugars, etc., which stabilizes the cell by acting as salt antagonists. They protect and stabilize the protein from denaturation (Chattopadhyay et al., 2004; Saum and Muller, 2008; Edbeib et al., 2016; Devi and Rasool, 2017).

10.3.2 MODIFICATION OF ENZYMES

A hypersaline environment has a severe effect on the stability and functionality of enzymes. A sharp distinction between halophilic and nonhalophilic proteins is that the surface of halophiles is occupied by the abundance of glutamate and aspartate (Zhang and Yi, 2013; Edbeib et al., 2016). For example, α-amylases, xylanases, proteases, etc., of halophiles have negatively charged surfaces which aids in competing with ions for water molecules as well as maintains its solubility (Britton et al., 2006; Karan et al., 2012; Edbeib et al., 2016). By genome sequencing of Halobacterium species NRC-1, the presence of an acidic proteome was reported (Ng et al., 2000; Das Sarma and Das Sarma, 2015; Edbeib et al., 2016).

10.4 ADAPTATION IN CYANOBACTERIA AND MICROALGAE

Variation in salt concentration affects cyanobacteria and microalgae resulting in osmotic stress, salt stress, and modification in cellular ion ratios (Gautam and Kapoor, 2022). So to acclimatize to the fluctuations in salinity following adaptations are demonstrated by microalgae-gathering compatible solutes to maintain cell osmolarity, adjusting membrane fluidity, discharge of extracellular polymeric substances (EPSs), regulation of photosynthesis to strike a balance between consumption and synthesis of energy, decrease in protein synthesis and interchanging between nonpolar storage lipids and polar structural lipids to alter and adapt to lipid metabolism (Borowitzka, 2018a, 2018b; Foflonker et al., 2018; Barati et al., 2018; Ivosevic De Nardis et al., 2019; Pavlinska et al., 2020; Misic Radic et al., 2021; Novosel et al., 2022). Polyunsaturated fatty acids (storage lipids) responsible for maintaining the rigidity of cell membranes (De Carvalho and Caramujo, 2018). Structural lipids are responsible for the change in the cellular environment, pathways of cell signaling, and membrane fluidity (Aratboni et al., 2019; Rogowska and Szakiel, 2020). Sterols control membrane permeability and fluidity in detection of osmotic changes (Novosel et al., 2022). In a hypersaline environment, marine species showed the transition from hydrophobic to hydrophilic characters at high salinities due to the quantity and proportion of sulfated and carboxylated polysaccharides, for example, *Cylindrotheca closterium* (Aquino et al., 2011; Arata et al., 2017). This behavior also determines the habit of microalgae as benthos or plankton (Griffiths and Harrison, 2009; Ozkan and Berberoglu, 2013; Novosel et al., 2021) With a sudden increase in salinity momentary entry of ions such as Na^+, Cl^-, and K^+ occur but the overall concentration in cyanobacteria is maintained quite low (Hagemann, 2013; Oren, 2015) block the high salt concentration due to Na^+/H^+ antiporter and H^+/ATPase pumps on the vacuole and cell membrane, respectively. Further reduction in size and yield is regulated by Na^+ channel blockers with the rising NaCl concentration (Taheri and Shariati, 2013; Ishika et al., 2018; Gautam and Kapoor, 2022). Several low molecular weight organic compounds or osmolytes enter the cell by transportation or can be synthesized de novo without affecting the cell's metabolic activities (Sirisattha et al., 2016). The salt concentration present on the exterior of the cell determines its concentration (Edbeib et al., 2016). Sucrose and trehalose as osmolytes provide protection at low salt concentrations in cyanobacteria. For example, *C. chthonoplastes* synthesizes trehalose at low concentrations while glucosylglycerol at high concentrations of salinity. Glucosylglycerol is also reported to be prevalent in *Coleofasciculus* and *Synechocystis*. Glycine betaine was found to be synthesized at high salt concentration from *A. halophytica*, *Synechocystis DUN52*, and *Gloeocapsa* strain ACMMN107. Glucosylglycerate from *Prochlorococcus* and *Synechococcus PCC7002*, trimethylamine-N-oxide from *Cyanothece* and *Gloeothece*, glutamate betaine from *Calothrix/Rivularia* and mycosporine-like amino acids (MAAs) isolated from cyanobacteria associated with saltern evaporation ponds (Volkmann et al., 2006; Klahn et al., 2010; Goh et al., 2010; Oren A2015).

10.5 ROLE IN ENVIRONMENTAL CLEAN-UP

10.5.1 SALINE SOIL RESTORATION

The large-scale enhancement of water-soluble salts in soil is referred to as soil salinization which affects crop production, pollutes freshwater sources, loss of biodiversity, damages to the economy and urban setup, deteriorates health, and poses a hazard to the ecosystem (Butcher et al, 2016; Shahid et al., 2018; Majeed and Muhammad, 2019; Zorb et al., 2019; Rocha et al., 2020; Pereira et al., 2020). Salt-impacted soil can be divided into saline, sodic, and saline–sodic, respectively. Cyanobacteria and microalgae due to their unique features can withstand extreme environments and provide a natural solution to the restoration of the ecosystem. Cyanobacteria and microalgae maintain the osmotic equilibrium by accumulation of compatible solutes (Kirsch et al., 2019; Shetty et al., 2019). The EPS production helps to stabilize the soil, forms aggregate, checks overflow, binds sodium ions, forms biofilm protecting plants from salt stress, and removes soluble sodium by biosorption (Singh et al., 2019; Li et al., 2019; Sadeghi et al., 2020). Cyanobacteria increases fertility and provides nutrition to plants by nitrogen fixation and carbon entrapment and requisitioning (Schipper et al., 2019). Uptake of contaminants from soil promotes salt movement to lower layers to prevent the crop from being damaged. Significant yield was observed in rice, wheat, and maize due to restoration of salt-affected soil by cyanobacteria and microalgae (Li et al., 2019). Cyanobacteria from biocrusts with their short–wet cycles restore the salinity in the desert ecosystem by the nutrient build-up and boost the soil conditions. Therefore, biocrusts have appeared as a prospective tool for the re-establishment of ecosystem operation (Antoninka et al., 2020). Inoculation of saline soil with pelletized cyanobacteria can open a new avenue for arid soil restoration (Roman et al.,

2020). Mine site restoration by carbon sequestration encourages seedling growth in such soil and remediates heavy metals and hydrocarbon from such polluted soil (Perera, 2018; Williams et al., 2019). Moreover, the dead and decomposed cyanobacteria have the capacity to increase soil biomass and water-holding properties (Rady et al., 2018; Rocha et al., 2020).

10.5.2 DESALINATION OF WATER

The challenge for the high demand for sustainable sources of water can be met by desalination of sea and brackish water. The present technologies are costly and require high energy expenditure with severe environmental threats. The adaptation of microalgae to hypersaline environment involves the regulation of the entry and exit of ions through the plasma membrane, accumulation of osmolytes synthesized photosynthetically, and turgor maintenance (Sahle-Demessie et al., 2019). The green algae *Scenedesmus* is reported to remove dissolved solids, Cl^-, PO^4, Na^+, NO^{3-}, and SO^{4-} efficiently from the seawater and utilize them for its growth and developmental processes (Gautam and Kapoor, 2022). *Scenedesmus obliquus* is reported to remove NaCl effectively by the process of adsorption rather than absorption which was evident from the synthesis of a single layer on the algal surface (Wei et al., 2020). *Dunaliella*, *Nannochloropsis oculata*, and *Chlorella vulgaris* successfully removed salt ions such as nitrogen, phosphate, and fluoride. With the increase in salinity, *S. obliquus* removed higher NaCl concentrations from brackish water as well as K^+, Ca^{2+}, and Mg^{2+} (Wei et al., 2020; Gautam and Kapoor, 2022). *Parelaphostrongylus tenuis* has the remarkable capacity to uptake salts into the vacuoles of its cells (Yensen, 2008; Gautam and Kapoor, 2022). Thus, desalination can emerge as a sustainable approach for a greener economy in the future.

10.5.3 WASTE WATER TREATMENT

The domestic, municipal, agricultural, and industrial effluents are designated as wastewater. The pollutants and contaminants present in the wastewater can be effectively removed by cultivation of microalgae and cyanobacteria in wastewater for generation of biomass. This approach is cost effective, and environment friendly with recycling of resources, low energy requirement, and low sludge formation, and the resulting biomass can be used for biodiesel production (Essa et al., 2018). The industries responsible for liberating highly saline wastewater are the pickling plants, and tanneries from oil and gas recovery processes (Oren, 2010; Gautam and Kapoor, 2022). Heavy metals, chemicals, and toxic petroleum compounds present in the wastewater of oil and gas industries have been designated as the produced water. Microalgae such as *Dunaliella*, *Chlorella*, *Scenedesmus*, *Nannochloropsis salina*, and *Parachlorella kessleri* could grow in produced water and successfully treat the same (Jabri et al., 2021; Rahman et al., 2021). Wastewater polluted with pharmaceutical and personal care compounds for example triclosan could be removed by extracellular or intracellular adsorption, accumulation, and degradation by microalgae (Xiong et al., 2018; Jabri et al., 2021).

TABLE 10.1 Selected Examples of Application of Halophilic Cyanobacteria and Microalgae for Bioremediation of Pollutants.

Bioremediation agent	Pollutant	Results	References
i) Cyanobacteria	Heavy metals and hydrocarbon from polluted mine site	Pollutants removed successfully	Perera, 2018; Williams et al., 2019
ii) Microalgae *Dunaliella*, *N. oculata* *C. vulgaris* *Scenedesmus* *N. salina*, *P. kessleri*	Nitrogen, phosphate, fluoride, K^+, Ca^{++}, and Mg^{++} Chloride, phosphate, sodium chloride, nitrate, sulfate Heavy metals, chemicals, and toxic petroleum compounds from produced water Pharmaceutical and personal care compounds such as triclosan	Pollutants effectively removed Pollutants effectively removed Pollutants successfully removed	Wei et al., 2020 Gautam and Kapoor, 2022 Gautam and Kapoor, 2022; Wei et al., 2020 Jabri et al., 2021 Rahman et al., 2021 Xiong et al., 2018 Jabri et al., 2021

10.6 PRODUCTION OF SECONDARY METABOLITES AND THEIR APPLICATIONS

In response to environmental stress microalgae and cyanobacteria synthesize certain organic compounds which are not involved in primary metabolic activities of the cell but are utilised in the secondary functions of the cell. Such, compounds are designated as secondary metabolites. In order to adapt to high salinity organic osmolyte glycerol is formed in high concentration in *Dunaliella* without hindering the functions of the enzymes. The consequence of high salinity shock reveals an association between glycerol synthesis and starch degradation in halophilic microalgae such as *D. salina* and *Chlamydomonas pulsatilla* (Skjanes et al., 2013). Glycerol produced as a compatible solute is highly soluble and nontoxic as it is inert and does not intervene with other metabolic pathways. Glycerol accumulation and increase in content occurs in *Chlorella autotrophica, Scenedesmus, Micrasterias* along with *Dunaliella* and *Chlamydomonas*. Under osmotic pressure, glycerol maintains the original volume of the cell (Ho et al., 2017; Shetty et al., 2019). Glycerol is used in different industries such as cosmetics, feedstock, pharmaceuticals, paints, textile, pulp and paper, automotive, etc., (Wang et al., 2001). Like glycerol, proline, a low molecular weight, neutrally charged osmolyte with high solubility increases with the increase in salinity in *Chlamydomonas reinhardtii* (Shetty et al., 2019). Around 24% NaCl concentration, *D. salina* accumulates carotenoids, which is the richest source of carotene (Das Sarma et al., 2010). Natural β-carotene produced by *Dunaliella* comprises of all trans and 9-cis-β-carotene while the synthetic ones consist of trans-β-carotene. The former is a better quencher of singlet oxygen and other free radicals and is assimilated in animal livers (Oren, 2010). These are the nonenzymatic antioxidants that act as scavengers of ROS upon UV exposure (Sirisattha et al., 2016). β-Carotene has a wide range of applications as a natural food-coloring agent, emulsion in confectionery and prepared food, source of provitamin A in animal feed, additive in cosmetics, health products, and multivitamin preparations, in pharmaceutical and medical fields as an antitumor and heart disease preventive mediators (Sirisattha et al., 2016; Coker, 2016). Microalgae *Scenedesmus*, and *Nannochloropsis* are involved in the production of zeaxanthin, a yellow-colored carotenoid that finds application in the pharmaceutical, cosmetics, and food industries (Encarnacao, 2015). Marine cyanobacteria and microalgae synthesize and build up MAAs in ample amounts in hypersaline environments. MAAs provide protection to the cell by absorbing UV radiation inhibiting ROS production and liberating energy in the form of heat. It acts as an antioxidant and checks and inhibits skin aging and skin cancer (Sirisattha et al., 2016; Singh, 2017). Immunomodulatory properties of sulfated polysaccharides of *A. halophytica*, a halophilic unicellular cyanobacterium found in a hypersaline environment have been reported. Enhanced salt stress indicates the production of trehalose in *Chlamydomonas*, *Chlorella*, and *Scytonema*. High lipid accumulation in response to salinity is noted in truly salt-adapted algae (Ho et al., 2017; Shetty et al., 2019).

10.7 PRODUCTION OF BIOFUELS

The noteworthy feature of halophilic microalgae is to create biomass and biofuel sustaining in hypersaline environments (Das Sarma et al., 2010). Exposure to high salinity showed enhanced lipid synthesis in microalgae. It has been reported that there is a sharp increase in the concentration of saturated and monounsaturated fatty acids in *Dunaliella* which has been considered as a potential candidate for the production of biodiesel. *D. salina* KSA-HS022 isolated from a lagoon was reported to augment the features of biodiesel with an increase of salinity from 50 to 150% (Abomohra et al., 2020). Further higher concentration of lipid build-up was noted in *S. obliquus* (Gautam and Kapoor, 2022). Biofuel production involves four steps- isolation, characterization, biomass production, and harvesting. Different types of biofuels such as production of biohydrogen, biogas, and biodiesel production by utilizing *D. salina* and other microalgae as a source of raw materials have been reported. Therefore, biofuel generation as a daily harvest of halophilic microalgae is an environment-friendly way to use nonarable lands, and nonpotable water and can reduce freshwater and land use (Das Sarma et al., 2010).

10.8 CONCLUSION AND FUTURE RESEARCH

The features such as rapid growth, simple culture methodology, high capacity of adaptation in adverse conditions, uptake of nutrients, removal of pollutants, and assimilation of carbon dioxide by halophilic cyanobacteria and microalgae indicates an enormous potential of these microorganisms in environmental clean-up. The desalination of water by halophilic microalgae and cyanobacteria acts as a safe alternative to obtain clean potable water and can meet the huge demand for water. These microalgae play a significant role in the restoration of salt-impacted soils and can serve as an effective tool in the field of biotechnology. Further, halophilic microalgae can be utilized as a feedstock for diesel which can be successfully performed by cultivating them in industrial waste water accompanied by lipid production. For a better understanding and knowledge of this vast diversity of halophilic microalgae and cyanobacteria and their mode of adaptation under stress conditions, further studies on their signaling pathways together with next generation sequencing strategies should be performed. Explorations and research are required to unearth the enormous capacities of halophilic microalgae in biofuel production. The harvested biomass is required to be valorized to augment the economic viability and environment sustainability. Thus, halophilic microalgae and cyanobacteria act as a reservoir of biomolecules and their application in various fields will result in the growth, sustenance, and protection of the ecosystem in the future.

ACKNOWLEDGMENT

I express my heartfelt thanks to the Department of Microbiology, Lady Brabourne College, Kolkata, West Bengal, India to provide me the opportunity to perform the research work.

KEYWORDS

- **salinization**
- **halophilic**
- **cyanobacteria**
- **microalgae**
- **environment clean-up**

REFERENCES

Abed, R. M. M.; Garcia-Pichel, F.; Herna´ndez-Marine, M. Polyphasic Characterization of Benthic, Moderately Halophilic, Moderately Thermophilic Cyanobacteria with Very Thin Trichomes and the Proposal of *Halomicronema excentricum* gen. nov., sp. nov. *Arch. Microbiol.* **2002,** *177*, 361–370.

Abomohra, A. E. F.; El-Naggar, A. H.; Alaswad, S. O.; Elsayed, M.; Li, M.; Li, W. Enhancement of Biodiesel Yield from a Halophilic Green Microalga Isolated under Extreme Hypersaline Conditions through Stepwise Salinity Adaptation Strategy. *Bioresour. Technol.* **2020,** *310*, 123462.

Al-Abri, K. Use of Molecular Approaches to Study the Occurrence of Extremophiles and Extremodures in Non-extreme Environments, PhD Thesis, University of Sheffield, UK, 2011.

Ali, I.; Prasongsuk, S.; Akbar, A.; Aslam, M.; Lotrakul, P.; Punnapayak, H.; Rakshit, S. K. Hypersaline Habitats and Halophilic Microorganisms, *Maejo Int. J. Sci. Technol.* **2016,** *10* (3), 330–345.

Antoninka, A.; Faist, A.; Rodriguez-Caballero, E.; Young, K. E.; Chaudhary, V. B.; Condon, L. A.; Pyke, D. A. Biological Soil Crusts in Ecological Restoration: Emerging Research and Perspectives. *Restor. Ecol.* **2020,** *28*, DOI: 10.1111/rec.13201.

Arata, P. X.; Alberghina, J.; Confalonieri, V.; Errea, M. I.; Estevez, J. M.; Ciancia, M. Sulfated Polysaccharides in the Freshwater Green Macroalga Cladophora surera not Linked to Salinity Adaptation. *Front. Plant Sci.* **2017,** *8*, 1972–1982

Aratboni, A. H.; Rafei, N.; García-Granados, R.; Alemzadeh, A.; MoronesRamirez, J. Biomass and Lipid Induction Strategies in Microalgae for Biofuel Production and Other Applications. *Microb. Cell Fact.* **2019,** *18*, 178–195.

Aquino, R. S.; Grativol, C.; Mourão, P. A. S. Rising from the Sea: Correlations Between Sulfated Polysaccharides and Salinity in Plants. *PLoS One* **2011,** *6*, e18862.

Australian Government. Keeping Salt Out of the Murray [Online] 2020. https://www.mdba.gov.au/publications/brochures-factsheets/keeping-salt-out-murray (accessed June 7, 2020).

Barati, B.; Gan, S. Y.; Lim, P. E.; Beardall, J.; Phang, S. M. Green Algal Molecular Responses to Temperature Stress. *Acta Physiol. Plant* **2019,** *41*, 26–45.

Bell, E. Life at Extremes: Environments, Organisms, and Strategies for Survival; CABI, 2012. DOI: 10.1079/9781845938147.0000

Bhatt, H. H.; Pasricha, R.; Upasani, V. N. Isolation and Characterization of a Halophilic Cyanobacterium Euhalothece SVLHO1 from Sambhar Saltlake India. *Int. J. Curr. Microbiol. App. Sci.* **2016,** *5* (2), 215–224.

Borowitzka, M. A. Biology of Microalgae. In Microalgae in Health and Disease Prevention; Levine, I. A., Fleurence, J., Eds.; Academic Press: London, 2018a; pp 23–72.

Borowitzka, M. A. The "Stress" Concept in Microalgal Biology—Homeostasis, Acclimation and Adaptation. *J. Appl. Phycol.* **2018b,** *30*, 2815–2825.

Briggs, S.; Taws, N. Impact of Salinity on Biodiversity—Clear Understanding or Muddy Confusion. *Aust. J. Bot.* **2003,** *51* (6). DOI: 10.1071/BT02114.

Britton, K. L.; Baker, P. J.; Fisher, M.; Ruzheinikov, S.; Gilmour, D. J.; Bonete, M. J.; Ferrer, J.; Pire, C.; Esclapez, J.; Rice, D. W. Analysis of Protein Solvent Interactions in Glucose Dehydrogenase from the Extreme Halophile Haloferax Mediterranei. *Proc. Natl. Acad. Sci. U. S. A.* **2006,** *103*, 4846–4851. DOI: 10.1073/pnas.0508854103

Burns, B. P.; Goh, F.; Allen, M.; Neilan, B. A. Microbial Diversity of Extent Stromatolites in the Hypersaline Marine Environment of Shark Bay Australia. *Environ. Microbiol.* **2004,** *6*, 1096–1101.

Butcher, K.; Wick, A. F.; DeSutter, T.; Chatterjee, A.; Harmon, J. Soil Salinity: A Threat to Global Food Security. *Agron. J.* **2016,** *108*, 2189–2200. DOI: 10.2134/agronj2016.06.0368.

Chaitanya, V.; Rama Krishna, C.; Ramana, V.; Beebi, S. K. Salinity Stress and Sustainable Agriculture-A Review. *Agric. Rev.* **2014,** *35* (1), 34–41.

Chattopadhyay, M. K.; Kern, R.; Mistou, M. Y.; Dandekar, A. M.; Uratsu, S. L.; Richarme, G. The Chemical Chaperone Proline Relieves the Thermosensitivity of a DnaK Deletion Mutant at 42 Degrees C. *J. Bacteriol.* **2004,** *186* (23), 8149–8152. DOI: 10.1128/JB.186.23.8149-8152.2004.

Cohen, Y.; Jørgensen, B. B.; Padan, E.; Shilo, M. Sulphide-Dependent Anoxygenic Photosynthesis in the Cyanobacterium *Oscillatoria limnetica*. *Nature* **1975a,** *257*, 489–492.

Coker, J. A. Extremophilesand Biotechnology: Current Uses and Prospects. *F1000Research* **2016,** 1–7.

Das Sarma, P.; Coker, J. A.; Huse, V.; Das Sarma, S. Halophiles, Industrial Applications. In *Encyclopedia of Industrial Biotechnology: Bioprocess, Bioseparatio, and Cell Technology*, 2010.

DasSarma, S.; DasSarma, P. Halophiles. *Encycl. Life Sci.* 2012. DOI: 10.1002/9780470015902.a0000394.pu

DasSarma, S.; DasSarma, P. Halophiles and their Enzymes: Negativity Put to Good Use. *Curr. Opin. Microbiol.* **2015,** *25*, 120–126. DOI: 10.1016/j.mib.2015.05.009

De Carvalho, C. C. C. R.; Caramujo, M. J. The Various Roles of Fatty Acids. *Molecules* **2018,** *23*, 2583–2619.

de Wit, R.; Falco´n, L. I.; Carpy-Roubaud, C. Heterotrophic Dinitrogen Fixation (acetylene reduction) in Phosphate-Fertilised *Microcoleus chthonoplastes* Microbial Mat from the Hypersaline Inland Lake 'la Salada de Chiprana' (NE Spain). *Hydrobiologia* **2005,** *534*, 245–253.

Devi, T.; Rasool, S. Adaptation of Halophiles at Hypersaline Environment, *Int. J. Adv. Res. Sci. Eng.* **2017,** *6* (9), 1444–1450.

Edbeib, M. F.; Wahab, R. A.; Huyop, F. Halophiles: Biology, Adaptation and their Role in Decontamination of Hypersaline Environments. *World J. Microbiol. Biotechnol.* **2016,** *32*, 135.

El- Swaify, S. A. Soiland Water Salinity. In *Plant Nutrient Management in Hawaii's Soils, Approaches for Tropical and Subtropical Agriculture*; Silva, J. A., Uchida, R., Eds.; College of Tropical Agriculture and Human Resources, University of Hawaii: Manoa, 2000.

Encarnacao, T.; Pais, A. A. C. C.; Campos, M. G.; Burrows, H. D. Cyanobacteria and Microalgae:A Renewable Source of Bioactive Compounds and Other Chemicals. *Sci. Prog.* **2015,** 145–168.

Essa, D.; Abo-Shady, A. M.; Khairy, H. M.; Abomohra, A. E. F.; Elshobary, M. E. Potential Cultivation of Halophilic Oleaginous Microalgae on Industrial Wastewater, *Egypt. J. Bot.* **2018,** *58* (2), 205–216.

Fofonker, F.; Mollegard, D.; Ong, M.; Su Yoon, H.; Bhattacharya, D. Genomic Analysis of Picochlorum Apecies Reveals how Microalgae may Adapt to Variable Environments. *Mol. Biol. Evol.* **2018,** *35*, 2702–2711.

Gautam, S.; Kapoor, D. Chapter8, Application of Halophilic Algae for Water Desalination. In *Handbook of Algal Biofuels, Aspects of Cultivation, Conversion and Biorefinery*, 2022; pp 167–179.

Goh, F.; Barow, K. D.; Burns, B. P.; Neilan, B. A. Identification and Regulation of Novel Compatible Solutes from Hypersaline Stromatolite-Associated Cyanobacteria. *Arch. Microbiol.* **2010,** *192*, 1031–1038.

Grifths, M. J.; Harrison, S. T. L. Lipid Productivity as a Key Characteristic for Choosing Algal Species for Biodiesel Production. *J. Appl. Phycol.* **2009,** *21*, 493–507.

Hagemann, M. Genomics of Salt Acclimation: Synthesis of Compatible Solutes among Cyanobacteria. *Adv. Bot. Res.* **2013,** *65*, 27–55.

Harmon, J. P.; Daigh, A. L. M. Attempting to Predict the Plant-Mediated Trophic Effects of Soil Salinity: A Mechanistic Approach to Supplementing Insufficient Information. *Food Webs* **2017,** *13*, 67–79. DOI: 10.1016/j.fooweb.2017.02.002

Ho, S. H.; Nakanishi, A.; Kato, Y.; Yamasaki, H.; Chang, J. S.; Misawa, N.; Yuu, H.; Jun, M.; Tomohisda, H.; Kondo, A. Dynamic Metabolic Profiling Together with Transcription Analysis Reveals Salinity-Induced Starch-to-Lipid Biosynthesis in alga Chlamydomonas sp. JSC4. *Sci. Rep.* **2017,** *7*, 45471.

Ionescu, D.; Lipski, A.; Altendorf, K.; Oren, A. M. Characterization of the Endoevaporitic Microbial Communities in a Hypersaline Gypsum Crust by Fatty Acid Analysis. *Hydrobiologia* **2007,** *576*, 15–26.
Ishika, T.; Bahri, P. A.; Laird, D. W.; Moheimani, N. R. The Effect of Gradual Increase in Salinity on the Biomass Productivity and Biochemical Composition of Several Marine, Halotolerant, and Halophilic Microalgae. *J. Appl. Phycol.* **2018,** *30* (3), 14531464.
Ivošević DeNardis, N.; Pečar Ilić, J.; Ružić, I.; Novosel, N.; Mišić Radić, T.; Weber, A.; Kasum, D.; Pavlinska, P.; Katalin Balogh, R.; Hajdu, B.; Marček Chorvátová, A.; Gyurcsik, B. Algal Cell Response to Laboratory-Induced Cadmium Stress: A Multimethod Approach. *Eur. Biophys. J.* **2019,** *48*, 231–248.
Jabri, H. A.; Das, P.; Khan, S.; Thaher, M.; Quadir, M. A. Treatment of Wastewaters by Microalgae and the Potential Application of the Produced Biomass- A Review. *Water* **2021,** *13*, 27.
Karan, R.; Capes, M. D.; DasSarma, S. Function and Biotechnology of Extremophilic Enzymes in Low Water Activity. *Aquat. Biosyst.* **2012,** *8*, 1. DOI: 10.1186/2046-9063-8-4
Kerkar, S. S. Ecology of Hypersaline Microorganisms, Department of Biotechnology, Goa University, Goa, 2004, pp 37–47.
Kirkwood, A. E.; Buchheim, J. A.; Buchheim, M. A.; Henley, W. J. Cyanobacterial Diversity and Halotolerance in a Variable Hypersaline Environment. *Microb. Ecol.* **2007,** *55*, 453–465. DOI: 10.1007/s00248-007-9291-5.
Kirsch, F.; Klähn, S.; Hagemann, M. Salt-Regulated Accumulation of the Compatible Solutes Sucrose and Glucosylglycerol in Cyanobacteria and Its Biotechnological Potential. *Front. Microbiol.* **2019,** *10*. DOI: 10.3389/fmicb.2019.02139. 120.
Kla¨hn, S.; Steglich, C.; Hess, W. R.; Hagemann, M. Glucosylglycerate: A Secondary Compatible Solute Common to Marine Cyanobacteria from Nitrogen-Poor Environments. *Environ. Microbiol.* **2010,** *12*, 83–94.
Kushner, D. J. Growth and Nutrition of Halophilic Bacteria. In *The Biology of Halophilic Bacteria*; Vreeland, R. H., Hochstein, L. I., CRC Press: Boca Raton, 1993; pp 87–103.
Lattemann, S.; Ho¨pner, T. Environmental Impact and Impact Assessment of Seawater Desalination. *Desalination* **2008,** *220* (1–3), 1–15.
Leogrande, R.; Vitti, C. Use of Organic Amendments to Reclaim Saline and Sodic Soils: A Review. *Arid Land Res. Manag.* **2019,** *33*, 1–21. DOI: 10.1080/15324982.2018.1498038.
Li, H.; Zhao, Q.; Huang, H. Current States and Challenges of Salt-Affected Soil Remediation by Cyanobacteria. *Sci. Total Environ.* **2019,** *669*, 258–272. DOI: 10.1016/j.scitotenv.2019.03.104.
Majeed, A.; Muhammad, Z. Salinity: A Major Agricultural Problem—Causes, Impacts on Crop Productivity and Management Strategies. In *Plant Abiotic Stress Tolerance*; Hasanuzzaman, M., Hakeem, R. K., Nahar, K., Alharby, H. F., Eds.; Springer: Cham, Switzerland, 2019; pp 83–99. ISBN 978-3-030-06117- 3.
Minhas, P. S.; Ramos, T. B.; Ben-Gal, A.; Pereira, L. S. Coping with Salinity in Irrigated Agriculture: Crop Evapotranspiration and Water Management Issues. *Agric. Water Manag.* **2020,** *227*. DOI: 10.1016/j.agwat.2019.105832.
Mišić Radić, T.; Čačković, A.; Penezić, A.; Dautović, J.; Lončar, J.; Omanović, D.; Juraić, K.; Ljubešić, Z. Physiological and Morphological Response of Marine Diatom *Cylindrotheca closterium* (Bacillariophyceae) Exposed to Cadmium. *Eur. J. Phycol.* **2021,** *56*, 24–36.
Muller, V.; Oren, A. In *Extremophiles*, 2003; vol 7, pp 261–266.

Ng, W. V.; Kennedy, S. P.; Mahairas, G. G.; Berquist, B.; Pan, M.; Shukla, H. D.; Lasky, S. R.; Baliga, N. S.; Thorsson, V.; Sbrogna, J. Genome Sequence of Halobacterium Species NRC-1. *Proc. Natl. Acad. Sci.* **2000,** *97*, 12176–12181. DOI: 10.1073/pnas.190337797

Novosel, N.; Ivošević DeNardis, N. Structural Features of the Algal Cell Determine Adhesion Behavior at a Charged Interface. *Electroanalysis* **2021,** *6*, 1436–1443.

Novosel, N.; Radic, T. M.; Zorinc, M. L.; Zemla, J.; Lekka, M.; Vrana, I.; Gasparovic, B.; Horvat, L.; Kasum, D.; Legovic, T.; Zutinin, P.; Udovic, M. G.; De Nardis, N. I. Salinity Induced Chemical, Mechanical, and Behavioral Changes in Marine Algae. *J. Appl. Phycol.* **2022,** *34*, 1293–1309.

Nu¨bel, U.; Garcia-Pichel, F.; Muyzer, G. The Halotolerance and Phylogeny of Cyanobacteria with Tightly Coiled Trichomes (Spirulina Turpin) and the Description of *Halospirulina tapeticola* gen. nov., sp. nov. *Int. J. Syst. Evol. Microbiol.* **2000b,** *50*, 1265–1277.

Oren, A. Diversity of Halophilic Microorganisms: Environments, Phylogeny, Physiology and Applications. *J. Ind. Microbiol. Biotechnol.* **2002,** *28*, 56–63.

Oren, A. Molecular Ecology of Extremely Halophilic Archaea and Bacteria. *FEMS Microbiol. Ecol.* **2002c,** *39*, 1–7. DOI: 10.1111/j.1574- 6941.2002.tb00900.x.

Oren, A. Industrial and Environmental Applications of Halophilic Microorganisms. *Environ. Technol.* **2010,** *31* (8-9), 825–834.

Oren, A. Cyanobacteria in Hypersaline Environments: Biodiversity and Physiological Properties. *Biodivers. Concerv.* **2015,** *24*, 781–798.

Ozkan, A.; Berberoglu, H. Physico-Chemical Surface Properties of Microalgae. *Colloids Surf B* **2013,** *112*, 287–293.

Paerl, H. W.; Pinckney, J. L.; Steppe, T. F. Cyanobacterial-Bacterial Mat Consortia: Examining the Functional Unit of Microbial Survival and Growth in Extreme Environments. *Environ. Microbiol.* **2000,** *2*, 11–26.

Patel, B. B.; Patel Bharat, B.; Dave, R. S. Studies on Infiltration of Saline–Alkali Soils of Several Parts of Mehsana and Patan Districts of North Gujarat. *J. Appl. Technol. Environ. Sanit.* **2001,** *1* (1), 87–92.

Pavlinska, Z.; Chorvat, D.; Mateasik, A.; Jerigova, M.; Velic, D.; Ivošević DeNardis, N.; Marcek Chorvatova, A. Fluorescence Responsiveness of Unicellular Marine Algae Dunaliella to Stressors under Laboratory Conditions. *J. Biotechnol.* **2020,** *324S*, 100018.

Pereira, P.; Barceló, D.; Panagos, P. Soil and Water Threats in a Changing Environment. *Environ. Res.* **2020,** *186*, 109501. DOI: 10.1016/j.envres.2020.109501

Perera, I.; Subashchandrabose, S. R.; Venkateswarlu, K.; Naidu, R.; Megharaj, M. Consortia of Cyanobacteria/Microalgae and Bacteria in Desert Soils: An Underexplored Microbiota. *Appl. Microbiol. Biotechnol.* **2018,** *102*, 7351–7363. DOI: 10.1007/s00253-018-9192-1.

Qadir, M.; Ghafoor, A.; Murtaza, G. Amelioration Strategies for Saline Soils: A Review. *Land Degrad. Dev.* **2000,** *11*, 501–521. DOI: 10.1002/1099-145X(200011/12)11:63.0.CO;2-S.

Rady, M. M.; Taha, S. S.; Kusvuran, S. Integrative Application of Cyanobacteria and Antioxidants Improves Common Bean Performance under Saline Conditions. *Sci. Hortic. Amst.* **2018,** *233*, 61–69. DOI: 10.1016/j.scienta.2018.01.047.

Rahman, A.; Pan, S.; Houston, C.; Selvaratnam, T. Evaluation of Galdieria sulphuraria and *Chlorella vulgaris* for the Bioremediation of Produced Water. *Water* **2021,** *13*, 1183.

Rocha, F, Lucas-Borja, M. E.; Pereira, P.; Munoz-Rojas, M. Cyanobacteria as a Nature Based Biotechnological Tool for Restoring Salt-Affected Soils. *Agronomy* **2020,** *10*, 1321.

Rogowska, A.; Szakiel, A. The Role of Sterols in Plant Response to Abiotic Stress. *Phytochem. Rev.* **2020,** *19*, 1525–1538.

Román, J. R.; Chilton, A. M.; Cantón, Y.; Muñoz-Rojas, M. Assessing the Viability of Cyanobacteria Pellets for Application in Arid Land Restoration. *J. Environ. Manag.* **2020,** *270*, 110795. DOI: 10.1016/j.jenvman.2020.110795. 1.

Roney, H. C.; Booth, G. M.; Cox, P. A. Competitive Exclusion of Cyanobacterial Species in the Great Salt Lake. *Extremophiles* **2009,** *13*, 355–361.

Sadeghi, S. H.; Kheirfam, H.; Zarei Darki, B. Controlling Runoff Generation and Soil Loss from Field Experimental Plots through Inoculating Cyanobacteria. *J. Hydrol.* **2020,** *585*, 124814. DOI: 10.1016/j.jhydrol.2020.124814.

Sahle- Demessie, E.; Hassan, A. A.; Badawy, A. E. Biodesalination of Brakish and Sea Water using Halophytic Algae. *Desalination* **2019,** *465*, 104–113.

Said-Al, A.; Omer, E. A. Medicinal and Aromatic Plants Production under Salt Stress: A Review. *Herba Polonica* **2011,** *57*, 72–87

Satyanarayana, T.; Johri, B. N.; Prakash, A. In *Microorganisms in Environmental Management: Microbes and Environment*; Springer: Berlin, 2012.

Saum, S. H.; Mu¨ller, V. Regulation of Osmoadaptation in the Moderate Halophile Halobacillus Halophilus: Chloride, Glutamate and Switching Osmolyte Strategies. *Saline Syst.* **2008,** *4*, 2014. DOI: 10. 1186/1746-1448-4-4.

Schipper, K.; Al Muraikhi, M.; Alghasal, G. S. H. S.; Saadaoui, I.; Bounnit, T.; Rasheed, R.; Dalgamouni, T.; Al Jabri, H. M. S. J.; Wijffels, R. H.; Barbosa, M. J. Potential of Novel Desert Microalgae and Cyanobacteria for Commercial Applications and CO2 Sequestration. *J. Appl. Phycol.* **2019,** *31*, 2231–2243. DOI: 10.1007/s10811- 019-01763-3.

Schleper, C.; Puhler, G.; Kuhlmorgen, B.; Zillig, W. Life at Extremely Low pH. *Nature* **1995,** *375*, 741–742.

Shahid, S. A.; Zaman, M.; Heng, L. Soil Salinity: Historical Perspectives and a World Overview of the Problem. In *Guideline for Salinity Assessment, Mitigation and Adaptation Using Nuclear and Related Techniques*; Springer International Publishing: New York, NY, USA, 2018; pp 43–53. ISBN 978-3-319-96190-3. 16.

Shankar, V.; Evelin, H. Strategies for Reclamation of Saline Soils. In *Microorganisms in Saline Environments: Strategies and Functions*; Soil Biology; Giri, B., Varma, A., Eds.; Springer: Cham, Switzerland, 2019; vol 56, pp 439–449. ISBN 978-3-030-18974-7.

Shetty, P.; Gitau, M. M.; Maroti, G. Salinity Stress Responses and Adaptation Mechanisms in Eukaryotic Green Microalgae. *Cells* **2019,** *8*, 1657.

Singh, S.; Kant, C.; Yadav, R. K.; Reddy, Y. P.; Abraham, G. Cyanobacterial Exopolysaccharides: Composition, Biosynthesis, and Biotechnological Applications. In *Cyanobacteria from Basic Science to Applications*; Mishra, A. K., Tiwari, D. N., Rai, A. N., Eds.; Academic Press: Cambridge, MA, USA, 2019; pp 347–358. ISBN 978-0-12-814667-5.

Singh, R.; Parihar, P.; Singh, M.; et al. Uncovering Potential Applications of Cyanobacteria and Algal Metabolites in Biology, Agriculture and Medicine: Current Status and Future Prospects. *Front. Microbiol.* **2017,** 1–37.

Sirisattha, R. W.; Kageyama, H.; Takabe, T. Halophilic Microorganism Resources and their Applications in Industrial and Environmental Biotechnology. *AIMS Microbiol.* **2016,** *2* (1), 42–54.

Skjanes, K.; Rebours, C.; Lindblad, P. Potential for Green Microalgae to Produce Hydrogen, Pharmaceuticals, and Other High Value Products in a Combined Process. *Crit. Rev. Biotechnol.* **2013,** *33* (2), 172–215.

Sommer, V.; Mikhailyuk, T.; Glaser, K.; Karsten, U. Uncovering Unique Green Algae and Cyanobacteria Isolated from Biocrusts in Highly Saline Potash Tailing Pile Habitats Using an Integrative Approach. *Microorganisms* **2020,** *8*, 1667.

Srivastava, P.; Wu, Q. S.; Giri, B. Salinity: An Overview. In *Microorganisms in Saline Environments: Strategies and Function*; Soil Biology, 2019; pp 3–18.

Taheri, R.; Shariat, M. Study of the Inhibitory Effect of the Media Culture Parameters and Cell Population to Increase the Biomass Production of Dunaliella tertiolecta. *Prog. Biol. Sci.* **2013,** *3* (2), 123133.

Vecchio, M. C.; Golluscio, R. A.; Rodríguez, A. M.; Taboada, M. A. Improvement of Saline-Sodic Grassland Soils Properties by Rotational Grazing in Argentina. *Rangel. Ecol. Manag.* **2018,** *71*, 807–814. DOI: 10.1016/j.rama.2018.04.010.

Vengosh, A. Salinization and Saline Environments. In *Salinization Treatise in Geochemistry*, Second Edition, 2014; vol 11, pp 325–378.

Volkmann, M.; Gorbushina, A. A.; Kedar, L.; Oren, A. The Structure of Euhalothece-362, A Novel Redshifted Mycosporine-Like Amino Acid, from a Halophilic Cyanobacterium (Euhalothece sp.). *FEMS Microbiol. Lett.* **2006,** *258*, 50–54.

Wei, L.; Gao, G.; Shen, G.; Yang, X.; Li, . The Role of Adsorption in Microalgae Biological Desalination: Salt Removal from Brackish Water using *Scenedesmus obliquus*. *Desalination* **2020,** *493*, 114616.

Wichern, F.; Islam, M. R.; Hemkemeyer, M.; Watson, C.; Joergensen, R. G. Organic Amendments Alleviate Salinity Effects on Soil Microorganisms and Mineralisation Processes in Aerobic and Anaerobic Paddy Rice Soils. *Front. Sustain. Food Syst.* **2020,** *4*. DOI: 10.3389/fsufs.2020.00030.

Williams W. D. Anthropogenic Salinization of Inland Waters. *Hydrobiology* **2001a,** *466*, 329–337.

Williams, W. D. Salinization: Unplumbed Salt in a Parched Landscape. *Water Sci. Technol.* **2001b,** *43*, 85–91.

Williams, W.; Chilton, A.; Schneemilch, M.; Williams, S.; Neilan, B.; Driscoll, C. Microbial Biobanking— Cyanobacteria-Rich Topsoil Facilitates Mine Rehabilitation. *Biogeosciences* **2019,** *16*, 2189–2204. DOI: 10.5194/bg16-2189-2019

Xiong, J. Q.; Kurade, M. B.; Jeon, B. H. Can Microalgae Remove Pharmaceutical Contaminants from Water? *Trends Biotechnol.* **2018,** *36*, 30–44.

Yensen, N. P. Halophyte Uses for the Twenty-First Century. In *Ecophysiology of High Salinity Tolerant Plants*; Springer, 2008; pp 367396.

Zhang, G.; Yi, L. Stability of Halophilic Proteins: From Dipeptide Attributes to Discrimination Classifier. *Int. J. Biol. Macromol.* **2013,** *53*, 1–6.

Zörb, C.; Geilfus, C. M.; Dietz, K. J. Salinity and Crop Yield. *Plant Biol.* **2019,** *21*, 31–38. DOI: 10.1111/plb.12884

CHAPTER 11

Structural Hierarchy Imparts Salinity Stress Tolerance Improvement in the Halophyte *Sesuvium portulacastrum* L.

VINAYAK H. LOKHANDE

Department of Botany, J.T.S.S.P.M., Shri Shiv Chhatrapati College of Arts, Commerce and Science, Bodkenagar, Junnar, Pune, Maharashtra, India

ABSTRACT

Structural organization in halophytes respond to salt stress at three different levels—cellular, tissue, and the whole plant; this has led to understand the salt stress tolerance mechanism in halophytes. *Sesuvium portulacastrum* L. (sea purslane) is a facultative, mangrove-associated halophyte. After two decades of extensive research, *S. portulacastrum* has now become one of the choices of model system among halophytes to understand the mechanism of stress tolerance at physiological, biochemical and molecular levels to array of various abiotic stresses such as salinity, drought, heavy metals, industrial effluents, etc. Present chapter deals with the discussion on growth, morphological, physiological and biochemical responses of *S. portulacastrum* to various concentrations of salinity stress at cellular, tissue and whole plant level. The efforts have been taken to understand the stress tolerance responses in *S. portulacastrum* with the help of combinations of in vitro and ex vitro growth culture systems. Interestingly, superior correlation has been found between the salt stress tolerance of callus, tissue and whole plant level. The stress responses recorded at cell in terms of callus and intact tissue like axillary shoots under in vitro conditions has further effectively helped to determine the salt stress responses at whole plant level under ex vitro conditions. Besides, the chapter also gives brief focus on clonal responses

Physiology of Halophytes: Signaling, Omics, and Tolerance Mechanisms.
Nivas Manohar Desai & Ganesh Chandrakant Nikalje (Eds.)

of *S. portulacastrum* to salinity stress which is of value from a scientific perspective for identification of traits associated with great stress tolerance and an applied perspective for identification of superior planting stock for specific restoration and reclamation needs.

11.1 INTRODUCTION

The plants being inherent to their sessile nature are always vulnerable to an array of different biotic (various types of pathogenic attacks) and abiotic (salinity, drought, temperature extremes, nutrient deficiencies, light intensity) environmental stresses (Ali and Yun, 2017). The consequences of these stresses results into limiting the crop productivity and average yield of economically important crop plants worldwide. Unfortunately, the human population is dependent for their food resources on salt-sensitive crop plants that come under the category of "glycophytes." The estimated rate of food requirement for the worldwide ever-growing human population by 2050 is the production of more than 70% of higher food from all sources as against the presently available cultivable land (Calone et al., 2021). However, the usable land availability for agriculture is astonishingly decreasing due to its contamination by toxic saline salts and other contaminants. It has been estimated that out of 5200 million hectares of land, more than 3600 million hectares of land is salinized, and this has led to losses of USD27.5 billion every year (Qadir et al., 2014; Rahman et al., 2021). These multifaceted adverse impacts of salinity associated with anthropogenic and natural calamities have been found responsible for the loss of 1.5 million hectares of cultivable land every year, thereby restricting the land availability by up to 50% for the cultivation of a variety of crops (Karakas et al., 2021). Therefore, with this increasing rate of pollution and its adverse impact on cultivable land due to toxic contaminants and saline salts, the salt sensitive crop plants belonging to glycophytes will not be able to accomplish the food demand of the ever-growing population. In this regard, it is the need of the hour to think over and find out the permanent remedies to the problems associated with salinity for sustainable agricultural development.

Mangroves or halophytes are salt-loving plants and grow under continuous saline conditions. Halophytes are extraordinary plants with the innate capacity to tolerate higher levels of salt concentrations which are otherwise toxic and cause damage to 99% of other plant species (Flowers and Colmer, 2008). Some of these plants have the capacity to grow in saline as well as in nonsaline habitats to cover the salt-contaminated barren lands for the achievement of multiple purposes of human beings. These plants are either called "salt accumulator" or "salt excluder" depending on their morphological, physiological, and anatomical structural organization in response to the surrounding climatic conditions. The excessive salts make the soil physiologically dry for the plants inhabiting such area, in which halophytes show their ability to grow, tolerate, and sustain this adverse environment (Munns and Tester, 2008). Consequently, halophytes have been used as a source of food, fodder, medicine, environmental protection, and for industrial purposes (Lokhande and Suprasanna, 2012; Lokhande et al., 2013; Hasanuzzaman et al., 2014; Sharma et al., 2016; Nikalje et al., 2019; Liu and Wang, 2021).

To comprehend the detailed mechanism of salinity stress tolerance, *Arabidopsis thaliana* from the glycophytes (salt-sensitive plant group) has been used as the model plant system due to the availability of wide information at genetic, proteomic as well as metabolomics levels. On the contrary, *Thellungiella salsuginea* (a halophytic relative of *A. thaliana*), previously known as *Thellungiella halophila* and now renamed as *Eutrema salsuginea* has been used as a decent genetic model plant system from the halophytes—a salt tolerant plant group (Ali and Yun, 2017). However, in the real world, the model plant systems which have shown the salinity stress tolerance at laboratory level all time will not become the possible solutions to extreme and adverse environmental constraints. Therefore, many other plant systems have also been investigated to reclaim the stress tolerance to various abiotic factors at field conditions (Lokhande et al., 2012; Nikalje et al 2019). Besides, the tolerance capacity of these plants has been investigated at cellular, tissue, and whole plant levels (Amor et al., 2005; Arzani 2008; Zhao et al., 2009, Aslam et al., 2011; Lokhande et al., 2010b,c; 2011a; 2012; Ali and Yun, 2017). It is important to study the mechanism operating at each level to develop the complete story of understanding plant salt stress tolerance (Aslam et al., 2011). In this regard, the present chapter highlights the use of *Sesuvium portulacastrum*, a halophytic nonmodel plant system to understand the salinity stress tolerance at cellular, tissue, and whole plant levels.

11.2 HALOPHYTES

Scientists from around the world working on halophytic plant systems have used different criteria to define halophytes in different ways (Hasanuzzaman et al., 2014; Liu and Wang, 2021). Schimper (1903) defined the halophytes as the plants capable of normal growth in saline habitats and also able to thrive on "ordinary" soil. The plants that show the capacity to bear salt concentrations above 0.5% at any stage of life could be grouped in the class of halophytes (Stocker, 1928). Dansereau (1957), simply defined the halophytes are the plants exclusively growing on saline soils. Greenway and Munns (1980) defined halophytes as "a kind of native flora of saline soils, which contain solutions with a Psi of at least 3.3 bar, being equivalent to 70mM monovalent salts." Thereafter, halophytes have been defined as plant species that grow and complete their life cycle in environments with NaCl concentrations above 200 mM (Munns and Tester, 2008).

On the basis of the plants growing under various ecological facets, halophytes are categorized into three types, namely, obligate, facultative, and habitat-indifferent halophytes in contrast to glycophytes (Sengbusch, 2003). Hasanuzzaman et al. (2014) have depicted the growth pattern of these all kinds of plants in response to their saline habitat. Obligate halophytes have the capacity to grow exclusively under saline conditions; the absence of a salty habitat severely hampers the growth of these halophytes. Facultative halophytes have chosen to survive in the presence or absence of a saline environment. Furthermore, by retaining their inherent capacity to tolerate extreme salt stress conditions, their growth has been found boosted up under nonsaline or soils with the presence of optimally low concentrations of salts (Lokhande et al., 2012). Habitat-indifferent halophytes actually are not true halophytes, because they usually grow in a salt-free environment. They grow on salty soils by competing with the salt-sensitive plant species. Their phenotypic and genotypic characteristic behaves differently in the presence and absence of a saline environment (Hasanuzzaman et al., 2014).

Halophyte species account for 1–2% of the world's land plants. There are approximately 235,000 plant species in the world, of which salt-tolerant and halophytes account for approximately 26,000 species (Liu and Wang, 2021). It has been reported that only the halophyte species are in the range of 2000–3000 and most of them are angiosperms. During1980s, James Aronson compiled a comprehensive database of 1554 halophyte species, termed HALOPH (Rahman et al., 2021). This has recently been converted to an interactive eHALOPH database (Santos et al., 2015) that also provides a bibliography of literature available for halophytes (Sharma et al., 2016).

11.3 SALINITY STRESS

Anything ""excess" or "less" to the plants from external conditions, refers to the stressful environment. Plant growth and productivity are adversely affected by nature's wrath in the form of various abiotic and biotic stress factors. Abiotic stresses such as salt, drought, chilling, freezing and high temperatures, flooding, heat, oxidative stress, and heavy metal toxicity cause US dollars losses worth hundreds of millions each year due to reduction in crop productivity and crop failure (Ahuja et al., 2010; Slama et al., 2015). Abiotic stresses cause various plant developmental processes such as seed germination, seedling growth and vigor, vegetative growth, flowering, and fruit set (Sairam and Tyagi, 2004; Rajalakshmi and Parida, 2012). Among the various types of abiotic stresses, salinity has been remaining a severe problem for all kinds of plants growing in variety of habitats. Therefore, there is an urgency to understand the mechanism of tolerance developed by the plants in response to these abiotic stresses, with the main emphasis on salt stress.

Salinity is one of the most important environmental constraints (Munns and Tester, 2008), that negatively affects crop productivity by affecting plant growth and development through water stress, cytotoxicity associated with excessive uptake of ions such as sodium (Na^+) and chloride, and nutritional imbalance and oxidative stress due to generation of reactive oxygen species (ROS) (Isayenkov and Maathuis, 2019). Salinity occurs due to inappropriate, poor-quality water irrigation affecting almost every aspect of a plant's physiology at the whole plant and cellular levels through osmotic stress in an earlier phase and ionic stress at a later stage. Generally, soils are classified as saline when the electrical conductivity of the saturated paste extract (ECe) is 4 dS m^{-1} or more which is equivalent to 40-mM NaCl and generates an osmotic pressure of approximately 0.2 MPa (Munns and Tester, 2008). To achieve salt tolerance, three interconnected aspects of plant activity are important: (i) damage must be prevented, (ii) homeostatic condition must be reestablished, and (iii) growth must resume. The plant responds to salinity based on its capability at physiological, biochemical, and molecular levels to tolerate salt stress conditions.

11.4 MECHANISM OF SALT STRESS TOLERANCE IN HALOPHYTE

In response to salt stress as a signal, halophytes have evolved with complex mechanisms for adaptation to these osmotic and ionic stresses. Osmotic adjustment under salt stress is an adaptation mechanism operated by halophytes to maintain their water balance (Flowers and Colmer, 2008). The best characterized biochemical response of plant cells to osmotic stress is an accumulation of some inorganic ions such as Na^+ and compatible organic solutes such as proline, glycine betaine, and soluble sugars (Flowers and Colmer, 2008; Slama et al., 2015; Isayenkov and Maathuis, 2019). These compatible solutes can accumulate to high levels without disturbing intracellular biochemistry (Bohnert and Jensen, 1996), protect subcellular structures, mitigate oxidative damage caused by free radicals (Attipali et al., 2004), and maintain the enzyme activities under salt stress (Jitesh et al., 2006).

Apart from the involvement of inorganic saline ions and organic osmolytes in the osmotic adjustment, potassium ion plays a major role in the biochemical and physiological processes of plants. Under typical physiological conditions, plant cells require high K^+ (100–200 mM) and lower Na^+ (less than 1mM) and accordingly the high cytosolic K^+/Na^+ ratio to maintain the osmotic balance (Tester and Davenport, 2003) for the proper functioning of the cell. To maintain a high K^+/Na^+ ratio in the cytosol, plant cell employs primary active transport mediated by channels and co-transporters for Na^+ extrusion and/or the intracellular compartmentalization of Na^+ into the vacuole (Blumwald, 2000). However, under salinity stress, ion ratios are altered by the uncontrolled influx of Na^+ through K^+ pathways. Halophytes being continuously exposed to high salinity conditions undergo osmotic adjustment preferably through higher accumulation of Na^+ and its efficient sequestration into the vacuole at low external water potential as well as low potassium use efficiency as compared to glycophytes.

In addition to osmotic adjustment through osmolyte synthesis and inorganic ions accumulation, one of the biochemical changes possibly occurring when plants are subjected to this harmful stress condition is the production of ROS such as superoxide (O^{2-}), hydrogen peroxide (H^2O^2), hydroxyl radicals (OH^-) and singlet oxygen ($^1O^2$) (Mittler, 2002; Ashraf, 2009; Miller et al., 2010). These cytotoxic ROS are responsible for the serious oxidative damage to membranes and other essential macromolecules such as lipids, proteins, and nucleic acids (Mittler, 2002; Ashraf, 2009; Miller et al., 2010) and the functioning of plant photosynthetic apparatus (Jitesh et al., 2006). These ROS produced as a result of salt stress need to be scavenged for maintaining normal growth. Plants have evolved a complex antioxidant system to scavenge these ROS, including antioxidant enzymes such as superoxide dismutase (SOD), catalase (CAT), guaicol peroxidase (GPX), ascorbate peroxidase (APX), monodehydroascorbate reductase (MDHAR) dehydroascorbate reductase (DHAR), and glutathione reductase (GR), (Mittler, 2002; Miller et al., 2010). In addition, plants have also evolved nonenzymatic antioxidants such as glutathione (GSH), ascorbate (AsA), carotenoids, and tocopherols (Ashraf 2009). The studies on antioxidant enzyme activities under salt stress conditions may provide information about possible oxidative damage generated by cytotoxic ROS and active role of antioxidant enzymes in scavenging these ROS molecules to maintain higher tolerance against salt stress in the halophytes.

11.5 SESUVIUM PORTULACASTRUM L.

The genus *Sesuvium* includes approximately 12 species with its distribution throughout the world growing in the coastal saline soils of tropical regions. However, *S. portulacastrum* L., commonly known as sea purslane and belongs to the family Aizoaceae, has been identified due to the presence of a unique character like uniflorous inflorescence and adventitious roots at nodal region (Lokhande et al., 2009a). It has been distributed among the eastern and western seashores of India as a mangrove-associated facultative halophyte, which has the capability to grow in highly saline as well as purely nonsaline soil unlike obligate halophytes or mangroves (Lokhande et al., 2009a, 2012). The plant shows perennial, psammophytic herbaceous habit with creeping stem producing adventitious roots at nodal regions (Lokhande et al., 2010). One of the important features of the plant is the presence of thick, fleshy leaves which are succulent in nature and arranged oppositely on the creeping stem. The stems and leaves have a tremendous capacity to store the excess amount of toxic salts and minerals in them, therefore *Sesuvium* comes under the category of 'salt accumulator' halophyte (Lokhande et al., 2009a, 2010, 2012; Wang et al., 2011). The creeping habit of the stem to run at a longer distance by developing and fixing the adventitious roots in the sand and succulent leaves at the nodal region helps to hold the sand or soil to a greater extent in the growing region of the plant. Therefore, *Sesuvium* has been widely considered as remedy for environmental protection and used as sand dune fixer as well as a saline soil stabilizer (Menzel and Leith, 1999, Lokhande et al., 2009a). Besides, the plant has shown extensive potential to remove excessive salts from the contaminated saline agricultural soil, to develop green vegetation on desert-contaminated lands and also have ornamental value in terms of landscaping (Lokhande et al., 2009a). The plant contains an insect molting hormone phytoecdysteroids—20 hydroxyecdysone hormones (Lokhande et al., 2009b), in addition to the potential to produce a variety of secondary metabolites which have medicinal value (Kapare et al., 2017). The extensive research conducted during the last two decades has shown that *Sesuvium* now becomes one of the choices of model system among halophytes to understand the mechanism of stress tolerance at growth, physiological, biochemical, molecular, proteomic, and metabolomic levels to an array of various abiotic stresses such as salinity, drought, heavy metals, industrial effluents, etc.

11.6 ABIOTIC STRESS TOLERANCE IN SESUVIUM PORTULACASTRUM

The *S. portulacastrum* has shown wide range of tolerance to various abiotic stress conditions such as salinity, drought, heavy metals, industrial effluents and textile dyes (Messedi et al., 2004; Ghnaya et al., 2005; Ramani et al., 2006; Slama et al., 2008; Rabhi et al., 2010; Lokhande et al., 2009a, 2009b, 2010a, 2010b, 2010c, Lokhande et al., 2011a, 2009b, 2009c; Zaier et al., 2010; Lokhande et al. 2012; Patil et al., 2012; Mariem et al., 2014; Lokhande et al., 2015; Mnasri et al., 2015; Srivastava et al., 2015; Wali et al., 2016; Muchate et al., 2016; Ayyappan et al., 2016; Feng et al., 2018; Nikalje et al., 2019; Lokhande et al., 2020; Kulkarni et al., 2021).

11.7 STRUCTURAL HIERARCHICAL RESPONSES TO SALINITY STRESS IN S. PORTULACASTRUM

11.7.1 EXPERIMENTAL DESIGN – SALT STRESS TREATMENTS, GROWTH, AND PHYSIOLOGICAL AND BIOCHEMICAL ANALYSES

The salinity stress responses with respect to structural hierarchy in *Sesuvium* were recorded at the cell in terms of callus and intact tissue-like axillary shoots under in vitro conditions. However, the salinity stress responses in terms of complex structural hierarchy were studied at the whole plant level under ex vitro conditions. The kind of plant material used for expressing structural hierarchy, growth medium, salt stress treatments and its exposure, growth, physiological and biochemical analyses performed at cellular, tissue, and whole plant level in *Sesuvium* are depicted in Table 11.1 (Lokhande et al., 2010a, 2010b, 2010c; 2011a, 2012).

TABLE 11.1 Salt Stress Treatments, Growth, Physiological, and Biochemical Analyses at Cellular, Tissue, and Whole Plant Level in *Sesuvium*.

Experimental conditions and analyses	Structural hierarchy			References
	Cell	Tissue	Whole plant	
Type of plant material used	Callus	Axillary shoots	Rooted plants	Lokhande et al. (2010b, 2011a, 2012)
Growth medium	MS + 10 μM 2,4-D + 5 μM BA	MS + 20 μM BA	Sand + ½ strength MS liquid medium	Lokhande et al. (2010b, 2011a, 2012)
Salt (NaCl) stress treatments (mM)	0, 100, 200, 400	0, 200, 400, 600	0, 200, 400, 600, 800	Lokhande et al. (2010b, 2011a, 2012)
Salt stress exposure	15 days	30 days	15 days	Lokhande et al. (2010b, 2011a, 2012)
Plant material used for analyses	Salt-stressed calli	Salt-stressed axillary shoots	The third leaf from salt-stressed whole plant	Lokhande et al. (2010b, 2011a, 2012)
Growth analyses in terms of relative growth rate (RGR)	Initial and final weight (g) of callus	Number of axillary shoots and leaves per explant, shoot, and third leaf length (mm)	Shoot and root length (cm) per plant	Lokhande et al. (2010b, 2011a, 2012)
Percent tissue water content (TWC %)	Fresh weight (FW) and dry weight (DW) of callus	FW and DW of axillary shoots	FW and DW of shoot and root	Lokhande et al. (2010b, 2011a, 2012)
Oxidative damage to the membrane	Relative electrolytic leakage (REL) and lipid peroxidation in terms of MDA content			Dionisio-Sese and Tobita (1998); Heath and Packer (1968)
Osmolytes accumulation	Proline, glycine betaine, and total soluble sugars (TSSs)			Bates et al. (1973), Grieve and Grattan (1983), Watanabe et al.(2000)
Antioxidant enzyme assays	CAT, APX, and SOD			Cakmak and Marschner (1992), Nakano and Asada (1981), Becana et al. (1986)

11.7.2 CELLULAR, TISSUE, AND WHOLE PLANTS RESPONSES TO SALT STRESS TREATMENTS

FIGURE 11.1 Growth and morphological responses of cell (callus), tissue (axillary shoots), and whole plants of *S. portulacastrum* to variable salt stress treatments; (a–d) effect of different concentrations of salt (0-, 100-, 200-, and 400-mM NaCl) on growth and proliferation of callus under in vitro conditions; (e–h) effect of different concentrations of salt (0-, 200-, 400-, and 600-NaCl) on growth and morphology of axillary shoots under in vitro conditions; (i–m) effect of different concentrations of salt (0-, 200-, 400-, 600- and 800-mM NaCl) on growth and morphology of whole plants of *Sesuvium* grown under ex vitro conditions.

11.7.2.1 GROWTH RESPONSES OF CALLUS, AXILLARY SHOOTS, AND WHOLE PLANTS OF SESUVIUM TO SALT STRESS

A correlation has been observed between the salt stress adaptations at cellular (callus) and whole plant levels in the plants belonging to halophytes and glycophytes (Arzani, 2008).

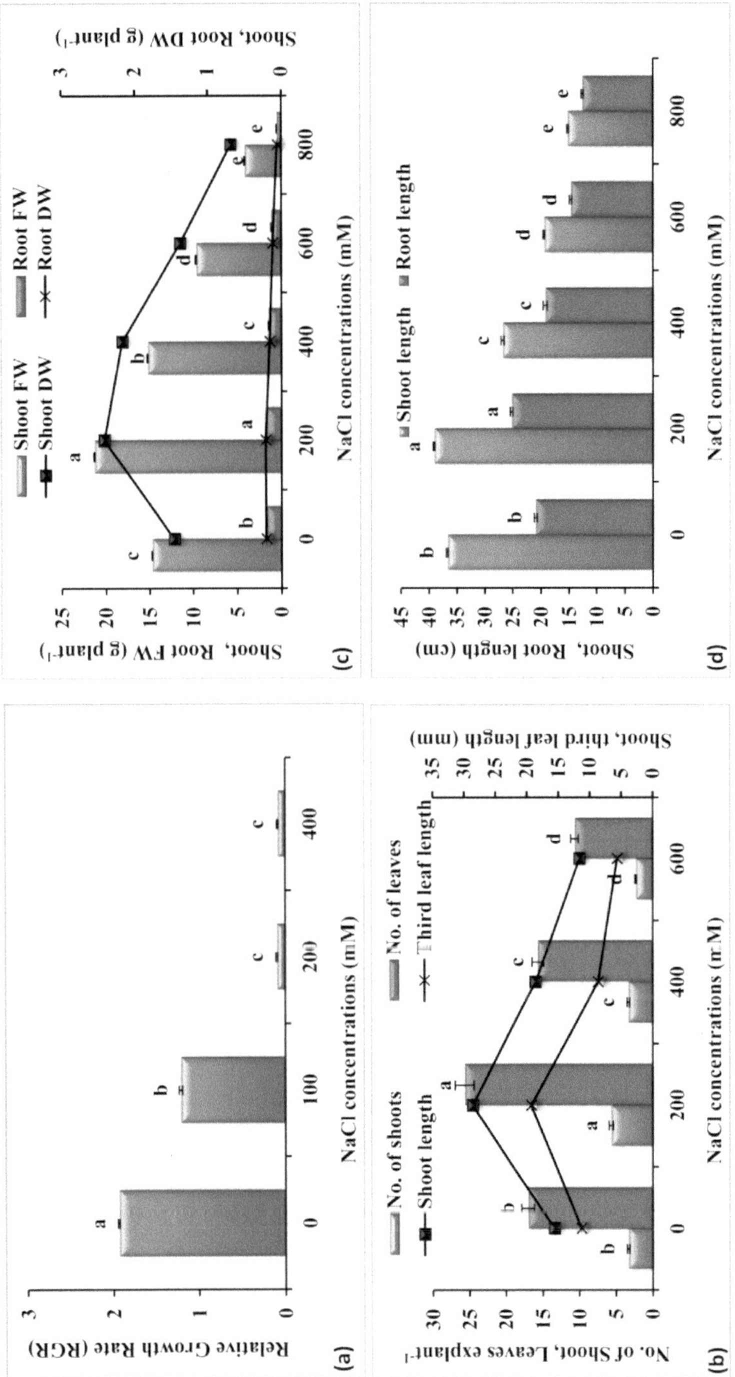

FIGURE 11.2 Effect of salt stress on growth and biomass at cellular, tissue, and whole plant level in *S. portulacastrum*; (a) RGR of callus; (b) number of shoots and leaves per explant, shoot, and third leaf length in axillary shoots; (c–d) shoot, root FW and DW, and shoot and root length of the whole plant in response to salt stress.

There are reports on understanding the mechanism of salt stress tolerance at the cellular level in mangroves under in vitro conditions (Cherian and Reddy, 2003; Amor et al., 2005; Zhao et al., 2009). The cellular integrity to form the tissue and structural organization of different tissues to form the whole plant body improves various growth, physiological, and biochemical activities of biological organisms. Earlier studies on salt stress responses are preferred in soil-less growth medium, that is, using tissue culture technique, supplemented with desired salt concentrations (Ahmad et al., 2007). In the present investigation, the concentrations of the salts in terms of NaCl were defined on the basis of the lethal toxicity of NaCl to the growth of the callus and tissues (axillary shoots) under in vitro conditions and whole plants under ex vitro conditions. Interestingly, the callus being the unorganized mass of the cells found to be very sensitive to higher salt concentrations; therefore, the higher dose of the NaCl for the callus was restricted to 400 mM. The observations recorded after 15 days of salt stress treatment to callus cultures showed a significant decrease in RGR (Figs. 11.1a-d; 11.2a), percent TWC % and FW/DW ratio of stressed calli with an increase in the salt levels from 0 to 400-mM NaCl (Lokhande et al., 2010b). It has been shown that reduced growth helps to save energy for defense purpose as well as limits the risk of heritable damage (May et al., 1998). Similarly, the cellular responses of the *Sesuvium* observed at the higher concentrations of NaCl have shown its survival capacity like other plants (Zhang et al., 2004; Patade et al., 2008; Zhao et al., 2009). Since the cells of the calli are not organized, the Na^+ ions might cause severe damage to the cells at earlier stages by interfering with the uptake of available essential nutrients by the cells from the medium (Patade et al., 2008); however, after perception of the salt stress response by the cells, they are found adopted to these stressed conditions and might have immediately activated the defense machinery for its survival.

Surprisingly, the improved structural hierarchy in terms of axillary shoots in comparison to the calli of the plant showed significant tolerance capacity. The improved growth of the axillary shoots in terms of the number of shoots and leaves per explant and shoot length and percent TWC % has been reported even under stressed conditions of 200-mM NaCl treatments in comparison to control (0 mM) and higher doses of NaCl exposures such as 400- and 600-mM NaCl (Figs. 11.1e–h; 11.2b) (Lokhande et al., 2011a). The improved growth and percent TWC at 200-mM NaCl treatments to axillary shoots has been correlated with the significant increase in FW/DW ratio over the control, a measure of succulence which is an inherent character of halophytes under optimum concentrations of NaCl (Lokhande et al., 2011a).

Moreover, the axillary shoots response in terms of growth, TWC %, and FW/DW ratio at 400-mM NaCl stress was found at par with control and lethal at 600-mM NaCl stress (Lokhande et al., 2011a). The higher osmotic potential of the saline medium affects water and nutrient uptake, which may in turn disturb the osmotic balance, and water status of the cell and inhibit the metabolic activities necessary for bud initiation and growth (Vijayan et al., 2003). Flowers and Launchli (1983) reported that, at low concentrations, NaCl exerts a significant positive effect on shoot proliferation in vitro due to the increased osmolarity; in contrast, higher levels implicated a negative effect on growth parameters which was also revealed in the present study. This has proved that the establishment of very strong cellular connections and transport machinery in the tissue (axillary shoots) has helped to distribute the burden of the excess Na^+ salts into different parts of the tissue in terms of Na^+ accumulation in the leaves of the axillary shoots, thereby improving the succulence of the tissue. The severe toxicity of the Na^+ at higher doses of NaCl (400 and 600 mM) might have helped the tissues to cope with such stress conditions by draining them more energy for their survival at the cost of reduced growth (Fig. 11.2b). Therefore, there are possibilities of reduced growth rates at higher salt concentrations. Besides, the inherent ability of the succulence in *Sesuvium* to provide the salt stress tolerance might have activated after the achievement of particular structural complexity which was not observed at the cellular level.

In the above cases, the cell (callus) and tissues (axillary shoots) of the plant are trying to cope and grow with the varying levels of salinity stress under in vitro (tissue culture) conditions, where the artificial growth conditions are provided for a longer period due to complete absence of (at cellular level) or development of incompetent photosynthetic machinery (at tissue level). Besides, the tissues are not with the presence of root system, which also supports to impart the salinity tolerance and plays an important role in restricting as well as partitioning the entry of excess amount of Na^+ ions in the plant. In such cases, the tolerance to salinity stress might be imparted less as compared to the whole plant response to stress conditions. However, when compared with the other crop systems, it is found very effective and significant tolerance capacity of the cell and tissues of the *Sesuvium* to the higher concentrations of NaCl.

Higher salinity is regarded as a stress factor for plants, even for most of the halophytes. To understand the physiological and biochemical mechanisms responsible for the salinity tolerance of the halophytes at the highly organized and structurally hierarchical and differentiated level of the plant, it is necessary to know whether their growth is being limited by the osmotic effect of the salt in the soil or the toxic effect of the salt within the plant. It has been extensively studied that in all types of plants exposed to salt stress, the simplest analysis of the response of plants to salinity stress is the reduction in growth (shoot and root length and biomass production) which occurs in two phases; a rapid response to the increase in external osmotic pressure, and a slower response due to the accumulation of Na^+ ions in leaves (Zhu 2001; Munns and Tester, 2008). In the present investigation, the story becomes more attentive at the whole plant level. The fully grown rooted plants when exposed to varying concentrations of salinity stress have shown improved growth performance (Lokhande et al., 2012) over the responses recorded at cellular and tissue levels (Figs. 11.1i-m; 11.2c–d). The survival of the plants at the expense of reduced growth rate was observed at significantly higher concentrations of the NaCl in the medium i.e. at 800-mM NaCl (Figs. 11.2c–d) (Lokhande et al., 2012). *Sesuvium* has shown its survival even more than 800-mM NaCl stress treatments (Lokhande et al., 2011). Similar to the results recorded under in vitro conditions for the improved growth performance of the axillary shoots without roots at 200-mM NaCl, the growth response of the well-rooted whole plants of *Sesuvium* was found many-folded and improved at 200-mM NaCl stress conditions in comparison to its control under ex vitro conditions (Figs. 11.1j; 11.2c–d). It is evident that the growth gets stimulated in halophytes at lower concentration of salt in the soil (Redondo-Gomez et al., 2009) which was also observed in the present investigation. The growth parameters recorded in terms of shoot and root length, biomass accumulation (FW and DW), and water status (TWC % and FW/DW ratio) of shoots and roots were found significant and astonishingly enhanced in the plants exposed to 200 mM NaCl treatments (Figs. 11.2c–d) (Lokhande et al., 2012). Although the growth characteristics of the whole plants found decreased at the higher levels of NaCl stress treatments i.e. at 400, 600, and 800 mM NaCl, the whole plants have shown significant tolerance and capacity to survive with slower growth rate without showing any negative phenotypic symptoms (Figs. 11.1k-m; 11.2c–d). The impediment in the growth of *Sesuvium* was correlated with the toxic concentrations of NaCl in the medium as well as severe disruption of nutrient uptake (Messedi et al., 2004). The improvement in stem diameter of the whole plants exposed to 200 mM NaCl stress has been correlated with the effective water use efficiency of the plant, perfect partitioning of the excessive Na^+ ions into its vacuoles and thereby significant increase in cortical cell size, cortex, and pith diameter of the stem. Besides, the thickening of the stem and leaves is associated with the inborn halosucculence nature of the plant (Lokhande et al., 2012).

To conclude with the growth performance of the *Sesuvium* at cellular, tissue, and whole plant levels in response to salinity stress, the elevated structural hierarchy of the plant (from cell to whole plant level) in terms of unprotected cells at callus level to protective epidermal coverings to all kinds of tissues at whole plant level against various abiotic factors; unorganized and poorly connected cell biomass (callus) to complex, strong and well-established cellular connections observed in whole plants; root-less system (axillary shoots) to well-established roots (in whole plants); the presence of abortive photosynthetic machinery in callus and axillary shoots as against effective photosynthetic machinery in whole plants; as well as improved and efficient water use efficiency has found responsible to establish the improved salinity stress tolerance in *Sesuvium* from cellular to whole plant levels.

11.7.2.2 PHYSIOLOGICAL AND BIOCHEMICAL RESPONSES OF CALLUS, AXILLARY SHOOTS, AND WHOLE PLANTS OF SESUVIUM TO SALT STRESS

The change in growth characteristics under the influence of salinity stress at cell, tissue, and whole plant levels are always correlated with the drastic alterations in metabolic pathways observed with the help of characterization of various salt stress-responsive physiological and biochemical key indicators. One of the severe attacks of salinity on plants is cell membrane damage caused by oxidative stress generated due to the presence of excessive Na^+ ions in the surrounding of the cell. The toxic concentration of Na^+ ions in the medium and shoots leads to the excess generation of ROS which are harmful to cell membranes and may cause damage to essential molecules like proteins, lipids, and nucleic acids (Sairam and Srivastava, 2002). Generally, the rate of cell membrane damage was measured in terms of REL and production of total malondialdehyde (MDA) content due to lipid peroxidation under stress conditions. Estimation of MDA content, which is a secondary end-product of polyunsaturated fatty acid oxidation due to disturbed osmotic balance caused by excess salt stress, is used widely to measure the extent of lipid peroxidation as an indicator of oxidative stress (Blokhina et al., 2003). The integrity of the cell membrane is always associated with the least REL and less amount of MDA production.

The present study revealed that, in general the REL was found significantly the highest at higher levels of NaCl stress treatments in cell, tissue, and whole plants of *Sesuvium.* The proof of the increased REL was achieved after the evaluation of total MDA content at cellular, tissue, and whole plant levels; which has been found to drastically increase at severe salt-stressed conditions. Surprisingly, the growth rate and morphological characteristics which were found to improve in axillary shoots and whole plants exposed to 200-mM NaCl stress treatments (Lokahnde et al., 2010c; 2012) have shown its significance in terms of reduced rate of REL and least production of MDA content, suggesting its role in the effective cellular membrane integrity due to a smaller amount of oxidative damage and establishment of salinity stress tolerance. In the present investigation, 200-mM NaCl treatment revealed lower oxidative damage to the membrane which might be associated with lower production of H^2O^2 and the finest osmotic balance developed by the accumulation of optimum concentrations of Na^+ ions, which is the cheapest source of osmoticum for halophytes under saline condition and its efficient sequestration into the vacuole. However, in the halophytes, suboptimal (control) or very high salt concentrations (more than 400-mM NaCl) drastically increase the level of H^2O^2, which disturbs the osmotic balance of the cell and thus induces more oxidative stress in plant tissues (Khan and Panda, 2007) and amplifying more damage to membrane lipids.

The prime negative impact of salt stress on the cell, tissues or whole plant is a drastic reduction in the water level of the cell, thereby altering various physiological and biochemical mechanisms of the plant to a severe extent. The immediate need of the cell to survive under such a stressful environment is, to maintain the osmotic equilibrium of the cell cytoplasm by fulfilling the requirement of water molecules in terms of avoiding excessive water loss by retaining the water level or synthesizing a similar kind of biomolecules for the replacement of unavailable water molecules in the cell. The biomolecules that are synthesized under stressful environments and enacted like the water molecules for maintenance of osmotic equilibrium of the cell are referred to as "osmolytes" or compatible solutes. However, the synthesis of such osmolytes is quite expensive for plants in terms of a lot of energy consumption, therefore it rewards the reduced growth of the cell, tissues, or whole plant. It is now a well-established fact that osmolytes accumulation increases significantly in all types of plants exposed to various abiotic stresses.

The osmotic stress generated in response to salinity in callus, tissues, and whole plants of the *Sesuvium* was found to be alleviated by the accumulation of water molecule compatible osmolytes such as proline, glycine betaine, and TSSs to maintain the osmotic balance of the cell. Accumulation of osmolytes under salt stress is an indication of oxidative damage which provides protection to cytosol from dehydration through the development of compatible cytoplasmic osmoticum. Both proline and glycine betaine have been reported to act as protective agents for cytoplasmic enzymes/proteins and as stabilizers of bio-membranes (Tripathi et al., 2007). Proline in addition is known to scavenge radicals (Matysik et al., 2002). In the present investigation, all these three types of osmolytes (proline, glycine betaine, and TSS) have shown drastic and progressive increases in cells, tissues, and whole plants of *Sesuvium* exposed to various concentrations of NaCl in comparison to control (without NaCl treatments) under in vitro and ex vitro conditions (Lokhande et al., 2010b, 2010c, 2012).

Proline and glycine betaine act as osmotica in the presence of low water content and plays a major role in the maintenance of osmotic balance, which was found to be higher in salt-tolerant species than in salt-sensitive species (Zhang et al., 2004). Besides, accumulation of TSS associated with an adaptation to different environmental conditions has been used as an indication for osmotic induction (Wang and Stutte, 1992). The unorganized mass of the cells, that is, callus exposed to salt stress has shown its survival at the reduced growth rate associated with the energy expenses in terms of adenosine triphosphate (ATP) for more and more amount of proline, glycine betaine (Fig. 11.3a) and TSS synthesis. However, the improved structural hierarchy observed at axillary shoots and whole plant level in line with the enhanced growth parameters, reduced REL and MDA content in the tissues and whole plants exposed to 200-mM NaCl treatments has further extended the importance of synthesis of lower amount of proline, glycine betaine (Figs. 11.3b and c) and TSS associated with stress tolerance (Lokhande et al., 2010b, 2010c; 2012).

In addition to the accumulation of organic osmolytes, excess Na^+ ion accumulation in the medium compensates for the loss of potassium ions, which have an added role of an osmoregulatory monovalent cation in the salt-stressed tissues (Munns and Tester, 2008). In contrast to the extremely energy energy-consuming osmolytes synthesis, the halophytes preferably use the accumulation of Na^+ ions in their vacuole at the expense of negligible or very less energy for the maintenance of osmotic balance. Therefore, Na^+ ions are not found to be very responsible for hampering the growth of the cell, tissue, or whole plants of halophytes even under salinity stress conditions. Instead, Na^+ ions are found responsible for improving the growth, stress tolerance, and survival of the plants under such conditions. Similarly, in the present investigation, the calli, axillary shoots, and whole plants of *Sesuvium* exposed to different levels of salt stress had shown an increased rate of Na^+ ion accumulation; however, the potassium and calcium ions accumulation was found less and not shown any impact on the growth and tolerance capacity of the plant. It might be possible because halophytes naturally

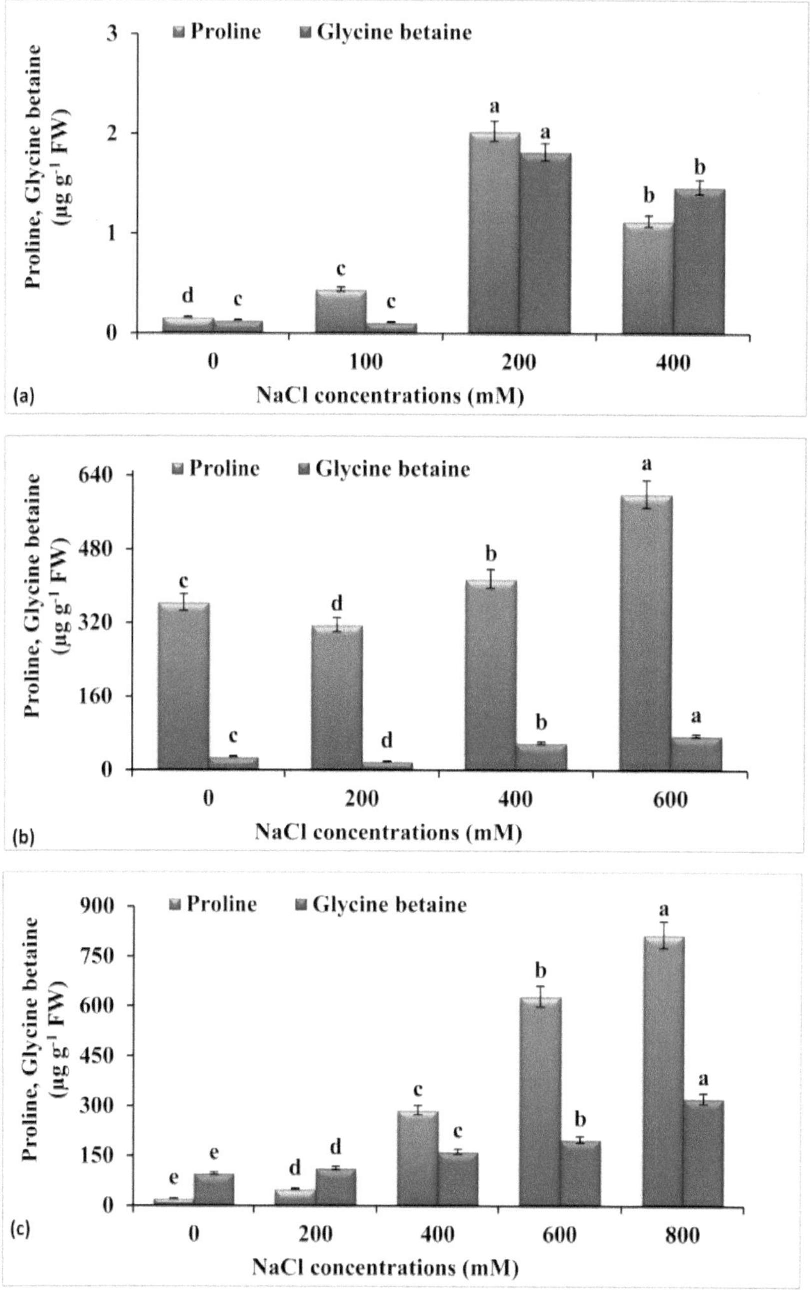

FIGURE 11.3 Stress tolerance achieved by the accumulation of osmolytes (proline and glycine betaine), (a) at cellular (callus), (b) tissue (axillary shoots), and (c) whole plant level in *S. portulacastrum* exposed to different concentrations of salts.

growing in saline environments mostly prefer Na^+ ions unlike potassium ions for the osmotic balance. The survival strategy of the cells (calli) was achieved by the significant accumulation of Na^+ ions as well as osmolytes at the cost of significant decrease in growth (Fig. 11.1). However, at 200-mM NaCl treatment to axillary shoots and whole plants, the coordinated action in terms of sufficient amount of Na^+ ion accumulation in the vacuoles and lower rate of proline, glycine betaine (Figs. 11.3b–c) and TSS synthesis with optimum use of the energy has found significantly effective in maintaining the osmotic balance of the axillary shoots and whole plants and therefore resulted into improved growth (Fig. 11.1) and salt stress tolerance (Lokhande et al., 2010b, 2010c; 2012).

Another foremost important feature imparting the survival and salt stress tolerance in calli, axillary shoots, and whole plants of *Sesuvium* is strongly correlated with the efficient and actively engaged enzymatic antioxidant machinery for the detoxification of excessively produced ROS in response to oxidative damage generated due to salinity stress. The presence of a strong antioxidative defense mechanism combined with physiological adaptations in the plant contributes to the difference in salt tolerance capacity between whole plants and in vitro structures like calli or axillary shoots (Cherian and Reddy, 2003). Several studies have demonstrated that salt-tolerant species increase their antioxidant enzyme activities and antioxidant content in response to salt stress whereas salt-sensitive species failed to do so (Yazici et al., 2007). However, not all the types of antioxidant enzymes will work at a time to scavenge the ROS such as H^2O^2. A number of enzymes isolated from halophytes are sensitive to salt when assayed in vitro at lower concentrations of NaCl than were present in their tissues as a whole (Flowers et al., 1977). In salt-stressed calli and axillary shoots of *Sesuvium*, APX instead of CAT was found more effective in scavenging the H^2O^2 (Lokhande et al., 2010b, 2010c). On the contrary, the whole plants exposed to different salt stress treatments showed significantly higher activities of CAT for scavenging of H^2O^2 (Lokhande et al., 2012). It clearly indicates that, the structural hierarchy has also found important for the activeness of the specific enzyme in the metabolic processes of the plant. Overall, in the present investigation, SOD enzyme has played a significant role in detoxifying the O^{2-} radicles in the calli, axillary shoots, and whole plants of *Sesuvium* exposed to different salt stress treatments (Lokhande et al., 2010b, 2010c; 2011a; 2012). An increase in the APX activity and concomitant increase in SOD activity has been found an adaptive strategy to remove the excess H^2O^2 generated due to an increase in SOD activity under excess salt stress conditions (Jitesh et al., 2006).

11.8 CLONAL BEHAVIOUR OF SESUVIUM TO SALINITY STRESS TOLERANCE

The clones of the *Sesuvium* collected from the geographically and climatically distant habitats i.e. from east and west coastal regions of India (Lokhande et al., 2009a) were analyzed for their responses to salinity stress tolerance (Lokhande et al., 2012). The earlier studies on the morphological and molecular diversity analysis of the clones of *Sesuvium* growing along the coastal regions of India presented a view that the morphological and molecular variations present among the *Sesuvium* clones were associated with the environmental conditions of the specific region and anthropogenic factors in addition to the genetic makeup of the clone within that region (Lokhande et al., 2009a). Therefore, the possibility of variations of *Sesuvium* clones to their salt stress tolerance in the different coastal regions of India has been documented by Lokhande et al. (2012). The study revealed that the growth, physiological, and biochemical changes that occurred among the *Sesuvium* clones showed a similar kind of behavior to salinity stress tolerance as observed earlier at tissue and whole plant level by Lokhande et al. (2010b, c). The potent clone from Maharashtra revealed a significantly higher degree of stress tolerance in comparison to other clones (Lokhande et al., 2012). Furthermore, stress tolerance achieved in terms of improved growth at 200- and 400-mM NaCl treatments for the stressed plants of a potent clone have been correlated with the significant accumulation of Na^+ ions, osmolytes, and better halosucculence nature of the plants as depicted in anatomical studies in terms of increased stem diameter, cortex thickness and pith region (Lokhande et al., 2012). The study conducted for the clonal behavior suggests the use of potent clone for solving the problems of environmental concern such as desalination, wetland restoration, and phytoremediation of heavy metals and toxic effluents.

11.9 CONCLUSION

In living organisms, the increased structural hierarchy and complexity from the cell to the whole organismic level show coordinated action of various metabolic processes thereby reflecting its strategies for survival and continuity of life in extreme surrounding environmental conditions. Plants inherent to sessile nature are always prone to variety of biotic and abiotic stresses. This has become severe in the plants growing in highly saline environments such as mangroves or halophytes. *S. portulacastrum*, which is one of the important mangroves associated halophytes has shown unique characteristics to grow in the complete absence of salts as well as having the capacity to survive even under lethal concentrations of the salts which are otherwise life determinant to other plants. This has reflected the facultative nature of the plant suitable to understand the salinity stress mechanism and correlates its tolerance strategies with the glycophytes, the plants sensitive to salinity stress.

In the present investigation, attempts have been made to understand the role and development of structural hierarchy and complexity in imparting salinity stress tolerance in cells (callus), tissue (axillary shoots), and whole plants of *Sesuvium.* It has been found that the unprotected and unorganized mass of the cells (callus) upon exposure to salinity stress revealed its survival at the expense of reduced relative growth rate by contributing more and more energy for the sequestrations of Na^+ ions in the vacuoles as well as synthesizing the energy consuming osmolytes like proline, glycine betaine, and TSS. Although cells (callus) are unable to show growth improvement, the cells have achieved the capacity to survive under salt stress conditions. The improved structural complexity observed at the tissue (axillary shoots) level has proved that well established cellular connections, the development of structural entities such as shoots and leaves, and the formation of an epidermal tissue system in axillary shoots for protection have helped to impart the salinity tolerance at the tissue level. This has been reflected in the establishment of succulence due to the retention of optimum osmotic balance through accumulation and sequestration of excessive Na^+ ions into the vacuoles as well as the synthesis of sufficient amount of osmolytes for retaining the cell turgor at optimum levels in salts (i.e., at 200-mM NaCl) in the medium. Furthermore, the extremely higher levels of complex tissue organization and differentiation in terms of root, stems and leaves at whole plant level have found fully responsible to impart the salinity tolerance even at supra-optimal concentrations of the salts in the growth medium. The structural complexity and hierarchy at the whole plant level might be responsible for efficiently distributing the effect of toxic saline ions into different parts of the plant through its sequestration into the vacuoles and/ or extrusion and hence showed better salt tolerance capacity which was found least at cellular level. In addition, restriction to the entry of excess toxic ions inside the plant at the root surface reduces the heritable damage to the whole plant. At whole plant level, the inbuilt and inherent capacity of Na^+ ion partitioning and mechanism of ion sequestration, development of succulence, fully efficient photosynthetic machinery, effective water use efficiency, elevated levels of osmolytes accumulation, and integrated responses of enzymatic antioxidants to scavenge the excessively generated ROS are the key factors to show salinity stress tolerance in *Sesuvium.* Furthermore, there might be possibly several other physiological, biochemical, and molecular factors operating at the whole plant level conferring salt tolerance which is lacking in the dedifferentiated and unorganized callus cultures. In general, coordinated salt tolerance behavior has been reported at cellular, tissue, and whole plant levels in *S. portulacastrum.* The study of the present investigation suggests tha, *Sesuvium*, having its facultative halophytic nature now become the choice of the plant for understanding the mechanism of stress tolerance against various abiotic stress factors at genomic, proteomic, and metabolomics levels. Besides, practically the plant has found remedies and eco-friendly solutions to various environment-related issues for making the journey towards sustainable development.

KEYWORDS

- **halophyte**
- **Sesuvium portulacastrum**
- **salinity stress**
- **structural hierarchy**
- **tolerance mechanism**

REFERENCES

Ahmad, M. S. A.; Javed, F.; Ashraf, M. Iso-Osmotic Effect of NaCl and PEG on Growth, Cations and Free Proline Accumulation in Callus Tissue of Two Indica Rice (*Oryza sativa* L.) Genotypes. *Plant Growth Regul.* **2007,** *53,* 53–63.

Ahuja, I.; de Vos, R. C. H.; Bones, A. M.; Hall, R. D. Plant Molecular Stress Responses Face Climate Change. *Trends Plant Sci.* **2010,** *15,* 664–674.

Ali, A.; Yun, D. J. Salt Stress Tolerance: What do We Learn from Halophytes? *J. Plant Biol.* **2017,** *60,* 431–439.

Amor, N. B.; Hamed, K. B.; Debez, A.; Grignon, C.; Abdelly, C. Physiological and Antioxidant Responses of the Perennial Halophyte *Crithmum maritimum* to Salinity. *Plant Sci.* **2005,** *168,* 889–899.

Arzani, A. Improving Salinity Tolerance in Crop Plants: A Biotechnological View. *In Vitro Cell Devel. Biol. – Plant.* **2008,** *44* (5), 373–383.

Ashraf, M. Y. Salt Tolerance Mechanism in Halophytes from Saudi Arabia and Egypt. *Res. J. Agri. Biol. Sci.* **2009,** *5* (3), 191–206.

Aslam, R.; Bostan, N.; Amen, N. E.; Maria, M.; Safdar, W. A Critical Review on Halophytes: Salt Tolerant Plants. *J. Med. Plant Res.* **2011,** *5* (33), 7108–7118.

Attipali, R. R.; Kolluru, V. C.; Munusamy, V. Drought Induced Responses of Photosynthesis and Antioxidant Metabolism in Higher Plants. *J. Plant Physiol.* **2004,** *161,* 1189–1202.

Ayyappan, D.; Sathiyaraj, G.; Ravindran, K. C. Phytoextraction of Heavy Metals by *Sesuvium portulacastrum* L. A Salt Marsh Halophyte from Tannery Effluent. *Int. J. Phytoremed.* **2016,** *18* (5), 453–459.

Bates, L. S.; Waldren, R. P.; Teare, I. D. Rapid Determination of Free Proline for Water Stress Studies. *Plant Soil.* **1973,** *39,* 205–208.

Becana, M.; Moran, J. F.; Iturbe-Ormaetxe, I. Iron-dependent Oxygen Free Radical Generation in Plants Subjected to Environmental Stress: Toxicity and Antioxidant Protection. *Plant Soil.* **1986,** *201,* 137–147.

Blokhina, O.; Virolainen, E.; Fagerstedt, K. V. Antioxidants, Oxidative Damage and Oxygen Deprivation Stress: A Review. *Ann. Bot.* **2003,** *91,* 179–194.

Blumwald, E. Sodium Transport and Salt Tolerance in Plants. *Curr. Opi. Cell Biol.* **2000,** *12,* 431–434.

Bohnert, H. J.; Jensen, R. G. Strategies for Engineering Water Stress Tolerance in Plants. *Trends Biotechnol.* **1996,** *14,* 89–97.

Cakmak, I.; Marschner, H. Magnesium Deficiency and High Light Intensity Enhance Activities of Superoxide Dismutase, Ascorbate Peroxidase and Glutathione Reductase in Bean Leaves. *Plant Physiol.* **1992,** *98,* 1222–1227.

Calone, R.; Bregaglio, S.; Sanoubar, R.; Noli, E.; Lambertini, C.; Barbanti, L. Physiological Adaptation to Water Salinity in Six Wild Halophytes Suitable for Mediterranean Agriculture. *Plants* **2021,** *10,* 309.

Cherian, S.; Reddy, M. P. Evaluation of NaCl Tolerance in the Callus Cultures of *Suaeda nudiflora* Moq. *Biol. Plant.* **2003,** *46,* 193–198.

Dansereau, P. Biogeography: An Ecological Perspective. Ronald Press: New York, NY, USA, 1957.

Dionisio Sese, M. L.; Tobita, S. Antioxidant Responses of Rice Seedlings to Salinity Stress. *Plant Sci.* **1998,** *135,* 1–9.

Feng, J.; Lin, Y.; Yang, Y.; Shen, Q.; Huang, J.; Wang, Zhu, X.; S.; Li, Z. Tolerance and Bioaccumulation of Cd and Cu in *Sesuvium portulacastrum. Ecotox. Environ. Saf.* **2018,** *147,* 306–312.

Flowers, T. J.; Colmer, T. D. Salinity Tolerance in Halophytes. *New Phytol.* **2008,** *179,* 945–963.

Flowers, T. J.; Launchli, A. Sodium Versus Potassium: Substitution and Compartmentation. In *Encyclopedia of Plant Physiology*; Springer: Berlin, 1983; 15, pp 651–681.

Flowers, T. J.; Troke, P. F.; Yeo, A. R. The Mechanism of Salt Tolerance in Halophytes. *Ann. Rev. Plant Physiol.* **1977,** *28,* 89–121.

Ghnaya, T.; Slama, I.; Messedi, D.; Grignon, C.; Ghorbel, M. H.; Abdelly, C. Cd-induced Growth Reduction in the Halophyte *Sesuvium portulacastrum* is Significantly Improved by NaCl. *J. Plant Res.* **2005,** *120,* 309–316.

Greenway, H.; Munns, R. Mechanisms of Salt Tolerance in Non-Halophytes. *Ann. Rev. Plant Physiol.* **1980,** *31,* 149–190.
Grieve, C. M.; Grattan, S. R. Rapid Assay for Determination of Water Soluble Quaternary Ammonium Compounds. *Plant Soil.* **1983,** *70,* 303–307.
Hasanuzzaman, M.; Nahar, K.; Alam, M. M.; Bhowmik, P. C.; Hossain, M. A.; Rahman, M. M.; Prasad, M. N. V.; Ozturk, M.; Fujita, M. Potential uses of Halophytes to Remediate Saline Soils. *Biomed. Res. Int.* **2014,** *589341.*
Heath, R. L.; Packer, L. Photoperoxidation in Isolated Chloroplasts. I–Kinetics and Stoiciometry of Fatty Acid Peroxidation. *Arch. Biochem. Biophy.* **1968,** *125,* 189–198.
Isayenkov, S. V.; Maathuis, F. Plant Salinity Stress; Many Unanswered Questions Remains. *Front. Plant Sci.* **2019,** *10,* 80.
Jitesh, M. N.; Prashant, S. R.; Sivaprakash, K. R.; Parida, A. K. Antioxidative Response Mechanism in Halophytes: Their Role in Stress Defence. *J. Genet.* **2006** *85* (3), 237–254.
Kapare, V.; Satdive, R.; Fulzele, D. P. Malpathak, N. Impact of Gamma Irradiation Induced Variation in Cell Growth and Phytoecdysteroid Production in *Sesuvium portulacastrum. J. Plant Growth Regul.* **2017,** *36,* 919–930.
Karakas, S.; Bolat, I.; Dikilitas, M. The use of Halophytic Companion Plant (*Portulaca oleracea* L.) on Some Growth, Fruit, and Biochemical Parameters of Strawberry Plants Under Salt Stress. *Horticulturae* **2021,** *7,* 63.
Khan, M. H.; Panda, S. K. Alterations in Root Lipid Peroxidation and Antioxidative Responses in two Rice Cultivars Under NaCl-salinity Stress. *Acta Physiol. Plant.* **2007,** *30,* 81–89.
Kulkarni, J.; Parab, H.; Srivastava, A. K.; Nikam, T. D.; Kumar, S. D.; Borde, M.; Suprasanna, P. Exposure to NaCl Enhances Cd^{2+} Biosorption Potential of *Sesuvium portulacastrum* (L.). *Environ. Technol. Innov.* **2021,** *23,* 101753.
Liu, L.; Wang, B. Protection of Halophytes and Their uses for Cultivation of Saline-Alkali Soil in China. *Biol.* **2021,** *10,* 353.
Lokhande, V. H.; Gor, B. K.; Desai, N. S.; Nikam, T. D.; Suprasanna, P. *Sesuvium portulacastrum,* A Plant for Drought, Salt Stress, Sand Fixation, Food and Phytoremediation. A Review. *Agron. Sustain. Dev.* **2013,** *33,* 329–348.
Lokhande, V. H.; Kudale, S.; Nikalje, G.; Desai, N.; Suprasanna, P. Hairy Root Induction and Phytoremediation of Textile Dye, Reactive green 19A-HE4BD, in a Halophyte, *Sesuvium portulacastrum* (L.) L. *Biotechnol. Rep.* **2015,** *8,* 56–63.
Lokhande, V. H.; Mulye, K.; Patkar, R.; Nikam, T. D.; Suprasanna, P. Biochemcial and Physiological Adaptations of the Halophyte *Sesuvium portulacastrum* (L.) L., (Aizoaceae) to Salinity. *Arch. Agr. Soil Sci.* **2012,** *59,* 1373–1391.
Lokhande, V. H.; Nikam, T. D.; Ghane, S. G.; Suprasanna, P. *In vitro* Culture, Plant Regeneration and Clonal Behaviour of *Sesuvium portulacastrum* (L.) L.: A Prospective Halophyte Studies on the Response of Growth Regulators for In Vitro Culture and Clonal Behavior of Six Clones of a Halophyte, *Sesuvium portulacastrum* L. *Physiol. Mol. Biol. Plants* **2010a,** *16* (2), 187–193.
Lokhande, V. H.; Nikam, T. D.; Patade, V. Y.; Ahire, M. L.; Suprasanna, P. Effects of Optimal and Supra-Optimal Salinity Stress on Antioxidative Defense, Osmolytes and *in vitro* Growth Responses in *Sesuvium portulacastrum* L. *Plant Cell Tiss. Org. Cult.* **2011a,** *104,* 41–49.
Lokhande, V. H.; Nikam, T. D.; Patade, V. Y.; Suprasanna, P. Morphological and Molecular Diversity Analysis Among the Indian Clones of *Sesuvium portulacastrum* L. *Genet. Res. Crop Evol.* **2009b,** *56,* 705–717.

Lokhande, V. H.; Nikam, T. D.; Suprasanna, P. Biochemical, Physiological and Growth Changes in Response to Salinity in Callus Cultures of *Sesuvium portulacastrum* L. *Plant Cell Tiss. Org. Cult.* **2010b,** *102,* 17–25.

Lokhande, V. H.; Nikam, T. D.; Suprasanna, P. Differential Osmotic Adjustment to Iso-Osmotic Salt and PEG Stress In Vitro in the Halophyte *Sesuvium portulacastrum* L. *J. Crop Sci. Biotechnol.* **2010c,** *13* (4), 251–256.

Lokhande, V. H.; Nikam, T. D.; Suprasanna, P. *Sesuvium portulacastrum* (L.) L., A Promising Halophyte: Cultivation, Utilization and Distribution in India. *Genet. Res. Crop. Evol.* **2009b,** *56,* 741–747.

Lokhande, V. H.; Patade, V. Y.; Srivastava, S.; Suprasanna, P.; Shrivastava, M.; Awasthi, G. Copper Accumulation and Biochemical Responses of *Sesuvium portulacastrum* (L.). *Mat. Today: Proceed.* **2020,** *31,* 4.

Lokhande, V. H.; Srivastava, A. K.; Srivastava, S.; Nikam, T. D.; Suprasanna, P. Regulated Alterations in Redox and Energetic Status are the Key Mediators of Salinity Tolerance in the Halophyte *Sesuvium portulacastrum* (L.) L. *Plant Grow Regul.* **2011c,** *65* (2), 287–298.

Lokhande, V. H.; Srivastava, S.; Patade, V. Y.; Dwivedi, S.; Tripathi, R. D.; Nikam, T. D.; Suprasanna, P. Investigation of Arsenic Accumulation and Tolerance in *Sesuvium portulacastrum* (L.) L. Chemosph. **2011b,** *82* (4), 529–534.

Lokhande, V. H.; Suprasanna, P. Prospects of Halophytes in Understanding and Managing Abiotic Stress Tolerance. In *Environmental Adaptations to Changing Climate: Metabolism, Productivity and Sustainability*; Springer: New York, 2012; pp 29–56.

Mariem, W.; Kilani, B. R.; Benet, G.; Abdelbasset, L.' Stanley, L.; Charlotte, P.; Abdely, C.; Tahar, G. How does NaCl Improve Tolerance to Cadmium in the Halophyte *Sesuvium portulacastrum* ? *Chemosph.* **2014,** *117,* 243–250.

Matysik, J.; Bhalu, A. B.; Mohanty, P. Molecular Mechanisms of Quenching of Reactive Oxygen Species by Proline Under Stress in Plants. *Review Curr. Sci.* **2002,** *82,* 525–532.

May, M. J.; Vernoux, T.; Leaver, C.; Van, M. M.; Inze, D. Glutathione Homeostasis in Plants: Implications for Environmental Sensing and Plant Development. *J. Exp. Bot.* **1998,** *49,* 649–667.

Menzel, U.; Leith, H. Annex 4: Halophyte Database Vers 2. In *Halophyte uses in Different Climates 1. Ecological and Physiological Studies*; Progress in Biometerology: Backhuys Publishers: Leiden, 1999; pp 159–258.

Messedi, D.; Labidi, N.; Grignon, C.; Abdelly, C. Limits Imposed by Salt to the Growth of the Halophyte Sesuvium Portulacastrum. *J. Plant Nut. Soil Sci.* **2004,** *167,* 720–725.

Miller, G.; Suzuki, N.; Ciftci-Yilmaz, S.; Mittler, R. Reactive Oxygen Species Homeostasis and Signaling During Drought and Salinity Stresses. *Plant Cell Environ.* **2010,** *33* (4), 453–467.

Mittler, R. Oxidative Stress, Antioxidants and Stress Tolerance. *Trends in Plant Sci.* **2002,** *7,* 405–410.

Mnasri, M.; Ghabriche, R.; Fourati, E.; Zaier, H.; Sabally, K.; Barrington, S.; Lutts, S.; Abdely, C.; Ghnaya, T. Cd and Ni Transport and Accumulation in the Halophyte *Sesuvium portulacastrum*: Implication of Organic Acids in These Processes. *Front. Plant Sci.* **2015,** *6,* 156.

Muchate, N. S.; Nikalje, G. C.; Rajurkar, N. S.; Suprasanna, P.; Nikam, T. D. Physiological Responses of the Halophyte *Sesuvium portulacastrum* to Salt Stress and Their Relevance for Saline Soil Bio-Reclamation. *Flora*, **2016,** *224,* 96–105.

Munns, R.; Tester, M. Mechanism of Salinity Tolerance. *Ann. Rev. Plant Biol.* **2008,** *59,* 651–581.

Murashige, T.; Skoog, F. A Revised Medium for Rapid Growth and Bioassays with Tobacco Tissue Cultures. *Physiol. Plant.* **1962,** *15,* 473–497.

Nakano, Y.; Asada, K. Hydrogen Peroxide is Scavenged by Ascorbate-Specific Peroxidase in Spinach Chloroplasts. *Plant Cell Physiol.* **1981,** *22,* 867–880.

Nikalje, G. C.; Kumar, J.; Nikam, T. D.; Suprasanna, P. FT-IR Profiling Reveals Differential Response of Roots and Leaves to Salt Stress in a Halophyte *Sesuvium portulacastrum* (L.) L. *Biotechnol. Rep.* **2019,** *23,* e00352.

Patade, V. Y.; Suprasanna, P.; Bapat, V. A. Effects of Salt Stress in Relation to Osmotic Adjustment on Sugarcane (*Saccharum officinarum* L.) Callus Cultures. *Plant Grow. Regul.* **2008,** *55* (3), 169–173.

Patil, A. V.; Lokhande, V. H.: Suprasanna, P.; Bapat, V. A.; Jadhav, J. P. *Sesuvium portulacastrum* (L.) L.: A Potential Halophyte for the Degradation of Toxic Textile Dye, Green HE4B. *Planta,* **2012,** *235* (5), 1051–1063.

Qadir, M.; Quillerou, E.; Nangia,V.; Murtaza,G.; Singh, M.; Thomas, R. J. Economics of Salt-Induced Land Degradation and Restoration. *Nat. Resour. Forum.* **2014,** *38,* 282–295.

Rabhi, M.; Ferchichi, S.; Jouini, J.; Hamrouni, M. H.; Koyro, H. W.; Ranieri, A.; Abdely, C.; Smaoui, A. Phytodesalination of a Salt-Affected Soil with the Halophyte *Sesuvium portulacastrum* L. to Arrange in Advance the Requirements for the Successful Growth of a Glycophytic Crop. *Biores. Tech.* **2010,** *101* (17), 6822–6828.

Rahman, M. M.; Mostofa, M. G.; Keya, S. S.; Siddiqui, M. N.; Ansari, M. M. U.; Das, A. K.; Rahman, M. A.; Tran, L. S. Adaptive Mechanisms of Halophytes and Their Potential in Improving Salinity Tolerance in Plants. *Int. J. Mol. Sci.* **2021,** *22,* 10733.

Rajalakshmi, S.; Parida, A. Halophytes as a source of genes for abiotic stress tolerance. *J. Plant Biochem. Biotechnol.* **2012,** *21* (Suppl 1), S63–S67.

Ramani, B.; Reeck, T.; Debez, A.; Stelzer, R.; Huchzermeyer, B.; Schmidt, A.; Papenbrok, J. *Aster tripolium* L. and *Sesuvium portulacastrum* L.: Two Halophytes, Two Strategies to Survive in Saline Habitats. *Plant Physiol. Biochem.* **2006,** *44,* 395–408.

Redondo-Gomez, S.; Mateos-Naranjo, E.; Figueroa, M. E.; Davy, A. J. Salt Stimulation of Growth and Photosynthesis in an Extreme Halophyte, *Arthrocnemum macrostachyumi. Plant Biol.* **2009,** *12,* 79–87.

Sairam, R. K.; Srivastava, G. C. Changes in Antioxidant Activity in Sub-Cellular Fractions of Tolerant and Susceptible Wheat Genotypes in Response to Long Term Salt Stress. *Plant Sci.* **2002,** *162,* 897–904.

Sairam, R. K.; Tyagi, A. Physiology and Molecular Biology of Salinity Stress Tolerance in Plants. *Curr. Sci.* **2004,** *86,* 407–421.

Santos, J.; Al-Azzawi, M.; Aronson, J.; Flowers, T. J. eHALOPH a Database of Salt-Tolerant Plants: Helping Put Halophytes to Work. *Plant Cell Physiol.* **2015,** *57,* e10.

Schimper, A. F. W. Plant Geography upon a Physiological Basis. Clarendon Press: Oxford, UK, 1903.

Sengbusch, P. V. "Halophytes" Botanik Online, University of Hamburg, 2003.

Sharma, R.; Wungrampha, S.; Singh, V.; Pareek, A.; Sharma, M. K . Halophytes as Bioenergy Crops. *Front. Plant Sci.* **2016,** *7,* 1372.

Slama, I.; Abdely, C.; Bouchereau, A.; Flowers, T.; Savoure, A. Diversity, Distribution and Roles of Osmoprotective Compounds Accumulated in Halophytes Under Abiotc Stress. *Ann. Bot.* **2015,** *115,* 433–447.

Slama, I.; Ghnaya, T.; Savouŕe, A.; Abdelly, C. Combined Effects of Long-Term Salinity and Soil Drying on Growth, Water Relations, Nutrient Status and Proline Accumulation of *Sesuvium portulacastrum*. *C. R. Biol.* **2008,** *331,* 442–451.

Srivastava, A. K.; Srivastava, S.; Lokhande, V. H.; D'Souza, S. F.; Suprasanna, P. Salt Stress Reveals Differential Antioxidant and Energetics Responses in Glycophyte (*Brassica juncea* L.) and Halophyte (*Sesuvium portulacastrum* L.). *Front. Environ. Sci.* **2015,** *3,* 19.

Stocker, O. Das Halophytenproblem. In *Ergebnisse der Biologie*, Springer: Berlin, Germany, 1928: pp 266–353.

Tester, M.; Davenport, R. Na^+ Tolerance and Na^+ Transport in Higher Plants. *Ann. Bot.* **2003,** *91,* 503–5027.

Tripathi, S. B.; Gurumurthi, K.; Panigrahi, A. K.; Shaw, B. P. Salinity Induced Changes in Proline and Betaine Contents and Synthesis in Two Aquatic Macrophytes Differing in Salt Tolerance. *Biol. Plant.* **2007,** *51,* 110–115.

Wali, M.; Gunsè, B.; Llugany, M.; Corrales, I.; Abdelly, C.; Poschenrieder, C.; Ghnaya, T. High Salinity Helps the Halophyte *Sesuvium portulacastrum* in Defense Against Cd Toxicity by Maintaining Redox Balance and Photosynthesis. *Planta*, **2016,** *244* (2), 333–346.

Wang, L.; Stutte, G. W. The Role of Carbohydrates in Active Osmotic Adjustment in Apple Under Water Stress. *J. Am. Soc. Hort. Sci.* **1992,** *117,* 816–823.

Wang, W.; Yan, Z.; You, S.; Zhang, Y.; Chen, L.; Lin, G. Mangroves: Obligate or Facultative Halophytes? A Review. *Trees,* **2011,** *25,* 953–963.

Yazici, I.; Turkan, I.; Sekmen, A. H.; Demiral, T. Salinity Tolerance of Purslane (*Portulaca oleracea* L.) is Achieved by Enhanced Antioxidant System, Lower Level of Lipid Peroxidation and Proline Accumulation. *Env. Expt. Bot.* **2007,** *61,* 49–57.

Zaier, H.; Ghnaya, T.; Lakhdar, A.; Baioui, R.; Ghabriche, R.; Mnasri, M.; Sghair, S.; Lutts, S.; Abdelly, C. Comparative Study of Pb-Phytoextraction Potential in *Sesuvium portulacastrum* and *Brassica juncea*: Tolerance and Accumulation. *J. Hazard. Mat.* **2010,** *183* (1–3), 609–615.

Zhang, F.; Yang, Y. L.; He, W. L.; Zhao, X.; Zhang, L. X. Effects of Salinity on Growth and Compatible Solutes of Callus Induced from *Populus euphratica*. *In Vitro Cell. Dev. Biol. – Plant.* **2004,** *40,* 491–494.

Zhao, X.; Tan, H. J.; Liu, Y. B.; Li, X. R.; Chen, G. X. Effect of Salt Stress on Growth and Osmotic Regulation in *Thellungiella* and *Arabisopsis* Callus. *Plant Cell Tiss. Organ Cult.* **2009,** *98,* 97–103.

Zhu, J. K. Plant Salt Tolerance. *Trends Plant Sci.* 2001, *6,* 66–71.

CHAPTER 12

Halophytes: Prospectives and Perspectives for Sustainable, Eco-friendly Environmental and Agricultural Development

VINAYAK H. LOKHANDE

Department of Botany, J.T.S.S.P.M., Shri Shiv Chhatrapati College of Arts, Commerce and Science, Bodkenagar, Junnar, Pune, Maharashtra, India

ABSTRACT

The ever-growing human population continuously generates pressure on natural resources and limits the fertility and productivity of agriculture. Almost one-third of every area farmed is significantly affected due to excess salt and the associated water logging, alkalinity, and heavy metals depositions in the fertile land. Unfortunately, most of the crop systems cultivated in agriculture come in the category of glycophytes that are unable to cope with adverse environmental abiotic stresses such as salinity, drought, heavy metals, etc. On the contrary, plants such as mangroves or halophytes are provided with the gift of nature to sustain under these adverse stresses. These are found to be the choice of plants to rescue the natural resources from these astonishingly increasing environmental pollutions and provide a way toward future sustainable agricultural development. The present era of agriculture is now struggling for the improvement of soil fertility and productivity as against the severely contaminated land and water resources. Halophytes are one of the solutions to make eco-friendly environment decontamination, phytoremediation, desalination, reclamation, and stabilization of saline soil from the agricultural area. The halophytes will find their way as a source of

Physiology of Halophytes: Signaling, Omics, and Tolerance Mechanisms.
Nivas Manohar Desai & Ganesh Chandrakant Nikalje (Eds.)

salt and heavy metals extractors or accumulators, supporting biocompost and fertilizer, soil conservators, etc. The present chapter emphasizes on applications of halophytes in resolving the serious environmental issues dealt with the agriculture and emphasis is given to their role in environmental clean-up using advanced strategies. The prospective and perspectives related to the role of halophytes are discussed in brief.

12.1 INTRODUCTION

Although water is considered a renewable resource, the limitations on the availability of water for human needs have been found correlated with its uneven distribution across the globe concomitant with its changing chemical and physical properties due to its worldwide pollution at harsh levels (Ventura and Sagi, 2013). Therefore, the problem of freshwater scarcity has been aggravated with time owing to its geographical distribution with limited access to the ever-growing human population. Furthermore, successful agriculture requires only nonpolluted fresh water for irrigation purposes. However, the use of available contaminated resources of fresh water for continuous irrigation of crop systems is responsible for increasing soil pollution to an extensive level by increasing the salinity of the soil due to the deposition of excess amounts of salts and toxic heavy metals thereby making the soil as well as the crops inefficient for water use and appropriate drainage. This synchronized interaction and continuous use of contaminated water and soil for agricultural purpose develops the cascade of many events for the failure of agriculture. Besides, the impact of globalization and urbanization has been found responsible for changing the environmental and climatic conditions to a severe extent and thus makes the environment unsuitable for the survival of biotic components. It is well-known fact that our food demand is satisfied through agricultural produce by the salt-sensitive glycophytes. To date, agricultural production is achieved by these glycophytes; however, in the course of development and severely polluted environmental conditions, these plants do not cope and accomplish the competition of food demand associated with their production. Glycophytes are found to be unsuccessful in producing economically under saline conditions because glycophytes are very sensitive to variable concentrations of the salts in the soil and water. Therefore, the accomplishment of global food demand for the ever-growing human population concomitant with the achievement of sustainable development is only possible with the availability of healthy agricultural soil, good quality water for irrigation, and favorable climatic and/or environmental conditions

(Koch et al., 2013; Zheng et al., 2018). Tactlessly, in the current situation, the progressive agricultural practices including inappropriate physical, chemical and biological operations adopted by the agrarian community to accomplish the goal of the highest food production were found unsustainable. In this context, the sustainable agriculture can be achieved through saline sea water irrigation to the halophytic crop systems which are being used as the source of cash crops (Lieth et al., 1999). In contrast to crop plants, these halophytes exist as specialists who can tackle adverse growth conditions and complete their lifecycle by residing in natural saline habitats. They have adopted distinctly at physiological, biochemical, and anatomical levels by counteracting the adverse effects of water deficiency and toxicity of salt ions (Koyro et al., 2008). The sustainable and favorable method of using halophytes is to explore them efficiently across saline regions which are otherwise not useful for conventional agriculture with the availability of low-quality water (Boer and Gliddon, 1998; Lieth et al., 1999; Koyro et al., 2008). Halophytes have shown their growth and productivity performance under extreme abiotic stress conditions and have remained the choice for solving the problems of environmental concern. The halophytes were found to succeed in managing and tolerating the combined and multidimensional abiotic stresses in comparison with glycophytes. The tolerance mechanism underlying in halophytes have been explored in glycophytes to adopt the abiotic stress tolerance in salt-sensitive crop plants. Besides, most of the halophyte systems have shown their potential as a cash crop in a saline environment, remedies to various diseases with a huge medicinal value, as well as in solving the problems of environmental concern (Lokhande and Suprasanna, 2012; Nikalje et al., 2017, 2019). Therefore, in the present chapter, the emphasis is given on the role of halophytes in resolving the serious environmental issues dealt with agriculture, and focus is given on their role in environmental clean-up using advanced but sustainable strategies.

12.2 ENVIRONMENTAL POLLUTION AND AGRICULTURE

Natural disasters (earthquakes, floods, drought, etc.) and anthropogenic activities leading to altered climatic conditions remain the main causes of environmental pollution. These activities account for rapid deforestation for industrial and urban development, change in landing patterns for advanced agriculture, extensive use of hazardous chemical fertilizers for maximizing the growth and productivity of the crop plants, generation of huge amounts of solid and liquid wastes concomitant with the use of nonbiodegradable cargos and their improper management, burning of agricultural waste and emission of toxic gases from industrial area and automobiles. Therefore, in the 21st century, the drastic change in the climatic conditions during this postindustrialization era has been correlated with the unethical planning and development toward industrialization, urbanization, widespread rate of fossil fuels burning, and increased levels of harmful greenhouse gases in the atmosphere (Chaudhry and Sidhu, 2021). These harmonized interactions of the ever-growing human population and multifaceted problems of environmental pollution have harshly distressed agricultural practices and crop patterns worldwide. Day by day, population growth and economic expansions are responsible for limiting the cultivable land and freshwater resources to a greater extent (Turcios et al., 2021), thereby imparting the failure requirement of essential needs of a growing population. It has been estimated that since the 1980s onward, withdrawal of groundwater from different sources has increased steadily by 1% per annum, and this problem has become serious with the aggravated climatic conditions (FAO, 2017; Turcios et al., 2021). Besides, the impact of urbanization and industrialization has rapidly acquired the available fertile land for its expansion, thereby limiting its availability for the qualitative and quantitative production of agricultural goods to fulfill the demand of the global population. To add more calamities, the abiotic environmental factors or stress conditions such as salinity, heavy metal contamination, and drought rapidly impart soil, as well as water pollution, and affect the processes of sustainable agricultural development. The rapid and injudicious urbanization and industrialization during the past few decades led to the release of large amounts of toxic wastes into the environment (Naikoo et al., 2019). This present scenario of environmental pollution in the context of soil, water, and air contamination has severely threatened the world of agricultural and agrarian communities. Therefore, it is an urgent need to make the world aware and attentive about saving the agricultural community by searching the permanent solutions to the problems of environmental pollution to save the wealth of the natural resources and made them available to future generations in a sustainable way for prolonged development.

12.3 SUSTAINABLE AGRICULTURE DEVELOPMENT—PROBLEMS AND SOLUTIONS

The invention of agriculture increased food production and led to a rapid growth in the human population, which reached 50 million by the bronze age, 250 million by the time of Christ, and 1 billion by the middle of the 18th century. Worldwide, the rate of human population has been endlessly augmented. Within 100 years, the population grew from 1 to 2 billion (in 1927), whereas it took only another 100 years to reach 7 billion (in 2020) (Tian et al., 2021). In line with the increased growth of population, the whole world is facing challenges to fulfill the food demand due to a drastic decrease in the availability of fertile cultivable lands allied with environmental pollution (Wallace et al., 2018). Although the green revolution leads to increased grain production and improvement of food shortages, exhaustive agriculture, also known as intensive farming and industrial agriculture, now becomes detrimental because it has been achieved with extensive use of chemical fertilizers, pesticides, and freshwater (Tian et al., 2021). With their exhaustive use, now the soil has become restless and nonproductive; in most of the arid and semi-arid regions, the agricultural lands have permanently lost their potential for fertility due to the comprehensive use of hazardous chemical fertilizers and pesticides, abrupt change in the soil pH, drastic depletion on the levels of soil organic carbon (SOC) and essential mineral nutrients, leaching of hazardous heavy metals on the ground surface and precipitation of salts due to improper irrigation practices from polluted water resources, improper or absence of soil aeration due to its compactness and water-logged conditions. Besides, microbial (bacterial and fungal) flora of the soil became monotypic or occasionally vulnerable which has affected or permanently lost the humus layer of the soil to a severe extent, a huge increase in the soil and water pollution levels associated with habitat losses led to reduction in cultivable land area across the world. Therefore, permanent solutions and strategies have to be developed to minimize the levels of environmental pollution and attain environmental sustainability. Sustainable agriculture has shown the potential to address some of the fundamental challenges facing agricultural practices worldwide. However, increasing agricultural productivity and food security in a sustainable manner remains a massive challenge with traditional agricultural practices. Therefore, planning and execution of sustainable agricultural practices should be in such a way that they will contribute to solving the issues related to the ever-growing human population, malnutrition, poverty, poor food quality and food insecurity, limited fertile land availability, associated with the adverse changes in climate such as global warming, elevated sea water levels, deforestation, environmental pollution, etc. Sustainable agricultural development can be achieved by systematic, strategic planning and designing of current well-cultivated crops to become the crops of future sustainable agriculture with the improvement in terms of increasing yield, improving nutritional quality, increasing agricultural resources use efficiency, and improving biotic and abiotic stress tolerance (Tian et al., 2021). Although susceptible glycophytic crop systems are coping in attaining the highest productivity under the circumstances of climate-changing patterns, they are unable to cope with the harshly available climatic conditions. Therefore, halophytes could play an important role and assist the glycophytes to come out of this danger with boosting the productivity and quality of the agriculture produce in a sustainable way. Sustainable agriculture development faces major challenges of environmental concern including abiotic stresses such as salinity, drought, and heavy metal contamination, depleted levels of SOC, and extensive use of chemical fertilizers and pesticides. Some of the issues are discussed here.

12.3.1 ABIOTIC STRESSES

Worldwide, the penalties of adverse and complex environmental conditions on agricultural development are due to abiotic stresses which include salinity, drought, heavy metals, freezing, hypoxia, strong light, ultraviolet (UV), and high and low temperatures (Hirayama and Shinozaki, 2010). The growth and yield of the plants are adversely affected by critical environmental conditions such as high temperatures and enhanced CO^2 concentrations (Raza et al., 2019). Plants being sessile organisms are found sensitive to these alterations in climatic conditions and therefore unable to adapt to changing environmental conditions (Lindner et al., 2010). Variations in temperature under natural conditions can affect plant growth and reproduction, and adverse climatic fluctuations damage the molecular mechanisms related to growth, thus affecting the development of plants (Bita and Gerats, 2013; Chaudhry and Sidhu, 2021). In addition, salinity being one of the major abiotic constraints, affects almost every physiological aspect of plants at the cellular and whole plant level (Hasegawa et al., 2000; Munns and Tester, 2008) which reflects the changes at phenotypic, genotypic, and physiobiochemical levels. It has been estimated that more than 800 million hectares of land considering 20% of all cultivated land and nearly half of irrigated land throughout the world is affected by salt (FAO, 2008; Turkan and Demiral, 2009) and this amount accounts for more than 6% of the world's total land area (Munns and Tester, 2008). Most of this salt-affected land has arisen from natural causes, from the accumulation of salts over a long period in addition to poor quality irrigation facilities and wrong agricultural practices performed by human beings in arid and semiarid zones of the world (Rengasamy, 2006). Unfortunately, most of the developing countries such as India and less-developed countries fall in these regions, and it is estimated that every year about 6 million ha arable land is lost from agricultural use in these developing countries. The population load in these countries is also high and agricultural land fails to fulfill the basic human needs and raw material supply to industries of their ever-growing population. The new land is unavailable for cultivation of crops and permanent solutions to salinity problems are not possible due to the poor economic status of these countries. In addition, freshwater resources are also becoming rapidly depleted due to the steady increase in the salinity levels in many regions. The rapid increase in population pressure is leading to greater utilization of remaining meager freshwater supplies for drinking, and living and even less for agriculture. In this complex scenario, it has become crucial to consider the development of nonconventional agricultural technologies which might more effectively utilize degraded, marginal, and saline lands for agriculture by using the "so-called" poor water (brackish water, seawater, and wastewater) (Khan and Duke, 2001).

Despite advances in increasing plant productivity and resistance to several pests and diseases, the improvement in the responses of crop plants to salinity stress remains elusive. Furthermore, limited success in increasing the yield stability of crop plants grown in saline soil might be too small in the minimal understanding of how salinity and other abiotic stresses affect the most fundamental processes of cellular functions including cell division, differentiation, and expansion which have a sustainable impact on plant growth and development (Hasegawa et al., 2000; Zhu 2001). In this context, it is need of the hour to broaden the search for studying the adverse effects of salt stress and mechanisms of salt tolerance developed by the plants at various levels of developmental stages for the selection of salt–tolerant species that will accomplish the demand of food, feed, fodder, and industrial raw material (Rengasamy, 2006). The problem can be resolved by enhancing the tolerance level to salt stress at the cost of reduced gain yield. The alternative and sustainable approach is to use salt–tolerant plants such as halophytes which perform better in productivity at extreme salinity levels. Almost all our modern crops are derived from glycophytes having a susceptible gene pool to salt tolerance and have received the most research attention. However, to date, enormous research papers exploring the salt tolerance mechanism of glycophytes have been published yet fewer than a dozen salt–tolerant cultivars were released, offering only slight improvement over the parent lines. It has been questioned whether any cultivar who is bred for salt tolerance has been commercially successful. In this regard, studies on halophytes provide information in the prospectives of (i) adaptations and survival mechanism acquired by the halophytes under saline conditions for making the efforts toward the development of salt tolerance in glycophytes (Bohnert and Jensen, 1996; Glenn and Brown, 1999); (ii) use of halophytes for evaluation of the possibility of saline agriculture in naturally available saline habitats (Glenn et al., 1997); and (iii) possibility of the use of halophytes as a source of new crops for sustainable development.

The most serious environmental concern comprehensively explored and researched in detail next to the salinity stress is abiotic stress occurred due to soil and water pollution caused by contamination with toxic and hazardous heavy metals. The rapid and injudicious urbanization and industrialization during the past few decades have been found responsible for the release of large amounts of toxic wastes into the environment (Naikoo et al., 2019). Contamination of agricultural lands by toxic heavy metals has become a serious environmental issue due to their adverse effects on ecosystems (Lokhande and Suprasanna, 2012). These heavy metals are widely spread in all types of natural resources and generating chronic toxic effects on plants, animals, and humans. Worldwide, every year, approximately USD35–50 billion is disbursed to clean the sites contaminated with toxic heavy metals. However, these clean-up strategies either eradicate, burn, or chemically treat contaminated soil at the cost of disrupting the microbial population of that site and damaging the environment (Gall and Rajakaruna, 2013). In the current scenario, there are cost-effective and environmental-friendly methodologies, which are easily available and are more sustainable, suitable, acceptable, and alternative to physical and chemical methods of heavy metal clean-up from contaminated sites. Fortunately, the methodology that has gained worldwide acceptance and popularity over the past few decades includes the sustainable use of biological systems (microbes and plants) for the removal of toxic pollutants and/or contaminants (heavy metals) from contaminated sites. This promising green technology is widely known as "phytoremediation" which uses plants to clean up environmental contaminants such as heavy metals, salt ions, organic compounds, and radioactive substances in soil or water (Dar et al., 2018; Naikoo et al., 2020). The concept of phytoremediation lies in the use of plants as natural systems to remove toxic contaminants (Lokhande and Suprasanna, 2012). Different plant systems have been explored for studying the approach of phytoremediation; however, the accumulation and detoxification of toxic heavy metals into less harmful or nontoxic forms have been correlated with the performance of the plant system used for phytoremediation purpose and variations occurred among their species. In the past few years, the emphasis has been given to sustainable agricultural development through the reclamation of contaminated regions of the world with the practical approach of exploring plants for the elimination of toxic heavy metals from contaminated soil and water bodies. In this context, to use the approach of phytoremediation, various plants have been discovered for their potential to accumulate or detoxify heavy metals; on that basis, the plants are categorized as "accumulators" or "hyperaccumulators." In this categorization, the halophytes as against the glycophytes have established a promising role in the phytoremediation of heavy metal-contaminated water and soil resources (Lokhande and Suprasanna, 2012; Liang et al., 2017; Nikalje et al., 2019; Naikoo et al., 2020). Recently there has been increased interest in using halophytes for phytoremediation, particularly in developing countries (Nouri et al., 2017; Naikoo et al., 2020).

12.3.2 DEPLETION IN SOIL ORGANIC CARBON

One of the major concerns or critical issues of today's agriculture and its failure to achieve the highest yield is the depleted value of SOC due to the uppermost threshold value of contaminated soil and the inaccessibility of the optimum level of essential nutrients from the soil. Besides, the improper implementation of agricultural practices such as extensive foliar applications of weedicides to control and remove the weeds from standing crops in the field, uncontrolled and overdosage of a variety of toxic pesticides in controlling the diseases, use of automated machinery in the field made the soil vulnerable to compactness and fertility. Therefore, maintaining or improving the soil properties has gained importance to sustain modern agriculture under increasing demands for preserving the biodiversity and quality of the environment. Although in the last few decades, the exploration of organic farming and the use of various types of nanoparticle-fortified biological-based organic fertilizers have played their role in improving soil fertility and enhancing the level of available organic carbon, its ratio found negligible when compared with the gross area of worldwide agriculture and its produce. Besides, there are some misconceptions and the unavailability of authentic products and procedures regarding the use of organically produced fertilizers in improving the level of soil organic carbon. Therefore, future agriculture should be precisely planned and focused in such a way that the available cultivable soil should be sustained and the barren contaminated land must get into cultivation to improve the total acreage production of crops with the usage of scientifically and rationally valid, economically cheap and environmentally pollution free advanced techniques and technologies.

SOC is a key component of functional ecosystems and is crucial for food, soil, water, and energy security (Stockmann et al., 2015). SOC mostly originates from the biological cycle between plants and soil and promotes the formation of soil aggregates and maintaining fertility (Deng et al., 2016; Yan et al., 2020). Plants are responsible for improving the soil's organic carbon by selectively absorbing the nutrients distributed in the deep layers of soil and carbon dioxide fixation from the atmosphere by the process of photosynthesis, thereby accumulating the organic carbon in the soil surface through the decomposition of plant litter on the soil. This is responsible for accelerating the process of soil development (Yan et al., 2020). Furthermore, the transformation of plant carbon to soil carbon is equally important, and it is chiefly performed by microbial decomposition and stabilization and is also mediated by the physical and geological properties of the soil (Lehmann and Kleber, 2015; Yan et al., 2020). This transformation process leads to the stabilization of plant-derived organic carbon in the soil for a longer period, followed by its conversion to SOC by releasing the carbon dioxide from the soil through the respiration process during microbial decomposition of plant litter and providing the output in terms of rich organic carbon levels of soil (Whalen et al., 2014). The contamination of soil, due to the environmental abiotic constraints, is accountable for limiting the transformation from plant-derived carbon to SOC. Besides, unethical and unsustainable agricultural practices have been found responsible for destroying the microbial flora of the soil; therefore, making the process of plant litter decomposition ineffective and resulting in a declining level of SOC. Although this transformation may happen to improve the SOC, the limited availability of fertile land area will not be able to cope with the gross annual production of the crops to meet the demand of the global population. In this context, the contaminated barren lands, waterlogged areas, as well as coastal saline land areas, should be acquired for increasing the overall productive land area with extensive manipulations and effective management of the water and mineral use efficiency of the crop plants in such areas. However, glycophytes cannot remain the choice for the cultivation of such contaminated lands, instead, halophytes could be explored in such areas to achieve the requirements of the globe. Besides, the halophytic systems can be introduced in the fertile but contaminated agricultural land areas to play their role. Halophytes by growing in contaminated fertile lands will effectively help to remove or extract the excessive saline salts and toxic minerals from the contaminated soil, reinstatement of diverse microbial flora for accelerating the process of decomposition of organic litter, thereby exploring its role in enhancing the SOC level. Such kinds of practices have been undertaken in wetland and saline areas (Yan et al., 2020) to increase the level of SOC. The sandy soils could also be explored for cultivation by improving their SOC level since sandy soils cover approximately 900 million hectares of area worldwide, particularly in arid and semi-arid regions (Yost and Hartemink, 2019). The halophytes thriving in saline habitats across the sandy lands of coastal areas, the plant species of halophytes having economic potential could be used as an alternative for developing novel agricultural lands with improved SOC levels to boost up the gross production and the world economy.

12.4 HALOPHYTES

Angiosperms, being the most advanced plant group have demonstrated the greatest diversity around the globe and are accountable for accomplishing all kinds of requirements of the under-developed, developing, and well-developed countries. Angiosperms cover the plants growing in diverse habitats, and based on their growth pattern related to the surrounding ecosystem, they are categorized as hydrophytes, xerophytes, mesophytes, halophytes, etc. due to their survival in extreme conditions such as aquatic, xeric, and salty conditions. However, plants are broadly categorized into two groups, "glycophytes" and "halophytes," based on their growth responses and capacity to grow under varying saline conditions (Munns and Tester, 2008). Unfortunately, worldwide the angiosperm crop plants which are sensitive to salts and any adverse conditions, growing across the agricultural zone come under the category of glycophytes. Halophytes constitute about 2% of all plant species and they are extraordinary plants that retain their survival and complete the life cycle in sodium salts environments which are usually found poisonous to other species of glycophytes and eradicate almost their entire population (Flowers and Colmer, 2008; Lokhande and Suprasanna, 2012). Halophytes are categorized as obligate halophytes which require a specific concentration of salts in the growing medium for their survival however, facultative halophytes can grow in salt-free environments and are also referred to as mangrove associates. Thus, the inbuilt capacity of the halophytes to grow in the presence or absence of salts takes advantage of glycophytes to manage the adverse abiotic stress conditions and therefore widely utilized in the program of productivity improvement of the crop plants (Lokhande and Suprasanna,

2012). Under the circumstances of unbalanced conditions of environmental factors, the remedy for future agriculture relies on the widespread use of halophytes because they could contribute to the growth of crop plants under stressed conditions and any genetic and biochemical modifications carrying risks to crop plants could easily be performed on halophytes which could improve their salt or drought-tolerance performances (Tipirdamaz et al., 2020).

In the context of widespread environmental pollution with its direct and adverse impact on growth, physiology, and productivity, even the survival of the crop plants, the sustainable use of halophytic plants provides a promising approach to exploring the contaminated zones of the world unsuitable for conventional agriculture (Boer and Gliddon, 1998; Koyro et al., 2008). Halophytic diversity has been used for economic interests to fulfill the needs of the human population or to resolve the problems of ecological concerns (sand dune fixation, soil desalinization, phytoremediation, and sequestration of carbon dioxide). Unfortunately, the exploration of the halophytes to a wide extent has not yet been reported (Koyro et al., 2008). Therefore, innovative solutions are needed worldwide with the widespread exploitation of halophytes to tackle the losses associated with gross agriculture production toward sustainable development (Custodio et al., 2021). Therefore, halophytes will be the answer to the problems of environmental pollution concerns such as salt desert and wasteland reclamation, besides being economically significant not only in India but also around the world. With this awareness, in 2016, India took a step ahead and made history by becoming the first country to set up the world's first Genetic Garden of Halophytes. It was set up at the coastal town of Vedaranyam in Tamil Nadu by MS Swaminathan Research Foundation (MSSRF). It has over 1600 plant species belonging to 550 genera and 117 families of halophytes. It also preserves halophytes that occur alongside the coasts of India comprising the Andaman and Nicobar Islands (Naikoo et al., 2020). The main aim of the foundation is the conservation and exploration of the halophytes for solving the problems of environmental concerns and developing the platform for lifelong sustainable development. Therefore, halophytes could be exploited to cover the underutilized large areas of saline wastelands and coastal belts due to the high saline environment. It will help to minimize the pressure on prime agricultural lands which have been otherwise overexploited and lost their productivity and fertility due to continuous cultivation. It is high time to make real efforts to identify, conserve, and use these plant species (Litalien and Zeeb, 2020; Naikoo et al., 2020).

12.5 PROSPECTIVES OF THE HALOPHYTES

12.5.1 ROLE OF HALOPHYTES IN ENVIRONMENTAL CLEAN-UP

Green environmental clean-up using halophytes, as an economic and environmental-friendly solution, has recently been more popular, particularly in developing countries. Halophytes remain one of the solutions to make the environment clean and eco-friendly through decontamination, phytoremediation, desalination, reclamation, and stabilization of saline soil from agricultural areas. The major constraints of environmental pollution are global warming, industrialization, and urbanization leading to generating excess load on natural resources and making the environment unsuitable for the survival of living organisms. Salinity and heavy metal contamination of soil and water bodies leftovers the leading environmental stresses and challenging issues to find remedies worldwide. Halophytes are routinely practiced to resolve these issues of environmental concern. During evolution, halophytes have adopted several strategies for tolerating the abiotic stresses in terms of succulence, partitioning of toxic metals and ions, synthesis of osmolytes, increase in activity of antioxidants, and synthesis of compatible solutes (Nikalje et al., 2019). In the last three decades, more emphasis has been given to using halophytes for environmental clean-up and to restore fertile cultivable lands and freshwater resources for irrigation purposes. Halophytes such as *Sesuvium*, *Medicago*, *Cackile*, *Salicornia*, *Atriplex*, and other species were well studied for their application in phytodesalination, phytoremediation, sand dune fixation, phytostabilization for attaining environmental sustainability (Lokhande and Suprasanna, 2012; Hasanuzzaman et al., 2014; Nikalje et al., 2019; Naikoo et al., 2020). Substantial research work has been carried out with thousands of research papers, as well as review articles and publications, on the use of halophytes for phytoremediation and desalination. Therefore, in the present chapter, no more discussion has been carried out on these topics. Instead, the role of halophytes in environmental clean-up for sustainable agriculture development has been discussed with a major emphasis on current environmental issues like the role of halophytes in improving the depleted levels of SOC and the use of halophyte compost as well as plant-growth-promoting rhizobacteria (PGPR) for extensive use in organic farming for improving the soil attributes of contaminated lands.

12.5.2 HALOPHYTES—BOON FOR ORGANIC FARMING

The universal food system has to be modified extensively with the intention of accomplishment and effective progress in sustainable agriculture that will help to supply adequate food to the world, contribute to supportable rural development, and offer livelihoods to the agrarian community without hazarding the natural resources. One of the ways of sustainable development to achieve these goals is the extensive use of organic farming. With the worldwide COVID-19 pandemic attack and their chronic risk factors as well as issues related to health, immunity, and safety, the human world is now in the demand of healthy and organically produced food. Therefore, organic agriculture adopted in developing countries on a small scale could remain the prime agenda of sustainable agriculture development policy for their overall growth. Organic farming illustrates benefits to agrarian communities, as it helps in reducing the environmental impacts of conventional agricultural concerns, increases the productivity of smaller fields, produces highly nutritious chemical-free, health-friendly products, provides cheap external inputs, and reduces the production cost, provides guaranteed and highest price premiums for organic produce.

Since the invention of agriculture to date, agricultural fields endlessly contributed to providing all kinds of necessities to the growing population which has ultimately affected its properties of fertility and productivity and polluted to the worst level. In the current scenario, the cultivable soil with inadequate agricultural fields is available with the presence of a full load of toxic contaminants/pollutants such as nonessential salts (sodium), heavy metals, remnants of chemical pesticides and fertilizers; devoid of organic carbon, constructive microflora, and water holding capacity. With such kind of contaminated and limited areas of so-called cultivable soil, the accomplishment of sustainable food production and its supply through the implementation of organic farming strategies and its success is fairly impossible. These contaminated farms with the available modified crop systems along with currently available hazardous over-use of chemicals, limited poor-quality organic fertilizers, and contaminated water resources utilized for the production of organic produce will not be considered sustainably factual organic products. Therefore, the foremost strategy adopted for successful organic farming is to rescue contaminated natural resources (soil and water) using eco-friendly methods which not only rescue the contaminated lands (soil) but also improve their wealth by using organically produced good quality biological fertilizers. Available crop systems predominantly glycophytes with their susceptible nature of abiotic stress tolerance are the ultimate source of food and raw materials for industrialization and urbanization. Although these plants are responsible for providing productivity on a large scale, these plants are not able to cope to the depth with the changing climatic and environmental conditions. In this regard, halophytes with their innate ability of abiotic stress tolerance could be used as the chief source to rescue and decontaminate the available contaminated cultivable lands. Organic farming through sustainable agriculture involves performing agriculture in an eco-friendly environment including chemical and contaminants soil, irrigation with unpolluted water resources, and controlling diseases using integrated pest management systems and biocontrol agents. Halophytes help in removing the toxic pollutants from the contaminated soil and water resources. The halophytic plant species, which act as hyperaccumulators or extractors, could be explored on a large scale to remove the contaminants permanently, meanwhile, they could convert the toxic polluted soil and water ecosystems into fertile and productive forms. The biomass of halophytes upon decomposition could be explored as biocompost with a rich source of organic acids and compounds, mineral salts, and nutrients and supplements for enriching the soil properties.

Organic farming is a complete manufacturing organization that encourages and augments environmental quality including biodiversity, bio–geo–chemical cycles, and soil floral and faunal activities. Various types of organic sources have been developed in the last few years to boon up organic farming including biogas slurry, blue–green algae (BGA), compost, green manure, farmyard manure, vermicompost, vermiwash, seaweed liquid fertilizers (Ayyapan and Ravindran, 2014). Due to the innate tolerance capacity of the halophytes, it has been shown that the halophytes are found superior the glycophytes in terms of nutrient diversity and availability of different plant-growth-promoting organic compounds (Cuevas, 1997; Balakrishnan et al., 2007). The halophytic compost is discovered as biofertilizer for improving the growth, yield, and productivity of many crop plants under saline agricultural soil conditions (Balakrishnan et al., 2007, 2010; Ayyapan et al., 2013; Ayyapan and Ravindran, 2014; Arumugam, 2018; Shah et al., 2020). The halophytic biocomposite has been recognized to enhance the soil microbial flora and their enzymatic activities to accelerate sustainable farming and therefore would be the substitute for chemical fertilizers (Balakrishnan et al., 2007).

The organic composts of plant origin are beneficiary for improving the nutrient availability for accelerating the survival and growth of the plants, soil aeration and retaining the water by soil particles, controlling weeds, and soil erosion. Besides, the foliar applications of these biological compost extracts are found beneficial in suppressing plant diseases (Arumugam, 2018). The halophytic compost in combination with the farmyard manure and consortium of phosphate sand sulfate solubilizing bacteria has improved the growth and yield attributes of crop plants. The study conducted by Balakrishnan et al. (2007) showed that the biocompost of three halophytes, namely, *Suaeda maritima* (L.) Dumort., *Sesuvium portulacastrum* L., and *Ipomoea pes-caprae* (L.) have been tested in combination with the farmyard manure and phosphate solubilizing bacteria for their impact on growth characteristics of one of the glycophytes, oil seed crop plant *Arachis hypogea.* Different combinations and permutations used during the study revealed that, among these three types of halophytic biocompost, the application of *S. maritima* biocompost (3.13 t/ha) combined with farmyard manure (3.13 t/ha) and phosphate solubilizing bacteria (*Bacillus megaterium*– 3 × 10^9/ha) significantly improved the soil microflora and their enzymatic activities in the soil cultivated with *Arachishypogaea.* Further studies conducted for the yield characteristics of *A. hypogea* have also been significantly increased with the same treatment conducted in coastal saline soil (Balakrishnan et al., 2010). Besides, the growth and yield characteristics of *Vignaradiata* grown along the lands on the coastal area in the presence of the addition of *Suaeda* biocompost + farmyard manure + phosphate solubilizing bacteria have shown significant improvement (Ayyapan et al., 2013). These studies revealed that the systematic and sustainable use of halophytic compost in appropriate combinations with farmyard manure and other nutrient-solubilizing bacteria could become one of the alternatives to chemical fertilizers, as the use of halophytic compost not only improves the growth, yield, and productivity attributes of glycophytes but also has significance in extraction of toxic salts of sodium and chloride from the contaminated saline agricultural lands (Ayyapan and Ravindran, 2014). Besides, the decomposition of these halophytic composts in the presence of effective soil microflora, these organic manures produce a variety of organic compounds, organic acids, and antibiotic substances which help to control certain fungal pathogens and nematode infestation in host crop plants. These by-products of decomposed halophytic compost adsorbed by the crop plants are responsible for increasing their resistance against some diseases and pests (Ayyapan and Ravindran, 2014).Arumugam (2018) reviewed the possible applications of halophytic plant-based compost in combination with compost tea for the improvement of soil health (physical and biological properties) as well as crop protection and production of different types of horticultural plants. The review provides information on the rigorous use of halophytic compost over chemical mineral fertilizers for enhancing crop productivity with minimum production cost, lowering the number of irrigation cycles for crops, and improving the organic matter of the soil. In concluding remarks, the review suggests that halophyte composting is technologically feasible, it will be a solution for environmental problems of the seashore and play a role in improving soil fertility of agricultural soil which is devoid of organic matter and nutrients. The use of halophytic compost in sandy saline soils could become a good amendment for the improvement of physical and nutritional characteristics of nutritionally deficient soils (Arumugam, 2018). The exploitation of organic compost from different plants as an unconventional source to diminish the use of chemical fertilizers is considered a decent sustainable agricultural practice. Shah and Ravindran (2020) in their studies revealed that the use of halophytic compost (*S. maritima*, 6.25 t/ha) in combination with farmyard manure (3.13 t/ha) and plant-growth-promoting nitrogen-fixing rhizobacteria, *Rhizobium* (2 kg/ha) in the soil showed significant improvement in development pattern of black gram in terms of plant height, number of compound leaves, leaf area, number of branches, FW and DW of the plant, number of root nodules, FW and DW of root nodules in comparison to other treatments having halophytic compost of *S. portulacastrum* L. The pH as well as electrical conductivity (EC) of the soil has been found reduced upon application of halophytic compost in comparison to control and has been correlated with the production of organic acids during the decomposition of organic compost in soil (Das and Singh, 2013; Shah and Ravindran, 2020). The study revealed consortium of halophytic compost could also achieve significant productivity of other crop plants in a sustainable way. Furthermore, Shah et al. (2020) explored the mixed administration of *S. maritima* compost at 6.25 t/ha + Farmyard manure at 3.13 t/ha + *Rhizobium* at 2 kg/ha in the soil showed a significant increase in soil microbial population and crop productivity in black gram. The continuous application of halophyte compost could improve the soil characteristics and biological fertility levels. Another approach has also been investigated, which includes the use of halophytic compost along with the farmyard manure and irrigation with sea water for exploring the halophytic crop systems for their nutritional, medicinal, forage, and wastewater treatment potentials (Ramasubramanian and Uyyavanthan, 2021). The halophyte, *Salicornia brachiata* has been grown across the seashore in coastal sandy loam soil supplemented with different degrees of farm yard manure and halophytic compost using vegetative parts of *Clerodendroninerme* L., *Ipomoea pes-caprae* (L.) followed by irrigation with seawater. The study discovered that the supplementation of these combined biological fertilizers has played a role in restoration of soil fertility, particularly in terms of the available nitrogen and organic carbon in saline soils and a significant increase in biomass production of *S. brachiata*. Therefore, the study carried out by different researchers to explore the potential of halophyte compost solely represents that, halophytes being innate properties of stress tolerance have shown significantly higher amounts of essential nutrients when compared to the green manure of glycophytes (Cuevas, 1997; Ayyapan and Ravindran, 2014) as depicted in Table 12.1.

TABLE 12.1 Nutrient Composition of Glycophytes Green Manure and Halophytes Compost.

Plant source	**Availability of essential nutrients (%)**		
	Nitrogen (N)	**Phosphorus (P)**	**Potassium (K)**
		Glycophytes	
Vignaradiate	0.72	0.18	0.53
Vignamungo	0.85	0.18	0.53
Vignacatajung	0.71	0.15	0.58
Crotalaria juncea	0.75	0.12	0.51
Halophytes	1.50–2.30	0.13–0.76	1.30–2.03

Source: Nutrient composition in green manure and halophytes (Dhana, 1997) from Ayyapan and Ravindran, 2014.

12.5.3 USE OF HALOPHYTES RHIZOSPHERE AS BIOFERTILIZERS IN SALINE AGRICULTURE LAND

Despite the capacity of salt–tolerant plants growing in saline and contaminated habitats has shown the potential to act as a source of organic manure, biological fertilizer, or compost to boost the growth and productivity of crop plants; the PGPR living in the vicinity of halophyte and mangrove rhizosphere have been explored the source of potential biofertilizers. Biofertilizers are the ingredients or preparations that contain living organisms (bacteria and fungi), which when applied as pure cultures or as a consortium in the soil to any crops or plants growing in the concerned fields are responsible for acclimatizing and colonizing in the given rhizosphere or internally get associated with the root systems of the plants [such as vesicular–arbuscular mycorrhiza (VAM) fungi]. After the establishment and growth of these microorganisms in the vicinity of the plant's roots, they perform the degradation of the complex nutrients through their enzymatic action and improve the availability and supply of essential nutrients for promoting the growth of the host plants (Malusa and Vassilev, 2014; Mahanty et al., 2017). The most important condition for the activeness and survival of these microorganisms is the maintenance of humid conditions and temperature. Under the influence of ideal growth conditions, these biofertilizers hasten the microbial processes and enhance the nutrient availability for their easy assimilation by the plants. Effective use of these biofertilizers is responsible for improving the health and wealth of the soil by contributing to its fertility through the fixation of atmospheric molecular nitrogen in the available forms as well as solubilizing the insoluble phosphates and sulfates and secreting plant-growth-promoting substances such as hormones, organic substances, and essential amino acids (Mazid and Khan, 2014). Operative biofertilizers can be prepared by harvesting the potent strains of bacteria and fungi from naturally available adverse and contaminated environments with their potential for nutrient mobilization, and synthesis of novel bioactive plant-growth-promoting compounds which will contribute to augmenting the crucial soil attributes like fertility, pH, water holding capacity, nutrient availability and ultimately leads to improve the yield and productivity of the crop plants (Pandey and Singh, 2012; Mahanty et al., 2017). Saline agricultural lands, natural wetlands as well as soil from mangroves and halophyte rhizosphere are some of the richest resources for the isolation, identification, and exploration of biologically important, abiotic stress tolerant microorganisms from these natural habitats for preparation and production of ideal biofertilizers because the microorganisms growing in such harsh conditions have gain natural tolerance against the toxic environment and potent capacity to survive under extreme conditions. Mangrove or halophytes-associated microflora has shown rich diversity and potential to act as biofertilizers for the organic farming of crop plants. Recently, the saline soils from the regions of Haryana state, India have been screened for the isolation of salt–tolerant PGPR (microbes), which have shown their significant capacity for phosphate solubilization, production of auxin indole-3-acetic acid (IAA), and ammonia production and suggested their use for alleviation of stresses and enhancing the yield of the crop plants under saline soil conditions (Sharma et al., 2021). Besides, Nikul and Goswami (2021) have isolated nitrogen-fixing bacteria from coastal areas and salinity-affected areas of Gir Somnath District, Gujarat, India and shown their potential to use as biofertilizers for the crops in the salinity-affected areas of the district. One of the interesting studies reported that the root powder of halophytes harboring the endophytic PGPR could be used as an alternative biofertilizer with a longer shelf life and serve as a carrier for preparation of effective biofertilizers for its use in saline soil agricultural lands along with other PGPR bioinoculants (Hassan et al., 2018). In this study, the dried root powder of Buffel grass (*Cenchrus ciliaris*

L.) along with the other five different halophytes (*Conyzanthus saqumatus*, *Solanum surattense* Burm. f., *Aerva javanica* (Burman f.) *Peganum harmala* L., and *Chrysopogonaucheri* L.) have been tested. Three endophytic plants PGPR, namely, *Bacillus cereus*, *Pseudomonad moraviensis*, and *Stenotrophomonas maltophilia* were isolated. The efficacy of sterilized root powder of *C. ciliaris* containing endophytic consortium of the PGPR on the growth, yield attributes and physiological status of the wheat under saline conditions have been studied which showed significant outcomes in terms of biofertilizers (Hassan et al., 2018). Furthermore, PGPR has been isolated from different types of halophytic plants and their effect on enhancing crop growth under saline conditions has been investigated (Redondo-Gómez et al., 2021). The effect of these halophytic PGPR consortiums on the growth of the glycophytic crops growing worldwide on salinized soils has been tested under varying conditions of carbon dioxides and temperatures. The study revealed that the application of biofertilizers neutralizes the adverse effect of salt stress on crop growth (Redondo-Gómez et al., 2021). This improved growth in crop plants under saline stress conditions due to halophyte-based PGPR biofertilizers fortification could be attempted because these PGPR directly provides the compounds synthesized by the bacterium such as N^2 fixation, phosphate solubilization, phytohormones: auxins, cytokinins, and gibberellins and lowering of the ethylene concentration, and compatible solutes production to the plants (Saenz-Mata et al., 2016). On the contrary, indirectly these PGPRs prevent adverse impacts of pathogen attack on plants by producing inhibitory substances for their growth (Mayak et al., 2004; Saenz-Mata et al., 2016). Although some of the halophyte's rhizosphere have been extensively screened for the isolation, characterization, and implementation of PGPR as biofertilizers, there is still an enormous opportunity to undertake the study on a large scale and establish the PGPR library or bank for its practical implementation with different types of consortia as biofertilizers in different forms for improving the growth, productivity of the crop plants as well as health and wealth of the agricultural saline soil.

12.5.4 ROLE OF HALOPHYTES IN SOIL ORGANIC CARBON (SOC) IMPROVEMENT

The fertility of the soil has now been counted in terms of the level of SOC in the soil. The quantity of terrestrial carbon is measured in the form of the level of available SOC. It has been estimated that the average content of

SOC is at the concentration of 2400 Pg to a soil depth of 2 m and has been compared to 3.2 times the atmospheric pool and 4.4 times the biotic pool (Sparks, 2003; Han et al., 2016). This has been correlated with a significant effect on the concentration of atmospheric CO^2 levels associated with the minute changes in the global SOC stock (Smith, 2008). The higher levels of CO^2 concentrations in the atmosphere could be minimized from 16 to 30% by increasing the levels of SOC concentrations from 5 to 15% in the soil at depths of up to 2 m (Baldock, 2007). In the current scenario, researchers from different disciplines, scientists, and environmentalists of the world along with the agrarian community are immensely working on resolving the problems of agricultural productivity associated with depleted levels of SOC in cultivable agricultural lands. Therefore, understanding the mechanisms underlying the storage and stability of organic carbon (OC) in soils has drawn considerable attention (Han et al., 2016). Halophytes, to some extent, have been utilized for improving the SOC level in soil-contaminated sites and their natural habitats residing along the coastal seashore.

The study has been conducted to compare the soil aggregate-associated SOC and aggregate-associated soil microbial biomass carbon (MBC) under four halophyte communities namely, *Karelinia caspia*, *Bassia dasyphylla*, *Haloxylon ammodendron*, and *Tamarix ramosissima* in China (Yang et al., 2016). The study revealed that aggregate-associated SOC concentrations ranged from 1.70 to 13.68 g kg^{-1}, and the aggregate-associated SOC was the highest under the halophytic communities of *H. ammodendron* and *T. ramosissima.* However, the aggregate-associated MBC ranged from 55.26 to 217.11 g kg^{-1}, and the aggregate-associated MBCs were higher under the *K. caspia* and *B. dasyphylla* communities. The study suggests that the *H. ammodendron* community could be beneficial for soil carbon storage in arid regions (Yang et al., 2016). Thereafter, the behavior of litter-derived organic carbon conversion of two halophytes in the soil and its influence on SOC stabilization of wetlands have been investigated (Yan et al., 2020). For this, in-field surveys and small-scale experiments have been carried out to identify the plant C turnover and evaluate using the carbon isotope technique to estimate the dynamics of SOC with the help of two halophytes, *Spartina alterniflora* and *Phragmites communis* (Yan et al., 2020). The study revealed that litter of the *S. alterniflora* undergoes rapid decaying processes and converts it into SOC in comparison to *P. communis*. The excess amount of converted SOC is lost due to soil respiration and the leaching process. On the contrary, the slower conversion of litter-derived carbon into SOC by *P. alterniflora* was found responsible for effective stabilization and an increase in the level of SOC in the soil. Therefore, the incorporation of *P. alterniflora* litter to increase the SOC level of poor original SOC regions along the coastal region could be adopted (Yan et al., 2020). Ding et al. (2020) conducted an experiment to understand the role of two *Atriplex* species (*Atriplex numularia* and *Atriplex amnicola*) of halophyte in combination with the glycophytes forage plants (pearl millet and cowpea) in enhancing the soil quality in metal contaminated soils. A significant increase in the levels of SOC (31–33% above the noncultivated contaminated soil) in the metal-contaminated sites cultivated with *Atriplex* plants was recorded in comparison to the SOC level recorded in contaminated soils cultivated with glycophytes. The study suggested the role of halophytes in stabilizing and improving the SOC level in contaminated lands along the coastal regions (Ding et al., 2020). The consequences of grassland overgrazing in the Flooding Pampa region in Argentina have been correlated with severe damage to vegetation followed by reduced levels of aboveground net primary productivity and depleted levels of SOC in that region. The simple technique of exclusion for more than 3 years has been implemented in degraded halophyte communities of the study area for improving the aboveground net primary productivity by up to 10 times and SOC level to 2 times in comparison to the levels recorded at the beginning of the study (Bolanos et al., 2016). This might have been achieved because of the continuous litter degradation and decomposition of the inhabitant halophytes in the study area and its conversion to SOC in a natural manner. Besides, the studies conducted on the application of halophyte compost as a tool for organic farming have been indirectly responsible for the increase in the levels of SOC in degraded and soil-contaminated lands.

12.6 CONCLUSION AND FUTURE PERSPECTIVES

Halophytes have numerous advantages and adaptations over glycophytes for cleaning up the salt-affected and heavy metal-contaminated environment of the world. They have developed the mechanisms of salts and heavy metal accumulation, excretion, sequestration, compartmentalization, chelation, detoxification, and stabilization. Despite only 2% diversity of halophytes as against the other plants, halophytes remain part of the solution to the problems of environmental concern. Since few decades, extensive research has been undertaken for phytoremediation and desalination of contaminated regions of the world, in the present chapter emphasis is given to the role of halophytes in mitigating the abiotic stresses through their use in the improvement of SOC and its exploration in organic farming for sustainable agricultural development. Future research on halophytes should consider the following perspectives: (1) use of halophytes for improving the economy of developing and under-developed countries with a major emphasis on undertaking the salt-affected barren lands for cultivation with suitable and sustainable agricultural techniques. An extensive search should be carried out to identify the cash crop halophyte species of economic returns. (2) Halophytes could be extensively screened and utilized for improving the SOC of contaminated cultivable agricultural lands; at the same time, the SOC restoration potential of the halophytes in their natural habitat should be explored. The large-scale cultivation of halophytes should be adopted on salt and heavy metal-contaminated lands across the world for a longer span and commercial production of halophyte compost as a prime source of biofertilizers should be developed to inculcate the concept of organic farming. The innovative and potent strains of PGPR can accelerate the efficiency of biofertilizers in saline agricultural lands as well as heavy metal-contaminated sites. (3) Integrated policies should be implemented to accelerate the efficacy of halophytes for phytoremediation and desalination of single or multiple-contaminated regions. The salt and heavy metal contaminated agriculture land rescue technology should be developed using large-scale exploration of the accumulator and hyperaccumulator halophyte species for converting the noncultivable lands into cultivable form. The alternative source of raw material to agricultural produce for industrial development should be assessed from the halophytes. (4) The introduction of salt and heavy metal tolerant PGPR consortia in combination with the intercropping of halophytes such as *Suaeda*, *Sesuvium*, etc. in glycophytes growing agriculture lands could be adapted for releasing the load of toxic contaminants during the growth and development of glycophytes. Besides, the modified halophyte liquid extracts or halophyte biosap with optimum concentrations of essential mineral nutrients and plant-growth-promoting organic compounds should be developed and used in foliar spray on glycophytes crop systems to improve the growth, yield, and productivity under adverse climatic conditions. (5) The growth and developmental strategies of halophytes in contaminated lands in the presence of elevated levels of carbon dioxide could be undertaken to explore their potential under extreme climatic conditions in the near future. The prospects and perspectives of halophytes for environmental clean-up and sustainable eco-friendly, environmental, and agricultural development are depicted in Figure 12.1.

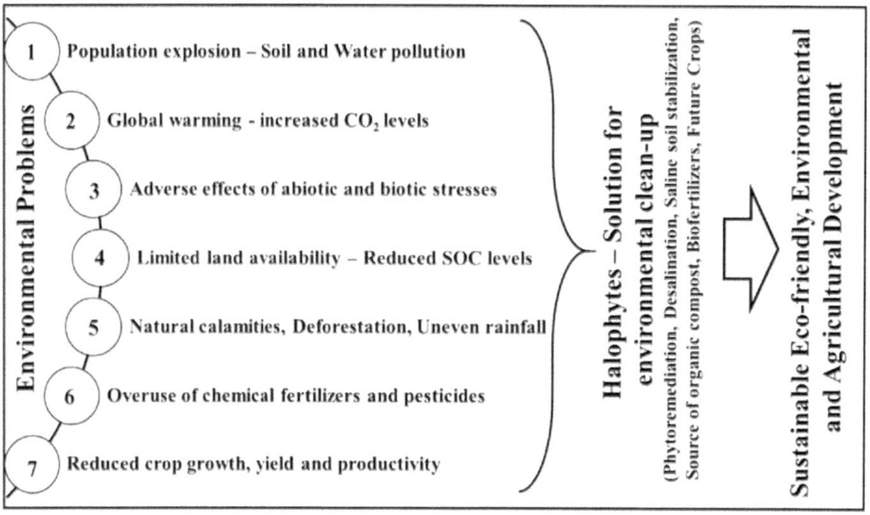

FIGURE 12.1 Use of halophytes for solving the problems of the environment for sustainable development.

KEYWORDS

- **halophytes**
- **environmental pollution**
- **agriculture**
- **sustainable development**
- **perspectives**
- **prospectives**

REFERENCES

Arumugam, R. Feasibility of Halophytic Plant-Based Compost and Compost Tea on Soil Health, Crop Protection and Production. *Int. J. Res. Anal. Rev.* **2018,** *5* (3), 355–365.

Ayyapan, D.; Ravindran, K. C. Utilization of Halophytes as A Source of Organic Manure. *Int. Lett. Nat. Sci.* **2014,** *12* (2), 162–166.

Ayyapan, D.; Sanjiviraja, K.; Balakrishnan, V.; Ravindran, K. C. Impact of Halophytic Compost on Growth and Yield Characteristics of *Vignaradiata* L. *Afric. J. Agri. Res.* **2013,** *8* (22), 2663–2672.

Balakrishnan, V.; Vdenkatesan K.; Sanjiviraja, K.; Indrajith A.; Ravindran, K. C. Application of and Organic Halophytic Manure on Yield Characteristics of *Arachis hypogea* Linn. *Int. J. Eng. Sci. Tech.* **2010,** *2* (1), 134–*143*.

Balakrishnan, V.; Venkatesan, K.; Ravindran, K. C. The Influence of Halophytic Compost, Farmyard Manure and Phosphobacteria on Soil Microflora and Enzyme Activities. *Plant Soil Environ.* **2007,** *53* (4), 186–192.

Baldock, J. A. Composition and Cycling of Organic Carbon in Soil, Nutrient Cycling in Terrestrial Ecosystems. In *Nutrient Cycling in Terrestrial Ecosystems,* Springer: Berlin, Heidelberg, 2007; p 1e35.

Bita, C.; Gerats, T. Plant Tolerance to High Temperature in a Changing Environment: Scientific Fundamentals and Production of Heat Stress-Tolerant Crops. *Front. Plant Sci.* **2013,** *4,* 273.

Boer, B.; Gliddon, D. Mapping of Coastal Ecosystems and Halophytes (Case Study of Abu Dhabi, United Arab Emirates). *Marine Freshwater Res.* **1998,** *49,* 297–301.

Bohnert, H. J.; Jensen, R. G. Strategies for Engineering Water Stress Tolerance in Plants. *Trends Biotechnol.* **1996,** *14,* 89–97.

Bolanos, V. R. A.; Vecchio, M. C.; Refi, R.; Golluscio, R. A. In *Zootechnical Value of the Community of Halophytes after Exclusion from Grazing in Flooding Pampa, Argentina.* Proceedings of the 10th International Rangeland Congress, 2016, pp 315–317.

Chaudhry, S.; Sidhu, G. P. S. Climate Change Regulated Abiotic Stress Mechanisms in Plants: A Comprehensive Review. *Plant Cell Rep.***2021,** *41 (1),* 1–31.

Cuevas, V. C. Rapid Composting Technology in the Phillippines: Its Role in Producing Good –Quality Organic Fertilizers. *Food Fert. Techn. Centre Extn. Bull.* **1997,** *444,* 1–13.

Custodio, M.; Lilebo, A.; Calado, R.; Villasante, S. Halophytes as Novel Marine Products - A Consumers' Perspective in Portugal and Policy Implications. *Marine Pol.* **2021,** *133,* 104731.

Dar, M. I.; Naikoo, M. I.; Green, I. D.; Sayeed, N.; Ali, B.; Khan, F. A.. Heavy metal hyperaccumulation and Hypertolerance in Brassicaceae. In *Plants Under Metal and Metalloid Stress,* Springer : Singapore, 2018; pp 263–276.

Das, I.; Singh, A. P. Effect of PGPR and Organic Manures on Soil Properties of Organically Cultivated Mungbean. *Bioscan.* **2013,** *9 (1)*, 27–29.

Deng, L.; Wang, K.; Tang, Z.; Shangguan, Z. Soil Organic Carbon Dynamics Following Natural Vegetation Restoration: Evidence from Stable Carbon Isotopes (δ13C). *Agric. Ecosyst. Environ.***2016,** *221,* 235–244.

Ding, Z.; Alharbi, S.; Almaroai, Y. A.; Eissa, M. A. Improving Quality of Metal-Contaminated Soils by Some Halophyte and Non-Halophyte Forage Plants. *Sci. Total Environ.* **2020,** *10,* 142885.

FAO.FAO-News Article: Exploring the Use of Wastewater in Agriculture [Online]. 2017, http://www.fao.org/news/story/en/item/463433/icode/ (accessed Mar 30, 2021).

FAO. FAO Land and Plant Nutrition Management Service [Online]. 2008, http://www.fao.org/ag/agl/agll/spush

Flowers, T. J.; Colmer, T. D. Salinity Tolerance in Halophytes. *New Phytol.* **2008,** *179,* 945–963.

Gall, J. E.; Rajakaruna, N. The Physiology, Functional Genomics, and Applied Ecology of Heavy Metal-Tolerant Brassicaceae. In *Brassicaceae: Characterization, Functional Genomics and Health Benefits*; Nova Science Publishers: New York, 2013; pp 121–148.

Glenn, E. P.; Brown, J. J. Salt Tolerance and Crop Potential of Halophytes. *Crit. Rev. Plant Sci.* **1999,** *18* (2), 227–255.

Glenn, E.; Miyamoto, M.; Moore, D.; Brown, J. J.; Thompson, T. L.; Brown, P. Water Requirements for Cultivating *Salicorniabigelovii*Torr. With Seawater on Sand in a Coastal Desert Environment. *J. Arid Environ.* **1997,** *36,* 711–730.

Han, L.; Sun, K.; Jin, J.; Xing, B. Some Concepts of Soil Organic Carbon Characteristics and Mineral Interaction from a Review of Literature. *Soil Biol. Biochem.* **2016,** *94,* 107–121.

Hasanuzzaman, M.; Nahar, K.; Alam, M. M.; Bhowmik, P. C.; Hossain, M. A.; Rahman, M. M.; Prasad, M. N. V.; Ozturk, M.; Fujita, M. Potential uses of Halophytes to Remediate Saline Soils. *Biomed. Res. Int.* **2014,** *589341.*

Hasegawa, P. M.; Bressan, R. A.; Zhu, J. K.; Bohnert, H. J. Plant Cellular and Molecular Responses to High Salinity. *Ann. Rev. Plant Physiol. Plant Mol. Biol.* **2000,** *51,* 463–499.

Hassan, T. U. I.; Bano, A.; Naz, I. Halophyte Root Powder: An Alternative Biofertilizer and Carrier for Saline Land. *Soil Sci. Plant Nutr.* **2018,** *64* (5), 653–661.

Hirayama, T.; Shinozaki, K. Research on Abiotic Stress Responses in the Post-Genome Era: Past, Present and Future. *Plant J.* **2010,** *61,* 1041–1052.

Khan, M. A.; Duke, N. C. Halophytes- A Resource for the Future. *Wetlands Ecol. Manag.* **2001,** *6,* 455–456.

Koch, A.; McBratney, A.; Adams, M.; Field, D.; Hill, R.; Crawford J.; ... Angers, D. Soil Security: Solving the Global Soil Crisis. *Glob. Policy* **2013,** *4,* 434–441.

Koyro, H. W.; Geissler, N.; Hussin, S.; Huchzermeyer, B. Mechanism of Cash Crop Halophytes to Maintain Yields and Reclaim Saline Soils in Arid Areas. In *Ecophysiology of High Salinity Tolerant Plants*; Springer: Berlin, 2008; pp 345–366.

Lehmann, J.; Kleber, M. The Contentious Nature of Soil Organic Matter. *Nature* **2015,** *528* (7580), 60–68.

Liang, L.; Liu, W.; Sun, Y.; Huo, X.; Li, S.; Zhou, Q. Phytoremediation of Heavy Metal Contaminated Saline Soils Using Halophytes: Current Progress and Future Perspectives. *Environ. Rev.* **2017,** *25,* 269–281.

Lieth, H.; Moschenko, M.; Lohmann, M.; Koyro, H. W.; Hamdy, A. Halophyte Uses in Different Climates I. Ecological and Ecophysiological Studies. In *Progress in Biometeorology*; Backhuys: Leiden, 1999; pp 258.

Lindner, M.; Maroschek, M.; Netherer, S.; Kremer, A.; Barbati, A.; Garcia-Gonzalo, J.; Seidl, R.; Delzon, S.; Corona, P.; Kolström, M.; Lexer, M. J. Climate Change Impacts, Adaptive Capacity, and Vulnerability of European Forest Ecosystems. *For Ecol. Manage.* **2010,** *259* (4), 698–709.

Litalien, A.; Zeeb, B. Curing the Earth: A Review of Anthropogenic Soil Salinization and Plant-Based Strategies for Sustainable Mitigation. *Sci. Total Environ.* **2020,** *698,* 134235.

Lokhande, V. H.; Suprasanna, P. Prospects of Halophytes in Understanding and Managing Abiotic Stress Tolerance. In *Environmental Adaptations to Changing Climate: Metabolism, Productivity and Sustainability*; Springer: New York, 2012; pp 29–56.

Mahanty, T.; Bhattacharjee, S.; Goswami, M.; Bhattacharyya, P.; Das, B.; Gosh, A.; Tribedi, P. Biofertilizers: A Potential Approach for Sustainable Agriculture Development. *Environ. Sci. Pollut. Res.* **2017,** *24* (4), 3315–3335.

Malusa, E.; Vassilev, N. A Contribution to Set a Legal Framework for Biofertilisers. *Appl. Microbiol. Biotechnol.* **2014,** *98* (15), 6599–6607.

Mayak, S.; Tirosh, T.; Glick, B. R. Plant Growth Promoting Bacteria Confer Resistance in Tomato Plants to Salt Stress. *Plant Physiol. Biochem.* **2004,** *42,* 565–572.

Mazid, M.; Khan, T. A. Future of Bio-Fertilizers in Indian Agriculture: An Overview. *Int. J. Agri. Food Res.* **2014,** *3* (3), 10–23.

Munns, R.; Tester, M. Mechanism of Salinity Tolerance. *Ann. Rev. Plant Biol.* **2008,** *59,* 651–581.
Naikoo, M. I.; Dar, M. I.; Khan, F. A. Trophic Transfer and Bioaccumulation of Lead Along Soil–Plant–Aphid–Ladybird Food Chain. *Env. Sci. Poll. Res.* **2019,** *26,* 23460–23470.
Naikoo, M. I.; Kafeel, U.; Naushin, F.; Khan, F. A. Halophytes in India and Their Role in Phytoremediation. In *Handbook of Halophytes*; Springer, Nature: Switzerland AG, 2020; p 115.
Nikalje, G. C.; Kumar, J.; Nikam, T. D.; Suprasanna, P. FT-IR Profiling Reveals Differential Response of Roots and Leaves to Salt Stress in a Halophyte *Sesuvium portulacastrum* (L.) L. *Biotechnol. Rep.***2019,** *23,* e00352.
Nikalje, G. C.; Srivastava, A. K.; Pandey, G. K.; Suprasanna, P. Halophytes in Biosaline Agriculture: Mechanism, Utilization and Value Addition. *Land Degr Dev.* **2017,** *29 (4),* 1081–1095.
Nikul, C.; Goswami, G. K. Isolation and Characterization of Nitrogen Fixing Bacteria from Saline Habitat (Gir-Somnath District): As Biofertilizers. *J. Adv. Sci. Res.* **2021,** 126–135.
Nouri, H.; Chavoshi B. S.; Nirola, R.; Hassanli, A.; Beecham, S.; Alaghmand, S.; Mulcahy, D. Application of Green Remediation on Soil Salinity Treatment: A Review on Halophytoremediation. *Process Safety Env. Prot.* **2017,** *107,* 94–107.
Pandey, J.; Singh, A. Opportunities and Constraints in Organic Farming: An Indian Perspective. *J. Sci. Res.* **2012,** *56,* 47–72.
Ramasubramanian, A.; Uyyavanthan, K. Growth, Productivity and Nutritional Status of *Salicornia brachiata* Roxb. Cultivated in Saline Soils Supplemented with Organic Manures Under Sea Water Irrigation. *J. Plant Stress Physiol.***2021,** *7,* 1–9.
Raza, A.; Razzaq, A.; Mehmood, S. S.; Zou, X.; Zhang, X.; Lv, Y.; Xu, J. Impact of Climate Change on Crops Adaptation and Strategies to Tackle its Outcome: A Review. *Plants* **2019,** *8* (2), 34.
Redondo-Gómez, S.; Mesa-Marín, J.; Pérez-Romero, J. A.; López-Jurado, J.; García-López, J. V.; Mariscal, V.; Molina-Heredia, F. P.; Pajuelo, E.; Rodríguez-Llorente, I. D.; Flowers, T. J.; Mateos-Naranjo, E. Consortia of Plant-Growth-Promoting Rhizobacteria Isolated from Halophytes Improve Response of Eight Crops to Soil Salinization and Climate Change Conditions. *Agronomy* **2021,** *11,* 1609.
Rengasamy, P. World Salinization with Emphasis on Australia. *J. Expt. Bot.* **2006,** *57,* 1017–1023.
Saenz-Mata, J.; Palacio-Rodriguez, R.; Sanchez-Galvan, H.; Balagurusamy, N. Plant Growth Promoting Rhizobacteria Associated to Halophytes: Potential Application in Agriculture. In *Sabkha Ecosystem V: The Americas;* Springer: Berlin, New York, 2016; pp 411–425.
Shah, M. A.; Ravindran, K. C. Optimzaiton of Halophytic Compost Along with Farmyard Manure and *Rhizobium* on Growth of *Vignamungo*L. *Plant Arch.* **2020,** *20* (2), 1398–1405.
Shah, M. Ravindran, K. C.; Elango, R. Synergistic Effect of Halophytic Compost in Combination with Farmyard Manure and Rhizobium on Soil Microflora and Yield Parameters of Black Gram (*Vignamungo* L.). *Int. J. Bot. Stud.* **2020,** *5* (3), 497–503.
Sharma, A.; Dev, K.; Sourrirajan, A.; Choudhary, M. Isolation and Characterization of salt-Tolerant Bacteria with Plant Growth Promoting Activities from Saline Agricultural Fields of Haryana, India. *J. Gent. Eng. Biotech.* **2021,** *19,* 99.
Smith, P. Land use Change and Soil Organic Carbon Dynamics. *Nutr. Cycling Agroecosyst.* **2008,** *81,* 169–178.
Sparks, D. L. *Environmental Soil Chemistry*; Academic Press: New York, 2003.

Stockmann, U.; Padarian, J.; McBratney, A.; Minasny, B.; Brogniez, D. D.; Montanarella, L.; Hong, S. Y.; Rawlins, B. G.; Field, D. J. Global Soil Organic Carbon Assessment. *Global Food. Sec.* **2015,** *6,* 9–16.

Tian, Z.; Wang, J. W.; Li, J.; Han, B. Designing Future Crops: Challenges and Strategies for Sustainable Development. *The Plant J.* **2021,** *105,* 1165–1178.

Tipirdamaz, R.; Karakas, S.; Dikilitas, M. Halophytes and the Future of Agriculture. In *Handbook of Halophytes*; Springer, Nature: Switzerland, 2020.

Turcios, A. E.; Miglio, Vela, R.; Sanchez, G.; Begier, T.; Bergier, A. W.; Cifuentes, J. I.; Pignataro, G.; Avellan, T.; Papenbrock, J. From Natural Habitats to Successful Application – Role of Halophytes in the Treatment of Saline Wastewater in Constructed Wetlands with a Focus on Latin America. *Env. Expt. Bot.* **2021,** *190,* 104583.

Turkan, I.; Demiral, T. Recent Developments in Understanding Salinity Tolerance. *Environ. Expt. Bot.* **2009,** *67,* 2–9.

Ventura, Y.; Sagi, M.. Halophyte crop cultivation: The Case for *Salicornia* and *Sarcocornia. Environ. Expt. Bot.* **2013,** *92,* 144–153.

Wallace, J. G.; Rodgers-Melnick, E.; Buckler, E. S. On the Road to Breeding 4.0: Unraveling the Good, the Bad, and the Boring of Crop Quantitative Genomics. *Annu. Rev. Genet.* **2018,** *52,* 421–444.

Whalen, J. K.; Gul, S.; Poirier, V.; Yanni, S. F.; Simpson, M. J.; Clemente, J. S.; Feng, X.; Grayston, S. J.; Barker, J.; Gregorich, E. G. Transforming Plant Carbon into Soil Carbon: Process-Level Controls on Carbon Sequestration. *Revue Canadienne De Phytotechnie.* **2014,** *94* (6), 271–279.

Yan, J.; Qian, L.; Fu, X.; Wu, J.; Tsang, Y. F.; Wang, L. Conversion Behaviors of Litter-Derived Organic Carbon of Two Halophytes in Soil and Their Influence on SOC Stabilization of Wetlands in the Yangtze River Estuary. *Sci. Total Environ.* **2020,** *716,* 137109.

Yang, H.; Wang, J.; Zhang, F. Soil Aggregation and Aggregate-Associated Carbon Under Four Typical Halophyte Communities in An Arid Area. *Environ. Soil Pollut. Res.* **2016,** *23,* 23920–23929.

Yost, J. L.; Hartemink, A. E. Soil Organic Carbon in Sandy Soils: A Review. *Adv. Agron.* 2019, *158,* 217–230.

Zheng, H.; Wang X.; Chen L.; Wang Z.; Xia Y.; Zhang Y.; Wang H.; Luo X.; Xing B. Enhanced Growth of Halophyte Plants in Biochar-Amended Coastal Soil: Roles of Nutrient Availability and Rhizosphere Microbial Modulation. *Plant Cell Environ.***2018,** *41,* 517–532.

Zhu, J. K. Plant Salt Tolerance. *Trends Plant Sci.* 2001, *6,* 66–71.

CHAPTER 13

Physiological, Proteomics, and Metabolomics Approaches for Elucidation of Salinity and Alkalinity Tolerance in Halophytes

ASHA KUMARI[1,2], MONIKA PATEL[1,2], and ASISH KUMAR PARIDA[1,2]

[1]*Plant Omics Division, CSIR—Central Salt and Marine Chemicals Research Institute (CSIR-CSMCRI), Gijubhai Badheka Marg, Bhavnagar, Gujarat, India*

[2]*Academy of Scientific and Innovative Research (AcSIR), Ghaziabad, Uttar Pradesh, India*

ABSTRACT

Soil salinization and alkalization are major abiotic stresses that negatively affect plant productivity worldwide, endangering the global food supply and leading to substantial economic penalties. Soil salinization can be defined as soluble salt accumulation in the soil, which disturbs soil fertility. Soil alkalization can be referred to as a rise in the soil pH. In nature, soil salinization and alkalization co-occur. Halophytes are native to saline environments and grow and flourish under high saline and alkaline habitats. Halophytes are known for their various medicinal properties, phytoremediation potential, and habitat protection by preserving ecological stability and restoration of salt/alkali-affected wastelands to develop sustainable agriculture. Great diversity in halophyte distribution has been reported under saline and alkaline habitats. For example, *Chloris virgata*, *Limonium bicolor*, *Kochia sieversiana*, *Puccinellia tenuiflora*, *Poa pratensis*, *Spartina alterniflora*, and *Tamarix hispida* have been reported for their tolerance to saline–alkaline stresses. In this

Physiology of Halophytes: Signaling, Omics, and Tolerance Mechanisms.
Nivas Manohar Desai & Ganesh Chandrakant Nikalje (Eds.)

context, halophytes have been exploited as principal species for remediation of salt-alkali affected soil for addressing the problem of growing unfertilized land. Growth stimulation under saline conditions (150–300-mM NaCl) has been reported in halophytes. However, in some halophytes, growth inhibition by low salinity has also been noted, for example, in the case of *Salicornia.* Dicotyledonous halophytes tend to grow under salinity up to 150-mM NaCl with reduced growth; for example, *Salicornia brachiata* (a model plant for dicotyledonous halophytes) showed a positive correlation between salinity and growth and survived up to 150-mM NaCl. In contrast, monocotyledon halophytes survive under a salinity of more than 1-M NaCl concentration. These halophytes tend to stimulate growth at low salinity levels. Learning the salt tolerance mechanism in monocotyledon halophytes is essential as all cereal crops are monocots. In the present chapter, the salinity and alkalinity tolerance mechanisms of different halophytic species are discussed with a focus on physiological, genomics, metabolomics, and proteomics aspects.

13.1 INTRODUCTION

Soil alkalization and salination are widespread environmental problems leading to a loss in agricultural productivity. Saline soils are composed of different types of soluble salt components differing in composition (Guo et al., 2017). Some saline-alkaline soils have low pH but high salinity, while some are noted with high pH but low salinity. Thus, making saline-alkaline soils very complex. Under saline–sodic conditions, the dissoluble mineral salt mainly consists of Na^+, K^+, Mg^{2+}, and Ca^{2+} as main cations, and HCO^3, SO^{42-}, Cl^-, NO^3, and CO^{32-} contribute to major anions (Li et al., 2009). NaCl and Na^2SO^4 are neutral salts contributing only toward soil salinity. Moreover, soil alkalinity is due to the presence of alkali salts (Na^2CO^{32-} and $NaHCO^3$). Under natural conditions, salt stresses are mostly from mixtures of neutral and alkali salts. However, existence of two diverse types of stresses (salt–alkali stress) has been reported since early (Guo et al., 2010a, 2015a; Gong et al., 2013; Kumar et al., 2018; Lu et al., 2021). It has been reported that additional occurrence of alkali salts elevates the soil pH resulting in a higher destructive effect on plants due to the combined damaging effects of both alkali and salts (Yang et al., 2009). However, less consideration is given to alkali stress. Thus, elucidating the holistic tolerance mechanism toward salt stress in plants needs research on both alkali and salt stresses. Salinity and alkalinity in the soil may be caused naturally (primary salinization) by atmospheric precipitation of sea and flood water, natural weathering of mineral rocks, and the addition of fossil salts. On the other hand, salinity may be introduced by anthropogenic activities (secondary salinization), including overuse of chemical fertilizers, brackish water irrigation, and introduction of industrial waste and sewage water (Hanin et al., 2016). Various remediation techniques for saline and saline sodic soils reclamation have been utilized (Srivastava, 2020), employment of salt–alkali tolerant crops may enticingly contribute for plunging the economic burden.

Salinity stress causes an increase in soil osmotic pressure, leading to physiological dryness for the plants. Besides, high salinity in the soil disturbs the ion metabolism, especially nitrogen, and indirectly spoils the soil structure resulting in decreased soil fertility. On the other hand, alkalinity also creates osmotic stress, but in addition is associated with carbonate ion toxicity, breaking in soil structure due to the extreme effect of Na^+, and inhibition of nutrient absorption (Zhang and Mu, 2009). Stress condition associated with high salt and alkali in plants is instigated by innumerable biochemical and physiological ailments that lead to reduced growth, decreased photosynthesis, less water availability, increased ion toxicity, high reactive oxygen species (ROS) production, and deficiency of essential mineral ions (Kumari et al., 2015; Abobatta, 2020). Figure 13.1 summarizes the effect of salinity and alkalinity in plants and their respective adaptive mechanisms. A comparative study of alkali and salt stress revealed additional damage caused by alkali stress due to high pH, which disrupts ionic and pH homeostasis in the plant tissue (Yang et al., 2007; Li et al., 2009; Zhang and Mu, 2009; Zhang et al., 2020b). Various defense mechanisms have been developed by plants to create osmotic, ionic, and ROS homeostasis toward salt–alkali stress (Zhang et al., 2020b). Plants tend to accumulate compatible osmolytes such as glycine betaine (GB), and proline for balancing osmotic potential, safeguarding membrane integrity, and enzyme activity to combat salt–alkali stress-induced osmotic stress (van Zelm et al., 2020). To prevent nutrient deficiency and cellular damage caused by disturbing ion homeostasis under salt–alkali stress, plants uphold a high K^+/Na^+ ratio to regulate ion homeostasis (Li et al., 2009). Furthermore, plants regulate ROS homeostasis by introducing various enzymatic and nonenzymatic antioxidants (Kumari et al., 2015). Plant retorts to salt–alkali stress various species to species (Munns et al., 2020). Slight acidic to neutral soil pH is considered feasible for the proper growth of plants. However, alkaline and highly alkaline condition or high acidic condition causes nutrient deficiency and are toxic to plants. Some plants are vulnerable to high soil salt content and thus can't survive in a saline environment. These plants are referred to as glycophytes, including most conventional crop plants (Munns et al., 2020). Unfortunately, under saline conditions, glycophytes decrease their average yield by 50% (Abobatta, 2020). Thus, interest in cultivars that can survive and produce good yields under saline and alkaline conditions has increased in the recent decade. Halophytes, also called salt-loving plants, emerges as a potential candidate for cultivation under abiotic stress conditions due to their ability to withstand high saline conditions.

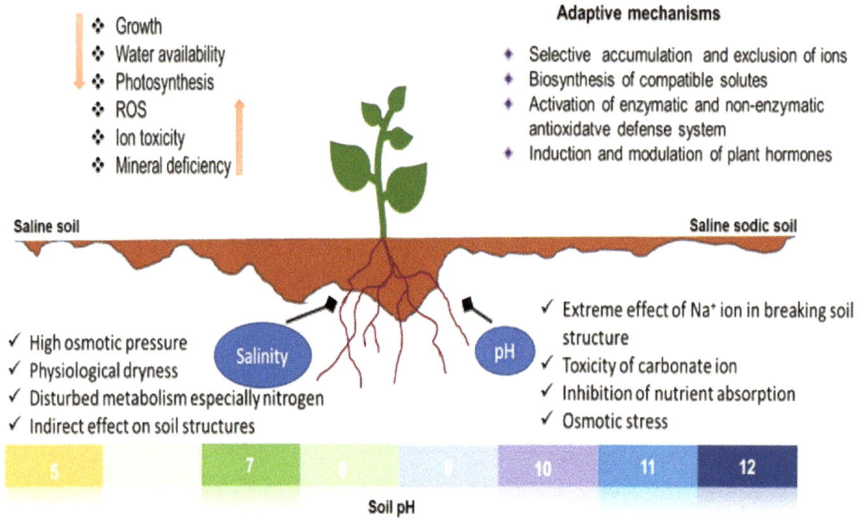

FIGURE 13.1 Diagrammatic presentation of effects of salinity and pH on plants and adaptive mechanism.

13.2 HALOPHYTES: POTENTIAL CANDIDATE FOR SALT-ALKALI TOLERANCE

Halophytes are the natural habitat of saline environments that can grow and thrive under high saline and alkaline conditions (Deinlein et al., 2014). Halophytes are known for their various medicinal properties, phytoremediation potential, and habitat protection by preserving ecological stability and restoration of salt/alkali-affected wastelands to develop sustainable agriculture (Kumari et al., 2015). Great diversity in halophyte distribution has been reported under saline and alkaline habitats. For example; *Chloris virgata*, *L. bicolor*, *Kochia sieversiana*, *Puccinellia tenuiflora*, *Poa pratensis*, *Spartina alterniflora*, and *Tamarix hispida* have been reported for their tolerance to saline–alkaline stresses (Yang et al., 2008; Li et al., 2009, 2010b; Wang et al., 2014a; Zhao et al., 2014). Halophytes are exploited as principal species for remediation of salt–alkali-affected soil to address the problem of growing unfertilized land (Reddy et al., 2008; Patel and Parida, 2020). Growth stimulation under saline conditions (150–300-mM NaCl) has been reported in halophytes (Mateos-Naranjo et al., 2010; Yu et al., 2013; Zhao et al., 2016; Yin et al., 2019). However, in some halophytes, growth inhibition by low salinity has also been noted, for example, in the case of *Salicornia* (Rozema and Schat, 2012). Thus, the differential responses of various halophytes make them suitable candidate for study of salt–tolerance mechanism and as source of salt responsive proteins, metabolites, and genes for development of salt–tolerant crops. The salt–tolerant capability of various halophytes is given in Table 13.1 below.

TABLE 13.1 Salinity Tolerance Capability of Different Halophytes.

S No.	Halophytes	Growth stimulating concentration (mM)	Maximum salinity tolerance (mM NaCl)	References
1	*Solanum chilense*	–	125	Gharbi et al. (2017)
2	*P. tenuiflora*	50	150	Yu et al. (2011)
3	*Hordeum brevisubulatum*	100	175	Zhang et al. (2020)
4	*Morus alba*	-	200	Huihui et al. (2020)
5	*Tagetes minuta*	50	200	Moghaddam et al. (2020)
6	*T. halophila*	150	250	Gong et al. (2005)
7	*A. tripolium*	-	250	Duarte et al. (2017)
8	*H. marinum*	100	300	Maršálová et al. (2016)
9	*P. patens*	-	350	Wang et al. (2008)
10	*S. salsa*	100	400	Qiu et al. (2007)
11	*Desmostachya bipinnata*	100	400	Asrar et al. (2017)
12	*B. gymnorhiza*	100	400	Li et al. (2008)
13	*Bruguiera cylindrica*	–	400	Sruthi and Puthur (2020)
14	*Bruguiera parviflora*	100	400	Parida and Das (2004)
15	*Salsola dendroides*	100	400	Heidari-Sharifabad and Mirzaie-Nodoushan (2006)
16	*K. sieversiana*	160	400	Yang et al. (2007)
17	*H. salicornicum*	200	400	Panda et al. (2019)
18	*Salicornia herbacea*	100	400	Lee et al. (2016)
19	*Cakile maritima*	100	500	Arbelet-bonnin et al. (2020)

TABLE 13.1 *(Continued)*

S No.	Halophytes	Growth stimulating concentration (mM)	Maximum salinity tolerance (mM NaCl)	References
20	*T. nitraria*	–	500	Cheng et al. (2015)
21	*Limonium delicatulum*	200	500	Aymen et al. (2016)
22	*S. maritima*	85	510	Moghaieb et al. (2004)
23	*S. brachiata*	200	600	Parida and Jha (2010)
24	*S. portulacastrum*	200	60	Peng et al. (2019)
25	*K. candel*	150	600	Wang et al. (2014)
26	*Atriplex halimus*	250	600	Khedr et al. (2011)
27	*S. altissima*	250	750	Meychik et al. (2013)
28	*A. lagopoides*	200	750	Sobhanian et al. (2010)
29	*Salsola drummondii*	200	800	Elnaggar et al. (2020)
30	*S. europaea*	200	800	Fan et al. (2011)
31	*S. fruticosa*	300	900	Shah et al. (2020)

13.3 HALOPHYTES DISTRIBUTION AND CLASSIFICATION

Earlier halophytes were demarcated as plant species that live permanently in saline environments. However, in the last decade, halophytes were defined as crops that can grow and reproduce in saline environment of greater or equal to 200-mM NaCl (Flowers and Muscolo, 2015). Around 1653, plant species have been identified as halophytes (Saslis-Lagoudakis et al., 2014). Halophytes comprise approximately 1% of world flora (Abobatta, 2020). Halophytes exist in 37 orders out of 65 plant orders. However, there are no separate families for halophytes. Some families, such as Chenopodiaceae, constitute the highest number of halophytes compared to other plant families. Under the Chenopodiaceae family, there are approximately 381 species of halophytes. Halophytes are distributed across 500 genera of flowering plants belonging to more than 71 families. Furthermore, both nonhalophytes and halophytes may belong to the same genus. For example, *Plantago*, *Glycine*, and *Solanum* (Abobatta, 2020). Based on various characteristics such as life habits, growth form, varying degrees of salt tolerance, and ecological niches, halophytes have been categorized into various groups which are discussed in the following:

a. Growth forms such as rhizomatous (*Juncus rigidus*), leafless succulent shrubs (*Suaeda aegyptiaca*), leafy nonsucculent shrubs (*Tamarix nilotica*), stoloniferous (*Aeluropus lagopoides*), nonsucculents herbs (*Tripteris amplexicaulis*), leafless trees (*Tamarix aphylla*), succulents herbs (*Nitraria retusa*), and succulent subshrubs (*Aptenia cordifolia*).

b. Life habits such as annual (*Apocynum venetum*), annual herb (*Acanthus* spp.), grasses (*Phacelurus latifolius*), perennials (*Suaeda fruticose*), perennial shrubs (*Atriplex portulacoides*), perennial rhizomes (*Apocynum pictum*), submerged plants (*Zannichellia palustris*), shrubs (*Lumnitzera racemose*), ferns (*Acrostichum* spp.), vines (*Flagellaria indica*), and trees (*Argania spinosa*).

c. Based on the varying degrees of actual salt tolerance, halophytes are grouped as (i) obligate halophytes (ii) facultative halophytes.

Obligate halophytes (euhalophytes or true halophytes) survive better under high saline conditions and require threshold salt in their environment for proper growth. Some examples of obligate halophytes include *Arthrocemum macrostachyum, J. rigidus,* and *A. portulacoides* (Zhao et al., 2011). Facultative halophytes are plant species that can grow both saline and nonsaline soils. They grow well on saline soils, but their growth is improved under low salt or nonsaline soils. Some examples of facultative halophytes include *Aster tripolium, Salvadora persica, Salvadora oleodies, Kali komarovii, and Chenopodium quinoa* (Reddy et al., 2008; Zhang et al., 2012; Adolf et al., 2013; Burundukova et al., 2017).

d. Based on ecological niche, halophytes are alienated into four groups:

   i. Extreme halophytes: These halophytes are well acclimated to high saline conditions and grow exclusively in a saline environment. For example, *Halimione pedunculata* and *Petrosimonia triandra* (Zhao et al., 2011).
   ii. Mesohalophytes: These halophytes comprise of intermediate anatomical adaptive features between glycophytes and extreme halophytes. Mesohalophytes grow in soil with moderate water content, such as semiarid habitats, salt meadows, and salt shrublands. *Atriplex*, *Aeuloropus*, *Limonlum*, and *Sueada* species are included in mesohalophytes (Zhao et al., 2011).

e. Xerohalophytes: Halophytes growing in salt deserts are called xerohalophytes. *Halocnemum*, *Tamarix*, *Haloxylon*, *Halostachys*, and

Zygophyyllum species are included in xerohalophytes (Abobatta, 2020).

f. Hydrohalophytes: Halophytes growing in places marshy, damp, and waterlogged areas are called hydrohalophytes. Hydrohalophytes are further subgrouped into

   i. Emergent hydrohalophytes that grows in salt marshes and coastal habitats. All the mangroves are grouped under hydrohalophytes. *Acanthus*, *Acrostichum*, *Bruguiera*, *Kandelia*, *Lumnitzera*, *Nypa*, *Rhizophora*, *Salicornia*, *Sonneratia*, and *Triglochin* are included in emergent halophytes.

   ii. Submerged hydrohalophytes: Halophytes that grow beneath the sea surface. *Cymodocae* spp., *Halophila* spp., *Phyllospadix japonica*, *Ruppia rostellata*, and *Zostera merina* are grouped in submerged hydrohalophytes (Abobatta, 2020).

13.4 HALOPHYTES VS GLYCOPHYTES

Salinity and alkalinity have an adverse effect on glycophytes and halophytes ionome, metabolome, and proteome profiles. The intricacy of a plant's response to stress is proportionate to the extent of stress. According to tolerance capacity, plants are grouped as glycophytes and halophytes. Cheeseman (2015) defined glycophytes as the plant species that can naturally survive in low sodium concentration and have the adaptation for maintaining low levels of sodium in the aerial tissues such as leaves. Glycophytes growth declines under salt–alkali stress, occasionally resulting in plant death because of hostile water relations, membrane ion leakage, and adverse effect of ion toxicity on plant metabolism. Most of the crops are glycophytes. Wheat and maize are most salt–tolerant crops. However, rice is the most salt-sensitive crop. According to salt–tolerant capacity, glycophytes are also divided into three groups: (i) Very sensitive crop plants include those unable to survive even 1 dS m^{-1} soil salinity. For example, *Ananas fregeria* (strawberry); (ii) Salt-sensitive crops are the crop plants that can tolerate ≤3 dS m^{-1} salinity in the soil. *Citrus paradise*, *Citrus sinensis*, *Daucus carota*, *Phaseolus vulgaris*, *Prunus domestica*, *Oryza sativa*, and *Solanum tuberosum* are all salt-sensitive crop plants; (iii) Intermediate salt-sensitive crops can survive up to 5 dS m^{-1} soil salinity, but their growth is highly reduced. Moderately/intermediate sensitive crop plants include *Saccharum officinarum*, *Vicia faba*, *Trifolium alexandrium*, *Cucumis sativus*, *Lactuca sativa*, and *Zea mays*; (iv) Moderately salt–tolerant crops: These crop plants can tolerate soil salinities ranging from 5–6.8 dS m^{-1}. *Glycine max*, *Sorghum bicolor*, and *Vigna unguiculate* are moderately salt–tolerant crops; (v) Salt tolerant crops embrace *Hordeum vulgare*, *Camelina sativa*, *Gossypium hirsutum*, *Carthamus tinctorius*, *Beta vulgaris*, and *Triticum aestivum* (Rozema and Schat, 2012).

Soil salinity and alkalinity induce severe growth reduction in glycophytes due to osmotic stress around the root and salt accumulation in the transpiring leaves. In contrast, halophytes prevent toxic ions accumulation in the leaves through different mechanisms. Additionally, halophytes can maintain a low Na^+/K^+ ratio for ion homeostasis. Nevertheless, glycophytes tend to have a greater relative growth rate, and germination rate under low salinity and nonsaline conditions having a competitive advantage over halophytes due to various ecological, phenological, and genetic factors that aid for earlier vegetative growth in glycophytes (Rozema and Schat, 2012). Halophytes flourish more in salty water and soil with low water potential than glycophytes (Munns and Tester, 2008). There is no absolute difference in the salinity tolerance of halophytes and glycophytes, but there are gradational variations in the degree of tolerance toward salinity stress from glycophytes and halophytes. A comparative proteomic study between *Arabidopsis thaliana* (glycophyte) and *Thellungiella halophila* (halophyte) revealed that halophytes have an advantage over glycophytes due to the well-organized regulation of basic biochemical tolerance mechanisms (Pang et al., 2010). Furthermore, many researchers have suggested that the salt tolerance capability of halophytes is major because of their capability to facilitate the low Na^+ concentration in the cytoplasm even under high saline and alkaline external environment as compared to glycophytes by efficiently excluding and compartmentalizing Na^+ (Sanchez et al., 2007; Pang et al., 2010; Sengupta and Majumder, 2010; Deinlein et al., 2014; Yin et al., 2019; Wang et al., 2020a)

Numerous reports on physiological and metabolic salt tolerance mechanisms in halophytes are available (Asrar et al., 2017; Pandit et al., 2017; Petropoulos et al., 2017; Wang et al., 2017; Kumari and Parida, 2018; Peng et al., 2019; Abobatta, 2020; Al-Shamsi et al., 2020; Zhao et al., 2020; Arif et al., 2020; Saddhe et al., 2020). Additionally, the potential of halophytes for sustainable agriculture, saline cultivation, and phytoremediation has been explored (Reddy et al., 2008; Rozema et al., 2013; Flowers and Muscolo, 2015). However, a smaller number of studies have undertaken the alkali or salt–alkali tolerance aspect of halophytes (Kumar et al., 2018; Yin et al.,

2019; Jia et al., 2020; Suo et al., 2020; Wang et al., 2020a; Lu et al., 2021; Ye et al., 2021). Thus, a better comprehension of the halophytes salt–alkali tolerance mechanism holds the key to developing tolerant crop varieties that can survive under high saline as well as highly alkaline environments (Pang et al., 2016a).

13.5 SALT AND ALKALI RESPONSIVE IONIC AND OSMOTIC HOMEOSTASIS IN HALOPHYTES

13.5.1 VACUOLAR COMPARTMENTALIZATION OF TOXIC IONS

Ionic imbalance is a common occurrence under salt–alkali stresses. Halophytes are experts in maintaining ion homeostasis under adverse conditions of high salt and pH for better survival. Halophytes maintain ion balance controlling ion influx and proper compartmentalization, which required for proper growth and survival of the halophyte in a saline environment. Three strategies are employed by halophytes, namely, (1) exclusion of Na^+ from the root; (2) sequestration of Na^+ into the vacuoles; and (3) controlling Na^+ and Cl^- loading into the xylem. Additionally, photosynthetically active mesophyll cells retrieve Na^+ from the xylem and recirculate via phloem into the shoot with HKT transporters' help for prevention of Na^+ accumulation in the cell (Zhao et al., 2020).

13.5.1.1 NA^+ EXCLUSION FROM THE ROOT

Na^+ is passively influxed in the root by a down electron gradient. Several ion transporters mediate the symplastic uptake of sodium by roots, including nonspecific K+ channels and nonselective cation channels (NSCCs) such as PIP that are permeable to monovalent cations and facilitate water uptake (Demidchik, 2018). It is normally presumed that approximately 95% of Na^+ taken by root is returned to the rhizosphere, and only 2–3% is transported into the shoot via xylem (Munns et al., 2020). Majorly there are two major types of ion channels which play a part in Na^+ exclusion from the root (a) NSCCs, such as cyclic nucleotide-gated channels (CNGCs) and glutamate receptor-like (GLRs), and (b) high-affinity potassium transporters (HKT^2). Other than that, high-affinity K^+ transporters (HAK^5), AKT^I (shaker type K^+ transporters), and LCT^I (low-affinity cation transporters) are also involved in Na^+ uptake through the root. In addition to this, aquaporins $PIP^{2;I}$ also help in uptake of Na^+ by root (Isayenkov and Maathuis, 2019). Furthermore, NSCC are also involved in the influx of Na^+ from the cell. NSCCs are weak voltage-dependent and voltage-independent cation channels. Based on voltage condition NSCCs can be of three types (1) depolarization-activated NSCCs (DA-NSCC), hyperpolarization-activated NSCCs (HA-NSCC), and voltage-insensitive NSCCs (VI-NSCC). VI-NSCCs are permeable to K^+ and help in retaining K^+ in the cell. It also participates in Na^+ influx in roots, thus restricting Na^+ to the root region. Zhao et al. (2016) reported that salt–alkali tolerance in halophyte *P. tenuiflora* was due to its ability to block apoplastic path of Na^+ entrance into the root endodermis with the help of plasma membrane located ion transporters (*PutPMP3-1* and *PutPMP3-2*), high-affinity K^+ transporter (*PutHKT2*), Na^+/H^+ antiporters (*PtNHA1*). Moreover, under high alkaline conditions, *P. tenuiflora* employs hydrogen-exporting adenosine triphosphatase (H+-ATPase) for pumping out excess H^+ to maintain pH homeostasis (Zhao et al., 2016). Salt overly sensitive (SOS) participates in maintenance of ion equilibrium under abiotic stress conditions. SOS1 that encodes for Na^+/H^+ antiporters also plays key part in the mitigation of ion toxicity caused due to salt–alkali stress by modulating Na^+ efflux from the cytosol and restricting the passage of toxic ions from root to shoot and other plant parts (Shabala et al., 2012). SOS2 and SOS3 are involved in calcium signaling for balancing Na^+ efflux under salinity stress. Therefore, it is a partial phenomenon for the salinity tolerance in halophytes.

13.5.1.2 VACUOLAR SEQUESTRATION OF NA^+

High Na^+ in the cytosol is toxic to the plant; therefore, halophytes tend to sequester excess Na+ toward older plant tissue or compartmentalize into vacuoles under high salinity and alkalinity condition. The vacuolar compartmentation of Na^+ takes place with the help of Na^+/H^+ antiporters (Gupta and Huang, 2014). The major proton pumps that participate in excess Na+ transports are V-PPase and H^+-ATPase (V-ATPase). V-ATPase equilibrates solute concentration, transportation, and plant growth (Kumar et al., 2019). Vacuolar sequestration of Na^+ is mediated by Na^+/H^+ exchangers (NHXs), such as NHX1 and NHX2 (Finazzi et al., 2015). Furthermore, HKT family proteins act as K^+/Na^+ symporters for the influx of K^+ and efflux of Na^+. K^+/Na^+ symporters are selective transporters for excluding excess sodium and maintaining high K^+ in the cell for conferring salinity tolerance (Arif et al., 2020). Upregulation of genes involved in ions compartmentalization under salt–alkali stress is documented in *P. tenuiflora* (Zhao et al., 2016; Lu et al., 2019). They reported that vacuolar membrane Ca^{2+}/H^+ antiporter encoded by *PutCAX1* played an essential role in the transportation of Zn^{2+}, Ba^{2+}, and Ca^{2+} and vacuolar Na^+/H^+ encoded by *PutNHX* was reported for its role in vacuolar sequestration of Na^+ in *P. tenuiflora*. In addition, Zhao et al. (2016) reported the important role of vacuolar Na^+/H^+ antiporter for pH regulation under alkaline salt conditions. Some research studies have suggested that efficient vacuolar sequestration of Na^+ is contributed not only by NHX transporters but also by the retention of Na^+ in the vacuole governed by competent regulation of tonoplast leak channels (Shabala et al., 2020; Zhao et al., 2020). The Na^+ permeable tonoplast channels SV (slow) and FV (fast) activating ion channels regulate the futile movement of Na^+ into and out of the tonoplast (Shabala et al., 2020). The current notion is that halophytes' salt tolerance mechanism also involves the exclusion of excess Cl^- from the shoot (Geilfus, 2018). This inference is supported by Cl^- accumulation (>1 M) in halophytes without any adverse effect on plant survival (Bazihizina et al., 2019). In this context, Cl^- uptake in the shoot has been reported in *Pennisetum giganteum*, relating to its salt tolerance ability (Hayat et al., 2020).

13.5.1.3 CONTROLLING OF NA^+ AND CL^- LOADING ONTO XYLEM

Regulation of Na^+ and Cl^- loading into the xylem is critically significant for salinity tolerance. Both active and passive transport systems are involved in the xylem loading of Na^+ under saline and alkaline conditions, but their respective role is defined by the exposure time of the plant to salinity (Ishikawa et al., 2018). Shabala et al. (2012) suggested that thermodynamically under most physiological conditions, Na^+ loading needs to be an active process. NSCCs play an important role in passive Na^+ loading into the xylem. Moreover, active loading is mediated by Na^+/H^+ antiporters encoded by the SOS[1] gene. Additionally, HKT[2] transporters (expressed highly in the stellar root tissues) also help in the xylem loading of Na^+, acting as a K^+/Na^+ symporter. The mechanism involved in the active loading of Na^+ is favored by the depolarization of parenchyma cells under saline–alkaline stress leading to passive movement of K^+ in the xylem, creating a driving force for xylem loading of Na^+ (Ishikawa et al., 2018). Furthermore, Cl^- loading into the xylem is mediated by cation–chloride cotransporters (CCC), which are favorably articulated at the xylem–symplast boundary. CCCs play a role in the symport of Na^+, K^+, and Cl^- and passively transport Cl^- into xylem. The passive movement of Cl^- into the xylem also adds to the driving force for active secondary loading of Na^+ into xylem (Zhao et al., 2020).

13.5.2 OSMOTIC ADJUSTMENT IN HALOPHYTES UNDER SALINE–ALKALINE STRESS

Osmotic adjustment is a primary strategy employed by plants for survival under abiotic stress like salinity and alkalinity. Halophytes achieve osmotic adjustment via two processes: de novo synthesis of compatible osmolytes such as proline and increasing uptake of inorganic ions (Guo et al., 2010; Zhao et al., 2016). For example, Guo et al. (2010) reported accumulation of organic acids and inorganic anions under saline–alkaline stress in *P. tenuiflora* for maintaining ionic and osmotic balance by harmonizing the colossal influx of cations. It has been reported that *P. tenuiflora* upheld osmotic homeostasis by accumulating betaine, soluble, and sugar proline in the protoplasm (Lu et al., 2021). Accumulation of inorganic ions is a basal strategy of halophytes for regulating osmotic balance in the leaf cells toward adverse salt–alkali stress. The relative input of different ions to osmotic homeostasis relies on membrane transport regulation: (a) transport energetics, kinetics, and selectivity; (b) negative feedback controls mechanism; and (c) membrane permeability. Different ions show diverse uptake rates, affecting ionic balance differently. For example, chloride ions contribute more toward the osmotic balance due to their easy absorption as compared to sulfate. Salinity is primarily caused when cation uptake rates exceed those of anion. Proline is the main osmolyte that is chiefly accumulated during ion homeostasis among other osmolytes (Kumar et al., 2018; Benjamin et al., 2019; Shuyskaya et al., 2020) such as GB (Kumari et al., 2015; Zhao et al., 2016; Behr et al., 2017; Yin et al., 2019); sugar (Hu et al., 2015) and polyols (Guo et al., 2015b; Lee et al., 2016; Benjamin et al., 2019). Accumulation of proline with increasing salinity has been reported in the roots and leaves of *Suaeda altissima* (Meychik et al., 2013). Meychik et al. (2013) suggested that 250-mM NaCl is the most favorable salinity level for *S. altissima. Atriplex leucoclada* accumulated greater proline contents than *Haloxylon salicornicum* and *Salicornica verginica* while minimum proline concentration was found in *Suaeda fruticosa* (Samiullah and Bano, 2011). In halophytes *S. fruticosa* and *A. leucoclada* the higher proline accumulation was found, while *H. salicornicum* and *S. verginica* have lesser proline accumulation. The major osmoprotectant proline accumulated 4–10 folds at

EC 35 dS m^{-1} and 6–8 folds at combined stress of pH 9.0 + EC 20 dS m^{-1} (Kumar et al., 2016). Tabot and Adams (2014) reported that proline accumulations were normally high across all salinity levels in the endemic halophyte (*Limonium linifolium*). Park et al. (2013) observed a marked increase in the proline levels at 500- and 700-mM NaCl in *Salicornia europaea* as well as an increase in GB linked to the salinity increment and corresponding induction of betaine aldehyde dehydrogenase (BADH) activity. *Atriplex nummularia* commonly known as saltbush showed a linear rise in proline content under saline stress conditions. For the levels of 450 and 600 mM of NaCl maximum contents were found at 45 days. In a report on *Atriplex prostrata*, Bueno (2016) reported from his study the proline being the major osmolyte showed considerable changes at 200- and 300-mM NaCl and the increase was greater (4 and 10 folds as compared to the control) followed by GB and in the case of *Plantago coronopus* the proline was the principal osmolyte, followed by sorbitol to a lower extent. Kumar et al. (2018) also found the similar results of proline accumulation and reported that proline accumulation was 4.8, 5.2, and 5.0 times larger at the sodic stress of pH 10.0, saline stress of electrical conductivity of a saturated soil extract (ECe) 35 dS m^{-1} and at combined sodic saline stress of pH 9.0 + ECe 20 dS m^{-1}, respectively, in *Sporobolus marginatus*, while the corresponding increases in *Urochondra setulosa* were 7.9, 10.59, and 6.85 folds, respectively. In another study on *Dichanthium*, Kumar et al. (2018b) noticed that under sodic stress proline showed 4.23 folds higher deposition, 5.62 folds at saline stress of ECe 35 dS m^{-1} and 8.07 folds at combined stress of pH 9.0 + ECe 20 dS m^{-1} as compared to control conditions. These results showed that proline accumulation plays an important part in cell survival under salinity and alkalinity conditions by protecting and stabilizing the cell membranes and also acts as a compelling osmoregulator that counteracts the harmful effects of toxic salt ions accumulation in cell vacuoles and alleviates the effect of NaCl on cell wall disruption. GB is chloroplastic in origin and plays an imperative role in osmotic homeostasis, thylakoid membrane protection, photosystem system II (PSII) against salt stress (Ashraf and Harris, 2004). Halophytes are hyperaccumulators of GB (>90 μmol dry weight) (Flowers and Colmer, 2015). Upregulation of BADH (*LcBDH1*, *LcBDH2*) and Choline monooxigenase (CMO) transcripts, responsible for the synthesis of betaine, have been reported in halophyte *Leymus chinensis* under salt stress (Wang et al., 2020a).

Sugars and polyols are some of the compatible solutes that are categorized as soluble carbohydrates, which are used for osmoprotection and osmotic adjustment. Tester and Devenport (2003) reported that maintaining osmotic homeostasis required augmentation of osmolytes in plant cells either by the production of metabolically compatible compounds or by uptake of soil solutes that can accumulate to high concentrations without influencing the metabolism of plants (Williamson et al., 2002). Organic osmolytes in this category are most commonly amino acids, protein, proline, and carbohydrates (such as sugar) documented by Yusuf (2008). Contribution of soluble carbohydrates, such as glucose, sucrose, fructose, and complex sugars like raffinose and trehalose have been reported toward adaptive mechanism under salt–alkali stress (Gong et al., 2005; Wang et al., 2014a; Hu et al., 2013; Rozentsvet et al., 2017). Bueno et al. (2020) reported a significant role of polyols in the retaliation to salt–alkali stress, compensating for the reduction in cellular water potential and as scavengers of ROS. The hydroxyl group in polyols establishes hydrogen bonds under cellular dehydration for supporting enzyme activities and protection of lipid membrane (Benjamin et al., 2019). Additionally, polyols act as signaling molecules, ROS scavenging, major cellular energy source, and precursors of metabolic compounds, (Edwards and Donoghue, 2013; Al-Shamsi et al., 2020) in many plant species, including halophytes. In all three species of *Salsola*, an increase in salinity reduces plants' soluble carbohydrate concentration (Li et al., 2010a; Kumar et al., 2019). Kumar et al. (2016) reported that sodic stress did not significantly influence total soluble sugar accumulation in *Aeluropus* and *Suaeda*. Nevertheless, the concentration was found to be drastically increased under conditions of saline stress alone and in combined stress with sodicity. In *Sporobolus* and *Urochondra*, TSS amounts significantly increased with the increasing magnitude of stress, irrespective of the nature of stress (i.e., either salinity or sodicity) in comparison to control plants (Kumar et al., 2018). Myoinositol is the utmost abundant isomer of inositol, playing important role in protecting oxidative stress and signaling (Sureshan et al. 2009).

13.6 EFFECTS OF SALT–ALKALI STRESSES AND SUBSEQUENT PHYSIOLOGICAL RESPONSES IN HALOPHYTES

Salinity and alkalinity impose several interdependent constraints on plant growth and development which include not only oxidative stress, physical dryness, and ion toxicity but also mineral ion imbalance and metabolic disorders (Guo et al., 2017). Salinity and alkalinity adversely impact the plant growth mostly crop plants. Under field conditions, crop plants barley are comparatively more tolerant to salinity (can tolerate up to 250-mM NaCl which is fatal for wheat and rice) and alkalinity compared to other cereals like rice, wheat, maize, and sorghum (Hanin et al., 2016). Various scientific approaches have contributed toward our mechanistic understanding of the effects of salinity and alkalinity on plant growth and development (Abobatta, 2020; Al-Shamsi et al., 2020; Arbelet-bonnin et al., 2020; Goussi et al., 2018; Guo et al., 2015; Kumar et al., 2019; Yin et al., 2019). The adverse effects of salinity and alkalinity on plant growth and development include two foremost constraints—ionic and osmotic stress (Zhao et al., 2020). Salinity and alkalinity also induce secondary stress like oxidative stress and organellar stress.

13.6.1 PLANT WATER RELATION AND SALINITY–ALKALINITY-INDUCED OSMOTIC STRESS

Salinity and alkalinity are known to cause a reduction in leaf osmotic potential, ionic potential, and turgor pressure (Arif et al., 2020). Accumulation of soluble salts tends to increase soil water potential, leading to the decline of water available for plant uptake, causing reduced turgor and water potential in above-ground plant parts (Betzen et al., 2019). Salinity and alkalinity-induced Na^+ accumulation in the leaf results in lower water potential of leaf, turgor, and osmotic potential, hampering various plant physiological traits. In this context decrease in the relative water content of the leaf is reported in pomegranate under salinity stress.

Salinity and alkalinity disturb the soil structure, leading to reduced soil water potential, ultimately decreasing leaf water potential, disturbing plant water relation and plant turgor pressure, leading to the generation of osmotic stress. Thus, osmotic stress can be referred to as lowering external water potential leading to a physiological dryness state for the plant. To adjust osmotically, plants increase inorganic ion intake, and resume growth but at a reduced growth. Thus, retardation of growth may be considered as a consequence of osmotic stress leading to reduced water uptake, declining in stomatal conductivity, decrease in leaf size, and retardation of cell division and elongation (Rozentsvet et al., 2017). In addition to osmotic stress, alkalinity causes the breaking of soil structure due to excessive retention of Na^+ and adds to the toxicity of carbonate ions, inhibiting the absorption of nutrients. Thus, soil alkalinization leads to injury to plants through salinity stress and high pH, making the natural saline-alkaline condition more complicated and severe for plant survival (Yang et al., 2009).

13.6.2 IONIC STRESS IN HALOPHYTES IMPOSED TO HIGH SALINITY AND ALKALINITY

Salinity and alkalinity imposed a second constrain is "ionic stress," also referred to as ionic toxicity. Under salinity stress, ionic stress is allied with an excess build-up of Na^+ and Cl^- ions in the cellular compartments. Salt stress imposes not only Na^+ toxicity but also is accompanied by Cl^- toxicity. The present notion is that accumulation of Cl^- in plant tissue causes severe physiological dysfunctions leading to loss of crop yield and product quality (Geilfus, 2018). Additionally, in association with high Na^+, excess chloride produces an antagonistic ion effect, hindering essential mineral ion uptake and leading to a nutrient deficiency under salt stress (Reich et al., 2017). In this regard, the adverse effect of excess Cl^- on nitrate, phosphate, and sulfate uptake has been documented in citrus, *Brassica rapa*, and rose (Reich et al., 2017). However, alkalinity stress imposes CO^{32-} toxicity in addition to Na^+ adversity. The most common explanation for the toxicity of Na^+ is its structural similarity with K^+. Under high saline conditions, Na^+ disrupts intake and translocation of K^+ and replaces K^+ in critical enzymatic reactions in the cytosolic compartments, disturbing the activities of important regulatory enzymes like enzymes associated with the Calvin cycle, glycolysis, starch, and polyamine biosynthesis (Zhao et al., 2020). In addition to this, alkali stress causes proton scarcity in the root region of the plant. It disrupts cross-membrane potential, inhibiting the captivation of essential mineral ions that alter plants' nutrient availability, ultimately distorting the ion and mineral balance greater than salinity alone (Yang et al., 2009). Drastic effects of salt stress on maize ionome status have been reported (Hessini et al., 2019).

13.6.3 SALINITY-ALKALINITY INDUCED OXIDATIVE STRESS AND ANTIOXIDATIVE DEFENSE

Plants require a threshold level of ROS for the effective functioning of various physiological and biochemical processes. ROS helps regulate various biological processes at lower concentrations, including growth and stimuli response to numerous biotic and abiotic stresses. However, ROS accumulation can result in lethal effect on plant growth and physiology, causing oxidative stress. Major ROS produced in plants under abiotic stress comprises singlet oxygen ($^1O^2$), hydroxyl radical ($OH^\bullet$), hydrogen peroxide (H^2O^2), and superoxide anion ($O^{2-\bullet}$). Deleterious ROS accumulation effects are attributed to their ability to cause peroxidation of the lipid membrane, denaturation of protein, DNA damage, oxidation of carbohydrates, and enzyme activity impairment (Asada, 2006). In addition, ROS accumulation under stress conditions also interrupts ionic homeostasis at the cellular level by activation of various ROS-sensitive ion channels (Demidchik, 2018). Hydrogen peroxide-sensitive Ca^{2+} ion channels have been reported in root epidermal cells and guard cells of stomata (Pel et al., 2000; Demidchik, 2018). Also, inhibition of the K^+ channel by H^2O^2 in stomata is reported (Pel et al., 2000). Furthermore, ROS-sensitive ion channels are also present in organellar membranes such as chloroplast and vacuolar membranes (Demidchik, 2018), which alters these organelles' operational capacity. For example, salinity induces swelling of the thylakoid membrane caused by the influx of Na^+ ions via H^2O^2-activated ion channels on the thylakoid membrane, resulting in decrease of photosynthesis (Barhoumi et al., 2007). Salinity and alkalinity induce over-production and accumulation of various ROS, causing abnormalities in organellar structures and electrolyte leakage (Arif et al., 2020). ROS production sites in plants include chloroplast, mitochondria, apoplast, and peroxisomes (Zhao et al., 2020). Salinity and alkalinity-induced osmotic stress led to the closure of stomata, and cytosolic Na^+ accumulation impairs photosynthetic machinery in plants leading to ROS formation in the chloroplast (Arif et al., 2020). Mitochondrial respiration is another foremost source of ROS production under salt–alkali stress. The main ROS produced through mitochondrial respiration is superoxide. Salinity causes excessive ubiquinone pool reduction, allowing leakage of electrons to molecular oxygen from the mitochondrial electron transport chain complexes I and II, resulting in superoxide production. Intracellular H^2O^2 is mostly produced in the peroxisome. Salinity stress enhances photorespiration resulting in a decreased ratio of CO^2/O^2 in the plant cell, causing the production of H^2O^2 in the peroxisomes. Peroxidase (POX), apoplastic diamine oxidase, plasma membrane-localized respiratory burst oxidase homologs, and polyamine oxidases are the primary mediators of ROS production in the apoplast. Upregulation of respiratory burst oxidase homologs AtRbohD and AtRbohF have been reported under salinity stress (Zhao et al., 2020). AtRbohD and AtRbohF are noticed to alleviate salinity stress. The main reason may be that the salinity-induced ROS production by AtRbohD and AtRbohF leads to the movement of potassium into the cytosol and increases the K^+/Na^+ ratio. AtRbohF restricts transpirational translocation of Na^+ to shoot from the root by limiting Na^+ distribution in xylem sap. Moreover, salinity-triggered propagation of long-distance signals is mediated by AtRbohD, suggesting a role of AtRbohD in systematic plant signaling under abiotic stress. Furthermore, ROS production in apoplast subsidizes lignin formation under salt stress (Arif et al., 2020).

13.6.3.1 ENZYMATIC DEFENSE SYSTEM AGAINST SALINITY AND ALKALINITY IN HALOPHYTES

High antioxidant properties of halophytes in comparison to glycophytes assist their great capability to cope with a high salt–alkali stress environment (Yu et al., 2011; Gong et al., 2013; Yin et al., 2019). Antioxidant defense system play a vital role in the detoxification of salinity- and alkalinity-induced ROS in the cellular environment. Halophytes have grown highly complex antioxidant defense mechanisms that involve enzymatic and nonenzymatic pathways to eradicate excessive ROS production while balancing a most favorable level of ROS for signaling. The enzymatic component consists of enzymes such as catalase (CAT), superoxide dismutase (SOD), dehydroascorbate reductase (DHAR), ascorbate peroxidase (APX), peroxidase (POX), glutathione reductase (GR), and ascorbate–glutathione cycle enzymes monodehydroascorbate reductase (MDHAR). Siddiqui et al. (2016) discovered that CAT and SOD activities of treated samples increased appreciably in *H. mucronatum* as compared to *C. ciliaris*. A study on a two-plant system (halophyte—*Sesuvium portulacastrum* and glycophyte—*Brassica juncea*) revealed that prevention of oxidative damage and proper utilization of energy are key determinants of efficient antioxidant defense system in halophytes for salt tolerance in comparison to glycophytes (Srivastava et al., 2015). In this context, *S. portulacastrum* was documented to avoid or delay oxidative damage caused by salinity as high ROS levels were only noted when the plant was exposed to salinity for a longer duration. However, inclination in ROS and MDA contents were reported in *B. juncea* under the short duration of salt treatment. Himabindu et al. (2016) stated that stress-induced ROS generation and oxidative stress regulation through specific genes/factors have also been concerned with stress tolerance. Proline accumulation, increase in salinity-induced activation of POX, and total proteins concentration in leaves of eu- and crynohalophytes (goosefoot, saltwort, and salt cedar) were found by Eva et al. (2016). In *S. portulacastrum*, the antioxidant enzymes such as SOD, CAT, POX, polyphenol oxidase, and increased up to the level of 600-mM NaCl concentration which is optimum and ahead of these levels the concentration decreased marginally found in their studies by Rajaravindran and Natarajan (2012). Zhou et al. (2017) notice that the antioxidant enzyme activity increased with the increase in salt treatment, which is consistent with the hypothesis that once the ROS was generated in plant tissues, SOD collaborated with other enzymes CAT, POD, and APX to attain the goal of ROS scavenging. SOD activity amplified with increasing sodicity. While two times high GR activity was noted under sodic stress in *S. marginatus*. The salt tolerance capability of plants is often related to an efficient antioxidative system. Neutral salt stress and low alkaline stress induced lower SOD and APX activity with no changes in CAT activity in *Lathyrus quinquenervius*. Moreover, high alkalinity induced reduced CAT activity and enhanced SOD and APX activity stated in *L. quinquenervius* (Zhang and Mu, 2009). Meanwhile, an increase in the activities of SOD, APX, and POX has been stated in halophyte *S. europaea* and halophyte alkali grass (Wang et al., 2009; Zhang et al., 2020c).

13.6.3.2 NONENZYMATIC DEFENSE MACHINERY: REDOX BUFFERS

Detoxification of $OH^{\bullet}$ in plants is majorly carried out by low molecular weight nonenzymatic antioxidant molecules, for example, ascorbate, tocopherols, carotenoids, glutathione, and polyphenols Sharma et al. (2018). The role of ascorbate–glutathione cycle for detoxification of ROS have been reported for improving salinity and alkalinity tolerance in halophytes (Adolf et al., 2013; Yu et al., 2013; Flowers and Muscolo, 2015; Kumari et al., 2015). Ascorbate and glutathione plays important role as redox buffers for neutralizing salt–alkali stress induced production of ROS. Ascorbate and glutathione play numerous roles in various physiological processes which include oxidative stress signaling, controlling of sulfate transport, and modulation of plant growth and development (Mir et al., 2018). Accumulation of methylglyoxal under salt–alkali stress is well known for instigating damage to lipid membranes and oxidative damage to cellular biomolecules. GSH-dependent glyoxalase converts methylglyoxal to D-lactate corroborating improved antioxidant system. Although halophytes employ various defensive strategies to cope with salt–alkali stress, maintenance of GSH:GSSG ratio and replenishing of oxidized glutathione for balancing redox buffer is one of them. For example, a maintained level of GSH:GSSG ratio is reported in halophytes *Cynodon dactylon*, *P. tenuiflora*, *Hordeum maritimum*, and *S. portulacastrum* (Lokhande et al., 2013; Suo et al., 2020; Ye et al., 2021). Ascorbate is a pervasive nonenzymatic antioxidant that acts as a reducing substrate for APX and guaiacol POX for H^2O^2 detoxification (Kumar et al.,

2018). Changes in the redox pool of glutathione and ascorbate are indicators of oxidative stress. Parida and Jha (2010) reported a decrease in an increase in AsA:DHA ratio at low salinity but a decline in the ratio with an increase in NaCl concentration in *Salicornia brachiata.* Furthermore, the induction of two key enzymes of the ascorbate–glutathione cycle (DHAR and GR) under salt–alkali stress is reported (Gong et al., 2013).

13.6.4 SALINITY AND ALKALINITY-INDUCED RESPONSES ON PHOTOSYNTHESIS, STOMATAL CONDUCTANCE, AND TRANSPIRATION

Photosynthesis is one of the foremost pivotal physicochemical processes for plants' survival; however, it is vulnerable to stress like salinity and alkalinity. Salt–alkali stress causes a decrease in photosynthesis either by stomatal limitations or nonstomatal limitations. Guard cells of stomata are sensitive to abiotic stress like salinity and alkalinity; it leads to stomatal closure under stress conditions. Furthermore, photosynthetic reaction enzymes are sensitive to salinity and alkalinity, leading to nonstomatal limitation of photosynthesis under stress conditions. Salinity and alkalinity cause reduced transpiration by decreasing stomatal conductance leading to low gaseous exchange and reduced net photosynthetic rate. Moreover, salinity and alkalinity also reduce photosynthetic pigments like chlorophyll a, b, and carotenoids (Ranganayakulu et al., 2013). Reduction in the number of grana, degradation of chloroplast structure, decreased photosynthetic pigments, and adversely affecting the electron transport chain have been reported in chickpeas (Pan et al., 2020). Furthermore, reduced mesophyll conductance due to decreased carboxylation under salinity stress has been reported in glycophytes such as pomegranate. Salinity-induced abnormalities in the ultrastructure of chloroplast and stomata and decreased photosynthetic pigments resulted in significantly reduced transpiration rate, net photosynthetic rate, stomatal conductance, and variable fluorescence to maximum fluorescence (Fv/Fm) ratio in *Eruca sativa* (L.) (Hniličková et al., 2017). Similarly, salinity induced a decrease in chlorophyll synthesis, reduced C^3 cycle utilization of NADPH, and Rubisco activity is reported in rice (Li et al., 2020). Likewise, under high salinity, reduced photosynthetic rate, chlorophyll fluorescence, impairment of chloroplast structure, and decreased activity of Rubisco have been documented in maize (Hessini et al., 2019). High salinity inhibited PSII activity in mustard, hampering net photosynthesis. Likewise, *Lisianthus* cultivars showed decreased gaseous exchange and transpiration (Cassaniti et al., 2013). Salinity-induced downregulation of photosynthesis due to stomatal limitation and reduced photochemical and carbon metabolism in sunflower cultivars (Hryvusevich et al., 2021). Thus, salinity induces a decline in net photosynthesis, transpiration rate, chlorophyll fluorescence, and stomatal conductance by hampering the ultrastructure of stomata and chloroplast interferes in the biosynthesis of photosynthetic pigments (Arif et al., 2020).

13.7 SALT- AND ALKALI-RESPONSIVE METABOLITES FROM HALOPHYTES: AN OVERVIEW

Metabolomics is an integral asset in explaining plant metabolic reactions to abiotic stresses (Janz et al., 2010). It centers around profiling the low atomic weight metabolites, which are the intermediates, initiators, or final results of vital physiological and biochemical pathways (Kumari and Parida, 2018). To determine biological responses caused by external factors, the existence and relative concentrations of metabolites would deliver information on the metabolic state of the plant. Metabolomic studies of plants' retaliation to salt–alkali stress and the comparison of sensitive and tolerant species' metabolic responses have revealed conserved and divergent patterns for stress tolerance in halophytes. Metabolomic studies have advantages over transcriptomic and proteomics studies as the content and composition of metabolites directly affect the phenotype because they are end-products of downstream processes of genes and protein activity (Feng et al., 2020). These metabolites might be essential defensive substances in plant reaction to abiotic stress (Cuadros-Inostroza et al., 2016). Enormous array of metabolites (primary and secondary metabolites) is hired by the plants in retaliation to salt–alkali stresses. Primary metabolites are universally present and are essential for plant growth and development. It has been reported that salinity and alkalinity tend to increase in amount of primary metabolites involved in nitrogen and amino acid metabolism, polyols, and carbohydrate metabolism (Sanchez et al., 2007). On the contrary, secondary metabolites are exploited more under stress conditions. Under salt–alkali-stress conditions, plants tend to accumulate compatible solutes, including an array of neutral and soluble organic metabolites like proline, GABA, etc. Salinity and alkalinity induce various changes in the levels of different primary and secondary metabolites involved in numerous biochemical and physiological reactions. These induced metabolites may act as osmoprotectants, signaling molecules, ROS scavengers, and transporters involved in tolerance plants' tolerance toward stress conditions (Parida et al., 2018). To cope with the heavy ounces of salinity and alkalinity, plants differentially modulate different metabolites involved in various cellular processes, including growth and energy metabolism, carbohydrate metabolism, photosynthesis, and antioxidant defense system. Changes in metabolome under saline–alkaline stress are well documented (Hu et al., 2015; Cuadros-Inostroza et al., 2016; Pang et al., 2016; Guo et al., 2017; Yang et al., 2017; Kumari and Parida, 2018; Lu et al., 2019; Feng et al., 2020). Most metabolites, primarily related to glycolysis and those which have a role as osmoprotectants, tend to increase under saline alkaline stress (Benjamin et al., 2019; Lu et al., 2019, 2021; Ye et al., 2021). Besides, alkali stress was reported to enhance tricarboxylic acid (TCA) cycle intermediates such as succinate, fumarate, and malate in contrast to salinity stress in soybean (Feng et al., 2020). This reveals that salinity stress alters the major activity of the TCA cycle for energy conservation diverted to the production of compatible solutes for osmotic adjustments. Moreover, metabolites related to photosynthesis such as fructose 6-phosphate, 3-phosphoglycerate, and glucose 6-phosphate have been documented to accumulate under salinity stress to cope with the adversity of reduced photosynthesis under stress conditions. On the contrary, the reduction of monophosphates has been noted down in *A. thaliana* under salinity stress for monophosphates in other pathways for amino acid metabolism.

13.7.1 METABOLITES FOR OSMOTIC ADJUSTMENT TOWARD SALT–ALKALI TOLERANCE

Metabolome analysis revealed salt–alkali stress-induced build-up of various amino acids and organic acids in halophytes majorly for osmotic homeostasis (Benjamin et al., 2019; Guo et al., 2010; Hu et al., 2015; Jia et al., 2020; Lu et al., 2019). Additionally, amino acids and organic acids also facilitate the protection of macromolecule structures and scavenge free radicals produced due to oxidative stress triggered by salt–alkali stress (Guo et al., 2010). Thus, the elevation of amino and organic acid contents has been directly linked to improved salt–alkali tolerance in halophytes (Flowers and Muscolo, 2015; Hu et al., 2015). Modulation of various amino acids, amide, and nonprotein amino acids has been reported under salt–alkali stress (Guo et al., 2010; Lokhande and Suprasanna, 2012). A decrease in amino acids such as arginine, serine, leucine, and threonine was reported in *P. pratensis*, which was associated with decreased growth of *P. pratensis* under salt–alkali stress (Hu et al., 2015). In contrast, accumulation of proline under salt–alkali stress is well documented (Benjamin et al., 2019; Bueno et al., 2020; Gao et al., 2014; Guo et al., 2010; Hayat et al., 2020; Hu et al., 2015; Kumar et al., 2018; Maršálová et al., 2016; Shuyskaya et al., 2020; Wang et al., 2014;

Yang et al., 2007; Yin et al., 2019). Proline not only plays an imperative part as osmoprotectant but also protects important enzymes of PSs and stabilizes protein complexes required for efficient photosynthesis under salt–alkali stress (Lehmann et al., 2010). However, proline toxicity has been reported under heat stress in glycophyte *A. thaliana* (Lv et al., 2011). Furthermore, accumulation of proline analog 4-hydroxy-N-methyl proline has been documented in halophyte *Melanleuca bracteata* under salinity stress for efficient regulation of various enzyme activity, vacuolar compartmentation of excess Na^+, and balancing the cost of energy production (Aslam et al., 2011). Besides proline, the vital role of amino acids such as tyrosine, valine, glycine, and isoleucine in mitigation of salinity stress has been documented in halophyte *Suaeda salsa* (Li et al., 2011). Differential regulation of malonic and 2-methyl-malic acid under salinity stress has been studied in *Lotus criticus*. Additionally, decrease in levels of succinic, citric, fumaric, and glycolic acid have been noted in *L. criticus* toward salt stress (Sanchez et al., 2011). Furthermore, organic acid contributes toward ionic homeostasis under salt–alkali stress. In this context, approximately 84.2% contribution of organic acid in sustaining ionic balance under salt–alkali stress in halophyte *K. sieversiana* has been documented (Yang et al., 2007).

Quaternary ammonium compounds also play an important role in the mitigation of salt–alkali stress in halophytes. Quaternary ammonium compounds include proline betaine, GB, choline sulfate, etc. GB is one of the key quaternary ammonium compounds that serves as a compatible solute for osmotic homeostasis (Lokhande and Suprasanna, 2012). Li et al. (2011) reported betaine as a key secondary metabolite for osmotic adjustment in *S. salsa*. Likewise, the build-up of GB in the cytoplasm for maintaining osmotic balance under salt–alkali stress was reported in halophyte *K. sieversiana* (Yang et al., 2007). Similarly, *Suaeda maritima* and *S. brachiata* showed a high accumulation of GB under salinity stress (Benjamin et al., 2019). Besides osmoprotection, GB is promptly involved in the protection of lipid membranes and photosynthetic apparatus (Lokhande and Suprasanna, 2012).

13.7.2 METABOLITES IN CARBOHYDRATE METABOLISM

Modulation of carbohydrate and energy metabolism is a vital strategy for assisting reduced growth under salt–alkali stress (Guo et al., 2017). Salinity and alkalinity induce inconsistent alteration in the carbohydrate pool of the plant. Yin et al. (2019) reported that halophyte *P. tenuiflora* modulates carbohydrate metabolism under alkali stress as an adaptive measure for maintaining threshold levels of soluble carbohydrates for energy supply and other metabolisms. Carbohydrates such as sugars and polyols (sugar alcohols) sever as important compatible solutes for maintaining cellular osmolarity conferring salinity and alkalinity tolerance in halophytes. Besides osmoprotection, these carbohydrates also have a role in ROS scavenging and carbon and energy storage (Wang et al., 2017). Accumulation of glucose, fructose, and sucrose reported in halophyte *T. halophila* toward salt stress (Sanchez et al., 2007). Likewise, high concentrations of trehalose and sucrose under salt–alkali stress were reported in *P. pratensis* (Hu et al., 2015). Furthermore, the accumulation of sucrose under salt–alkali stress for osmotic adjustment and energy production is documented in *Malus halliana* (Jia et al., 2020). Sugar alcohols such as myoinositol, mannitol, pinitol, and glycerol are kept in hand with sugars for osmotic tuning under salt–alkali stress. In addition to their role as compatible solutes, polyols also act as low molecular weight chaperons and help in the mitigation of oxidative stress induced by the production of ROS under a stress environment. High concentrations of pinitol in mangroves reported under salt stress (Kumari et al., 2015). Likewise, the accumulation of IMT1 and INO1 gene (responsible for pinitol biosynthesis) homologs is documented in *Porteresia coarctata* (Sengupta and Majumder, 2009). Myoinositol is an important osmolyte for osmotic homeostasis under stress conditions. It facilitates Na^+ symport in the shoot, reducing Na^+ toxicity. Upregulation of MIPS, an enzyme responsible for myoinositol biosynthesis, was reported in *Mesembryanthemum crystallinum* (Sengupta et al., 2012).

13.7.3 METABOLITES LINKED TO ENERGY METABOLISM

Modulation of energy metabolism is a vital strategy of halophytes for survival under adverse conditions. An increase in energy demand under stress conditions is a well-known fact. Hence, upregulation of metabolites linked to ATP generation such as adenosine, adenine, and adenosine monophosphate, have been reported in different halophytes such as *P. tenuiflora*, *A. lagopoides*, *Tangut nitraria*, *M. halliana*, and *S. salsa* (Wang et al., 2008; Cheng et al., 2015; Guo et al., 2015a; Maršálová et al., 2016; Peng et al., 2019; Jia et al., 2020; Shabala et al., 2020). Alkali stress-induced augmentation in the levels of enzymes of glycolysis like fructokinases (FRK), phosphoglycerate kinase, enolase, and triosephosphate isomerase is conveyed in *P. tenuiflora* (Yu et al., 2013). FRK is a key enzyme of glycolysis that catalyzes nonreversible phosphate group transfer from ATP to fructose, controlling the influx of sugar in glycolysis reactions. Hence, FRK accumulation under alkali stress modulates the breakdown of glucose for fulfilling the energy demand under salt–alkali stress (Yu et al., 2013). Moreover, salt–alkali stress-induced aggregation of malate dehydrogenase in halophytes *P. tenuiflora* and xerohalophyte *H. salicornicum* (Yu et al., 2013; Panda et al., 2020). Additionally, induction of ATP synthase homologs has been reported in *P. tenuiflora* under alkali stress (Yu et al., 2013). Hence, the capacity to maintain threshold levels of ATP generation in response to osmotic imbalance and decreased photosynthesis is a basal survival strategy of halophytes for tolerating high salt–alkali stress.

13.7.4 METABOLIC MODULATION OF GLYCOLYSIS AND TCA CYCLE

Fulfilling the increased energy demand under salt–alkali stress is mediated by differential regulation of various glycolytic enzymes such as glyceraldehyde-3-phosphate dehydrogenase, fructose bisphosphate aldolase, and fructokinases, etc. (Maršálová et al., 2016). Increase in the expression of fructose bisphosphate aldolase and glyceraldehyde-3-phosphate dehydrogenase have been stated in *Halogeton glomeratus* and *Bruguiera gymnorhiza* (Tada and Kashimura, 2009; Wang et al., 2015) Also, decreased expression of glycolysis intermediates such as pyruvate have been reported in *A. lagopoides,* fulfilling the demand of proline for osmotic adjustment (Sobhanian et al., 2010). Reduced activity of the TCA cycle is known under salt–alkali stress for gratifying the supply and demand of amino acids. In this context, decreased expression of TCA cycle enzymes malate dehydrogenase and succinyl-coenzyme A (CoA) ligase subunit β have been stated in *Hordeum marinum* under salt stress (Maršálová et al., 2016). Sobhanian et al. (2010) stated decreased expression of TCA cycle metabolites like succinate, fumarate, citrate, and ketoglutarate in halophyte *A. lagopoides*. However, increased expression of TCA cycle enzymes such as citrate synthase, succinate dehydrogenase, and isocitrate dehydrogenase were reported in xerohalophyte *H. salicornicum* (Panda et al., 2020).

13.8 SALINITY- AND ALKALI-INDUCED PROTEOMIC RESPONSES IN HALOPHYTES

13.8.1 MODULATION OF CALVIN CYCLE PROTEINS FOR EFFICIENT SALT–ALKALI TOLERANCE

Differential regulation of various Calvin cycle enzymes such as RuBisCo small and large subunit, carbonic anhydrase, RuBisCO activase, ferredoxin-thioredoxin reductase, and phosphoribulokinase (PRK) is reported under salt–alkali stress (Sengupta and Majumder, 2009; Yu et al., 2011, 2013; Zhang et al., 2012; Wang et al., 2015; Panda et al., 2020). Reduced photosynthesis under salt–alkali stress may be attributed to the downregulation of enzymes associated with the Calvin cycle and light-harvesting complexes (Yu et al., 2011, 2013). RuBisCo is the key enzyme for photosynthetic fixation of CO^2. Thus, the photosynthetic rate is proportionate to the amount of RuBisCo in the leaves. Reduced levels of carbonic anhydrase, RuBisCO large and small subunit, PRK, and RuBisCO activase in *P. tenuiflora* and *T. nitraria* under salt–alkali stress have been documented (Yu et al., 2011, 2013; Cheng et al., 2015). On the contrary, increased expression of RuBisCo large and small subunits and RuBisCo activase in *S. europaea, H. glomeratus*, and *H. salicornicum* have been stated (Wang et al., 2009, 2015; Panda et al., 2020). Likewise, PRK increased expression under salinity stress reported in *S. europaea* and *P. coarctata* (Sengupta and Majumder, 2009; Wang et al., 2015). Furthermore, upregulation of Calvin cycle regeneration phase enzymes, such as protein transketolase, have been stated in mangrove *Kandelia candel* (Wang et al., 2014b).

13.8.2 REGULATION OF PHOTOSYSTEM ASSOCIATED PROTEINS

Proteomic studies have revealed differential expression of various proteins associated with PS under salt–alkali stress. Modulation of different PS-related proteins under salinity stress include oxygen-evolving enhancer protein (OEE), cytochrome b6-f complex iron–sulfur subunit, ATP synthase, D1/2 protein, light-harvesting complex proteins, chl a/b binding protein (Wang et al., 2008; Tada and Kashimura, 2009; Sobhanian et al., 2010; Yu et al., 2011; Cheng et al., 2015). OEE is key component of PSII that facilitates light induced oxidation of water during photosynthesis. There are three subunits of OEE, namely, OEE1, OEE2, and OEE3. Upregulation of OEE1 and OEE2 under the influence of salinity has been stated in halophytes *B. gymnorhiza*, *S. portulacastrum*, and *Festuca arundinacea* (Tada and Kashimura, 2009; Pawłowicz et al., 2018; Peng et al., 2019). Likewise, increased expression of phosphoenol pyruvate, D1/D2 protein, and OEE have been reported in *A. lagopoides* (Sobhanian et al., 2010). On the contrary, a decrease in the expression of OEE1 and PSII assembly/stabilization D1/D2 protein was documented in *P. tenuiflora* (Yu et al., 2011). Similarly, PSI-LHC complex protein CAB2 downregulation under salinity stress have been stated in halophytes *T. nitraria*, *Physcomitrella patens*, and *P. tenuiflora* (Wang et al., 2008; Yu et al., 2011; Cheng et al., 2015).

TABLE 13.2 Salinity and Alkalinity Induced Functional Metabolites in Halophytes.

Osmolytes		Function	Halophyte Species	Reference
Amino acid	Proline	Osmotic adjustment; ROS scavenging Protein stabilization; Protection of membrane and subcellular structure	*T. halophila* *Juncus acutus*, *Juncus maritimus*, *H. marinum*	Gong et al. (2005); Rozentsvet et al. (2017); Maršálová et al. (2016)
	Alanine, valine, serine	Osmoprotection	*S. portulacastrum*	Lokhande et al. (2013)
Quaternary ammonium compounds	GB	Osmotic adjustment; ROS scavenging stabilization of membrane and macromolecule	*K. sieversiana* *P. tenuiflora*, *S. maritima*	Yang et al. (2007); Yin et al. (2019); Benjamin et al. (2019)
	β-A betaine; Hydroxyproline	Osmoprotectant	*S. salsa*	Sui et al. (2010)
	Polyamine	Osmotic adjustment; Reduce lipid peroxidation; Ichannel modulation; activation of antioxidant enzymes	*P. coronopus*, *Frankenia pulverulenta*, *A. prostrata*, *S. portulacastrum*, *S. maritima*, *S. brachiata*	Bueno et al. (2020) Benjamin et al. (2019)
Polyols	Mannitol	Osmotic adjustment; macromolecular structure stabilization; ROS scavenger; Protection of photosynthetic apparatus	*A. prostrata P. coronopus*, *F. pulverulenta*, *T. halophila*, *P. pratensis*	Bueno et al. (2020); Gong et al. (2005); Hu et al. (2015)
	Glycolipids	Osmotic adjustment	*S. portulacastrum*	Nikalje et al. (2018)
	D-P	Osmoprotection	*H. salicornicum*	Panda et al. (2021)
	Myoinositol	Osmotic adjustment; Sustainability of cell division	*M. crystallinum* *H. salicornicum*	Panda et al. (2021); Bueno et al. (2020)

TABLE 13.2 *(Continued)*

Osmolytes		Function	Halophyte Species	Reference
Soluble s	Sucrose	Osmotic adjustment; Redox homeostasis; Signal molecule	*A. halimus*	Rozentsvet et al. (2017)
	Glucose; Fructose	Osmotic adjustment; C energy reserve; protecting lipid bilayer	*A. halimus*	Rozentsvet et al. (2017)
Complex Sugars	Raffinose	Osmotic adjustment	*T. halophila*, *P. pratensis*	Gong et al. (2005); Hu et al. (2015)
	Trehalose	Osmotic adjustment; Sof membrane and protein	*T. hispida*, *P. pratensis*	Hu et al. (2015); Wang et al. (2014)
TCA cycle intermediates	Malic acid	Generation of reducing power; Osmoregulation	*Nitraria tangutorum*, *S. corniculate* *Lotus* sp.	Pang et al. (2016); Sanchez et al. (2011)
	Succinic acid	Generation of reducing power; Osmoregulation	*N. tangutorum*	Ni et al. (2015)
	Asparagine; glutamine,	Generation of reducing power, Osmoregulation	A. lagopoides	Sobhanian et al. (2010)

TABLE 13.3 Key Proteins Reported to be Involved in Salt–Alkali Tolerance in Halophytes

S. No.	Protein category	Proteins	Molecular Function	Halophytic species	References
1.	Calvin cycle proteins	RuBisCo small and large subunits	Metal ion binding; RNA binding; Structural molecule activity	*S. europaea*	Wang et al. (2009)
				H. glomeratus	Wang et al. (2009)
				H. salicornicum	Panda et al. (2020)
				P. tenuiflora	Yu et al. (2011)
				T. nitraria	Cheng et al. (2015)
		Carbonic anhydrase	Bidirectional conversion of carbon dioxide (CO^2) and water (H^2O) into bicarbonate (HCO^{3-}) and protons (H^+)	*P. tenuiflora*	Yu et al. (2011)
				T. nitraria	Cheng et al. (2015)
		RuBisCO activase	Photosynthetic fixation of CO^2	*S. europaea*	Wang et al. (2009)
				H. glomeratus	Wang et al. (2015)
				H. salicornicum	Panda et al. (2020)
				P. tenuiflora	Yu et al. (2011)
				T. nitraria	Cheng et al. (2015)
		Ferredoxin-thioredoxin reductase	Catalyzes the light-dependent activation of several photosynthesis enzymes	*H. salicornicum*	Panda et al. (2020)
		PRK	ATP-dependent phosphorylation of ribulose 5-phosphate	*S. europaea*	Sengupta and Majumder, (2009)
				Porteresia coarctata	Wang et al. (2015)
				P. tenuiflora	Yu et al. (2011)
				T. nitraria	Cheng et al. (2015)
		Protein transketolase	Catalyzes the cleavage of carbon–carbon bonds to transfer two ketol carbon units from donor ketose sugars	*K. candel*	Wang et al. (2014b)

TABLE 13.3 *(Continued)*

S. No.	Protein category	Proteins	Molecular Function	Halophytic species	References
2.	PS-associated proteins	OEE protein	Key component of PSII that facilitates light-induced oxidation of water during photosynthesis	*B. gymnorhiza*	Tada and Kashimura, (2009)
				S. portulacastrum	Pawłowicz et al. (2018)
				F. arundinacea	Peng et al. (2019)
		Cytochrome b6-f complex iron–sulfur subunit	Metal ion binding	*P. patens*	Wang et al. (2008)
		ATP synthase	Catalytic activity; transporter activity	*B. gymnorhiza*	Tada and Kashimura, (2009)
		D1/D2 protein	Involved in the repair of PSII components during stress condition	*A. lagopoides*	Sobhanian et al. (2010)
		Light-harvesting complex proteins	Binding chlorophyll (Chl) and carotenoid molecules that absorb light and transfer the energy to the reaction center Chl of the PS	*P. tenuiflora*	Yu et al. (2011)
		Chl a/b binding protein	Metal ion binding; protein binding	*T. Nitraria*	Cheng et al. (2015)
		Phosphoenol pyruvate carboxylase	Catalytic activity	*A. lagopoides*	Sobhanian et al. (2010)
		Chlorophyll a/b binding protein 2	Captures and delivers excitation energy to PSs with which it is closely associated	*T. nitraria*	Wang et al. (2008)
				P. patens	Yu et al. (2011)
				P. tenuiflora	Cheng et al. (2015)
3.	Antioxidant proteins	APX	Antioxidant activity; Catalytic activity	*P. tenuiflora*	Yin et al. (2019)
				T. halophila	Pang et al. (2010)
				S. corniculata	Zang et al. (2021)

TABLE 13.3 *(Continued)*

S. No.	Protein category	Proteins	Molecular Function	Halophytic species	References
				H. salicornicum	Panda et al. (2020)
				T. Nitraria	Cheng et al. (2015)
		GST	Maintain redox homeostasis in the cell	*P. tenuiflora*	Yin et al. (2019)
				T. Nitraria	Cheng et al. (2015)
		MDHAR	Catalytic activity; Nucleotide binding	*P. tenuiflora*	Yin et al. (2019)
				T. Nitraria	Cheng et al. (2015)
		GR	Antioxidant activity; Catalytic activity	*P. tenuiflora*	Yin et al. (2019)
				T. halophila	Pang et al. (2010)
		Glyoxalase I	Increased the tolerance of methylglyoxal and maintained high GSH levels	*P. tenuiflora*	Yin et al. (2019)
		DHA	Detoxification of H^2O^2	*T. halophila*	Pang et al. (2010)
				H. salicornicum	Panda et al. (2020)
		PR5	Induced synthesis of ethylene under stress condition	*T. halophila*	Pang et al. (2010)
		Chloroplast 2-cys peroxiredoxin	Maintain redox homeostasis in the cell	*T. halophila*	Pang et al. (2010)
		Prx (Prx, Prx10, Prx13, Prx40, PrxCB);	Antioxidant activity; catalytic activity	*S. corniculata*	Zang et al. (2021)
		Glutathione S-transferase U17	Maintain redox homeostasis in the cell	*S. corniculata*	Zang et al. (2021)
		SOD	Antioxidant activity; Catalytic activity; Metal ion binding	*S. corniculata*	Zang et al. (2021)
				H. salicornicum	Panda et al. (2020)
		Peroxiredoxin Q	Antioxidant activity; Catalytic activity; Protein binding	*H. salicornicum*	Panda et al. (2020)

TABLE 13.3 *(Continued)*

S. No.	Protein category	Proteins	Molecular Function	Halophytic species	References
		POX	Antioxidant activity; Catalytic activity	*H. salicornicum*	Panda et al. (2020)
				T. Nitraria	Cheng et al. (2015)
		GPX	Antioxidant activity; Catalytic activity	*H. salicornicum*	Panda et al. (2020)
				T. Nitraria	Cheng et al. (2015)
		CAT	Detoxification of H^2O^2	*T. Nitraria*	Cheng et al. (2015)
		Oxygen-evolving enhancer protein 1	Responsible for water oxidation during photosynthesis	*T. Nitraria*	Cheng et al. (2015)
		Peroxiredoxin	Detoxification of H^2O^2	*T. Nitraria*	Cheng et al. (2015)
		Thioredoxin POX	Detoxification of H^2O^2	*T. Nitraria*	Cheng et al. (2015)
		Ferredoxin-NADP reductase	Make available reductants needed for CO^2 fixation during the dark reaction	*T. Nitraria*	Cheng et al. (2015)
		NADH ubiquinone oxidoreductase	Catalytic activity; Nucleotide binding	*T. Nitraria*	Cheng et al. (2015)
4.	Protein synthesis-related proteins	Membrane-associated 30 kDa protein	Protein destination and transport	*P. tenuiflora*	Yin et al. (2019)
		CSP	Binds and cleaves RNA, particularly in stem loops	*P. tenuiflora*	Yin et al. (2019)
		Regulator of chromosome condensation	Play a role as chromosomal passenger complex (CPC) implicated in all cell cycle phases	*P. tenuiflora*	Yin et al. (2019)
		RRM-containing protein	Protein synthesis	*T. halophila*	Pang et al. (2010)
		Translationally controlled tumor protein			
		40S ribosomal protein S7			

TABLE 13.3 *(Continued)*

S. No.	Protein category	Proteins	Molecular Function	Halophytic species	References
		40S ribosomal protein S24	Protein destination and transport		
		Eukaryotic translation initiation factor 3A	Protein synthesis		
		Ribosomal protein S15A			
		Elongation factor 1-α			
		Ribosomal protein S3			
		Cytosolic tRNA-Ala synthetase			
		50S ribosomal protein L3			
		Ribosomal protein S2	RNA binding; Structural molecule activity		
		60S ribosomal protein L5			
		60S ribosomal protein L13A			
		Elongation factor-Tu	Catalytic activity; Metal ion binding; Nucleotide binding;	*H. salicornicum*	Panda et al. (2020)
			RNA binding		
		Initiation factors 1 and 2	Nucleotide binding; Protein binding; RNA binding	*H. salicornicum*	Panda et al. (2020)
		Rpo cluster C and B	RNA binding; Structural molecule activity	*H. salicornicum*	Panda et al. (2020)
			49 60S acidic ribosomal protein P0-3 P57691 1.29 0.79 structural		

TABLE 13.3 *(Continued)*

S. No.	Protein category	Proteins	Molecular Function	Halophytic species	References
5.	Transport related proteins	Membrane-ATPase transporter	Catalytic activity; Metal ion binding; Nucleotide binding;	*H. salicornicum*	Panda et al. (2020)
				T. halophila	Vera-Estrella et al. (2015)
			Transporter activity	*A. tripolium*	Ramani et al. (2006)
				P. maritima	Vera-Estrella et al. (2015)
		Vacuolar ATPase transporter		*H. salicornicum*	Panda et al. (2020)
		Germin-like protein subfamily 2 member 4	Metal ion binding	*H. salicornicum*	Panda et al. (2020)
		V-Type proton ATPase subunit a3	Catalytic activity; protein binding; transporter activity	*H. salicornicum*	Panda et al. (2020)

13.8.3 ROS DEFENSE-RELATED PROTEINS IN HALOPHYTES

Salt–alkali stress causes ion imbalance, hyperosmotic stress, ROS production and oxidative damage. The proteomics data of Yin et al., (2019) showed that the abundances of APX, GST, MDHAR, and GR were enhanced in the leaves of the halophyte *P. tenuiflora* under NaHCO^3 stress. Moreover, the overexpression of glyoxalase I in *P. tenuiflora* increased the tolerance of methylglyoxal and maintained high GSH levels under salinity stress (Yin et al., 2019). In *Thellungiella*, relatively small changes in the levels of APX, a DHA, a PR5, a chloroplast 2-cys peroxiredoxin, and a GR were spotted under salt stress (Pang et al., 2010). ROS scavenging-related proteins such as APX1 and Prx (Prx, Prx10, Prx13, Prx40, and PrxCB) were highly accumulated while SOD was downregulated in *Salicornia corniculata* under both salt–alkali stress. Moreover, glutathione S-transferase U17 was significantly downregulated by alkali and upregulated by salt stress in *Suaeda corniculate* (Zang et al., 2021). Salinity stress significantly upregulates APX, peroxiredoxin Q, POX, GPX, SOD in *H. salicornicum* (Panda et al., 2020). The expression of MDHAR was significantly upregulated by salt stress in *H. salicornicum*. Higher activity of this enzyme led to increased ascorbate accumulation and efficient ROS detoxification in *H. salicornicum* under high salinity (Panda et al., 2020). Differential expression of APX, POD, MDHAR, GPX, CAT, SOD, POX precursor, oxygen-evolving enhancer protein 1, peroxiredoxin, thioredoxin POX, glutathione transferase, ferredoxin-NADP reductase, and NADH ubiquinone oxidoreductase (UOR) 24 kDa subunit implicate redox homeostasis in dicotyledonous halophyte *T. Nitraria* in response to salt stress (Cheng et al., 2015).

13.8.4 PROTEIN SYNTHESIS-RELATED PROTEINS

Salinity and alkali inhibit transcription and protein synthesis. In *P. tenuiflora*, three transcription-related proteins membrane-associated 30 kDa protein, chloroplast stem-loop binding protein (CSP), and regulator of chromosome condensation were decreased under alkali stress (Yin et al., 2019). RNA recognition motif (RRM)-containing protein, and translationally controlled tumor protein was downregulated in *T. halophila* by salt stress (Pang et al., 2010). Moreover, 40S ribosomal protein S7, 40S ribosomal protein S24, eukaryotic translation initiation factor 3A, and ribosomal protein S15A were upregulated and nine proteins elongation factor 1-α, elongation factor EF-2, ribosomal protein S3, cytosolic tRNA-Ala synthetase, 50S ribosomal protein L3, ribosomal protein S2, 60S ribosomal protein L5, and 60S ribosomal protein L13A were downregulated by salt in *T. halophila* (Pang et al., 2010). Salinity-induced significantly changed proteins of the ribosomal pathway including ribosomal protein components such as elongation factor-Tu (EF-Tu), initiation factor 1 and 2 (IF1, 2), Rpo cluster C and B in *H. salicornicum* (Panda et al., 2020).

13.8.5 TRANSPORT-RELATED PROTEINS

Salinity tolerance in plants may be achieved by either exclusion of toxic ions from the shoot and root or tolerance of the tissue to these ions and osmotic adjustments. Several ion transporters and channels are involved in the specific transport of particular ions for exclusion and compartmentalization of toxic ions. Halophytes use Na^+ as the osmoticum and they have the efficient adaptation mechanisms for excretion and/or compartmentalization of excess Na^+. Halophytes have strong NHX efficiency due to the induced activity of H^+-pumps in response to salinity. The membrane-ATPase transporter activity of both plasma membrane and vacuoles was significantly upregulated by high salinity in *H. salicornicum* (Panda et al., 2020). Likewise, an increase in the membrane-ATPase transporter activity has been reported in *T. halophila* leaves (Vera-Estrella et al., 2015). On the contrary, a decrease in the membrane-ATPase transporter activity has also been documented in *A. tripolium* and *Prunus maritima* (Ramani et al., 2006; Vera-Estrella et al., 2015). The abundance of vacuolar ATPase subunit B decreased under NaCl treatment in *S. portulacastrum* (Ding et al., 2022). Moreover, salinity significantly altered proteins having transporter activity includes germin-like protein subfamily 2 member 4, and V-type proton ATPase subunit a3 in *H. salicornicum* under salinity (Panda et al., 2020).

13.9 CONCLUSION AND FUTURE PROSPECTIVE

Soil salinization and alkalinization are the widespread abiotic stresses reducing the crop yield worldwide, endangering the global food supply. Under natural conditions, salt stresses are mostly from mixtures of neutral and alkali salts. Thus, holistic research on the salt tolerance mechanism requires a comparative study on salt–alkali stress. Plant growth and biomass yield are stated to reduce under salt–alkali stress. Salt–alkali stress negatively affect seed germination as they tend to alter the seed's physiological activity. Saline–alkaline stress also increases free radicals, which damage membranes and disrupt selective membrane permeability. Ion toxicity due to saline–alkaline stress is an important factor that limits plant growth, development, and yield. Salt–alkali stress limits the photosynthesis rate either by stomatal limitations or nonstomatal limitations. Plant tolerance to salt–alkali stress depends on diverse adaptive strategies, including regulation of morphology, physiology, biochemistry, and molecular genetics. Halophytes are native to saline environments and grow and flourish under high saline and alkaline habitats. Halophytes have been studied as the principal species for exploiting and recovering salt-alkalinized soil to address the growing problem of increased salinized and alkalized lands. Thus, understanding the underlying mechanisms of salinity and alkalinity tolerance in halophytes implicates noteworthy aspects of exploiting this pioneer halophyte in the future. In the present chapter tolerance ability of halophytes toward salt–alkali stress, and mechanistic insights at physiological, biochemical, metabolic, and proteomic levels have been documented for a better understanding of halophytes' salt–alkali tolerance mechanisms to develop tolerant crop varieties that can survive under high saline and high alkaline environment. Our study will add to the pool of information on the mechanisms of salinity/alkalinity tolerance in various halophytes, which may accord with the development of salinity/alkalinity resistant crops in the future.

ACKNOWLEDGMENTS

AK and MP are thankful to the Department of Science and Technology (DST), New Delhi, for providing financial support in the form of the INSPIRE Fellowship. Financial support of DST (Grant No. SERB/SB/SO/PS-14/2014), Government of India, New Delhi, to AKP, is duly acknowledged. This manuscript was assigned CSIR–CSMCRI Communication No. 068/2020.

KEYWORDS

- **alkalinity**
- **antioxidant**
- **halophytes**
- **ion transporters**
- **malic acid**
- **manitol**
- **myo-inositol**
- **metabolites**
- **polyamine**
- **oxidative stress**
- **salinity**
- **trehalose**

REFERENCES

Abobatta, W. F. Plant Responses and Tolerance to Extreme Salinity: Learning from Halophyte Tolerance to Extreme Salinity. In *Salt and Drought Stress Tolerance in Plants, Signaling and Communication in Plants*, Hasanuzzaman, M., Tanveer, M., Eds.; Springer: Switzerland, AG, 2020; pp 177–210.

Adolf, V. I.; Jacobsen, S. E.; Shabala, S. Salt Tolerance Mechanisms in Quinoa (*Chenopodium quinoa* Willd.). *Environ. Exp. Bot.* **2013,** *92,* 43–54.

Al-Shamsi, N.; Hussain, M. I.; El-Keblawy, A. Physiological Responses of the Xerohalophyte *Suaeda Vermiculata* to Salinity in its Hyper-Arid Environment. *Flora Morphol. Distrib. Funct. Ecol. Plants* **2020,** *273,* 151705.

Arbelet-bonnin, D.; Blasselle, C.; Palm, E. R.; Redwan, M.; Ponnaiah, M.; Laurenti, P.; et al. Metabolism Regulation During Salt Exposure in the Halophyte *Cakile maritima. Environ. Exp. Bot.* **2020,** *177,* 104075.

Arif, Y.; Singh, P.; Siddiqui, H.; Bajguz, A.; Hayat, S. Salinity Induced Physiological and Biochemical Changes in Plants : An Omic Approach Towards Salt Stress Tolerance. *Plant Physiol. Biochem.* **2020,** *156,* 64–77.

Asada, K. Production and Scavenging of Reactive Oxygen Species in Chloroplasts and Their Functions. *Plant Physiol.* **2006,** *141,* 391–396.

Ashraf, M. P. J. C.; Harris, P. J. C. Potential Biochemical Indicators of Salinity Tolerance in Plants. *Plant Sci.* **2004,** *166,* 3–16.

Aslam, R.; Bostan, N.; Maria, M.; Safdar, W. A Critical Review on Halophytes : Salt Tolerant Plants. *J. Med. Plants Res.* **2011,** *5,* 7108–7118.

Asrar, H.; Hussain, T.; Hadi, S. M. S.; Gul, B.; Nielsen, B. L.; Khan, M. A. Salinity Induced Changes in Light Harvesting and Carbon Assimilating Complexes of *Desmostachya bipinnata* (L.) Staph. *Environ. Exp. Bot.* **2017,** *135,* 86-95.

Barhoumi, Z.; Djebali, W.; Chaïbi, W.; Abdelly, C.; Smaoui, A. Salt Impact on Photosynthesis and Leaf Ultrastructure of *Aeluropus littoralis*. *J. Plant Res.* **2007,** *120,* 529–537.

Bazihizina, N.; Colmer, T. D.; Cuin, T. A.; Mancuso, S.; Shabala, S. Friend or Foe? Chloride Patterning in Halophytes. *Trends Plant Sci.* **2019,** *24,* 142–151.

Behr, J. H.; Bouchereau, A.; Berardocco, S.; Seal, C. E.; Flowers, T. J.; Zörb, C. Metabolic and Physiological Adjustment of *Suaeda maritima* to Combined Salinity and Hypoxia. *Ann. Bot.* **2017,** *119,* 965–976.

Benjamin, J. J.; Lucini, L.; Jothiramshekar, S.; Parida, A. Metabolomic Insights into the Mechanisms Underlying Tolerance to Salinity in Different Halophytes. *Plant Physiol. Biochem.* **2019,** *135,* 528–545.

Betzen, B. M.; Smart, C. M.; Maricle, K. L.; MariCle, B. R. Effects of Increasing Salinity on Photosynthesis and Plant Water Potential in Kansas Salt Marsh Species. *Trans. Kans. Acad. Sci.* **2019,** *122,* 49–58.

Bueno, M.; Lendínez, M. L.; Calero, J.; Del Pilar Cordovilla, M. Salinity Responses of Three Halophytes from Inland Saltmarshes of Jaén (Southern Spain). *Flora Morphol. Distrib. Funct. Ecol. Plants* **2020,** *266,* 151589.

Burundukova, O. L.; Shuyskaya, E. V.; Rakhmankulova, Z. F.; Burkovskaya, E. V.; Chubar, E. V.; Gismatullina, L. G.; et al. *Kali komarovii* (Amaranthaceae) is a Xero-Halophyte with Facultative NADP-ME Subtype of C^4 Photosynthesis. *Flora - Morphol. Distrib. Funct. Ecol. Plants* **2017,** *227,* 25–35.

Cassaniti, C.; Romano, D.; Hop, M. E. C. M.; Flowers, T. J. Growing Floricultural Crops with Brackish Water. *Environ. Exp. Bot.* **2013,** *92,* 165–175.

Cheeseman, J. M. Tansley Review The evolution of Halophytes, Glycophytes and Crops, and its Implications for Food Security Under Saline Conditions. *New Phytol.* **2015,** *206,* 557–570.

Cheng, T.; Chen, J.; Zhang, Ji.; Shi, S.; Zhou, Y.; Lu, L.; et al. Physiological and Proteomic Analysis of Leaves from the Halophyte *Tangut nitraria* Reveals Diverse Response Pathways Critical for High Salinity. *Front. Plant Sci.* **2015,** *6,* 1–13.

Cuadros-Inostroza, A.; Ruíz-Lara, S.; González, E.; Eckardt, A.; Willmitzer, L.; Peña-Cortés, H. GC-MS Metabolic Profiling of *Cabernet Sauvignon* and *Merlot* Cultivars During Grapevine Berry Development and Network Analysis Reveals a Stage- and Cultivar-Dependent Connectivity of Primary Metabolites. *Metabolomics* **2016,** *12,* 1–17.

Deinlein, U.; Stephan, A. B.; Horie, T.; Luo, W.; Xu, G.; Schroeder, J. I. Plant Salt-Tolerance Mechanisms. *Trends Plant Sci.* **2014,** *19,* 371–379.

Demidchik, V. ROS-activated Ion Channels in Plants: Biophysical Characteristics, Physiological Functions and Molecular Nature. *Int. J. Mol. Sci.* **2018,** *19,* 17–21.

Ding, G.; Yang, Q.; Ruan, X.; Si, T.; Yuan, B.; Zheng, W.; Xie, Q.; Souleymane, O. A.; Wang, X. Proteomics Analysis of the Effects for Different Salt Ions in Leaves of True Halophyte *Sesuvium portulacastrum*. *Plant Physiol. Biochem.* **2022,** *170,* 234–248.

Edwards, E. J.; Donoghue, M. J. Is It Easy to Move and Easy to Evolve? Evolutionary Accessibility and Adaptation. *J. Exp. Bot.* **2013,** *64,* 4047–4052.

Feng, Z.; Ding, C.; Li, W.; Wang, D.; Cui, D. Applications of Metabolomics in the Research of Soybean Plant Under Abiotic Stress. *Food Chem.* **2020,** *310,* 125914.

Finazzi, G.; Petroutsos, D.; Tomizioli, M.; Flori, S.; Sautron, E.; Villanova, V.; et al. Ions Channels/Transporters and Chloroplast Regulation. *Cell Calcium* **2015,** *58,* 86–97.

Flowers, T. J.; Colmer, T. D. Plant Salt Tolerance: Adaptations in Halophytes. *Ann. Bot.* **2015,** *115,* 327–331.

Flowers, T. J.; Muscolo, A. Halophytes in a Changing World. *AoB Plants* **2015,** *7,* 1–5.

Gao, Z.; Han, J.; Mu, C.; Lin, J.; Li, X.; Lin, L.; et al. Effects of Saline and Alkaline Stresses on Growth and Physiological Changes in Oat (*Avena Sativa* L.) Seedlings. *Not. Bot. Horti Agrobot. Cluj-Napoca* **2014,** *42,* 357–362.

Geilfus, C. M. Review on the Significance of Chlorine for Crop Yield and Quality. *Plant Sci.* **2018,** *270,* 114–122.

Gong, B.; Wen, D.; VandenLangenberg, K.; Wei, M.; Yang, F.; Shi, Q.; et al. Comparative Effects of NaCl and $NaHCO^3$ Stress on Photosynthetic Parameters, Nutrient Metabolism, and the Antioxidant System in Tomato Leaves. *Sci. Hortic. (Amsterdam)* **2013,** *157,* 1–12.

Gong, Q.; Li, P.; Ma, S.; Indu Rupassara, S.; Bohnert, H. J. Salinity Stress Adaptation Competence in the Extremophile *Thellungiella halophila* in Comparison with its Relative *Arabidopsis thaliana. Plant J.* **2005,** *44,* 826–839.

Goussi, R.; Manaa, A.; Derbali, W.; Cantamessa, S.; Abdelly, C.; Barbato, R. Comparative Analysis of Salt Stress, Duration and Intensity, on the Chloroplast Ultrastructure and Photosynthetic Apparatus in *Thellungiella Salsuginea. J. Photochem. Photobiol. B Biol.* **2018,** *183,* 275–287.

Guo, C. Y.; Wang, X. Z.; Chen, L.; Ma, L. N.; Wang, R. Z. Physiological and Biochemical Responses to Saline–Alkaline Stress in Two Halophytic Grass Species with Different Photosynthetic Pathways. *Photosynthetica* **2015a,** *53,* 128–135.

Guo, L. Q.; Shi, D. C.; Wang, D. L. The key Physiological Response to Alkali Stress by the Alkali-Resistant Halophyte *Puccinellia tenuiflora* is the Accumulation of Large Quantities of Organic Acids and into the Rhyzosphere. *J. Agron. Crop. Sci.* **2010a,** *196,* 123–135.

Guo, R.; Shi, L.; Ding, X.; Hu, Y.; Tian, S.; Yan, D.; et al. Effects of Saline and Alkaline Stress on Germination, Seedling Growth, and Ion Balance in Wheat. *Agron. J.* **2010b,** *102,* 1252–1260.

Guo, R.; Shi, L.; Yan, C.; Zhong, X.; Gu, F.; Liu, Q.; et al. Ionomic and Metabolic Responses to Neutral Salt or Alkaline Salt Stresses in Maize (*Zea mays* L.) Seedlings. *BMC Plant Biol.* **2017,** *17,* 41.

Guo, R.; Yang, Z.; Li, F.; Yan, C.; Zhong, X.; Liu, Q.; et al. Comparative Metabolic Responses and Adaptive Strategies of Wheat (*Triticum aestivum*) to Salt and Alkali Stress. *BMC Plant Biol.* **2015b,** *15,* 1–13.

Gupta, B.; Huang, B. Mechanism of Salinity Tolerance in Plants: Physiological, Biochemical, and Molecular Characterization. *Inter. J. Genom.* **2014**

Hanin, M.; Ebel, C.; Ngom, M.; Laplaze, L.; Masmoudi, K. New Insights on Plant Salt Tolerance Mechanisms and their Potential Use for Breeding. *Front. Plant Sci.* **2016,** *7,* 1–17.

Hayat, K.; Zhou, Y.; Menhas, S.; Bundschuh, J.; Hayat, S.; Ullah, A.; et al. *Pennisetum giganteum:* An Emerging Salt Accumulating/Tolerant Non-conventional Crop for Sustainable Saline Agriculture and Simultaneous Phytoremediation. *Environ. Pollut.* **2020,** *265,* 114876.

Hessini, K.; Issaoui, K.; Ferchichi, S.; Saif, T.; Abdelly, C.; Siddique, K. H. M.; et al. Interactive Effects of Salinity and Nitrogen forms on Plant Growth, Photosynthesis and Osmotic Adjustment in Maize. *Plant Physiol. Biochem.* **2019,** *139,* 171–178.

Himabindu, Y.; Chakradhar, T.; Reddy, M. C.; Kanygin, A.; Redding, K. E.; Chandrasekhar, T. Salt-Tolerant Genes from Halophytes are Potential Key Players of Salt Tolerance in Glycophytes. *Environ. Exp Bot.* **2016,** *124,* 39–63.

Hniličková, H.; Hnilička, F.; Martinková, J.; Kraus, K. Effects of Salt Stress on Water Status, Photosynthesis and Chlorophyll Fluorescence of Rocket. *Plant, Soil Environ.* **2017,** *63,* 362–367.

Hryvusevich, P.; Navaselsky, I.; Talkachova, Y.; Straltsova, D.; Keisham, M.; Viatoshkin, A.; et al. Sodium Influx and Potassium Efflux Currents in Sunflower Root Cells Under High Salinity. *Front. Plant Sci.* **2021,** *11.*

Hu, L.; Zhang, P.; Jiang, Y.; Fu, J. Metabolomic Analysis Revealed Differential Adaptation to Salinity and Alkalinity Stress in kentucky Bluegrass (*Poa pratensis*). *Plant Mol. Biol. Rep.* **2015,** *33,* 56–68.

Hu, L.; Zhang, P.; Jiang, Y.; Fu, J. Metabolomic Analysis Revealed Differential Adaptation to Salinity and Alkalinity Stress in *Kentucky bluegrass* (*Poa pratensis*). *Plant Mol. Biol. Rep.* **2015,** *33,* 56–68.

Hu, T.; Agriculture, S.; Botanical, W.; Yi, H. Stomatal and Metabolic Limitations to Photosynthesis Resulting from Nacl Stress in Perennial Ryegrass Genotypes Differing in Salt Tolerance. *J. Am. Soc. Hort. Sci.* **2013,** *138,* 350–357.

Isayenkov, Stanislav V.; Frans, J. M. Maathuis. Plant Salinity Stress: Many Unanswered Questions Remain. *Front. Plant Sci.* **2019,** *10,* 80.

Janz, D.; Behnke, K.; Schnitzler, J.-P.; Kanawati, B.; Schmitt-Kopplin, P.; Polle, A. Pathway Analysis of the Transcriptome and Metabolome of Salt Sensitive and Tolerant Poplar Species Reveals Evolutionary Adaption of Stress Tolerance Mechanisms. *BMC Plant Biol.* **2010,** *10,* 150.

Jia, X.; Zhu, Y.; Zhang, R.; Zhu, Z.; Zhao, T.; Cheng, L.; et al. Ionomic and Metabolomic Analyses Reveal the Resistance Response Mechanism to Saline-Alkali Stress in *Malus halliana* Seedlings. *Plant Physiol. Biochem.* **2020,** *147,* 77–90.

Kumar, A.; Kumar, A.; Lata, C.; Kumar, S.; Mangalassery, S.; Singh, J. P.; et al. Effect of Salinity and Alkalinity on Responses of Halophytic Grasses *Sporobolus marginatus* and *Urochondra setulosa. Indian J. Agric. Sci.* **2018,** *88,* 1296–1304.

Kumar, A.; Mann, A.; Lata, C.; Kumar, N.; Sharma, P. Salinity-Induced Physiological and Molecular Responses of Halophytes. In *Research Developments in Saline Agriculture*; Dagar, J. C., Ed.; Springer Nature Singapore, 2019; pp 331–356.

Kumari, A.; Parida, A. K. Metabolomics and Network Analysis Reveal the Potential Metabolites and Biological Pathways Involved in Salinity Tolerance of the Halophyte *Salvadora persica. Environ. Exp. Bot.* **2018,** *148,* 85–99.

Kumari, A.; Das, P.; Parida, A. K.; Agarwal, P. K. Proteomics, Metabolomics and Ionomics Perspectives of Salinity Tolerance in Halophytes. *Front. Plant Sci.* **2015,** *6,* 1–20.

Lee, S. J.; Jeong, E. M.; Ki, A. Y.; Oh, K. S.; Kwon, J.; Jeong, J. H.; et al. Oxidative Defense Metabolites Induced by Salinity Stress in Roots of *Salicornia herbacea. J. Plant Physiol.* **2016,** *206,* 133–142.

Lehmann, S.; Funck, D.; Szabados, L.; Rentsch, D. Proline Metabolism and Transport in Plant Development. *Amino Acids* **2010,** *39,* 949–962.

Li, C.; Fang, B.; Yang, C.; Shi, D.; Wang, D. Effects of Various Salt-Alkaline Mixed Stresses on the State of Mineral Elements in Nutrient Solutions and the Growth of Alkali Resistant Halophyte *Chloris virgata. J. Plant Nutr.* **2009,** *32,* 1137–1147.

Li, Q.; Ma, C.; Tai, H.; Qiu, H.; Yang, A. Comparative Transcriptome Analysis of Two Rice Genotypes Differing in their Tolerance to Saline–Alkaline Stress. *PLoS ONE* **2020,** *15,* 1–15.

Li, W.; Zhang, C.; Lu, Q.; Wen, X.; Lu, C. The Combined Effect of Salt Stress and Heat Shock on Proteome Profiling in *Suaeda salsa. J. Plant Physiol.* **2011,** *168,* 1743–1752.

Lokhande, V. H.; Suprasanna, P. Prospects of Halophytes in Understanding and Managing Abiotic Stress Tolerance. In *Environmental Adaptations and Stress Tolerance of Plants in the Era of Climate Change*; Ahmad, P., Prasad, M. N. V., Eds.; Springer: New York, 2012; pp 29–56.

Lokhande, V. H.; Gor, B. K.; Desai, N. S.; Nikam, T. D.; Suprasanna, P. *Sesuvium portulacastrum,* a Plant for Drought, Salt Stress, Sand Fixation, Food and Phytoremediation. A Review. *Agron. Sustain. Dev.* **2013,** *33,* 329–348.

Lu, X.; Chen, Q.; Cui, X.; Abozeid, A.; Liu, Y.; Liu, J.; et al. Comparative Metabolomics of Two Saline–Alkali Tolerant Plants *Suaeda Glauca* and *Puccinellia tenuiflora* Based on GC-MS Platform. *Nat. Prod. Res.* **2019,** 1–4.

Lu, X.; Chen, Q.; Cui, X.; Abozeid, A.; Liu, Y.; Liu, J.; et al. Comparative Metabolomics of Two Saline–Alkali Tolerant Plants *Suaeda Glauca* and *Puccinellia tenuiflora* Based on GC-MS Platform. *Nat. Prod. Res.* **2021,** *35,* 499–502.

Lv, W. T.; Lin, B.; Zhang, M.; Hua, X. J. Proline Accumulation is Inhibitory to Arabidopsis Seedlings During Heat Stresss. *Plant Physiol.* **2011,** *156,* 1921–1933.

Maršálová, L.; Vítámvás, P.; Hynek, R.; Prášil, I. T.; Kosová, K. Proteomic Response of *Hordeum vulgare* cv. Tadmor and *Hordeum marinum* to Salinity Stress: Similarities and Differences Between a Glycophyte and a Halophyte. *Front. Plant Sci.* **2016,** *7,* 1–19.

Mateos-Naranjo, E.; Redondo-Gómez, S.; Álvarez, R.; Cambrollé, J.; Gandullo, J.; Figueroa, M. E. Synergic Effect of Salinity and CO^2 Enrichment on Growth and Photosynthetic Responses of the Invasive Cordgrass *Spartina densiflora. J. Exp. Bot.* **2010,** *61,* 1643–1654.

Meychik, N. R.; Nikolaeva, Y. I.; Yermakov, I. P. Physiological Response of Halophyte *Suaeda altissima* L. Pall. and Glycophyte *Spinacia oleracea* L. to Salinity. *Am. J. Plant Sci.* **2013,** *04,* 427–435.

Mir, M. A.; John, R.; Alyemeni, M. N.; Alam, P.; Ahmad, P. Jasmonic Acid Ameliorates Alkaline Stress by Improving Growth Performance, Ascorbate Glutathione Cycle and Glyoxylase System in Maize Seedlings. *Sci. Rep.* **2018,** *8,* 2831.

Munns, R.; Tester, M. Mechanisms of Salinity Tolerance. *Annu. Rev. Plant Biol.* **2008,** *59,* 651–681.

Munns, R.; Day, D. A.; Fricke, W.; Watt, M.; Arsova, B.; Barkla, B. J.; et al. Energy Costs of Salt Tolerance in Crop Plants. *New Phytol.* **2020,** *225,* 1072–1090.

Nikalje, G. C.; Variyar, P. S.; Joshi, M. V.; Nikam, T. D.; Suprasanna, P. Temporal and Spatial Changes in Ion Homeostasis and Accumulation of Flavanoids and Glycolipid in a Halophyte *Sesuvium portulacastrum* (L.) L. *PLoS One* **2018,** *13* (4), e0193394. DOI: 10.1371/journal.pone.0193394.

Pan, T.; Liu, M.; Kreslavski, V. D.; Zharmukhamedov, S. K.; Nie, C.; Yu, M.; et al. Technology Non-Stomatal Limitation of Photosynthesis by Soil Salinity. *Crit. Rev. Environ. Sci. Technol.* **2020,** *3,* 1–35.

Panda, A.; Rangani, J.; Parida, A. K. Comprehensive Proteomic Analysis Revealing Multifaceted Regulatory Network of the Xero-Halophyte *Haloxylon salicornicum* Involved in Salt Tolerance. *J. Biotechnol.* **2020,** *324,* 143–161.

Pandit, P. R.; Fulekar, M. H.; Karuna, M. S. L. Effect of Salinity Stress on Growth, Lipid Productivity, Fatty Acid Composition, and Biodiesel Properties in *Acutodesmus obliquus* and *Chlorella vulgaris*. *Environ. Sci. Pollut. Res.* **2017,** *24,* 13437–13451.

Pang, Q. Y.; Chen, S. X.; Dai, S. J.; Chen, Y. Z.; Wang, Y.; Yan, X. F. Comparative Proteomics of Salt Tolerance in *Arabidopsis thaliana* and *Thellungiella halophila*. *J. Proteome Res.* **2010,** *9,* 2584–2599.

Pang, Q.; Zhang, A.; Zang, W.; Wei, L.; Yan, X. Integrated Proteomics and Metabolomics for Dissecting the Mechanism of Global Responses to Salt and Alkali Stress in *Suaeda corniculata*. *Plant Soil* **2016,** *402,* 379–394.

Parida, A. K.; Jha, B. Antioxidative Defense Potential to Salinity in the Euhalophyte *Salicornia brachiata*. *J. Plant Growth Regul.* **2010,** *29,* 137–148.

Parida, A. K.; Panda, A.; Rangani, J. Metabolomics-Guided Elucidation of Abiotic Stress Tolerance Mechanisms in Plants. In *Plant Metabolites and Regulation Under Environmental Stress*; Ahmad, P., Ahanger, M. A., Singh, V. P., Tripathi, D. K., Alam, P., Alyemeni, M. N., Eds.; Elsevier Inc., 2018; pp 89–131.

Park, K. W.; An, J. Y.; Lee, H. J.; Son, D.; Sohn, Y. G.; Kim, C. G.; Lee, J. J. The Growth and Accumulation of Osmotic Solutes of the Halophyte Common Glasswort (*Salicornia europaea*) Under Salinity Conditions. *J. Aquat. Plant Manag.* **2013,** *51,* 103–108.

Patel, M.; Parida, A. K. Salinity Alleviates the Arsenic Toxicity in the Facultative Halophyte *Salvadora persica* L. by the Modulations of Physiological, Biochemical, and ROS Scavenging Attributes. *J. Hazard. Mater.* **2020,** *401,* 123368.

Pawłowicz, I.; Waśkiewicz, A.; Perlikowski, D.; Rapacz, M.; Ratajczak, D.; Kosmala, A. Remodeling of Chloroplast Proteome Under Salinity Affects Salt Tolerance of *Festuca arundinacea*. *Photosynth. Res.* **2018,** *137,* 475–492.

Peng, C.; Chang, L.; Yang, Q.; Tong, Z.; Wang, D.; Tan, Y.; et al. Comparative Physiological and Proteomic Analyses of the Chloroplasts in Halophyte *Sesuvium portulacastrum* Under Differential Salt Conditions. *J. Plant Physiol.* **2019,** *232,* 141–150.

Petropoulos, S. A.; Levizou, E.; Ntatsi, G.; Fernandes, Â.; Petrotos, K.; Akoumianakis, K.; et al. Salinity Effect on Nutritional Value, Chemical Composition and Bioactive Compounds Content of *Cichorium spinosum* L. *Food Chem.* **2017,** *214,* 129–136.

Ramani, B.; Reeck, T.; Debez, A.; Stelzer, R.; Huchzermeyer, B.; Schmidt, A.; Papenbrock, J. *Aster tripolium* L. and *Sesuvium portulacastrum* L.: Two Halophytes, Two Strategies to Survive in Saline Habitats. *Plant Physiol. Biochem.* **2006,** *44,* 395–408.

Ranganayakulu, S. G.; Veeranagamallaiah, G.; Sudhakar, C. Effect of Salt Stress on Osmolyte Accumulation in Two Groundnut Cultivars (*Arachis hypogaea* L.) with Contrasting Salt Tolerance. *Afr. J. Plant Sci.* **2013,** *7,* 586–592.

Reddy, M. P.; Shah, M. T.; Patolia, J. S. *Salvadora persica*, A Potential Species for Industrial Oil Production in Semiarid Saline and Alkali Soils. *Ind. Crops Prod.* **2008,** *28,* 273–278.

Reich, M.; Aghajanzadeh, T.; Helm, J.; Parmar, S.; Hawkesford, M. J.; De Kok, L. J. Chloride and Sulfate Salinity Differently Affect Biomass, Mineral Nutrient Composition and Expression of Sulfate Transport and Assimilation Genes in *Brassica rapa*. *Plant Soil* **2017,** *411,* 319–332.

Rozema, J.; Schat, H. Salt Tolerance of Halophytes, Research Questions Reviewed in the Perspective of Saline Agriculture. *Environ. Exp. Bot.* **2012,** *92,* 83–95.

Rozema, J.; Muscolo, A.; Flowers, T. Sustainable Cultivation and Exploitation of Halophyte Crops in a Salinising World. *Environ. Exp. Bot.* **2013,** *92,* 1–3.

Rozentsvet, O. A.; Nesterov, V. N.; Bogdanova, E. S. Structural, Physiological, and Biochemical Aspects of Salinity Tolerance of Halophytes. *Russ. J. Plant Physiol.* **2017,** *64,* 464–477.

Saddhe, A. A.; Manuka, R.; Nikalje, G. C.; Penna, S. Halophytes as a Potential Resource for Phytodesalination. *Handb. Halophytes* **2020,** 1–21.

Samiullah, A.; Bano, A. Evaluation of Oxidative Stress Tolerance, Growth Responses, Osmolytes and Phytohormonal Estimation in Selected Halophytes. *IPCBEE* **2011,** *16.*

Sanchez, D. H.; Siahpoosh, M. R.; Roessner, U.; Udvardi, M.; Kopka, J. Plant Metabolomics Reveals Conserved and Divergent Metabolic Responses to Salinity. *Physiol. Plant.* **2007,** *132,* 209–219.

Saslis-Lagoudakis, C. H.; Hua, X.; Bui, E.; Moray, C.; Bromham, L. Predicting Species' Tolerance to Salinity and Alkalinity Using Distribution Data and Geochemical Modelling: A Case Study Using Australian Grasses. *Ann. Bot.* **2014,** *115,* 343–351.

Sengupta, S.; Majumder, A. L. Insight into the Salt Tolerance Factors of a Wild Halophytic Rice, *Porteresia coarctata*: A Physiological and Proteomic Approach. *Planta* **2009,** *229,* 911–929.

Sengupta, S.; Majumder, A. L. *Porteresia coarctata* (Roxb.) Tateoka, a Wild Rice: A Potential Model for Studying Salt-Stress Biology in Rice. *Plant Cell Environ.* **2010,** *33,* 526–542.

Sengupta, S.; Mukherjee, S.; Goswami, L. Manipulation of Inositol Metabolism for Improved Plant Survival Under Stress : A "Network Engineering Approach." *J. Plant Biochem. Biotechnol.* **2012,** *21,* 15–23.

Shabala, L.; Mackay, A.; Tian, Y.; Jacobsen, S. E.; Zhou, D.; Shabala, S. Oxidative Stress Protection and Stomatal Patterning as Components of Salinity Tolerance Mechanism in Quinoa (*Chenopodium quinoa*). *Physiol. Plant.* **2012,** *146,* 26–38.

Shabala, S.; Chen, G.; Chen, Z. H.; Pottosin, I. The Energy Cost of the Tonoplast Futile Sodium Leak. *New Phytol.* **2020,** *225,* 1105–1110.

Sharma, D. K.; Shah, K. R.; Dave, R. S. A Review on the Pharmagnostic Evaluation of Meswak, *Salvadora persica. Biosci. Biotechnol. Res. Commun.* **2018,** *11,* 734–742.

Shuyskaya, E. V.; Rakhmankulova, Z. F.; Toderich, K. N. Role of Proline and Potassium in Adaptation to Salinity in Different Types of Halophytes. In *Handbook of Halophytes*; Springer International Publishing, 2020; pp 1–23.

Sobhanian, H.; Motamed, N.; Jazii, F. R.; Nakamura, T.; Komatsu, S. Salt Stress Induced Differential Proteome and Metabolome Response in the Shoots of *Aeluropus lagopoides* (Poaceae), a Halophyte C^4 Plant. *J. Proteome Res.* **2010,** *9,* 2882–2897.

Srivastava, N. Reclamation of Saline and Sodic Soil Through Phytoremediation. In *Environmental Concerns and Sustainable Development: Volume 2: Biodiversity, Soil and Waste Management*; Shukla, V., Kumar, N., Eds.; Springer: Singapore, 2020; pp 279–306.

Suo, J.; Zhang, H.; Zhao, Q.; Zhang, N.; Zhang, Y.; Li, Y.; et al. Na^2CO^3-Responsive Photosynthetic and ROS Scavenging Mechanisms in Chloroplasts of Alkaligrass Revealed by Phosphoproteomics. *Genom., Proteom. Bioinforma.* **2020,** *18,* 271–288.

Sureshan, K. M.; Murakami, T.; Watanabe, Y. Total Syntheses of Cyclitol Based Natural Products from Myo-Inositol: Brahol and Pinpollitol. *Tetrahedron* **2009,** *65,* 3998–4006.

Tabot, P. T.; Adams, J. B. Salt Secretion, Proline Accumulation and Increased Branching Confer Tolerance to Drought and Salinity in the Endemic Halophyte *Limonium linifolium. S. Afr. J. Bot.* **2014,** *94,* 64–73.

Tada, Y.; Kashimura, T. Proteomic Analysis of Salt-Responsive Proteins in the Mangrove Plant, *Bruguiera gymnorhiza. Plant Cell Physiol.* **2009,** *50,* 439–446.

van Zelm, E.; Zhang, Y.; Testerink, C. Salt Tolerance Mechanisms of Plants. *Annu. Rev. Plant Biol.* **2020,** *71,* 403–433.

Vera-Estrella, R.; Barkla, B. J.; Garcia-Ramirez, L.; Pantoja, O. Salt Stress in *Thellungiella halophila* Activates Na^+ Transport Mechanisms Required for Salinity Tolerance. *Plant Physiol.* **2015,** *139,* 1507–1517.

Wang, Gao, C.; Wang, L.; Zheng, L.; Yang, C.; Wang, Y. Comprehensive Transcriptional Profiling of $NaHCO^3$-Stressed *Tamarix hispida* Roots Reveals Networks of Responsive Genes. *Plant Mol. Biol.* **2014a,** *84,* 145–157.

Wang, H.; Xiang, Y.; Li, L. H.; Bhanbhro, N.; Yang, C. W.; Zhang, Z. Photosynthetic Response and Transcriptomic Profiling Provide Insights into the Alkali Tolerance of Clone Halophyte *Leymus chinensis. Photosynthetica* **2020a,** *58,* 780–789.

Wang, J.; Meng, Y.; Li, B.; Ma, X.; Lai, Y.; Si, E.; et al. Physiological and Proteomic Analyses of Salt Stress Response in the Halophyte *Halogeton glomeratus. Plant, Cell Environ.* **2015,** *38,* 655–669.

Wang, L. S.; Li, W. L.; Qi, X. W.; Ma, L.; Wu, W. L. Physiological and Proteomic Response of *Limonium bicolor* to Salinity. *Russ. J. Plant Physiol.* **2017,** *64,* 349–360.

Wang, L.; Liu, X.; Liang, M.; Tan, F.; Liang, W.; Chen, Y.; et al. Proteomic Analysis of Salt-Responsive Proteins in the Leaves of Mangrove *Kandelia candel* During Short-Term Stress. *PLoS One* **2014b,** *9.*

Wang, X.; Fan, P.; Song, H.; Chen, X.; Li, X.; Li, Y. Comparative Proteomic Analysis of Differentially Expressed Proteins in Shoots of *Salicornia europaea* Under Different Salinity. *J. Proteome Res.* **2009,** *8,* 3331–3345.

Wang, X.; Yang, P.; Gao, Q.; Liu, X.; Kuang, T.; Shen, S.; et al. Proteomic Analysis of the Response to High-Salinity Stress in *Physcomitrella Patens. Planta* **2008,** *228,* 167–177.

Williamson, J. D.; Jennings, D. B.; Guo, W. W.; Pharr, D. M.; Ehrenshaft, M. Sugar Alcohols, Salt Stress, and Fungal Resistance: Polyols-Multifunctional Plant Protection? *J. Am. Soc. Hortic. Sci.* **2002,** *127,* 467–473.

Yang, C. W.; Wang, P.; Li, C. Y.; Shi, D. C.; Wang, D. L. Comparison of Effects of Salt and Alkali Stresses on the Growth and Photosynthesis of Wheat. *Photosynthetica* **2008,** *46,* 107–114.

Yang, C.; Chong, J.; Li, C.; Kim, C.; Shi, D.; Wang, D. Osmotic Adjustment and Ion Balance Traits of an Alkali Resistant Halophyte *Kochia sieversiana* During Adaptation to Salt and Alkali Conditions. *Plant Soil* **2007,** *294,* 263–276.

Yang, C.-W.; Xu, H.-H.; Wang, L.-L.; Liu, J.; Shi, D.-C.; Wang, D. L. Comparative Effects of Salt-Stress and Alkali-Stress on the Growth, Photosynthesis, Solute Accumulation, and Ion Balance of Barley Plants. *Photosynthetica* **2009,** *47,* 79–86.

Yang, D. Shuang; Zhang, J.; Li, M. Xia; Shi, L. Xuan. Metabolomics Analysis Reveals the Salt-Tolerant Mechanism in *Glycine Soja. J. Plant Growth Regul.* **2017,** *36,* 460–471.

Ye, T.; Wang, Y.; Feng, Y. Q.; Chan, Z. Physiological and Metabolomic Responses of Bermudagrass (*Cynodon dactylon*) to Alkali Stress. *Physiol. Plant.* **2021,** *171,* 22–33.

Yin, Z.; Zhang, H.; Zhao, Q.; Yoo, M. J.; Zhu, N.; Yu, J.; et al. Physiological and Comparative Proteomic Analyses of Saline-Alkali $NaHCO^3$-Responses in Leaves of Halophyte *Puccinellia tenuiflora. Plant Soil* **2019,** *437,* 137–158.

Yu, J.; Chen, S.; Wang, T.; Sun, G.; Dai, S. Comparative Proteomic Analysis of *Puccinellia tenuiflora* Leaves Under Na^2CO^3 Stress. *Int. J. Mol. Sci.* **2013,** *14,* 1740–1762.

Yu, J.; Chen, S.; Zhao, Q.; Wang, T.; Yang, C.; Diaz, C.; et al. Physiological and Proteomic Analysis of Salinity Tolerance in *Puccinellia tenuiflora*. *J. Proteome Res.* **2011,** *10,* 3852–3870.

Yusuf, M.; Hasan, S. A.; Ali, B.; Hayat, S.; Fariduddin, Q.; Ahmad, A. Effect of Salicylic Acid on Salinity-Induced Changes in *Brassica juncea*. *J. Integr. Plant Biol.* **2008,** *50,* 1096–1102.

Zang, W.; Miao, R.; Zhang, Y.; Yuan, Y.; Pang, Q.; Zhou, Z. Metabolic and Molecular Basis for the Salt and Alkali Responses of *Suaeda corniculata*. *Environ. Exp. Bot.* **2021,** *192,* 104643.

Zhang, H.; Feng, H.; Zhang, J.; Ge, R.; Zhang, L.; Wang, Y.; et al. Emerging Crosstalk Between Two Signaling Pathways Coordinates K^+ and Na^+ Homeostasis in the Halophyte *Hordeum brevisubulatum*. *J. Exp. Bot.* **2020a,** *71,* 4345–4358.

Zhang, H.; Han, B.; Wang, T.; Chen, S.; Li, H.; Zhang, Y.; et al. Mechanisms of Plant Salt Response: Insights from Proteomics. *J. Proteome Res.* **2012,** *11,* 49–67.

Zhang, J.-T. T.; Mu, C.-S. S.; Effects of Saline and Alkaline Stresses on the Germination, Growth, Photosynthesis, Ionic Balance and Anti-Oxidant System in an Alkali-Tolerant Leguminous Forage *Lathyrus quinquenervius*. *Soil Sci. Plant Nutr.* **2009,** *55,* 685–697.

Zhang, K.; Tang, J.; Wang, Y.; Kang, H.; Zeng, J. The Tolerance to Saline–Alkaline Stress was Dependent on the Roots in Wheat. *Physiol. Mol. Biol. Plants* **2020b,** *26,* 947–954.

Zhang, W.; Liu, J.; Zhang, Y.; Qiu, J.; Li, Y.; Zheng, B.; et al. A High-Quality Genome Sequence of Alkaligrass Provides Insights into Halophyte Stress Tolerance. *Sci. China Life Sci.* **2020c,** *63.*

Zhao, C.; Zhang, H.; Song, C.; Zhu, J.-K.; Shabala, S. Mechanisms of Plant Responses and Adaptation to Soil Salinity. *Innov.* **2020,** *1,* 100017.

Zhao, K.; Song, J.; Feng, G.; Zhao, M. Species, Types, Distribution, and Economic Potential of Halophytes in China. *Plant Soil* **2011,** *342,* 495–509.

Zhao, Q.; Suo, J.; Chen, S.; Jin, Y.; Ma, X.; Yin, Z.; et al. Na^2CO^3-Responsive Mechanisms in Halophyte *Puccinellia tenuiflora* Roots Revealed by Physiological and Proteomic Analyses. *Sci. Rep.* **2016,** *6,* 1–23. doi:10.1038/srep32717.

Zhao, Y.; Lu, Z.; He, L. Effects of Saline–Alkaline Stress on Seed Germination and Seedling Growth of *Sorghum bicolor* (L.) Moench. *Appl. Biochem. Biotechnol.* **2014,** *173,* 1680–1691.

Zhou, C.; Shen, W.; Lu, C.; Wang, H.; Xiao, Y.; Zhao, Y.; et al. Effects of Salinity on the Photosynthesis of Two Poaceous Halophytes. *Clean - Soil, Air, Water* **2017,** *43,* 1660–1665.

CHAPTER 14

Reactive Oxygen Species and Oxidative Damage in Halophytes Under Salt Stress

MALI RAM HINDALA, RIDHI JOSHI, and VIDYA PATNI

Department of Botany, University of Rajasthan, Jaipur, Rajasthan, India

ABSTRACT

Generation of reactive oxygen species (ROS) is an inexorable consequence of aerobic life on the earth, which are produced through normal cellular anabolic (constructive) and catabolic (destructive) biological processes. Halophytic plant species are adopting similar mechanisms and strategies to develop antioxidant protection. The level of ROS is enhanced under various stress states, such as salt, owing to an earnest disproportion between assembly and elimination. These are highly toxic. Excluding proper defense strategies, it can raise cellular oxidative damage to DNA, RNA, lipids, proteins, etc., starting in the redox (reduction as well as oxidation) state. Besides the functioning of toxic by-products of stress, ROS place a leading emphasis on signal transduction elements in growth, development, and response to different cellular stressors. Halophytic plants regulate the concentration of ROS through the sequential cycles of the nonenzymatic and enzymatic antioxidant pathways. The regulated antioxidative defensive mechanisms in between nonenzymatic and enzymatic action and tolerance have been extensively discussed in relation to salt stress. In the present chapter, we discuss several functional roles of reactive oxygen species, biochemical targets (at the elemental and molecular level), antioxidant enzymes, scavengers, sites, synthesis, mechanisms of ROS production, adverse effects of oxidative stress, and different routes in halophytic species in terms of salt stress and oxidative damage.

Physiology of Halophytes: Signaling, Omics, and Tolerance Mechanisms.
Nivas Manohar Desai & Ganesh Chandrakant Nikalje (Eds.)

14.1 INTRODUCTION

Globally, soil salinization has become a broad threat to the diversity of the halophytic species population and is a serious environmental concern (Zhang et al., 2018). Soil salinity restricts worldwide food production owing to the diminishing impact of salt on the vegetative and reproductive growth phases of crops (Yang and Guo, 2018). Soil salt stress is a renowned abiotic constraint that greatly influences plant growth, development, and metabolism in halophytes (Attia et al., 2020). In halophytes, roughly 20 and 50% of crops and agricultural land worldwide are under salt stress, respectively, so different defensive strategies are essential for better crop improvement. Land is limited worldwide, although the population is growing rapidly (Kumar et al., 2019). Halophytic plants occupy roughly 60% of total arable land worldwide (Shi and Sheng, 2005). Photosynthesis is a biophysicochemical, anabolic (synthetic) process, and the principle controller of plant growth and development. They actively participate in the carbon concentration and assimilation processes (Munns et al., 2016). Photosynthesis is the most sensitive anabolic tool to ecological barriers. The salt stress effect includes (i) photoinhibition, (ii) decreasing gross photosynthetic efficiency, affecting carbon assimilation, and (iii) reduced plant growth and development (Shelke et al., 2017; Sahin et al., 2018). In plant tissues, excess of Na^+ and other ions such as Ca^{2+}, Mg^{2+}, and Cl^- generates a physiological disturbance or imbalance, ultimately causing cellular oxidative destruction in relation to photosynthetic activity (Ahanger et al., 2018). Soil salinity induces many physiological obstacles, including photosynthetic organization, photosystem inactivation, and photoinhibition in halophytes, ultimately causing a reduction in carbon assimilation (Li et al., 2020).

The aforementioned is primarily expressed through: (i) decreasing the apparent quantum (single packet of matter or energy) efficiency of the photosynthetic process, inhibiting photosystem-I and II functions, and inducing dark respiration (Zhang et al., 2018); (ii) overexcitation of PS-II, which enhances much energy consumption through nonradiative mechanisms (Meloni et al., 2003); and (iii) salt stress that causes the storing potential of ROS and lipid peroxidation that impede standard oxidoreduction state along with decreased photosynthetic effectiveness (Khan et al., 2014). Interaction with ion assimilation and ion uptake reveals plant water depletion as well as oxidative destruction (Abass et al., 2018). Subsequently, hydrogen peroxides, as well as superoxides, will gather in plants under salinity (Attia et al., 2020). The halophytes possess the inherent capacity to tolerate high salt concentrations. The abundance of salt concentrations in soil is a great concern for halophytes. This stress reduces the productivity and total yields of crops through osmotic and ionic effects. For the purpose of water assimilation in plants, soluble salts at the soil-water interface are essential to maintaining the osmotic adjustments. Halophytic plants retain the diversity in their salt sustaining strategies to reduce the toxic effects of high salt concentrations (English and Colmer, 2013). The word, reactive oxygen species (ROS) has been devised as an advanced category of intrinsic or endogenous, extremely reactive oxygen- and nitrogen-containing molecules. In the atmosphere, the level of molecular oxygen is increasing (around 2.5 years ago) (Lesser, 2006) because of the oxygenic activity of cyanobacteria (blue green algae) generating the evolution of aerobic respiration, sequentially leading to the development of eukaryotic (advanced) organisms (Lesser, 2006).

The terms RNS and RNI have been designated to explain an evolving category of intrinsic, extremely O^2- and N^2-containing compounds. The word ROS explains that chemical species are synthesized due to partial or incomplete reduction of molecular oxygen, namely, $OH^{\cdot}$, $O^{2\cdot-}$, H^2O^2, and singlet oxygen; meanwhile, ROS specifies both ROI as well as ozone (O^3), which is to a lesser extent in plant cells. The words ROS and ROI are sometimes apparently identical and may include hypobromous (HOBr), hypochlorous (HOCl), hypoiodous (HOI), and organic hydroperoxide (ROOH), which are synthesized through peroxidase-catalyzed oxidation of halides. The prime reactive nitrogen intermediates (RNI) that affect ROS are nitrogen-oxygen free radicals, nitrogen dioxide nitrite, and peroxynitrite (Nathan and Ding, 2010). ROS are also further classified into free nonradical and free radical species. All O^2 radicals are ROS, but not all ROS are O^2 radicals (Halliwell, 2006) (Table 14.1).

Oxidative stress signaling and ROS are essential elements in halophytic plants to overcome the salt-stressed condition. Without protective covering or strategy, ROS can be toxic and spawn oxidative injuries to cellular DNA, proteins, and lipids. To endure multiple salt stresses, they retain different structures and adaptations to overcome imposed salt stress. In terms of defensive approach, the principle antioxidative catalysts, namely, catalase, ascorbate peroxidase, SOD, and peroxidase, are elaborated in the protection of the cellular machinery of halophytes. Thus, regulation of ROS is an essential approach to reduce the salinity stress (Oprică and Vochița, 2021). Halophytes are those plants that possess peculiar adaptations like salt glands (secretary in nature) and structures like vacuoles, cytoplasm, and cellular organelles to overcome the salinity stress. Halophytes are used as a plant model organism in terms of salt tolerance and adaptation (Hamed, 2022).

TABLE 14.1 Some Selected Reactive Oxygen Species, Free Radicals, and Nonradicals in Halophytes.

Free radicals	**Nonradicals**
ROS includes:	ROS includes:
Singlet, O^{21}	Hypochlorous acid, HOCl
Carbonate, $CO^{\bullet 3-}$	Nitrosoperoxycarbonate, $ONOOCO^{2-}$
Peroxyl, $RO^{\bullet 2}$	Peroxomonocarbonate, $HOOCO^{2-}$
Hydroxyl, OH•	Peroxynitrous acid, ONOOH
Hydroperoxyl, $HO^{\bullet 2}$	Peroxynitrate, $O^{2}NOO^{-}$
Superoxide, $O^{2\bullet -}$	Peroxynitrite, $ONOO^{-}$
Superoxide, $O^{\bullet -2}$	Hypobromous acid, HOBr
Alkoxyl, RO•	Organic peroxides, ROOH
	Singlet, O^{21}
	Hydrogen peroxide, $H^{2}O^{2}$
	Ozone, O^{3}
Reactive nitrogen species (RNS):	Reactive nitrogen species (RNS):
• $NO^{\bullet 3}$	Peroxynitrite, $ONOO^{-}$
• NO•	Peroxyacetyl nitrate, $CH^{3}C(O)OONO^{2}$
• $NO^{\bullet 2}$	Nitrous acid, HNO^{2}
	Dinitrogen tetroxide, $N^{2}O^{4}$
	Nitryl chloride, $NO^{2}Cl$
	Nitrosyl cation, NO^{+}
	Dinitrogen trioxide, $N^{2}O^{3}$
	Alkyl peroxynitrites, ROONO
	Nitroxyl anion, NO^{-}
	Nitronium cation
	Peroxynitrous acid, ONOOH
	Peroxynitrate, $O^{2}NOO^{-}$
Reactive chlorine species:	Reactive chlorine species (RCS):
• Atomic chlorine, Cl•	Chloramines,
	Hypochlorous acid (HOCl),
	Chlorine (Cl^{2}),
	Nitryl chloride, $NO^{2}Cl$,
	Bromine chloride,
	Chlorine dioxide (ClO^{2}),
Reactive bromine species	Reactive bromine species (RBS):
• Atomic bromine	Bromine chloride (BrCl),
	Hypobromous acid,
	Bromine (Br^{2})

Source: Halliwell, 2006.

ROS actively participate in oxidative damage, causing negative effects on central components of DNA, RNA, protein, fatty acid, and another cellular fraction. Oxidative damage is developed owing to unevenness in ROS synthesis as well as restricted antioxidant defenses in cells. This effect ultimately caused many malfunctions or barriers to normal development. Mitochondria (energy-generating organelle) level during cellular respiration is keenly associated with the aging process. Reduction of excessive ROS is crucial in the aging process. Free radical chemistry (the concept of aging) is linked with an aging and degenerative disorder, mainly contributing to the negative effects of antioxidants on cellular fractions as well as on connective tissues. The origin of free radicals is a sequential chain reaction involving molecular oxygen that is catalyzed by many oxidative enzymes. Multifarious trace micronutrient elements, namely, manganese, cobalt, and iron, play a crucial role in reaction initiations (Halliwell and Gutteridge, 2015). The regulation of the redox or dual functioning of oxidants or reductants particularly involves the limited formation of reactive species (both O^2 and N^2). These species react with a key functional group of proteins of interest (mainly the iron and sulfur group, cysteine), which leads to a protein modification (covalent bonding). ROS as a secondary messenger are very crucial for many TFs' expressions as well as signal transduction mechanisms, for instance, nuclear factor and heat shock protein (HSP) factors. They have directly participated in the regulatory system of gerontological processes or aging, cell adhesions, and redox-dependent amplifications of immune responses in organisms (both in plants and animals) (Nathan, 2003).

To explain the highly complex and universal impacts of ROS in halophytes, multidisciplinary strategies are indeed needed, where chemists may construct a new approach or mechanisms to generate and detect ROS and their by-products with high accuracy and spatiotemporal regulation (Winterbourn, 2008). The concept of oxidative damage in halophytes is universal, notorious, and nonspecific. It involves transferring electrons to any electron acceptor where electronic contact facilitates (Imlay, 2003).

14.1.1 OXIDATIVE DAMAGE TO DNA

Many studies have suggested that halophytic species such as *Hordeum marinum* (sea barley-grass), *Sesuvium portulacastrum*, and *Cakile maritima* have more natural and potent antioxidative defense mechanisms than the glycophytic species (Seckin et al., 2010; Ellouzi et al., 2011; Srivastava et al.,

2015). Barriers in halophytic plant (*Puccinellia tenuiflora*) photosynthesis would increase ROS synthesis and production in cells, ultimately causing oxidative damage to DNA, proteins, and lipids. Thus, it directly influences the integrity or organization of the cellular membrane, enzymatic activity, and multifarious functions of photosynthetic machinery (Yu et al., 2011). In addition, the induction of high salts in halophyte *Salvadora persica* in photosynthesis has caused a significant increase in ROS levels and oxidation in plants. They have reported that the constitutive level of all the antioxidative enzymes in *S. persica* has been much high in relation to glycophytes (Nazar et al., 2011; Lee et al., 2013; Rejeb et al., 2015).

Production of ROS could stimulate multifarious injuries in DNA and cause mutational procedures like deletion, addition, duplication, or other genetic defects in halophytes. Both sugar and base pairs mismatching are sensitive to oxidation states, arising from single-strand base pairs' breakdown, and cross-linking of proteins and base pairs (Imlay, 2003). Hydrogen peroxide toxicity in *Escherichia coli* contributes to DNA injuries, basically by a Fenton (chemical) reaction. This interaction produces a native form of radicals from the substrate hydrogen peroxide (H^2O^2), DNA-bound ions, and the fixed inception of reducing equivalents. In vivo DNA injuries' production steps through H^2O^2, including the reducing state of nicotinamide adenine dinucleotide (NADH), the ferrous form of iron. To overcome the abnormal impacts of hydrogen peroxide radicals, cells consuming scavengers of these radicals and DNA damage control enzymes are important. In vitro hydrogen peroxide cannot cause strands to be cut down under standard physiological processes, together with their abnormal or negative impacts. In vivo is the probable outcome of the Fenton reaction in the presence of a catalyst (ferrous ions). The prokaryotic and eukaryotic organisms (both) possess DNA repair enzymes; for the DNA damage cells, these are important steps that determine the fate of peculiar cells (Beyer et al., 1991). ROS are synthesized over a few threshold limits; they could destruct internal compartments, namely, DNA, RNA, proteins, and lipids. Damage in these components can cause abnormal functions in halophytic plant cells (Sunkar et al., 2003).

14.1.2 OXIDATIVE DAMAGE TO PROTEINS

Salinity-induced modulation in photosynthetic pigment, relative water level, lipid peroxidation, photosynthesis, photosystem-II effects, and alterations in the activities of many antioxidative enzymes have been studied in noncrop plant halophyte (*Salvadora persica)*. In the absence of plant pigment degradation, reduction of water quantity, regulation of water use efficiency, and photosystem-II from salinity-induced oxidative damage in antioxidative enzymes are crucial factors for salt tolerance in *S. persica* (Rangani et al., 2016).

Peptide chain fragmentation, clustering of cross-linked products, variable electrical charge, particular location of amino acid alteration, and being prone to deletion and degradation are prime causes of oxidative damage to proteins. Peptide bridges in amino acids vary in sensitivity to damage in proteins, and multiple states of ROS are also variable in their powerful reactivity. Protein configurations are essential in determining sensitivity of every peptide bond to active attack through ROS (Halliwell and Gutteridge, 2015). For various types of enzymes, the oxidation through O^{2-} of the ferrous iron cluster generates enzyme inactivation (Hyslop, et al., 1988), as well as many more amino acids, namely, arginine, cysteine, proline, lysine, and histidine form carbonyl groups when oxidized (Stadtman, 1986). A broad array of proteins and their peptide bonds are damaged and degraded by ROS (Davies, 1987). Reactive compounds such as catechols and hydrogen peroxides are formed upon oxidative radical attack on proteins. These metabolic products and otherwise damaged proteins are stored in biological systems. In cells, the storage of such proteins has been hypothesized to be part of a gerontological mechanism (Dean et al., 1993).

14.1.3 OXIDATIVE DAMAGE TO LIPIDS

In halophytic species, the production of ROS is highly toxic. Without the involvement of protection mechanisms, it can cause oxidative damage to lipids and other cellular components, leading to changes in redox potential to maintain their homeostasis. In addition, it can cause further damage to plant halophyte cells (Ozgur et al., 2013).

HO·-dependent ROS, along with lipids, are the prime biological mechanisms of cellular damage and the order of membrane fluidity. The functional role of lipids is affected by oxidative damage (Halliwell and Gutteridge, 2015). The mitochondrial lipid peroxidation mechanism is primarily cytotoxic, with multifarious effects on cellular enzyme activity, the synthesis of ATP, and the initiation of programmed cell death (Green and Reed, 1998).

The lipid peroxidation by-products are hydrocarbons (such as ethylene and ethane) and aldehydes (e.g., malondialdehydes). Both vitamins (C and E) mutually interact to rescue lipids and their membranes, in contrary to the peroxidation reaction. Conceptually, vitamin E (aqueous state) and vitamin C (membranous state) are mainly lipid-soluble small-molecule antioxidants and terminal water-soluble small molecules, respectively. Although vitamin C possesses the inherent property of recycling the vitamin E, it thus helps in repairing mechanisms of tocopheroxyl free radicals of vitamin E (tocopherols). Therefore, allowing vitamin E to act against the chain breaking of free radical antioxidants (Freeman and Crapo, 1982; Gutteridge and Halliwell, 1990; Buettner, 1993).

In the present chapter, we discuss the various sources, redox potential, electronic configurations, biochemical targets, chemical characteristics, sites, synthesis, pathway or route of ROS, oxidative enzymes, injuries, and some causes, particularly in halophytes, and some general discussion about ROS. Additionally, we have also explained different defensive strategies (antioxidants, scavengers, etc.) to protect from ROS in halophytic plants in relation to salt stress.

14.2 CHEMICAL CHARACTERISTICS OF REACTIVE OXYGEN SPECIES

To explain the biological and chemical aspects of ROS or intermediates, fundamental information about their origin, type, and reactivity (particularly redox potentials and electronic configuration) is essential.

14.2.1 REDOX POTENTIALS

The thermodynamics of free active radical reaction is essential toward the knowledge steps in the direction of electron transfer (from an outstanding knowledge point of view). The thermodynamics or biophysical parameters can be preowned to determine the pecking sequence, or order, or hierarchy of free radicals and antioxidants. The prime kinetic and thermodynamic properties are keenly interconnected with chain-breaking antioxidants. The redox potentials of many ROS, which are engaged in the depletion of molecular O^2 to H^2O, are depicted in Table 14.2.

TABLE 14.2 Some Selected Potentials (Adapted and Modified in Relation to Standard Hydrogen Electrodes).

Sr no.	Redox couple	Redox potential (E0/V)
1	Ubiquinone, H^+/semiubiquinone	−0.036
2	Dehydroascorbic/ascorbate$^{\cdot -}$	−0.17
3	$O^2/O^{2\cdot -}$	−0.33
4	Methyl viologen (MV^{2+})/$MV^{\cdot +}$	−0.45
5	O^2, $H^+/HO^{2\cdot}$	−0.46
6	RSSR/RSSR$^{\cdot -}$	−1.50
7	Semiubiquinone, H^+/ubiquinol ($CoQ^{\cdot -}$, $2H^+/CoQH^2$)	0.20
8	Ascorbate$^{\cdot}$, H^+/ascorbate monoanion	0.28
9	H^2O^2, H^+/H^2O, $^{\cdot}OH$	0.32
10	Trolox C ($TO^{\cdot}$, H^+/TOH)	0.48
11	α-Tocopheroxyl	0.50
12	Catechol-$O^{\cdot}$, H^+/catechol-OH	0.53
13	$^1O^2/O^{2\cdot -}$	0.65
14	$O^3/O^{3\cdot -}$	0.89
15	$RS^{\cdot}/RS^-$	0.92
16	$O^{2\cdot -}$, $2H^+$/Hydrogen peroxide	0.94
17	$ROO^{\cdot}$, H^+/ROOH	1.00
18	$HOO^{\cdot}$, H^+/Hydrogen peroxide	1.06
19	$RO^{\cdot}$, H^+/ROH	1.60
20	$O^{3\cdot -}$, $2H^+/H^2O + O^2$	1.80
21	$HO^{\cdot}$, H^+/H^2O	2.33

Source: Buetter, 1993.

In addition, few redox values are organized from a high reducing to a high oxidizing state, the single redox potential of several ROS of biological value, as well as the single electron of the redox potential of ROS or intermediate scavengers (Buettner, 1993). During the redox reaction, the di-/bi-radical properties of O^2 limit it to accepting a single electron at a time, with spin paired. The molecular O^2 (redox potential with –0.16 volts; pH—7; oxygen concentration of 1M; pressure 1 bar) is somehow a weak or poor universal electron acceptor (Imlay, 2003).

In addition, molecular oxygen is a fairly nontoxic, benign, and weak oxidant for several biomolecules. The unpaired electrons of molecular oxygen cooperate with the unpaired electrons of organic radicals as well as transition metals (intermediate effects between metal and nonmetal).

The single reduction of electron molecular O^2 to the superoxide anion is thermokinetically unfavorable compared to its direct $2e^-$ reduction to hydrogen peroxide (H^2O^2, with a redox potential of +0.30 V). Subsequently, $2e^-$ required of the latter is unfavorable (Valko et al., 2004). The synthesis of a superoxide radical anion but not H^2O^2 is a peculiar feature of the auto-oxidative mechanism. Because of its anionic charge, the superoxide radical possesses a natural, restricted reaction with electron-rich clusters or centers (D'Autréaux and Toledano, 2007).

Alongside O^{2-} ions' protonation, perhydroxyl radicals are produced. A novel candidate of redox potential has better increased reduction potentials (having +1.06 V) and is a good oxidant (Buettner, 1993). Owing to its low concentration, perhydroxyl radicals are, however, considered to play a minor role in the physiological state. Beyond its too high reduction potential of +0.94 V, oxygen concentration varies with biological compounds. A single electron reduction of O^{2-} ions gives rise to H^2O^2 production. Hydrogen peroxide possesses a single electron reduction potential of +0.32 (Buettner, 1993) to +0.38 V (Valko et al., 2004), depending upon the source of organisms (both plants and animals) and even more supported or favorable toward $2e^-$ reduction potentials (+1.35V) (Valko et al., 2004). Conceptually, it is more stable (less energy) under a physiological state (Rodriguez-Muñiz et al., 2014).

The redox potential of the ROS is unique for each species. It determines biochemical activation by altering their electronic ionic charges. Each radical and nonradical ROS and other relevant species functions are affected via various nonenzymatic and enzymatic biochemical mechanisms in halophytes to provide extra protection cover from ROS. Different antioxidants are activated to counteract the adverse effects that are produced upon ROS attack in cellular functional components in salt-tolerant plants.

14.2.2 DIFFERENT ELECTRONIC STRUCTURES IN REACTIVE OXYGEN SPECIES

The ground-level molecular oxygen state is paramagnetic bi/di-radicals, which include two electron pairs with parallel spins. Mostly, nonradicals are diamagnetic with opposite spin rotation. This spin allows molecular O^2 to actively assist in redox reactions with other atoms or molecules from the reductant that possess a similar spin in nature; thus, it can accommodate in the open or free space in pi orbitals of O^2. Therefore, it cannot perform efficient oxidized molecular synthesis through the addition steps of $2e^-$ at the same time. The rotation of spin pairs in a redox reaction results in O^2 accepting an electron at the same time. Hence, molecular O^2 can function faster with respect to other radicals through a single electron shift. They function with nonradical ions consisting of unpaired electrons; for instance, iron is a transition metal in ferrous-sulfur (Fe-S) cluster centers. A single reduction of electrons puts steps toward the construction of superoxide ions ($O^{2.-}$), and a further one electron reduction of $O^{2.-}$ results in the formation of some ROS, namely, H^2O^2. Furthermore, a reduction in H^2O^2 is converted into the production of $OH^.$ radicals, which subsequently undergo reduction to obtain water (hydroxyl ions) (Imlay, 2003). Molecular oxygen in its ground state can be turned into a more active form. The energy supply of oxygen results into the synthesis of a more active form of oxygen ($^1O^2$). The singlet form of oxygen possesses opposite spin rotation; therefore, spin restriction is removed, enhancing the oxidizing ability of $^1O^2$ with respect to the ground state of oxygen (Imlay, 2003).

14.3 ORIGIN OF ROS IN HALOPHYTES

Fundamentally, ROS can be generated in two ways, namely, intrinsic or endogenously and extrinsic or exogenously in halophytes.

14.3.1 INTRINSIC (ENDOGENOUS) ORIGIN OF ROS PRODUCTION IN HALOPHYTES

Halophytes with intrinsic and efficient potential, contrary to salinity, can be used as models for enhancing the response of crops in terms of salinity stress. As crops, glycophytes and halophytes have major molecular differences, which may lead to potent candidature of guiding and optimizing the transformation of glycophytes along with halophytic tolerance genes (Barros et al., 2021). Mostly in crops, glycophyte species have a lower threshold of salinity; whereas, halophytes face salinity efficiently (Barros et al., 2021). Recently, the existing knowledge concerning ROS homeostasis and functions relies on the crops' seeds and model plants. The information about the seeds of noncrops, for instance, halophytic plants, is restricted to just a few studies. In addition, mechanisms underlying multifarious roles like alternative routes, downstream targets, and crosstalk with other molecules are still obscure (Rasheed et al., 2019).

Under extreme environmental conditions such as salinity, drought, and high temperature, morphophysiological adaptations in xerohalophytes could provide much information for producing more resilient cash crops (Rasheed et al., 2022).

ROS can be generated through intracellular or extracellular modes, depending on various sources. These are generated in a broad array through biophysicochemical processes. Different types of enzymes are actively participating in the synthesis of ROS. Cytosolic enzymatic action is attributed to ROS production and multifarious isoforms of NOXs (NADPH oxidase, location—transmembrane).

The NOXs are a superoxide-producing enzymatic system (Nathan and Cunningham-Bussel, 2013; Lambeth, 2004). Cytosolic (inner membrane) regions of NOx shift an electron from nicotinamide adenine dinucleotide phosphate, (oxygen serves as a substrate) to FAD (flavin adenine dinucleotide, a type of flavoprotein derived from vitamin B2-riboflavin) biomolecule factor. From there, the electron returns to a heme group, where it receives the O^2 on the outer side of the membrane, thus producing O^{2-} (Nathan and Cunningham-Bussel, 2013). Various types of nicotinamide adenine dinucleotide phosphate oxidase are manifested in distinct cells separately. It performs contrasting functions in a diverse manner on cellular components. NADPH oxidase enzyme superfamilies have shown the specificity of ROS and their special emphasis on homeostasis in relation to cellular signaling (Nathan and Cunningham-Bussel, 2013).

The cell organelle, mitochondria, is a prime generation site (intracellularly or endogenously) of ROS. When a single electron (e^-) is transferred to oxygen at two discrete locations of ETC, such as complex-I (location—matrix) and complex-III (location—inner mitochondrial space), mitochondrial superoxide radicals occur (Turrens, 2003; Finkel, 2011). Under normal pressure (normobaric) and high pressure (hyperbaric) in vitro conditions, these locations in the electron transport chain of mitochondria convert roughly 1–2% of total consumed oxygen to superoxide molecules (Boveris and Chance, 1973; Schaffer and Suleiman, 2010; Skulachev, 1997).

The rate of superoxide radical production in isolated mitochondria is high in vitro with respect to in vivo conditions (Finkel and Holbrook, 2000). Somehow, single electron reaction (ETC) is primarily predominant. A $2e^-$ superoxide production reaction is also existing, which permits the direct reduction of molecule O^2 to H^2O^2. Both complexes I and III of ETC are superoxide radical production sticks. Mitochondria is also the additional intracellular source of ROS nonenzymatically (predominantly) that occurs as a transitional or intermediate phase in the redox reaction of Co-Q (Q^{10}). Once a radical is formed, $Q^{\cdot-}$ can move electrons to molecular O^2 with the subsequent production of superoxide radicals (Finkel and Holbrook, 2000). ROS synthesis is a function of metabolic rate. Not only NADPH oxidase, mitochondria, but also other intracellular sources of ROS generation, include oxidases for amino acids and polyamines, NO synthases, ERO^1 (flavoenzyme, source location—endoplasmic reticulum), cytochrome P^{450}, lipoxygenase, cyclooxygenase, demethylase (flavin-dependent), etc. (Nathan and Cunningham-Bussel, 2013; Finkel, 2011).

These radical-generating oxidants function through normal enzymatic action. As a result of the Fenton reaction, copper and iron ions are released from ferrous iron clusters/centers, and the heme group (a metal storage protein) can change $O^{2\cdot-}$ and H^2O^2 to $OH^{\cdot}$. In addition, lipid superoxide holds for the synthesis of alkoxyl (a peroxyl) radicals in the lipid membrane (Halliwell, 2006). Both superoxide and nitric oxide (NO) are involved in free radicals' conversion processes, either directly or indirectly, through nonenzymatic antioxidants (Dröge, 2002).

14.3.2 EXTRINSIC (EXOGENOUS) ORIGIN OF ROS PRODUCTION IN HALOPHYTES

Halophytes have adopted many strategies to survive in salty or harsh environments. However, some microbiota could help overcome harsh environmental conditions. Bacterial endophytes in halophytic species are a potential source of beneficial microbes for sustainable agriculture (Christakis et al., 2020).

ROS can also be produced by the host through an exogenous mode. Various environmental pollutants, including ionizing (x-ray and gamma rays) and nonionizing radiations (ultraviolet irradiation), are prime induction factors that promote ROS extracellularly. Paraquat, or methyl viologen, is the herbicide that reacts to form free radicals such as ozone, peroxide, etc., through ozonation and peroxidation enzymatic reactions. Bipyrimidiulium (closer to paraquat herbicide) and nitroaromatics are organic pesticides of biochemical nature that promote superoxide production. The amino phenols, phenols, polyhalogenated, and alkanes are the aromatic and aliphatic chemical groups, respectively, which are metabolized to free radicals. Chemicals that produce free iron and copper could promote hydroxyl radical formation (Finkel and Holbrook, 2000; Winterbourn, 2008; Halliwell and Gutteridge, 2015).

ROS could be produced or generated easily with the help of the radiolysis of water by treating irradiations (ultraviolet, X-ray, gamma rays, etc.,) of H^2O^2 (Leach et al., 2001). The production of secondary products through radioionization mechanisms can strongly produce or promote the primary ionization events. The theoretical values have shown that superoxide or H^2O^2 is produced in very minute concentrations by ionization events (Leach et al., 2001).

Furthermore, sensitizer ultraviolet ray irradiation leads to the production of singlet oxygen. Non-DNA-based reactive carcinogens (environmental factors) can generate ROS in cells in two ways, namely, (i) through promoting an endogenous source of ROS and (ii) through the metabolism of primary radical intermediates or species (Klaunig et al., 2011). Various exposures to xenobiotic of altered structures (variable) and activity, the promotion or induction of oxidative damage and stress have been observed or reflected. Metal ions, phorbol ester, chlorinated compounds, and radiations are some peroxide-proliferating compounds that induce oxidation damage and stress in vivo and in vitro (Klaunig, 2011).

ROS-mediated programmed cell death or apoptosis effects have been encountered in the mode of action of multifarious chemotherapeutic cancer-causing factors. Cisplatin and adriamycin (an antitumor drug) promote ROS in excess, leading to DNA injuries and apoptosis (Szatrowski and Nathan, 1991). Bactericidal antibiotic drug target counteraction, induces collapse in the iron control dynamic mechanism, stimulating the synthesis of harmful $OH^{\cdot}$ radicals via Fenton components (substrates: Fe^{2+}, H^2O^2; products: hydroxide (OH^-) and hydroxyl radicals) in gram-positive and gram-negative bacteria, ultimately leading to PCD (Kohanski et al., 2007).

In halophytes, there are extrinsic factors that affect their solute accumulation and their emphasis on osmoprotection and osmoregulation. Osmolyte contents in halophytes are responsible for improving growth and yield under salinity (Slama et al., 2015). Glycine-betaine is an important osmolyte that provides protection from many stresses, including ion toxicity, dehydration, and temperature or heat, through stabilizing macromolecules. It also protects chloroplasts and photosystem-II by altering their thermodynamic properties (Subbarao et al., 2001). Glycine-betaine accumulations in chloroplasts provide safety valves to the photosynthetic and oxygen-evolving complex of photosystem-II activities, hindering high salt-stimulated breakdown of exogenous proteins (Yang., 2008).

Different intrinsic and extrinsic sources of ROS can be helpful to overcome the detrimental effects developed from cellular oxidative damage in halophytes. Halophytes have adopted many strategies to survive in salty or harsh environments. Molecular interaction between endogenous and exogenous free radicals and nonradicals may provide the backbone to explore the serious diseases in halophytes.

14.4 BIOCHEMICAL TARGETS OF ROS AND OXIDATIVE DAMAGE IN HALOPHYTES

ROS, or intermediates, are oxidants that could be functional through single electron oxide and electron double oxidation (nonradical) (Halliwell, 2006). Radical O^2 species are concerned thermodynamically as enzyme activity barriers for single electron reactions. Radical ROS are primarily chain reaction propagators such as free radical-mediated autooxidation of polyunsaturated fatty acids (PUFA) (Yin et al., 2011). More toxic reactivity is concerned with ROS being nonspecific in choosing biological targets to react with substrates. Signaling responses in cells are a better choice than well-arranged biochemical targets. There is a great anomaly in the selection of biological targets. Highly reactive ROS (OH·) possess a wide range of nonspecific or nonselective targets. ROS, which is marked by H^2O^2 or $O^{2.-}$, will be particular. Substrate-based availability and the relative rate of reaction constants are crucial factors in determining the biological targets. Some atomic elements within target molecules are facilitated by low reduction signaling. In addition, subsets of atoms in ROS lead to covalent protein modifications (D'Autreaux and Toledano, 2007). To explain the better concept of biochemical targets' machinery in ROS, first details about atomic targets are given prior to the molecular address.

14.4.1 ATOMIC TARGETS OF ROS IN HALOPHYTES

Amino acids like methionine and cysteine and sulfur cooperate with ROS, and in the case of selenocysteine, selenium is the most commonly encountered atomic target in ROS (Nathan and Cunningham-Bussel, 2013). Redox potentials can be bidirectional or reversal (forward as well as backward) in the case of sulfur and selenium or in the case of arginine, threonine, cysteine, and proline (Nathan and Ding, 2010), as well as carbon acts as atomic targets in PUFA (Yin et al., 2011). In addition, both in the ferrous-sulfur cluster and heme, ferrous (Fe^{++}) interacts with ROS as an atomic target site (Imlay, 2003). Hydrogen peroxide acts together with the thiol group. An anionic form is too crucial to perform the reaction better. Hence, the thiol form is more reactive. Superoxide radical anion favors reaction with iron sulfur center, where susceptibility of such group is basically owing to electrostatic interaction with superoxide anions (Imlay, 2003). Rational reactions of H^2O^2 and $O^{2.-}$ emanate biological targets of their transcriptional agents, OxyR (cysteine remains) and SoxR (Fe-S chains), in *E. coli* (Pomposiello and Demple, 2001; Green and Paget, 2004). These transcriptional factors are promoted by the superoxide anion and hydrogen peroxide, respectively. They are actively participating in the execution of oxidation-preventing catalysts, the OxyR response to H^2O^2, and the SoxR response to $O^{2.-}$ (Hidalgo et al., 1997).

14.4.2 MOLECULAR TARGETS OF DEOXYRIBONUCLEIC ACID IN TERMS OF OXIDATIVE DAMAGE

Major targets of ROS are sources of mitochondrial DNA within cells. This target brings on a compromised mitochondrial role. ROS reactions with DNA (without the involvement of proteins) itself may function to promote the protein synthesis (Nathan and Cunningham-Bussel, 2013). Aging cells have the highest level of ROS in damaged nuclear DNA (Finkel and Holbrook, 2000).

14.4.3 MOLECULAR TARGETS OF LIPIDS IN HALOPHYTIC SPECIES

The salt stress state promotes the halophytic plants to increased production of ROS in plant tissues. Halophytes have an inherent property to reduce the toxic ROS impacts using antioxidant systems such as enzymatic and nonenzymatic constituents, for instance, *Prosopis strombulifera.* Oxidative damage in tissues has been determined by H^2O^2 and malondialdehyde fractions. In halophytic plant tissues, there is a considerable accumulation of phenolic and flavonoidic compounds under sodium sulfate salt stress. Under these circumstances, it exerts strong antioxidant activity. In addition, it plays a pivotal role for these phenolic and flavonoidic compounds in counteracting many salts in salinized soils, altering morphophysiological and biochemical responses in *Prosopis strombulifera* (Reginato et al., 2014). The role of glycine-betaine has also been observed in reducing lipid peroxidation in halophytic plants (Cushman, 2001).

A lipid molecule takes part in repairing damaged cells in live cells and membrane systems. Autooxidation of PUFA fractions is promoted by a radical, for instance, a hydroxyl free radical. When hydroxyl radicals react with fatty acids, they produce carbon-lipid-centered radicals (Cosgrove et al., 1987). Under different physiological conditions, carbon-lipid-centered radicals readily trap the molecular oxygen to synthesize lipid peroxyl radicals (Hasegawa and Patterson, 1978; Niki and Noguchi, 2004), chain carriers in the autooxidation mechanism of lipids. In the oxidation process, the cis-configuration of the fatty-acyl chain is changed to form a trans-configuration (Chatgilialoglu et al., 2000; Chatgilialoglu and Ferreri, 2005) or may fragment into alpha-betta unsaturated aldehyde (electrophilic) (Sun and Salomon, 2004; Esterbauer et al., 1991). The breakdown of cis double bonds and peroxidation may result in a reduction in membrane fluidity. Autooxidation of PUFA fractions produces an array of secondary deleterious cell destruction by-products that are responsible for many diseased impacts (Kagan et al., 2000). PUFA within the inner mitochondria is keenly sensitive to ROS, which gives rise to oxidative damage (Barnham et al., 2004). Oxidative signaling mechanisms are developed from the synthesis of alpha-betta unsaturated (single bond) aldehydes that further react with nucleotides (West and Marnett, 2006). In addition, many oxidation signals have been reported in mitochondria, where oxidation and peroxidation processes are completed (Kagan et al., 2005).

14.4.4 MOLECULAR TARGETS OF PROTEINS IN HALOPHYTES

Aquaporin protein plays an important role, particularly in abiotic stress such as salt, drought, and cold, for example, *Sesuvium portulacastrum* (a well-known halophytic species). With respect to glycophytes, halophytes have excellent salt and drought tolerance. The plant halophytes could increase salt tolerance (SpAQP1 gene) in transgenic tobacco by enhancing the antioxidant properties (Chang et al., 2016).

Many protein functions are affected by reactive O^2 species or intermediates. ROS attack is responsible for changing the conformational structures that regulate protein activity. Cysteine (disulfide bond) residue in cysteine amino acid (building blocks for protein synthesis) functions as a redox switch. Oxidation of the Cys-fragment may cause structural conformational alterations of proteins. In the phosphatase enzyme, cysteine residue oxidation is a crucial molecular target at the protein level in biological systems, as it influences protein phosphorylations. Thus, it affects the overall functioning of the proteome (total protein constitution) in a cell. Oxidative damage and stress in proteins lead to the synthesis of carbonyl derivatives with a basic skeleton. Such molecular targets at the protein level are preowned as an indicator for common cellular injuries in halophytic plant cells (Finkel and Holbrook, 2000).

14.5 REACTIVE OXYGEN SPECIES OR INTERMEDIATES, SCAVENGERS, AND ANTIOXIDANTS IN HALOPHYTES

Different oxidative ions and their generated stress in halophytic species include K^+ retention, sodium secretion and compartmentalization, osmotic adjustment, and ROS homeostasis (Hamed, 2022). To cope with salinity stress, different antioxidant protective strategies have been reported in terms of various salt stresses in halophytes (Kumar et al., 2019). An extrinsic or exogenous supply of salicylic acid, a signaling molecule, increases oxidative damage in barley (*Hordeum vulgare*) under saline and drought conditions. Strong, efficient scavenging properties of ROS have been detailed in *Hordeum vulgare's* functions as a signaling molecule. The role of salicylic acid includes (i) an increase in antioxidant enzymatic function and (ii) the maintenance of photosynthetic activity (Habibi, 2012).

In halophytes, such as *Vigna radiata*, glycine-betaine and ethylene are responsible for the mitigation of salinity-induced photosynthesis and inhibition of growth through 2-hydroxybenzoic acid (Khan et al., 2014). In cells, enzymatic and nonenzymatic biochemical mechanisms of antioxidants are required to regulate the amount of ROS and related intermediate species (Fridovich, 1978). The discovery of the SOD enzyme remains a strong historical invention in the arena of free antioxidant biology (Imlay, 2011). The SOD enzyme indicated that ROS are digested, whereby they will cause high oxidative damage in cells (Imlay and Fridovich, 1991). Enzymatic and nonenzymatic antioxidant processes have evolved with successive time, namely, NADH peroxidase, glutathione, and catalase (Imlay, 2008). Ascorbic acid, glutathione, and alpha-tocopherol (dissolving in fat) are the best examples of nonenzymatic antioxidants (Lesser, 2006; Halliwell and Gutteridge, 2007; Thomas, 1995). Vitamins C, E, and glutathione are free radicals' scavengers. The glutathione and thiol-free radicals may react with proteins or lipids. It is responsible for the synthesis of disulfide bonds (S–S bonds) with the thiolate of secondary glutathione molecules. This bonding is further scavenged by O^2, a disulfide (neutral or inert) (Jonsson et al., 1993).

Much antioxidant activity of reduced glutathione and a low level of ROS have been reported in halophytes of the *Hibiscus* genera under salt stress. There are two types of cultivars of *Hibiscus*, namely, sensitive and tolerant. In the former cultivar, the degree of photoinhibition in photosystem-II was compared to the latter cultivar under salt conditions (Lu et al., 2022).

Vitamin C yields ascorbyl radicals that are freely digested, so new secondary or additional free radicals' products are synthesized. Alpha-tocopherol (a prime member of vitamin E) is a natural, cosmopolitan, and most active antioxidant. Its effects have been observed predominantly in autooxidation (free radical chain initiator) reactions. Alpha-tocopherol produces tocopheroxyl radicals that react freely through hydrogen transfer. Alpha-tocopherol terminates the two-chain reaction (Jonsson et al., 1993), and radicals may also scavenge ascorbate sites where ascorbate ions are considered as ROS sink (Kagan et al., 2005). Plants will avoid the abnormal consequences of surplus ROS by enhancing antioxidant enzyme action. Oxidation preventing molecules such as ascorbate and glutathione possess the potential to control the ROS concentration in terms of salt tolerance (Dumanović et al., 2021).

14.6 COMMON ROUTES OR PATHWAYS FOR ROS IN HALOPHYTES

Various pathways have been accepted for ROS generation through different enzymatic systems. The whole mechanism possesses many steps to generate and regulate ROS breakdown and synthesis. ROS are considered as by-products of aerobic metabolism in halophytes and are generated through many cellular counterparts such as chloroplasts (Dietz et al., 2016), mitochondria (Huang, et al., 2016), and peroxisomes (Sandalio and Romero-Puertas, 2015). All steps have been narrated in simplified form (Fig. 14.1, adapted and modified, from Nathan and Ding, 2010).

Step 1: In the cytosol, the conversion of superoxide anions $O^{2\cdot-}$ to H^2O^2 using the copper zinc superoxide dismutase enzymatic system is done by reducing the level of $ONOO^-$. Step 2: In mitochondria, manganese superoxide of eukaryotic cells and ferrous and nickel superoxide in prokaryotes break down a similar reaction. Step 3: In the case of hydrogen peroxide, if the amount is high, then it favors the catalysis of H^2O^2 to water and oxygen. If the amount is high, then it follows catalytic function and functions as peroxide and more reactive substrates. Step 4: The glutathione redox mechanism is powered by NADPH, and they possess glutathione peroxidase, glutaredoxin, glutathione reductase, and cysteinyl tripeptide glutathione. This process reduces the disulfide bond of proteins, donates a single electron to a few peroxiredoxins or ascorbates, alters the S-nitroso bonds through glutathioxydation of cysteinyl sulfhydryls, and removes the xenobiotics. Step 5: The Trx redox pathway is powered by NADP. Trx peroxidase, Trx reductase, and oxidorectase Trx actively participate in the mechanism. This cycle reduces the disulfide bond of proteins and donates a single electron to ribonucleo-trireductase, methionine sulfo-oxidoreductase, and peroxiredoxin. Trx binds and controls many signals. Step 6: Secreted thioredoxin is chemotactic in nature, reduces disulfide bonds of proteins, and retains the cytokine (signaling molecule) like activity. Step 7: Secreted thio-redoxin types (1 to 6) reduce hydrogen peroxide, organic products, or peroxyl nitril anions ($ONOO_{-}$). Step 8: In the cytosol, mitochondria are engaged in peroxiredoxin catalysis. In the principle methionine (sulfur containing amino acids), sulfoxide reductase, the methione-S-sulfoxide and the methionine-R- sulfoxide changed to methinine amino acids. This is the most important step for the consumption of the cost of NADPH. Step 9: Alpha-keto acids include alpha-ketoglutaryl pyruvate. Pyruvate reacts faster, stoichiometrically and nonenzymatically, to digest (and vice versa) via the process of oxidative decarboxylation.

Step 10: Superoxide anions ($O^{2.-}$) convert into more reactive products and less reactive nitric oxide (NO) forms. Step 11: Ascorbic acid breaks down the hydrogen peroxide. And finally, Step 12: Tocopherol (vitamin E) reacts with lipid radicals and induces ROS (Nathan and Ding, 2010). No signaling participates in salt resistance among halophytes, and it improves various stress tolerances in terms of high salinity, drought, waterlogging, and mechanical wounding (Hamed, 2022). In addition, nitric oxide (salt-induced ROS) is a crucial regulator in relation to salt stress and performs the principle emphasis downstream of abscisic acid. The interaction between NO and ABA in relation to salt stress has been explored (León et al., 2014). ROS and ethylene are crucial biochemical signaling components that perform several biological functions, including cell fate determination, root and root hair development, and involving biotic and abiotic responses (Steffens, 2014).

ROS production in halophytes is a systematic and sequential reaction, which targets different substrates depending on the oxidative stages. Each and every step has enzyme specificity in terms of production.

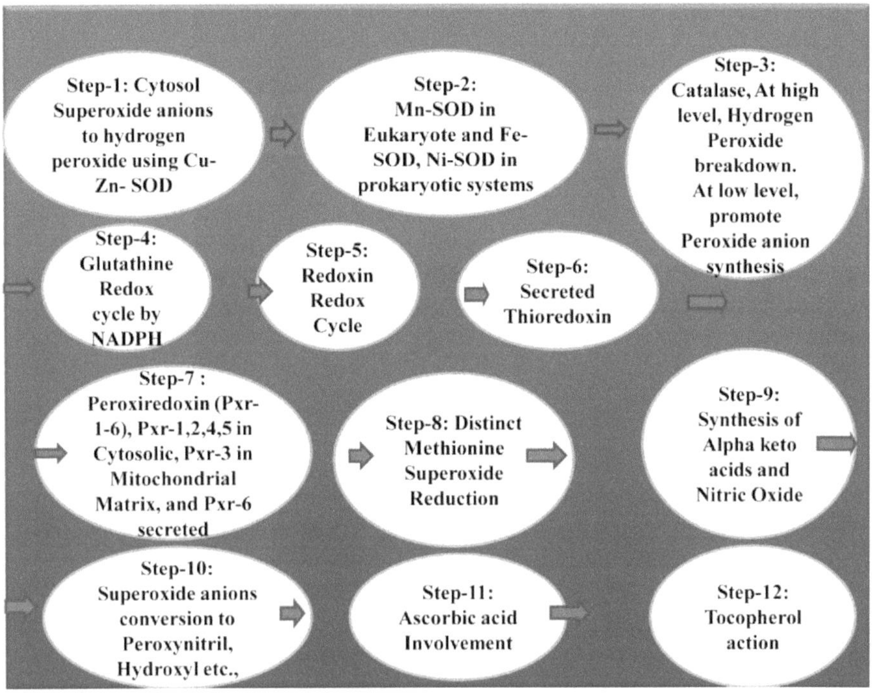

FIGURE 14.1 Common routes for ROS generation through enzymatic systems.

In halophytes, many antioxidative enzymes are involved in ROS homeostasis. Some selected comparative studies of antioxidant enzymes in glycophytes and halophytes have been discussed in tabulated form that are directly (intrinsic or endogenously) or indirectly (extrinsic) in halophytic ROS regulatory pathways (Table 14.3).

14.7 ROS SYNTHESIS, SITES, AND MECHANISMS IN HALOPHYTES

Under normal conditions, ROS are synthesized as a final source of anabolic and catabolic reactions in halophytes. Oxidative damage and salt stress cause adverse effects on cellular metabolism. In spite of the fact that, under normal conditions, oxidative stress and damage can also occur as secondary components of other stress, for instance, salinity, the uncoupling of light reactions with products of dark reactions, is the principle source of ROS production. Mechanisms of synthesis of ROS are similar in all plants, and halophytic plants are not unique to this. In general, the antioxidant defense system retains

TABLE 14.3 Some Selected Comparative Studies of Antioxidant Enzymes in Glycophytes and Halophytes.

S. no.	Glycophytic species	Halophytic species	Valuable findings in terms of antioxidative enzyme activity	Reference(s)
1.	*Arabidopis thaliana*	*Cakile maritima*	Increased levels of POX activity in halophytes	Ellouzi et al., 2011
2.	*Hordeum vulgare*	*Hordeum marinum*	GR5 (novel isozymes) found first time and activities of GR1, 3, 6, 7 have been reported in halophytes	Seckin et al., 2010
3.	*Hordeum vulgare*	*Hordeum marinum*	Increases POX activity in halophytes	Seckin et al., 2010
4.	*Plantago media*	*Plantago maritima*	APX activity increased in halophytes but remain constant in glycophytes at 0.2M	Sekmen et al., 2007
5.	*Plantago major*	*Thellungiella salsuginea*	Constitutively lower in glycophytes, but higher in halophytes	Radyukina et al., 2007
6.	*Lycopersicon esculentum* (*Solanum lycopersicum*	*Lycopersicon pennellii*	Higher GPX activity in halophyte mitochondrial fractions in relation to glycophytes	Mittova et al., 2003
7.	*Arabidopis thaliana*	*Thellungiella salsuginea*	In halophytes, excessive production of CAT 3 isoforms.	Wang et al., 2004
8.	*Hordeum vulgare*	*Hordeum marinum*	Increased CAT activity and constitutively higher expression in halophytes	Seckin et al., 2010
9.	*Lycopersicon esculentum*	*Lycopersicon pennellii*	Constitutive higher CAT activity in peroxisome. In addition, further increased in halophytes under salt stress	Shalata et al., 2001
10.	*Arabidopis thaliana*	*Cakile maritime*	Higher SOD activity in halophytes with respect to glycophytes	Ellouzi et al., 2011

the potential to cope with these ROS stress conditions. Oxidative injuries or damage and the removal or detoxification of abnormal cellular components of ROS are involved in plant cell, tissue growth, development, and stress tolerance (Nikalje et al., 2018). Halophytes are saline soil flora. The vacuole, the cell organelle, is a storing site of salt cations and anions in halophytes. They adjust the osmotic pressure under salt salinity. In the cytoplasm, many organic solutes are stored to minimize the abnormal metabolic effects.

In halophytes, sodium chloride toxicity and the cellular basis of salt tolerance have been reported (Flowers et al., 2015). There are three cellular organelles (chloroplast, mitochondria, and peroxisomes) that are actively participating in ROS production. They also participate in scavenging ROS effects in halophytic plants cooperatively. In chloroplasts, ROS ($^{1}O^{2}$, $O^{2\cdot -}$, $H^{2}O^{2}$, and $HO\cdot$) are produced through biochemical catalysis. Different sources include Fenton or Haber–Weiss reaction products, photosystem-I Mehler reactions of photosynthesis components, and photosystem-II overexcitation. $^{1}O^{2}$ can also synthesize in PS-II of photosynthesis when the electron transport chain of mitochondria is overcharged (overloaded). The $H^{2}O$-$H^{2}O$ cycle is a sensitive tool to stop much reactive $^{1}O^{2}$ form that can be overcome through nonenzymatic reactions (Sunkar et al., 2003). When stomata are closed, carbon fixation inhibitory action generates different forms of ROS, particularly $^{1}O^{2}$, superoxide anions, and $H^{2}O^{2}$ under a salinity-stressed state (Kumar et al., 2019).

$^{1}O^{2}$ and $HO\cdot$ free radicals are the most reactive oxidative damage molecules in ROS. The latter molecule cannot be removed enzymatically. The detoxification process is based on nonenzymatic antioxidants like β-carotene and alpha-tocopherol. $^{1}O^{2}$ and $HO\cdot$ free radicals are initial substrates for lipid peroxidation (Sunkar et al., 2003). This is exhibited by oxidative decarboxylation of PUFA in the membrane. This process can be stimulated in many ways, namely, (i) autooxidation, (ii) photooxidation, and (iii) lipooxygenase or cyclooxygenases (Feussner and Wasternack, 2002). This is a nonenzymatic autooxidation reaction mechanism owing to ROS. The abundance of lipid peroxidation affects the stereospecificity of membranes and permeability, ultimately resulting in the leakage of ions and the synthesis of a few other metabolites (Amor et al., 2006). The threshold salt concentration at which lipid peroxidation increases has been reported (Table 14.4).

TABLE 14.4 Examples of Halophytic Species and Their Effective Salt Concentrations in Oxidative Damage Under Salt Stress.

S. no.	Halophytic species	Effective salt concentration causing oxidative damage	References
	Salicornia europaea;	0.2–0.5M	Mohammadi and Kardan, 2016
	Atriplex leucoclada;	0.2–0.5M	
	Kochia scoparia	0.2–0.5M	
	Gypsophila oblanceolata	0.150M	Sekmen et al., 2012
	Suaeda europaea	0.2M	Aghaleh et al., 2009
	Cakile maritima (Jerba)	0.2M	Ben Amor et al., 2006
	Tecticornia indica	2.0M	English and Colmer, 2013
	Atriplex portulacoides	0.1M	Benzarti et al., 2012
	Tecticornia pergranulata	2.0M	English and Colmer, 2013

TABLE 14.4 *(Continued)*

S. no.	Halophytic species	Effective salt concentration causing oxidative damage	References
	Sesuvium portulacastrum	0.4–0.6M	Lokhande et al., 2011
	Hordeum marinum	0.3M	Seckin et al., 2010
	Salicornia branchiate	0.4M	Parida and Jha, 2010
	Crithmum maritimum	*0.2M*	Ben Amor et al., 2006
	Suaeda persica	0.3M	Aghaleh et al., 2009
	Spartina alterniflora	0.2M	Song and Wang, 2015
	Prosopis strombulifera	1M	Reginato et al., 2012
	Crithmum maritimum	0.3M	Ben Hamed et al., 2007

Many physiological as well as morphological responses may be affected by salt stress (variable concentrations); for instance, *Salicornia europaea* possesses high shoots and dry root matter with respect to a few other halophytic species. Under these circumstances, up to 0.2M photosynthetic pigment (PP) production is more. Thereafter, it reduces the level of pigment drastically. The photosynthetic pigment in halophytic species is found to be high in *Atriplex leucoclada* with respect to others. The highest and lowest Na^+ cations among halophytes have been reported in *Salicornia europaea* and *Kochia scoparia* sequentially. The sodium/potassium cation proportions could be an index for salt detection. Protein and nitrogen contents, metabolizable energy, and digestibility parameters have been found more in common glasswort in relation to other halophytic species under salt-stressed conditions (Mohammadi and Kardan, 2016).

In addition, hydroxyl radicals and nonenzymatic lipid peroxidation products can alter the protein configuration and their biological activity. Major harmful components, lipid peroxidation, or aldehyde, and detoxifying the aldehyde are very crucial to prevent further oxidative damage. The enzymes dehydrogenase and aldose reductase play a key role in the detoxification or removal process of their components. Overexpression of enzymes such as aldehyde or aldose reductase and dehydrogenase stimulated salinity tolerance in *Arabidopsis*. It also provides protection against lipid peroxidation and oxidative injuries (Kotchoni et al., 2006). Glycine-betaine aldehyde dehydrogenase, a class of aldose dehydrogenase, is the principle catalyst in halophytes. Thus, this enzyme actively participates in the production of glycine-betaine (Fitzgerald et al., 2009).

ROS can alter the membrane protein configuration through the oxidation of a few amino acid residues. Cysteine and methionine amino acids are very sensitive to oxidation. The thiol residues of cysteine can be sulfonic acid, sulfinic acid, or disulfide bonds that produce oxidized sulfinic acid upon oxidation of the thiol residues of cysteine. Disulfide and sulfinic acid products are reversible in a redox state and assist in protein regulation. The sulfinic acid can be removed via sulfiredoxins. Sulfinic acid oxidation is irrevocable (Rinalducci et al., 2008). Protein carbonylation has been reported to be much higher in mitochondria with respect to chloroplasts and peroxisomes under salt stress. The mitochondrial membrane is a very sensitive and principle site for oxidative damage in halophytic species (Bartoli et al., 2004).

Oxidative modification of proteins in halophytes is too restricted; for instance, under salt stress, lipid peroxidation levels increased, but protein carbonylation remained the same in *Hordeum marinum* (Hafsi et al., 2010). Many reports have shown that in cells, lipids and proteins are crucial targets of oxidative damage in ROS, predominantly. An estimation of the oxidation state of proteins bound to thiol groups can be a crucial factor to reflect the oxidative state of cellular components (Ben Hamed et al., 2007).

In halophytes, the TBARS assay is important in the determination of lipid peroxidation levels under salt stress or oxidative injury (Hayakawa and Agarie, 2010). Oligosaccharides and beta pigments (mainly anthocyanin and betacyanin, a water-soluble pigment) can be estimated in the TBARS assay. Many molecules can accumulate various kinds of metabolic products that interact with a probe (Lugan et al., 2010).

In addition, the TBARS assay, an electrolyte efflux, and candid membrane injuries via enhanced permeability are used to estimate the level of oxidative damage in the membranes of halophytic plants (Ben Hamed et al., 2007). Oxidation of proteins is detected through the resolution of carbinylate amino acid residues, which is irreversible oxidative damage. TBARS doesn't interact with the thiol residues of proteins, which are greatly altered by the oxidative surplus load in cells. Measurement of the oxidative state of the thiol group of proteins may be essential to estimate the redox potential of cells (Ozgur et al., 2013)

Many enzymes, such as alpha-tocopherol, β-carotene, SOD, Apx, CAT, Gpx, peroxiredoxin, etc., act as scavengers. Antioxidative enzymes are engaged to prevent cellular oxidative damage in halophytes. In mitochondria, ROS ($O^{2\cdot-}$, H^2O^2, $HO_\cdot$) are produced through complex-I and III of ETC, dismutation of superoxide anions, and Fenton or Haber–Weiss reaction products. The ROS source includes multifarious catalysts such as SOD, Apx, glutathione peroxidase, and peroxiredoxin. These sources are participating in the detoxification of cellular abnormal products (Ozgur et al., 2013).

In the peroxisome, ROS (H^2O^2, HO·) are generated in the channel of photorespiration and the Fenton, or Haber–Weiss reaction. A few sources of ROS in peroxisomes include CAT and APX, which catalyze the biochemical functioning to detoxify cellular products. Upon the action of many chemical signals or in stress-related ROS, $O^{2\cdot-}$, H^2O^2, and HO· are produced in the plasma membrane apoplastic site through the catalytic action in halophytic plants (figure adapted and modified from Ozgur et al., 2013; Figs. 14.2 and 14.3, adapted and modified from Nathan and Ding, 2010).

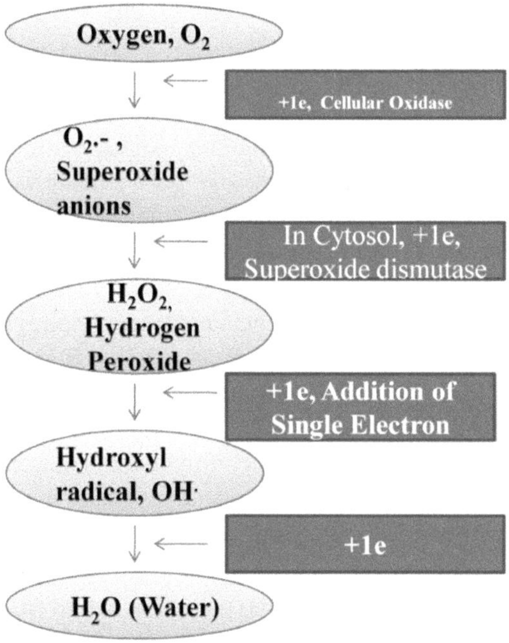

FIGURE 14.2 ROS synthesis and breakdown.

Under salinity conditions, stomata are closed to reduce the amount of water, which limits the gaseous conversion. Owing to the restriction of gaseous exchange, several redox potential states are seen in plant cells. Lesser amounts of CO^2 can be sensational in relation to ROS synthesis when photosynthetic active radiation (PAR) is too high (Bauwe et al., 2010). Due to stomata closing, it reduces carbon dioxide and oxygen ratios within leaf tissues and thus inhibits the carbon fixation mechanism. This inhibitory action leads to a greater reduction in complex-I and III components of the ETC of mitochondria (Kumar et al., 2019).

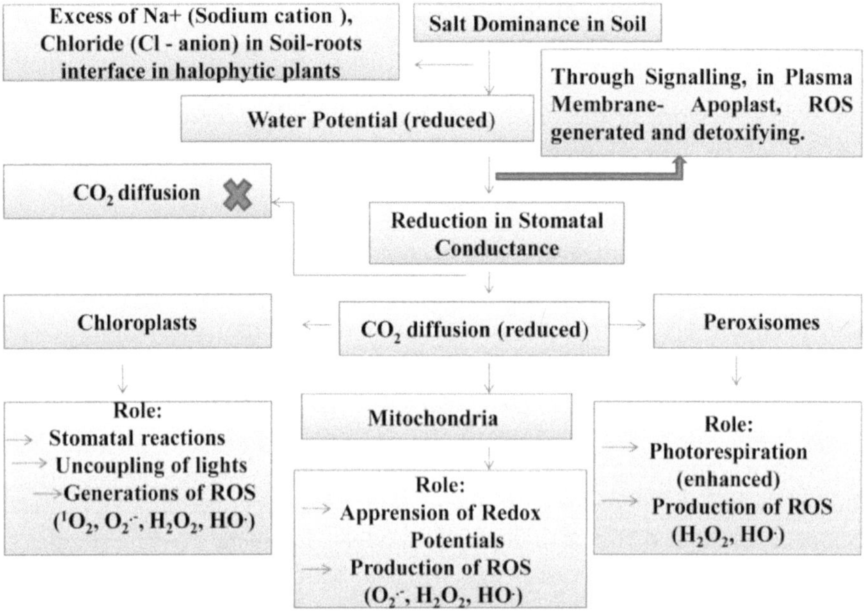

FIGURE 14.3 Synthesis, sites, and mechanisms of ROS under salinity in halophytes.

Under salt stress, the ROS-producing mechanism is caused by restricted gaseous conversion, which facilitates photorespiration. In peroxisomes, H^2O^2 is produced through glycolate oxidase. Under oxidative damage and salt stress states, any decreased level of CO^2 increases the rate of photo-respiration. Subsequently, an inadequate antioxidant protection system in photorespiration can augment oxidative consignment, ultimately resulting in cellular injuries. Notwithstanding, this is a very crucial step in halophytic plants through C4-type photosynthesis, owing to the CO^2 concentration system around RUBISCO, a dual-functioning enzyme, oxygenation (oxygenase) as well as carboxylation (carboxylase) (Bauwe et al., 2012).

Coordination between cellular organelles such as chloroplasts, mito-chondria, and peroxisomes is essential to produce ROS and other highly reactive species. As excessive ROS production causes adverse effects, an antioxidative strategy is required to overcome such effects under salt stress. In order to overcome this, halophytes perform in different enzymatic and nonenzymatic modes to complete the mechanisms of ROS production.

14.8 DISCUSSIONS, CONCLUDING REMARKS, AND FUTURE PROSPECTS

An extensive work of the biochemical and physiological adaptation of the xerohalophyte *Salsola drummondii* to seasonal alterations in salty areas has been explored and it could be used as a plant model organism in an arid saline environment in the near future (Rasheed et al., 2022). In halophytes, broadening prospectives in research for the purpose of cultivation, commercialization, and utilization of salt-tolerant crop species have been shown effectively (Barros et al., 2021). In salt-tolerant plants, studies on ROS to date have been centered on the detoxification process in relation to salt stress. However, the regulatory and signaling functions of ROS are too restricted. The roles of redox-active components such as peroxisomes, thioredoxin, and glutaredoxin in halophytes are still not obvious. Thus, further studies are indeed required to explore the molecular and biochemical evidence through different halophytic models, for instance, *Thellungiella.* Data extracted from *Arabidopsis* will be beneficial to explore similar ROS production and defense strategies in terms of salt stress in halophytic species. The whole genome sequence analysis of *Thellungiella parvula* and *T. salsuginea* (both are glycophytes) has been explored to retain higher salt (NaCl) concentration. This genome sequence could be beneficial to understand the defense mechanism in relation to salt concentration (Dassanayake et al., 2011; Wu et al., 2012b). This is a great exercise to assess and enhance fundamental comprehension in terms of salinity in halophytes. Furthermore, valuable information is obtained by applying pyrosequencing methods, which provides extra backup with proteomics and the transcriptomic concept of halophytes. It is very essential to elucidate whole knowledge about the halophytic salt tolerance strategies. The production and mechanism of ROS in monocotyledonous halophytic model plant are still unexplored (Byrt and Munn's., 2008). Although dicotyledonous halophytes model seems to be a crucial source in terms of salt tolerance, the monocotyledonous halophytic model plant is also required to elucidate the difference between monocotyledonous and dicotyledonous salt-tolerant species. In the present chapter, there is reasonable information about the constituents of antioxidants (role of antioxidant enzymes and contents of nonenzymatic, for instance, glutathione and ascarbate). In salt-tolerant plants, research on other constituents of antioxidant components and the link between ROS and signaling is restricted with respect to salt stress. Further, multidisciplinary omic approaches such as proteomics, phenomics, transcriptomics, and metabologenomics are needed to assess the ROS or antioxidant interaction. This approach would provide further details on the biochemical and molecular facets of ROS, antioxidants, and salt stress in halophytes. A special emphasis on aldehydes reductase or dehydrogenase in relation to high salinity is to be a part of further investigation in halophytes. Similarly, in halophytes, a biotechnological approach can be used to establish a clear-cut relationship between ROS and oxidative damage in the near future.

KEYWORDS

- **antioxidant**
- **halophytes**
- **enzymatic and nonenzymatic**
- **metabolism**
- **oxidative damage**
- **reactive oxygen species**
- **redox potential**
- **scavenger**
- **salt stress and tolerance**

REFERENCES

Aghaleh, M.; Niknam, V.; Ebrahimzadeh, H.; Razavi, K. Salt Stress Effects on Growth, Pigments, Proteins and Lipid Peroxidation in *Salicornia persica* and *S. europaea*. *Biol. Plant.* **2009,** *53* (2), 243–248.

Ahanger, M. A.; Alyemeni, M. N.; Wijaya, L.; Alamri, S. A.; Alam, P.; Ashraf, M.; Ahmad, P. Potential of Exogenously Sourced Kinetin in Protecting *Solanum lycopersicum* from NaCl-induced Oxidative Stress Through Up-Regulation of the Antioxidant System, Ascorbate-Glutathione Cycle and Glyoxalase System. *PLoS One* **2018,** *13* (9), e0202175.

Amor, N. B.; Jiménez, A.; Megdiche, W.; Lundqvist, M.; Sevilla, F.; Abdelly, C. Response of Antioxidant Systems to NaCl Stress in the Halophyte *Cakile maritima*. *Physiol. Plant.* **2006,** *126* (3), 446–457.

Barnham, K. J.; Masters, C. L.; Bush, A. I. Neurodegenerative Diseases and Oxidative Stress. *Nat. Rev. Drug Discov.* **2004,** *3* (3), 205–214.

Barros, N. L. F.; Marques, D. N.; Tadaiesky, L. B. A.; de Souza, C. R. B. Halophytes and other Molecular Strategies for the Generation of Salt-Tolerant Crops. *Plant Physiol. Biochem.* **2021,** *162*, 581–591.

Bartoli, C. G.; Gómez, F.; Martinez, D. E.; Guiamet, J. J. Mitochondria are the Main Target for Oxidative Damage in Leaves of Wheat (*Triticum aestivum*L.). *J. Exp. Bot.* **2004,** *55* (403), 1663–1669.
Bauwe, H.; Hagemann, M.; Fernie, A. R. Photorespiration: Players, Partners and Origin. *Trends Plant Sci.* **2010,** *15* (6), 330–336.
Bauwe, H.; Hagemann, M.; Kern, R.; Timm, S. Photorespiration has a Dual Origin and Manifold Links to Central Metabolism. *Curr. Opin. Plant Biol.* **2012,** *15* (3), 269–275.
Ben Hamed, K.; Castagna, A.; Salem, E.; Ranieri, A.; Abdelly, C. Sea Fennel (*Crithmum maritimum* L.) Under Salinity Conditions: A Comparison of Leaf and Root Antioxidant Responses. *Plant Growth Regul.* **2007,** *53* (3), 185–194.
Benzarti, M.; Ben Rejeb, K.; Debez, A.; Messedi, D.; Abdelly, C. Photosynthetic Activity and Leaf Antioxidative Responses of *Atriplex portulacoides* Subjected to Extreme Salinity. *Acta Physiologiae Plantarum* **2012,** *34* (5), 1679–1688.
Beyer, W.; Imlay, J.; Fridovich, I. Superoxide Dismutases. *Prog. Nucleic Acid Res.* **1991,** *40,* 221–53.
Boveris, A.; Chance, B. The Mitochondrial Generation of Hydrogen Peroxide. General Properties and Effect of Hyperbaric Oxygen. *Biochem. J.* **1973,** *134* (3), 707–716.
Buettner, G. R. The Pecking Order of Free Radicals and Antioxidants: Lipid Peroxidation, α-Tocopherol, and Ascorbate. *Arch. Biochem. Biophys.* **1993,** *300* (2), 535–543.
Byrt, C. S.; Munns, R. Living with Salinity. *New Phytol.* **2008,** 903–905.
Chang, W.; Liu, X.; Zhu, J.; Fan, W.; Zhang, Z. An Aquaporin Gene from Halophyte *Sesuvium portulacastrum*, SpAQP1, Increases Salt Tolerance in Transgenic Tobacco. *Plant Cell Rep.* **2016,** *35* (2), 385–395.
Chatgilialoglu, C.; Ferreri, C. Trans Lipids: The Free Radical Path. *Acc. Chem. Res.* **2005,** *38* (6), 441–448.
Chatgilialoglu, C.; Ferreri, C.; Ballestri, M.; Mulazzani, Q. G.; Landi, L. Cis–Trans Isomerization of Monounsaturated Fatty Acid Residues in Phospholipids by thiyl Radicals. *J. Am. Chem. Soc.* **2000,** *122* (19), 4593–4601.
Christakis, C. A.; Daskalogiannis, G.; Chatzakis, A.; Markakis, E. A.; Sagia, A.; Rizzo, G. F.; Sarris, P. F. Halophytic Bacterial Endophytome: A Potential Source of Beneficial Microbes for a Sustainable Agriculture. *BioRxiv* 2020.
Cosgrove, J. P.; Church, D. F.; Pryor, W. A. The Kinetics of the Autoxidation of Polyunsaturated Fatty Acids. *Lipids* **1987,** *22* (5), 299–304.
Cushman, J. C. Osmoregulation in Plants: Implications for Agriculture. *Amer. Zool.* **2001,** *41* (4), 758–769.
Dassanayake, M.; Oh, D. H.; Haas, J. S.; Hernandez, A.; Hong, H.; Ali, S.; Cheeseman, J. M. The Genome of the Extremophile Crucifer *Thellungiella parvula*. *Nat. Genet.* **2011,** *43* (9), 913–918.
Davies, K. J. A. Protein Damage and Degradation by Oxygen Radicals. *J. Biol. Chem.* **1987,** *262,* 9895–9901.
D'Autréaux, B.; Toledano, M. B. ROS as Signalling Molecules: Mechanisms that Generate Specificity in ROS Homeostasis. *Nat. Rev. Mol. Cell Biol.* **2007,** *8* (10), 813–824.
Dean, R. T.; Gieseg, S.; Davies, M. J. Reactive Species and Their Accumulation on Radical-Damaged Proteins. *TrendsBiochem. Sci.* **1993,** *18*, 437–441.
Dietz, K. J.; Turkan, I.; Krieger-Liszkay, A. Redox-and Reactive Oxygen Species-Dependent Signalling into and out of the Photosynthesizing Chloroplast. *Plant Physiol.* **2016,** *171* (3), 1541–1550.

Dröge, W. Free Radicals in the Physiological Control of Cell Function. *Physiol. Rev.,* 2002.

Dumanović, J.; Nepovimova, E.; Natić, M.; Kuča, K.; Jaćević, V. The Significance of Reactive Oxygen Species and Antioxidant Defense System in Plants: A Concise Overview. *Front. Plant Sci.* **2021,** *11,* 552969.

Ellouzi, H.; Ben Hamed, K.; Cela, J.; Munné-Bosch, S.; Abdelly, C. Early Effects of Salt Stress on the Physiological and Oxidative Status of *Cakile maritima* (halophyte) and *Arabidopsis thaliana* (glycophyte). *Physiologia Plantarum* **2011,** *142* (2), 128–143.

English, J. P.; Colmer, T. D. Tolerance of Extreme Salinity in two Stem-Succulent Halophytes (*Tecticornia species*). *Funct. Plant Biol.* **2013,** *40* (9), 897–912.

Esterbauer, H.; Schaur, R. J.; Zollner, H. Chemistry and Biochemistry of 4-Hydroxynonenal, Malonaldehyde and Related Aldehydes. *Free Radic. Biol. Med.* **1991,** *11* (1), 81–128.

Finkel, T. Signal 岡田 Duction by Reactive Oxygen Specil巴 s. よ Cell. *Biol* **2011,** *194,* 7–15.

Finkel, T.; Holbrook, N. J. Oxidants, Oxidative Stress and the Biology of Ageing. *Nature* **2000,** *408* (6809), 239–247.

Fitzgerald, T. L.; Waters, D. L.; Henry, R. J. Betaine Aldehyde Dehydrogenase in Plants. *Plant Biol.* **2009,** *11* (2), 119–130.

Feussner, I.; Wasternack, C. The Lipoxygenase Pathway. *Annu. Rev. Plant Biol.* **2002,** *53,* 275.

Flowers, T. J.; Munns, R.; Colmer, T. D. Sodium Chloride Toxicity and the Cellular Basis of Salt Tolerance in Halophytes. *Ann. Bot.* **2015,** *115* (3), 419–431.

Fridovich, I. The biology of Oxygen Radicals: The Superoxide Radical is an Agent of Oxygen Toxicity; Superoxide Dismutases Provide an Important Defense. *Science* **1978,** *201* (4359), 875–880.

Green, D. R.; Reed, J. C. Mitochondria and Apoptosis. *Science* **1998,** *281,* 1309–1312.

Green, J.; Paget, M. S. Bacterial Redox Sensors. *Nat. Rev. Microbiol.* **2004,** *2* (12), 954–966.

Habibi, G. Exogenous Salicylic Acid Alleviates Oxidative Damage of Barley Plants Under Drought Stress. *Acta BiologicaSzegediensis* **2012,** *56* (1), 57–63.

Hafsi, C.; Romero-Puertas, M. C.; Gupta, D. K.; Luis, A.; Sandalio, L. M.; Abdelly, C. Moderate Salinity Enhances the Antioxidative Response in the Halophyte *Hordeummaritimum* L. under Potassium Deficiency. *Environ. Exp. Bot.* **2010,** *69* (2), 129–136.

Halliwell, B. Oxidative Stress and Neurodegeneration: Where are We Now? *J. Neurochem.* **2006,** *97* (6), 1634–1658.

Halliwell, B.; Gutteridge, J. M. C. *Free Radicals in Biology and Medicine*; Oxford University Press: Oxford, 2015.

Hamed, K. B. Responses of Halophytes to Nitric Oxide (NO). In *Nitric Oxide in Plant Biology*; Academic Press, 2022, pp 391–406.

Hassan, F.; Al-Yasi, H.; Ali, E.; Alamer, K.; Hessini, K.; Attia, H.; El-Shazly, S. Mitigation of Salt-Stress Effects by Moringa Leaf Extract or Salicylic Acid Through Motivating Antioxidant Machinery in Damask Rose. *Can. J. Plant Sci.* **2020,** *101* (2), 157–165.

Hasegawa, K.; Patterson, L. K. Pulse Radiolysis Studies in Model Lipid Systems: Formation and Behavior of Peroxy Radicals in Fatty acids. *Photochem. Photobiol.* **1978,** *28* (4–5), 817–823.

Hidalgo, E.; Ding, H.; Demple, B. Redox Signal Transduction via Iron-Sulfur Clusters in the SoxR Transcription Activator. *Trends Biochem. Sci.* **1997,** *22* (6), 207–210.

Huang, S.; Van Aken, O.; Schwarzländer, M.; Belt, K.; Millar, A. H. The Roles of Mitochondrial Reactive Oxygen Species in Cellular Signaling and Stress Response in Plants. *Plant Physiol.* **2016,** *171* (3), 1551–1559.

Hyslop, P. A.; Hinshaw, D. B.; Halsey, W. A. Jr.; Schraufs¨atter, I. U.; Sauerheber, R. D.; et al. Mechanisms of Oxidant-Mediated Cell Injury. *J. Biol. Chem.* **1988,** *263,* 1665–1675.

Hashida, S. N.; Takahashi, H.; Uchimiya, H. The role of NAD Biosynthesis in Plant Development and Stress Responses. *Ann. Bot.* **2009,** *103* (6), 819–824.

Hayakawa, K.; Agarie, S. Physiological Roles of Betacyanin in a Halophyte, *Suaeda japonica* Makino. *Plant Prod. Sci.* **2010,** *13* (4), 351–359.

Imlay, J. A. Pathways of Oxidative Damage. *Annu. Rev. Microbiol.* **2003,** *57,* 395–418.

Imlay, J. A. Cellular Defenses Against Superoxide and Hydrogen Peroxide. *Ann. Rev. Biochem.* **2008,** *77,* 755.

Imlay, J. A. Redox pioneer: Professor Irwin Fridovich. *Antioxid. Redox Signal.* **2011,** *14* (3), 335–340.

Imlay, J. A.; Fridovich, I. Assay of Metabolic Superoxide Production in *Escherichia coli. J. Biol. Chem.* **1991,** *266* (11), 6957–6965.

Jonsson, M.; Lind, J.; Reitberger, T.; Eriksen, T. E.; Merenyi, G. Free Radical Combination Reactions Involving Phenoxyl Radicals. *J. Phys. Chem.* **1993,** *97* (31), 8229–8233.

Kagan, V. E.; Fabisiak, J. P.; Shvedova, A. A.; Tyurina, Y. Y.; Tyurin, V. A.; Schor, N. F.; Kawai, K. Oxidative Signaling Pathway for Externalization of Plasma Membrane Phosphatidyl Serine During Apoptosis. *FEBS Lett.* **2000,** *477* (1–2), 1–7.

Kagan, V. E.; Tyurin, V. A.; Jiang, J.; Tyurina, Y. Y.; Ritov, V. B.; Amoscato, A. A.; Borisenko, G. G. Cytochrome c Acts as a Cardiolipin Oxygenase Required for Release of Pro-Apoptotic Factors. *Nat. Chem. Biol.* **2005,** *1* (4), 223-232.

Khan, M. I. R.; Asgher, M.; Khan, N. A. Alleviation of Salt-Induced Photosynthesis and Growth Inhibition by Salicylic Acid Involves glycinebetaine and Ethylene in Mungbean (*Vigna radiata* L.). *Plant Physiol. Biochem.* **2014,** *80,* 67–74.

Klaunig, J. E.; Wang, Z.; Pu, X.; Zhou, S. Oxidative Stress and Oxidative Damage in Chemical Carcinogenesis. *Toxicol. Appl. Pharmacol.* **2011,** *254* (2), 86–99.

Kohanski, M. A.; Dwyer, D. J.; Hayete, B.; Lawrence, C. A.; Collins, J. J. A common mechanism of cellular death induced by bactericidal antibiotics. *Cell* **2007,** *130* (5), 797–810.

Kotchoni, S. O.; Kuhns, C.; Ditzer, A.; Kirch, H. H.; Bartels, D. Over-Expression of Different Aldehyde Dehydrogenase Genes in *Arabidopsis thaliana* Confers Tolerance to Abiotic Stress and Protects Plants against Lipid Peroxidation and Oxidative Stress. *Plant, Cell Environ.* **2006,** *29* (6), 1033–1048.

Lambeth, J. D. NOX Enzymes and the Biology of Reactive Oxygen. *Nat. Rev. Immunol.* **2004,** *4* (3), 181–189.

Leach, J. K.; Van Tuyle, G.; Lin, P. S.; Schmidt-Ullrich, R.; Mikkelsen, R. B. Ionizing Radiation-Induced, Mitochondria-Dependent Generation of Reactive Oxygen/Nitrogen. *Cancer Res.* **2001,** *61* (10), 3894–3901.

Lee, M. H.; Cho, E. J.; Wi, S. G.; Bae, H.; Kim, J. E.; Cho, J. Y.; Chung, B. Y. Divergences in Morphological Changes and Antioxidant Responses in Salt-Tolerant and Salt-Sensitive Rice Seedlings after Salt Stress. *Plant Physiol. Biochem.* **2013,** *70,* 325–335.

León, J.; Castillo, M. C.; Coego, A.; Lozano-Juste, J.; Mir, R. Diverse Functional Interactions Between Nitric Oxide and Abscisic Acid in Plant Development and Responses to Stress. *J. Exp. Bot.* **2014,** *65* (4), 907–921.

Lesser, M. P. Oxidative Stress in Marine Environments: Biochemistry and Physiological Ecology. *Annu. Rev. Physiol.* **2006,** *68,* 253–278.

Li, P.; Zhu, Y.; Song, X.; Song, F. Negative Effects of Long-Term Moderate Salinity and Short-Term Drought Stress on the Photosynthetic Performance of Hybrid *Pennisetum*. *Plant Physiol. Biochem.* **2020,** *155*, 93–104.

Lokhande, V. H.; Nikam, T. D.; Patade, V. Y.; Ahire, M. L.; Suprasanna, P. Effects of Optimal and Supra-Optimal Salinity Stress on Antioxidative Defence, Osmolytes and In Vitro Growth Responses in *Sesuvium portulacastrum* L. *Plant Cell Tissue Organ Cult.* **2011,** *104* (1), 41–49.

Lugan, R.; Niogret, M. F.; Leport, L.; Guégan, J. P.; Larher, F. R.; Savouré, A.; Bouchereau, A. Metabolome and Water Homeostasis Analysis of *Thellungiella salsuginea* Suggests that Dehydration Tolerance is a Key Response to Osmotic Stress in This Halophyte. *Plant J.* **2010,** *64* (2), 215–229.

Lu, W.; Wei, G.; Zhou, B.; Liu, J.; Zhang, S.; Guo, J. A comparative Analysis of Photosynthetic Function and Reactive Oxygen Species Metabolism Responses in two *Hibiscus* Cultivars Under Saline Conditions. *Plant Physiol. Biochem.* 2022.

Meloni, D. A.; Oliva, M. A.; Martinez, C. A.; Cambraia, J. Photosynthesis and Activity of Superoxide Dismutase, Peroxidase and Glutathione Reductase in Cotton Under Salt Stress. *Environ. Exp. Bot.* **2003,** *49* (1), 69–76.

Mohammadi, H.; Kardan, J. Morphological and Physiological Responses of Some Halophytes to Salinity Stress. In *Annales Universitatis Mariae Curie–Sklodowska, sectio C–Biologia,* **2016;** Vol. 70, No. 2, p 31.

Munns, R.; James, R. A.; Gilliham, M.; Flowers, T. J.; Colmer, T. D. Tissue tolerance: An Essential but Elusive Trait for Salt-Tolerant Crops. *Funct. Plant Biol.* **2016,** *43* (12), 1103–1113.

Nathan, C. Specificity of a Third Kind: Reactive Oxygen and Nitrogen Intermediates in Cell Signalling. *J. Clin. Invest.* **2003,** *111* (6), 769–778.

Nathan, C.; Cunningham-Bussel, A. Beyond Oxidative Stress: an Immunologist's Guide to Reactive Oxygen Species. *Nat. Rev. Immunol.* **2013,** *13* (5), 349–361.

Nathan, C.; Ding, A. SnapShot: Reactive Oxygen Intermediates (ROI). *Cell* **2010,** *140* (6), 951–951.

Nazar, R.; Iqbal, N.; Syeed, S.; Khan, N. A. Salicylic Acid Alleviates Decreases in Photosynthesis Under Salt Stress by Enhancing Nitrogen and Sulfur Assimilation and Antioxidant Metabolism Differentially in Two Mungbean Cultivars. *J. Plant Physiol.* **2011,** *168* (8), 807–815.

Kumar, N.; Lamba S.; Kumar, A.; Kumar, P.; Mann, A.; Devi, S.; Rani, B. Antioxidant Defence in Halophytes Under High Salinity. In *Halophytes and Climate Change: Adaptive Mechanisms and Potential Uses;* Wallingford UK: Cabi, 2019; pp 196–208.

Mittova, V.; Tal, M.; Volokita, M.; Guy, M. Up-regulation of the leaf mitochondrial and peroxisomal antioxidative systems in response to salt-induced oxidative stress in the wild salt-tolerant tomato species *Lycopersicon pennellii. Plant, Cell Environ.* **2003,** *26* (6), 845–856.

Nikalje, G. C.; Mirajkar, S. J.; Nikam, T. D.; Suprasanna, P. Multifarious role of ROS in Halophytes: Signaling and Defense. In *Abiotic Stress mediated Sensing and Signaling in Plants: An Omics Perspective*; Springer: Singapore, 2018; pp 207–223.

Niki, E.; Noguchi, N. Dynamics of Antioxidant Action of vitamin E. *Acc. Chem. Res.* **2004,** *37* (1), 45–51.

Oprică, L.; Vochiţa, G. Enzymatic Activity in Halophytes. *Handbook of Halophytes: From Molecules to Ecosystems towards Biosaline Agriculture*; 2021, pp 1877–1899.

Ozgur, R.; Uzilday, B.; Sekmen, A. H.; Turkan, I. Reactive oxygen species regulation and antioxidant defence in halophytes. *Funct. Plant Biol.* **2013,** *40* (9), 832–847.

Parida, A. K.; Jha, B. Antioxidative Defense Potential to Salinity in the Euhalophyte *Salicornia brachiata*. *J. Plant Growth Regul.* **2010,** *29* (2), 137–148.

Pomposiello, P. J.; Demple, B. Redox-Operated Genetic Switches: The SoxR and OxyR Transcription Factors. *Trends Biotechnol.* **2001,** *19* (3), 109–114.

Rangani, J.; Parida, A. K.; Panda, A.; Kumari, A. Coordinated Changes in Antioxidative Enzymes Protect the Photosynthetic Machinery from Salinity Induced Oxidative Damage and Confer Salt Tolerance in an Extreme Halophyte *Salvadora persica* L. *Front. Plant Sci.* **2016,** *7*, 50.

Rasheed, A.; Koyro, H. W.; Hameed, A.; Gul, B. Physiological Responses of the Xero-Halophyte *Salsola drummondii* to Seasonal Alterations of Environmental Conditions in a Salt Desert. *Ecol. Res.* 2022.

Rejeb, K. B.; Benzarti, M.; Debez, A.; Bailly, C.; Savouré, A.; Abdelly, C. NADPH Oxidase-Dependent H2O2 Production is Required for Salt-Induced Antioxidant Defense in Arabidopsis Thaliana. *J. Plant Physiol.* **2015,** *174*, 5–15.

Reginato, M. A.; Abdala, G. I.; Miersch, O.; et al., Changes in the Levels of Jasmonates and Free Polyamines Induced by Na^2SO^4 and NaCl in Roots and Leaves of the Halophyte *Prosopis strombulifera* . *Biologia* **2012,** 67, 689–697.

Reginato, M. A.; Castagna, A.; Furlán, A.; Castro, S.; Ranieri, A.; Luna, V. Physiological Responses of a Halophytic Shrub to Salt Stress by Na2SO4 and NaCl: Oxidative Damage and the Role of Polyphenols in Antioxidant Protection. *AoB Plants* **2014,** *6*.

Rinalducci, S.; Murgiano, L.; Zolla, L. Redox Proteomics: Basic Principles and Future Perspectives for the Detection of Protein Oxidation in Plants. *J. Exp. Bot.* **2008,** *59* (14), 3781–3801.

Rodriguez-Muñiz, G. M.; Gomis, J.; Arques, A.; Amat, A. M.; Marin, M. L.; Miranda, M. A. Hydroxyl Radical as an Unlikely Key Intermediate in the Photodegradation of Emerging Pollutants. *Photochem. Photobiol.* **2014,** *90* (6), 1467–1469.

Sahin, U.; Ekinci, M.; Ors, S.; Turan, M.; Yildiz, S.; Yildirim, E. Effects of Individual and Combined Effects of Salinity and Drought on Physiological, Nutritional and Biochemical Properties of Cabbage (*Brassica oleracea var. capitata*). *Scientia Horticulturae* **2018,** *240*, 196–204.

Sandalio, L. M.; Romero-Puertas, M. C. Peroxisomes Sense and Respond to Environmental Cues by Regulating ROS and RNS Signalling Networks. *Ann. Bot.* **2015,** *116* (4), 475–485.

Schaffer, S. W.; Suleiman, M. S. (Eds.). *Mitochondria: The Dynamic Organelle*. Springer Science & Business Media, 2010; Vol. 2.

Seckin, B.; Turkan, I.; Sekmen, A. H.; Ozfidan, C. The role of Antioxidant Defense Systems at Differential Salt Tolerance of *Hordeum marinum* Huds.(sea barleygrass) and *Hordeum vulgare* L.(cultivated barley). *Environ. Exp. Bot.* **2010,** *69* (1), 76–85.

Sekmen, A. H.; Turkan, I.; Tanyolac, Z. O.; Ozfidan, C.; Dinc, A. Different Antioxidant Defense Responses to Salt Stress During Germination and Vegetative Stages of Endemic Halophyte *Gypsophila oblanceolata* Bark. *Environ. Exp. Bot.* **2012,** *77*, 63 76.

Sekmen, H. A.; Türkan, İ.; Takio, S. Differential Responses of Antioxidative Enzymes and Lipid Peroxidation to Salt Stress in Salt-Tolerant *Plantago maritima* and Salt-Sensitive Plantago Media. *Physiol. Plant.* **2007,** *131* (3), 399–411.

Shalata, A.; Mittova, V.; Volokita, M.; Guy, M.; Tal, M. Response of the Cultivated Tomato and its Wild Salt-Tolerant Relative *Lycopersicon pennellii* to Salt-Dependent Oxidative Stress: The Root Antioxidative System. *Physiologia Plantarum* **2001,** *112* (4), 487–494.

Shelke, D. B.; Pandey, M.; Nikalje, G. C.; Zaware, B. N.; Suprasanna, P.; Nikam, T. D. Salt Responsive Physiological, Photosynthetic and Biochemical Attributes at Early Seedling Stage for Screening Soybean Genotypes. *Plant Physiol. Biochem.* **2017,** *118,* 519–528.

Shi, D.; Sheng, Y. Effect of various salt–Alkaline Mixed Stress Conditions on Sunflower Seedlings and Analysis of Their Stress Factors. *Environmental and experimental Botany*, *54* (1), 8–21. (*Sesuvium portulacastrum* L.). *Front. Environ. Sci.* **2005,** *3,* 19.

Song, J.; Wang, B. Using Euhalophytes to Understand Salt Tolerance and to Develop Saline Agriculture: *Suaeda salsa* as a Promising Model. *Ann. Bot.* **2015,** *115* (3), 541–553.

Srivastava, A. K.; Srivastava, S.; Lokhande, V. H.; D'Souza, S. F.; Suprasanna, P. Salt Stress Reveals Differential Antioxidant and Energetics Responses in Glycophyte (*Brassica juncea* L.) and Halophyte, 2015.

Stadtman, E. R. Oxidation of Proteins by Mixed-Function Oxidation Systems: Implication in Protein Turnover, Aging and Neutrophil Function. *Trends Biochem. Sci.* **1986,** *11,* 11–12.

Steffens, B. The Role of Ethylene and ROS in Salinity, Heavy Metal, and Flooding Responses in Rice. *Front. Plant Sci.* **2014,** *5,* 685.

Skulachev, V. P. Membrane-linked systems preventing superoxide formation. *Biosci. Rep.* **1997,** *17* (3), 347–366.

Sunkar, R.; Bartels, D.; Kirch, H. H. Overexpression of a Stress-Inducible Aldehyde Dehydrogenase Gene from *Arabidopsis thaliana* in Transgenic Plants Improves Stress Tolerance. *Plant J.* **2003,** *35* (4), 452–464.

Sun, M.; Salomon, R. G. Oxidative Fragmentation of Hydroxy Octadecadienoates Generates Biologically Active γ-Hydroxyalkenals. *J. Am. Chem. Soc.* **2004,** *126* (18), 5699–5708.

Szatrowski, T. P.; Nathan, C. F. Production of Large Amounts of Hydrogen Peroxide by Human Tumor Cells. *Cancer Res.* **1991,** *51* (3), 794–798.

Thomas, M. J. The Role of Free Radicals and antioxidants: How do We Know that They are Working?. *Crit. Rev. Food Sci. Nutr.* **1995,** *35* (1–2), 21–39.

Turrens, J. F. Mitochondrial Formation of Reactive Oxygen Species. *J. Physiol.* **2003,** *552* (2), 335-344.

Valko, M.; Izakovic, M.; Mazur, M.; Rhodes, C. J.; Telser, J. Role of Oxygen Radicals in DNA Damage and Cancer Incidence. *Mol. Cell. Biochem.* **2004,** *266* (1), 37 56.

Wang, Z. L.; Li, P. H.; Fredricksen, M.; Gong, Z. Z.; Kim, C. S.; Zhang, C.; Zhang, H. Expressed Sequence Tags from *Thellungiella halophila*, A New Model to Study Plant Salt-Tolerance. *Plant Sci.* **2004,** *166* (3), 609–616.

West, J. D.; Marnett, L. J. Endogenous reactive intermediates as modulators of cell signaling and cell death. *Chem. Res. Toxicol.* **2006,** *2* (19), 173–194.

Wu, H. J.; Zhang, Z.; Wang, J. Y.; Oh, D. H.; Dassanayake, M.; Liu, B.; Xie, Q. Insights into Salt Tolerance from the Genome of *Thellungiella salsuginea*. *Proc. Nat. Acad. Sci.* **2012,** *109* (30), 12219–12224.

Yang, Y.; Guo, Y Elucidating the Molecular Mechanisms Mediating Plant Salt-stress Responses. *New Phytol.* **2018,** *217,* 523– 539. Wiley Online Library.

Yin, H.; Xu, L.; Porter, N. A. Free Radical Lipid Peroxidation: Mechanisms and Analysis. *Chem. Rev.* **2011,** *111* (10), 5944–5972.

Yu, J.; Chen, S.; Zhao, Q.; Wang, T.; Yang, C.; Diaz, C.; Dai, S. Physiological and Proteomic Analysis of Salinity Tolerance in *Puccinellia tenuiflora*. *J. Proteom. Res.* **2011,** *10* (9), 3852–3870.

Zhang, Y.; Fang, J.; Wu, X.; Dong, L. Na+/K+ Balance and Transport Regulatory Mechanisms in Weedy and Cultivated rice (*Oryza sativa* L.) Under Salt Stress. *BMC Plant Biol.* **2018,** *18* (1), 1–14.

CHAPTER 15

Halophyte Adaptation Is Nature's Prodigy to Environmental Changes

LAXMISHREE S. CHENGALA

Department of Botany, G. M. Momin Women's College, Bhiwandi, Dist. Thane, India

ABSTRACT

Soil salinity is among the most disturbing consequences of environmental change. Factors like low/scanty rainfall, contamination of water supply and ground water, poor irrigation systems, and salt ingression increasingly contribute to soil salinity. The ability of halophytes to thrive in extremely saline soil conditions can aid in solving the problems associated with salt stress. The halophyte germplasms have the potential to lead the agricultural revolution to meet the increasing food demand in times when agricultural lands are increasingly under salt stress. Various contributors to the stress response of halophytes include genetic factors, antiporters, ion channels, antioxidation potential, and most importantly the ability of halophytes to modulate plant physiology according to the salt stress. Besides, the seed morphology of halophytes, their germination physiology as well as their response and recovery show a multitude of strategies and mechanisms to adapt to salt stress. Another rewarding characteristic of halophytes is their ability to remediate polluted lands by accumulation of heavy metals. Their bioaccumulation capacity can prove to be tremendously useful in restoration of deforested and barren lands. Furthermore, studies on genetic, molecular, and biochemical aspects can help in the selection of responsive genes to increase salt tolerance in regular plants through genetic engineering. In this chapter, the characteristic adaptation strategies of halophytes are described

Physiology of Halophytes: Signaling, Omics, and Tolerance Mechanisms.
Nivas Manohar Desai & Ganesh Chandrakant Nikalje (Eds.)

in detail with the help of up to date published literature to shed light on ecophysiological perspectives for their survival in extreme conditions of salt stress.

15.1 INTRODUCTION

Environmental changes have accompanied the industrial and technological advancements with time affecting soil, water, and atmosphere. Most significantly, the continuous and uninterrupted use of dry lands for agricultural purposes, with simultaneous use of pesticides and chemical fertilizers has altered the soil ecosystem (Shabala, 2013). It is estimated that approximately 50% of agricultural lands have become unavailable due to an increase in soil salinity and upper layer of soil is lost in over 20% of land area due to erosion or soil degradation (Vineeth et al., 2020; Food and Agriculture organization, 2022). Consequently, salinity has affected 831 million hectares (Mha) of global soil and threatens around 1.5 Mha of land each year (Vineeth et al., 2020; Hasanuzzaman et al., 2014). Inappropriate anthropogenic activities also contribute to soil salinity and threaten sustainable agriculture and future food security (Food and Agriculture organization, 2022). Besides, large areas of agricultural lands are converted into urban cities. Moreover, there is an estimated increase in food demand of more than 50% by 2050 (Kumar and Sharma, 2020). To make matters worse, more climatic changes are expected over the next decades which can decrease annual precipitation in subtropical regions. This will further increase our dependence on non-potable water for irrigation, and contribute to our existing problem (Drought and Climate Change, 2022). Thus, the burden on agriculture to meet food demands is increasing exponentially and major breakthrough in the possible reuse of disrupted lands is an immediate necessity.

The losses in agricultural production due to increased salinity alone accounts to the annual loss of over 12 billion USD every year (Shabala, 2013). In such a scenario, halophyte plant species are our best option to utilize saline soil to reduce agricultural losses and climatic changes to a certain extent. Hence, understanding the mechanisms of stress tolerance and the involvement of associated genes can significantly help in introducing salinity tolerance traits in food crops; thereby reducing the burden on agricultural lands (Rahman et al., 2021). Further, these traits can also be introduced in plants propagated for biofuel production, which are increasingly acquiring fertile agricultural lands to overcome the energy crisis (Xian-Zhao et al.,

2012). Undoubtedly, the propagation of halophytes in saline soil can prove to be an ideal solution for problems associated with increasing salinity. This is because, in addition to coping with saline soils, they can overcome the current water crisis that will necessitate the use of (apparently abundant) saline water for irrigation in near future. Various applications of halophyte species that can be exploited for human and environmental benefits are presented in Table 15.1.

Halophytes can either be obligate or facultative requiring absolute necessity of saline water for proliferation or exhibiting tolerance to increased saline concentration in agricultural soil, respectively. Based on a different classification system, halophytes are further subclassified as hydrohalophytes and xero-halophytes depending on their habitat. As the name suggests, the former are adapted to aquatic and wet soil conditions (along the coastline, marshlands and mangroves) whereas the latter grow in dry habitats with saline soils (in semiarid desert regions). Precisely, the xero-halophytes are obligate and characterized as extremophiles since they require more than 500 mM NaCl concentration for seed germination and propagation (Rahman et al., 2021; Mishra and Tanna, 2017). In this chapter, we discuss the factors favoring the utilization of halophytes to overcome climatic changes. The physiological and molecular characteristics of halophyte species are described in detail and an up-to-date review of current progress in application of halophytes in saline soil is presented.

15.2 CHARACTERISTIC FEATURES OF HALOPHYTES THAT FAVOR ADAPTATION TO SALT STRESS

Halophytes are plants that can tolerate or even benefit from substantial amounts of salt in soil and account for ~2% of the plant flora, worldwide. Considering the gradual increase in saline soil conditions due to climatic change, the adaptation of wide range of plant families to salt stress during evolution is a conceivable hypothesis. In agricultural crops, soil salinity adversely affects seed germination as well as plant growth mainly by disrupting osmotic balance and photosynthetic activity (Shabala, 2013). Consequently, these plants suffer from failed seed germination, nutritional and metabolic imbalance, ion toxicity, and increased production of reactive oxygen species (ROS) in response to salt stress (Hasanuzzaman et al., 2014; Rahman et al., 2021). Hence, during evolution, the halophytes have undergone physiological and molecular adaptations and characteristically

TABLE 15.1 Potential Characteristics and Uses of Halophyte Species.

Sr. No.	Applications	Example of halophyte Species	Common name	Family	Characteristics	Reference
1.	Prevention of soil erosion as well as restoration of eroded and barren lands	*Suaeda fruticosa* L.	Shrubby seablite	Amaranthaceae	They survive and proliferate in soil with high pH (up to 10.5), salinity (up to 65 dS m^{-1}), and shortage of water.	(Wungrampha et al., 2019)
2.	Carbon sequestration	*Eucalyptus tereticornis* Sm.	Forest red gum	Myrtaceae	They can sequester up to 400 metric tons per hectare of carbon in water clogged saline lands during their life cycle.	(Kumar et al., 2019)
3.	Biofuel production	*Salicornia virginica* L.	American glasswort, pickleweed	Amaranthaceae	They contain high amounts of saturated oils in seeds and their biomass is rich in lignocellulose. Thus, they are rich sources of hydrocarbons that can be used as biofuels.	(Bomani et al., 2022)
4.	Phytoremediation	*Tamarix africana* Poiret	Tamarisk, salt cedar, taray	Tamaricaceae	The enormous and deep rooted system of halophytes allows the accumulation of heavy metal from a large area of soil and renders them harmless through phytoextraction or associated halophilic microbial metabolic activity.	(Ayyappan et al., 2016)
5.	Nutraceuticals	*Cynodon dactylon* L.	Bermuda grass, dhoob, dūrvā grass, ethana grass, dubo, dog's tooth grass, Bahama grass, devil's grass, couch grass	*Poaceae*	They are rich sources of phytochemicals and exhibit potential bioactivity against several chronic diseases.	(Faustino et al., 2019)

TABLE 15.1 *(Continued)*

Sr. No.	Applications	Example of halophyte Species	Common name	Family	Characteristics	Reference
6.	Source of proteins	*Chenopodium quinoa* Willd.	Quinoa	Amaranthaceae	The protein content of quinoa ranges from 12 to 17% depending on variety, environment, and crop management practices. Hence, it is an essential source of protein among vegetarians. They are also rich in lysine and methionine—essential amino acids absent in most cereals and pulses.	(Hussain, et al., 2018)
7.	Restoration of saline sodic soils	*Leptochloa fusca* (L.) Kunth	Sprangletops	Poaceae	They are highly adaptive, proliferative, and improve carbon sequestration. The roots of these plants are extremely deep which helps in absorbing nutrients from well below the soil level.	(Kaur et al., 2002)
8.	Antibiotics	*Cynometra ramiflora* L.	Wrinkle Pod Mangrove	Fabaceae	The seed oil is a rich source of bioactive agents and is indicated as a potent medicine for skin diseases.	(Abeysinghe, 2010)
9.	Agar production	*Gracilaria* spp.	Warm water seaweeds	Gracilariaceae	It is a carrageenophyte (red sea weed) commercially used as a gelling agent that is used in food, pharmaceutical, and microbiological industries.	(Vuai, 2022)
10.	Agroforestry	*Terminalia catappa* L.	Tropical almond or Indian almond	Combretaceae	Biosaline agriculture is a relatively new field with a prospective view for the utilization of halophytes for food and fodder; thereby converting the present salinity crisis into an opportunity.	(Arora et al., 2013)

developed mechanisms for growth regulation in response to salt stress with the help of antiporters, ion channels, and antioxidation potential (Shabala, 2013). Another favorable characteristic of halophiles is their di- or heteromorphic nature and ability to modulate plant physiology according to the salt stress (Rahman, 2021). Predominantly, families Chenopodiaceae and Amaranthaceae dominate the saline soil environments globally followed by Poaceae. They comprise of annual or perennial herbs, shrubs, and small trees (Lucia et al., 2018). In this section, we have described the characteristics of halophyte seeds and different proliferation phases of life cycle favoring their adaptation.

15.2.1 SEED CHARACTERISTICS AND DORMANCY PHASE

In order to completely adapt to any extreme environmental conditions, the plants are required to develop mechanisms to ensure seed germination in such habitats and maintain seed viability for extended duration; and thus, complete their life cycle. Halophytes exhibit several strategies like seed heteromorphism, short recovery time, dormancy, and rapid germination to ensure their propagation in diverse environment (González, 2022). Obviously, these factors are primarily controlled by genetic factors. However, environmental factors significantly dominate the adaptive response of plants to external stress. For instance, the genetic expression allows seed germination under high salinity as observed commonly in *Suaeda salsa* L. indicating its tolerance to high salt concentration. Conversely, the seeds of *Borszczowia aralocaspica* (Bunge) remain dormant in saline conditions and germinate only when soil conditions return to normal, indicating adaptive environmental response (Zhang et al., 2015). A recent study reported the ability of the seeds of *Tetraena qatarense* (Zygophyllaceae), *Suaeda aegyptica* (Amaranthaceae), and *Caroxylon imbricatum* (Amaranthaceae) to germinate and proliferate in soil irrigated with saline water containing 600 mM salts (Alhaddad et al., 2021). Another interesting observation reported in their study was higher germination rate of *Limonium axillare* under saline conditions as compared to distilled water.

The seed coat of halophytes provides added protection under extreme conditions. Song et al. (2017) reported abundance of waxes in seed coat of black seeds of *Suaeda salsa* as compared to brown seed. These waxes were suggested to provide protection to embryos against ion toxicity, maintain seed viability, and exhibit protective action against ROS-induced cellular damage. The seed color is another characteristic feature of halophytes (Zhang et al., 2015). Typically, they are either brown or black. Several studies have indicated that brown seeds have soft outer seed coat which allows absorption of water more readily and hence have a higher germination rate. The genetic traits of brown seeds enable tolerance of highly saline environments and are transferred to the progeny halophytes. Phytohormones like gibberellin, abscisic acid, indole-3-acetic acid, phosphatidylglycerol, glycerophospholipid, and zeatin riboside are produced in larger quantities in brown seeds. Gibberellin also aids in accumulation of storage proteins, allowing greater membrane fluidity under salt stress. In contrast, the black seeds have significantly low germination rate due to their hard and smooth outer coat which prevents water absorption but they show extreme salt tolerance. Black seeds are reported to germinate in soil containing up to 600 mM salts. Thus, the production of heteromorphic seeds is a key characteristic of halophytes to adapt to extreme environments. However, these characteristics are not always transferred to the progeny, as reported in *Suaeda aralocaspica*. Instead, they disappear during further stages of plant growth (Song et al., 2017; Cao et al., 2016). This is one of the important features of halophytes that allow their growth in saline or normal soil.

Dormancy and seed bank (production of enormous amount of seeds) are other favorable strategies of halophytes that enable their adaptation to environmental changes. For instance, *Kalidium gracile* was reported to have a seed bank of ~7030 seeds per meter square of desert habitat, among which only 1.4% seeds germinated and almost 28% seeds turned dormant and remained buried under saline soil (Cao, et al., 2014). The larger size of seeds also supports germination under saline conditions. Halophytes produce small seeds that lay dormant in extremely saline soil and germinate when the situation is conducive for plant survival (González, 2022). Endogenous concentration of hormones like gibberellin and abscisic acid is much higher in small dormant seeds as compared to large seeds of halophytes; whereas large seeds contain higher concentration of ethylene. Hence, they are linked to effective regulation of dormancy and germination. However, the exact role of these phytohormones is not completely elucidated in halophytes, but they are suggested to have a species-specific activity (Pujol, et al., 2000).

15.2.2 PHYSIOLOGICAL CHANGES OCCURRING DURING GERMINATION AND GROWTH PHASE UNDER SALT STRESS

Some halophytes show presence of salt glands (also known as salt bladders) that can selectively exclude or accumulate ions. These halophytes are known as recretohalophytes. The salt glands are epidermal cells but with much larger size (up to 10 times of epidermal cells) that can sequester 1000-fold more Na^+ compared to epidermal cells. Besides Na^+, these cells also regulate the concentration of other ions like K^+, Ca^{2+}, Mg^{2+}, Fe^{2+}, Mn^{2+}, Zn^{2+}, Cl^-, Br^-, I^-, SO^{42-}, PO^{43-}, and NO^{3-}. The salt glands are surrounded by a thickened cuticle and enclose multiple plasmodesmata, mitochondria, and small vesicle. These cells are devoid of chloroplast (Yuan et al., 2016; Bamidele and Malloch, 2006). The monocotyledonous halophytes mostly show the presence of 2 salt glands whereas dicotyledonous halophytes may contain 4–40 salt glands (Rahman, 2021). Their density increases with increased salinity levels and their appearance is similar to that of trichomes exhibiting glandular hair, thorns, and surface glands (Bamidele and Malloch, 2006). They also play a role in reducing water loss and preventing excessive UV damage when exposed to salinity. Salt glands are represented by bicellular microhairs in the epidermal layers of the vascular strands in leaves of most halophyte grasses and few cereal crops. Hence, it is believed that with suitable modification, the salt tolerance in cereal crops can be increased. However, very few insights related to typical pathways of salt glands are available at present to derive practical conclusions (Yuan et al., 2016). The phytohormones like gibberellin and abscisic acid regulate trichome initiation and growth in salt bladders for salt exclusion (Lu et al., 2021). Morris et al. (2019) characterized the salts excreted from *Distichlis spicata* and *Spartina pectinata* that appeared as sap-like excretion (of 31–45 μm diameters) on leaves. They indicated that haloconduction (mobilization through air) of salts helps in their dispersal to far areas of nonsaline soil regions where they act as nutrients instead of pollutants.

The root system of most hydrohalophytes has also evolved by development of distinct layers of cell barrier in endodermis and exodermis. These barriers are known as the casparian bands and suberin lamella composed of lignin and long-chain fatty acids, respectively. The casparian band promotes apoplastic transport into the strictly regulated symplastic system whereas the suberin lamella aids in ion and water uptake regulation. Together, they result in effective exclusion of salts from the roots and prevent accumulation of toxic ions in plant shoots (Bing et al., 2021). It was reported recently that the roots of *Suaeda salsa* accumulated higher concentrations of Na^+, Cl^-, and K^+ ions through elongation of its root length under salt stress (Wang et al., 2021).

In semiarid regions, halophytes that occur in succulent forms survive by diluting high ion concentration in plant tissues. These plants are characterized by large size of mesophyll cells and smaller intercellular space in leaves and stems. They also possess larger size of cellular organelles including mitochondria and vacuoles (Herrera, 2020). It is hypothesized that enlarged mitochondria helps in higher ATP production which is required for salt excretion (through salt glands) or salt exclusion (through barriers in apoplastic root transport system) (Winter, 2019). However, the complex physiological interconnection of structural and functional aspects in halophytes poses challenges in clear understanding of the functional biology of succulent halophytes. Other physiological strategies in overcoming salt stress include regulation of stomatal opening to retain high water levels and activation of antioxidant enzymes (Batanouny, 2001). In *Halocnemum strobilaceum* and *Zygophyllum aegyptium*, the succulence traits increased with increased soil salinity (Serag, 1999). Other succulents like *T. medusa* and *T. auriculata* have shown survival in soil containing up to 2000 mM salts (Moir-Barnetson et al., 2016). In both the studies, the plants exhibited mechanisms of salinity tolerance through osmotic adjustments, compartmentalization of ions in cell organelles, and production of glycine betaine (osmolyte).

15.2.3 EFFECT OF LIGHT AND TEMPERATURE ON HALOPHYTE SEED GERMINATION

Light and temperature are key signaling elements of environment that regulate seed germination as well as plant growth. While most agricultural seeds prefer dark conditions for germination, halophytes tend to germinate in light (González, 2022). One such example of light-induced germination of heteromorphic seed is observed in *Salsola ferganica*. It typically grows in cold deserts and heavy saline-alkaline wastelands. The dormant seeds of this plant are extremely sensitive to light and temperature. Hence, to overcome this sensitivity, *S. ferganica* produces large amount of seeds which vary in size and shape. This ensures efficient dispersal ability (based on size and weight) of seeds to large areas and their germination under favorable conditions (Ma et al., 2018). The photoperiod of 12 h light and 12 h darkness and temperature of 15°C–20°C has been reported to be optimum for germination of *Halothamnus iraqensis* Botsch. Its germination rate reduces on extended storage (Bhatt et al., 2019). In a succulent shrub *Halocnemum strobilaceum*, highest germination rate (almost 100%) was observed at conditions of constant fluorescent light (400–700 nm) and 35°C. However, most of these seedlings were distorted and died shortly after. These seedlings were also sensitive to increase in salinity (Qu et al., 2008). It is suggested that the light conditions induce phosphorylation of protein kinases and regulate the activities of membrane channels in seeds of halophyte plants. This increases the permeability of ions like K^+ and Ca^2, which acts as important osmotic agents in seed germination (Ma et al., 2018; Qu et al., 2008).

15.3 PHYSIOLOGICAL CHARACTERISTICS OF SOIL MICROENVIRONMENT IN SALT STRESS

Fertile soil promotes plant growth due to its suitable structure that allows holding optimum levels of organic nutrients, water, and minerals. It also supports the growth of microflora that produces a range of bioenzymes and acts synergistically with plant roots to enhance their nutrient uptake. As the salinity of soil increases, several alterations occur in soil structure and consequently limit its capacity to support plant growth. High salt concentrations increase the density of soil and causes surface crusting. It also results in extreme soil compaction (Otlewska et al., 2020). These factors collectively impede the hydraulic conductivity leading to low water infiltration and loss of beneficial microflora.

Microbial activities are responsible for decomposition of organic matter, mineralization of soil, and stabilization of soil aggregates. Thus, soil microflora is an essential element of nature that plays a critical role in building fertile soil structure. Alteration in environmental conditions disturbs the micro-ecosystems in soil by either killing beneficial microorganisms or altering the soil microbiome. In either instance, nutrients in soil are made unavailable to plants. In saline soil, most microorganisms are unable to maintain the osmotic balance and die as a result of plasmolysis. In absence of beneficial microbiome, soil is severely deprived of biochemical enzymes like ureases, dehydrogenases, alkaline phosphatases, and β-glucosidases that break down organic matter into elemental form that can be utilized by plants (Zvinavashe et al., 2019). Halophytes overcome the challenges of hydraulic conductivity by compartmentalization of mineral ions. In turn, the intake of these ions creates a microenvironment surrounding the roots that encourage growth of microorganisms. During the course of evolution, the halophytes have developed genetic mechanisms (expression of genes including SOS1 and HKT1 described in Section 15.4 of this chapter) in response to reduced hydraulic conductivity due to saline salt conditions for their survival (Otlewska et al., 2020; Zvinavashe et al., 2019).

Besides microbial activity, high salinity interrupts imbibition in plant seeds and hence affects germination. As the osmotic potential of the soil is lowered, the enzyme activities and protein metabolism of seeds are disrupted leading to hormonal imbalance and distortion in ultrastructure of cells, tissues, and organs in seedlings. The interrupted germination and plant growth in saline conditions induces significant stress on plant leading to production of several ROS including OH, O^{2-}, H^2O^2, and $^1O^2$. This further adds to disruptions caused in protein expression and genomic stability. The ROS species also damage the photosynthetic apparatus (Liu et al., 2021). The seed characteristics of presence of outer coats, seed heteromorphism, dormancy phase, and seed banks together ensure germination of halophyte seeds in extreme environments.

The plants transplanted from normal soil to saline soil undergo reduction in plant growth and biomass depending on the level of salinity in soil. In highly saline soils, plants cells shrink within seconds due to loss of water and ultimately wither. High salinity due to Na^+ and Cl^- also restrict the uptake of macronutrients like Ca^{2+} and Mg^{2+}. In turn, this disrupts signal transduction, membrane stability, chlorophyll synthesis, stomatal conductance, and stomatal opening. The decreased production of photosynthetic pigments decreases carbon assimilation by suppression of ribulose-1, 5-bisphosphate carboxylase/oxygenase (RuBisCO). Soil compaction results in decreased respiration in plants which is primarily indicated by lower growth of new leaves and increased necrosis. Halophytes primarily deal with these conditions by activation of salt exclusion or salt excretion mechanisms (described briefly in Section 15.4.1) (Drought and Climate Change, 2022; González, 2022; Qu, et al., 2008; Otlewska et al., 2020).

15.4 SALT TOLERANCE MECHANISMS

Halophytes have evolved a range of adaptations to tolerate higher salinity levels in soil. These adaptations are extremely complex and involve combinations of physiological, cellular, biochemical as well as molecular mechanisms. The physiological modifications occurring in leaves and roots, and presence of salt glands is described in Section 15.2 of this chapter. These physiological aspects are governed by cellular and molecular mechanisms which lay the foundation of halophyte life cycle (Fan, 2020). Ideally, adaptive responses are observed in halophytes exposed to lower salinity levels. These plants modulate simple functions like decreased transpiration, salt excretion, and increased ion dilution to maintain osmotic balance in order to survive extreme conditions. For this reason, the genetic expression, morphology, and seed germination in same plant species studied from different regions show different characteristics among halophytes (Zhao et al., 2020; Hassan et al., 2017). Generally, halophytes grow at a concentration of around 50 mm NaCl for monocots, and between 100 and 200 mm for dicots. For adaptation to higher salt concentrations, the genetic structure of obligate halophytes has shown similar levels of expression in different habitat (Oh et al., 2010; Wang et al., 2015). This understanding is overwhelming in the context of halophyte's volunteering response in soil mitigation. Simply stated, halophyte plants are nature's way of repairing the damage caused by human interference through irresponsible agricultural and industrial practices. In this section, the mechanisms of salt tolerance are described in brief.

15.4.1 SALT EXCRETION AND SALT EXCLUSION

Salt excretion and salt exclusion are two active mechanisms of halophytes to tolerate ion stress. As described earlier, salt excretion occurs through specialized excretory structures known as salt bladders that are selectively present in recretohalophytes. These cells collect the salt ions transported by mediator mesophyll cells and excrete them on the surface of leaves in the form of microcrystals. The absence of salt glands in mutant *Mesembryanthemum crystallinum* causes loss of halophyte characteristics, indicating the dependence of recretohalophytes on this strategy to overcome salt stress (Agarie et al., 2007). In addition, several studies have reported activation of genes in salt bladders for production of organic osmolytes, increased metabolism for energy generation, and stress response on exposure to high salinity in *M. crystallinum*. However, the exact gene sequence and its associated roles for above processes are still under study (Barkla et al., 2016; Barkla and Vera-Estrella, 2015).

Salt exclusion is a term used to emphasize the strategies to reduce the uptake of salt ions or its compartmentalization after uptake by roots to prevent organ damage in halophytes. The uptake of key functional ions like Na^+, Cl^-, and K^+ are regulated by maintenance of osmotic balance by roots. Osmoprotectors like proline, glycine-betaine, sugars, and polyols further help in regulating the osmotic pressure (Serag, 1999; Moir-Barnetson et al., 2016). Even during transport to leaves, the H^+ pump encourages compartmentalization of these ions between vacuoles, cytosol, and apoplast with the help of tonoplast Na^+/H^+ antiporters (Chen et al., 2018). It is notably reported that the barriers in apoplastic root transport system strengthen with increase in soil salinity (Batanouny, 2001). This is indicated by increased production of lignin and suberin which are major components of casparian band and suberin lamella, respectively, in cell barriers of halophyte roots (Chen et al., 2018). During germination, the black seeds of halophytes are reported to undergo compartmentalization of ions in the pericarps, and significantly lower the exposure of embryos toward these ions; thus preventing ion toxicity (González, 2022). Comparatively lower, yet significant degree of compartmentalization is also observed in bracts, seed coat, vacuoles, and older tissues (Zhang et al., 2015).

15.4.2 ANTIOXIDATIVE MECHANISMS

Under salt stress, plants can overcome the harmful effects of ROS through an efficient antioxidation system. Antioxidants are enzymatic or nonenzymatic reactive molecules and endogenous phytohormones that neutralize the ROS and thus maintain homeostasis. They play a central role in detoxification of ROS in cells of all living things. In absence of active antioxidation system, the ROS species react with macromolecules including proteins, DNA and RNA irreversibly causing significant damage to cells. A number of antioxidant enzymes are identified in almost all halophytes studied and reported so far (Jithesh et al., 2006). These enzymes include monodehydroascorbate reductase, glutathione transferases, ascorbate peroxidases, and superoxide dismutases. Moreover, the expression of these genes is proportional to the degree of salt stress (Jithesh et al., 2006). *Cakile maritime* (a succulent halophyte), *Thellungiella halophila,* and *Suaeda salsa* are reported to increase the expression of antioxidants when exposed to higher salinity (Soundararajan et al., 2019).

In halophytes, the level of ROS including H^2O^2 and advanced glycation products like malondialdehyde are reported to be much higher than glycophytes (salt sensitive plants). For this reason, the antioxidant systems of halophytes are more evolved as compared to glycophytes. Interestingly, the cloning of an antioxidant metallothionein gene sbMT-2 from *Salicornia brachiate* into mutant tobacco plant lacking the ROS detoxification genes resulted in significant enhancement of salt tolerance (Jithesh et al., 2006). This suggests that antioxidants alone can effectively manage salt stress in absence of other mechanisms adopted by halophytes. However, considering the high levels of accumulation of ROS in halophytes, they are not always neutralized efficiently even on activation of antioxidant enzymes. Moreover, higher activities of enzymatic and nonenzymatic antioxidants are directly related to low solute leakage and membrane damage. Hence, the halophytes simultaneously improve compartmentalization and transient localization of ROS on increase in soil salinity (Soundararajan et al., 2019).

15.4.3 GENETIC EXPRESSION

The genetic expression of salt tolerance genes has been studied in halophytes with the help of transcriptomic, proteomic, and metabolomic analyses. Practically, a functional gene has been identified for all physiological processes described in this chapter. Hemabindu et al. (2016) have enlisted most of these genes and their associated functions in detail. As described earlier, the halophytes undergo several modifications to overcome ion stress. Consequently, several genes with diverse functions are activated under salt stress and their expression increases with increase in ionic concentration. For instance, *Limonium bicolor* simultaneously activates the genes and transcription factors responsible for ion transport, vesicles formation, antioxidant production, and salt tolerance under salt stress (Kashyap et al., 2020). Few examples of salt tolerance genes identified in halophytes are represented in Table 15.2

Given the increasing pressure of food crops on agricultural lands, the introduction of halophytic genes in glycophytes to utilize lands with higher salinity is a promising approach not only to meet food demands but also to mitigate climate changes (Hafeez et al., 2021). Hence, identification of genes responsible for various mechanisms adapted by halophytes to overcome salt stress is of extreme importance. Interestingly, (salt tolerance) gene homologs of halophytes are identified in glycophytes in nonfunctional forms (Kashyap et al., 2020). If these genes are triggered and expressed, it may be easy to achieve our goal of utilizing more land for plant growth. More encouraging insights are observed from studies that provide evidence that many glycophytes containing these homologs confer moderate salt tolerance when exposed to increased soil salinity (Hemabindu et al., 2016). Besides, considering the evolution of halophytes that suggest accumulation of adaptive physiological and biochemical mechanisms over time, it can be possibly assumed that the glycophytes with homologous nonfunctional halophyte genes may soon become adapted to salt stress on continuous exposure by natural selection. Similarly, various ortholog genes are also identified among halophytes and glycophytes. For example, a reduced expression of a gene ortholog of SOS1 in *Eutrema salsugineum* compared to *Arabidopsis thaliana* significantly reduces its tolerance to salt (Oh et al., 2009).

TABLE 15.2 Genes Responsible for Adaptation of Halophytes and Their Role in Plant Survival.

Sr. no.	Genes	Plant	Role
1	Salt overly sensitive 1 (*SOS1*)	*Avicennia marina, Salicornia dolichostachya*	Increased exclusion of Na^+ by roots through Na^+/H^+ antiporter
2	*HKT1*	*Suaeda salsa, Salicornia dolichostachya, Leptochioa fusca, Aeluropus logopoides*	High-affinity K^+ transporter
3	*AKT1, KUP/HAK/KT,* or *UP/HAK/KT*	*Suaeda maritime, Eutrema salsugineum, Reaumuria trigyna, Limonium bicolor*	Inward-rectifying K^+ Channel 1
4	*NHX*	*Limonium gmelinii, Karelinia caspica, Salicornia brachiate, Aeluropus littoralis*	Vacuolar-type Na^+/H^+ antiporter that pumps excessive Na^+ into the vacuole to avoid toxic Na^+ level in the cytoplasm
5	*CLC*	*Mesembryanthemum crystallinum, Limonium bicolor*	Chloride channel
6	*AQP, PIP,* and *TIP*	*Sesuvium portulacastrum, Suaeda salsa, Avicennia officinalis*	Increased aquaporin expression
7	*AoCYP86B1*	*Avicennia officinalis*	Regulation of suberin biosynthesis

Source: Adapted from Hafeez et al. (2021) and Mishra and Tanna (2017).

15.5 POTENTIAL OF HALOPHYTES IN PHYTOREMEDIATION

Like increasing soil salinity, soil pollution is another serious environmental concern. Industrial chemicals, heavy metals, pesticides, chemical fertilizers, and many other xenobiotics enter the soil and water bodies causing pollution and health hazards. Most of these chemicals are manufactured in the form of reagent- or water-soluble salts. As a result, in addition to pollution, they also cause increase in the salinity of soil. Hence, the use of halophytes is an ideally sustainable approach for revegetation as well as phytoremediation of polluted lands with high salinity. The intrinsic salt tolerance characteristics exhibited by halophytes can be extremely beneficial for economical desalination and hence restoration of saline soils (Caparrós et al., 2022; Manousaki and Kalogerakis, 2011). Phytoremediation, in general, is not a novel approach; instead it has been practiced since a very long time for remediation of sewage sludge polluted lands. Phytoremediation of saline soil is not a new idea either. Among the first reported studies, Chaudhri et al. (1964) indicated the potential of halophytes to remove up to 4 tons of salt per hectare of land in a single harvest of annual aerial parts of the plant.

Halophytes practice remediation mainly by means of phytostabilization and phytoextraction of heavy metals, xenobiotics, and other chemicals. During phytostabilization, the halophytes can grow in polluted soil and show tolerance to heavy metals or other pollutants. However, they are not capable of translocating and accumulating the metals in the aerial parts of the plant. Thus, they are used as vegetative cover over polluted saline soil. In these regions, the halophytes stabilize the soil and prevent the metals from leaching into the groundwater. On the other hand, some plant species are capable of accumulating the heavy metals and other xenobiotic compounds in their tissues. This process is referred to as "phytoextraction" which enables harvesting and removal of organic pollutants. In either case, halophytes continuously remove salts from the soil to the point that the land is reclaimed to its balanced and productive state, making them a promising phytodesalinization tool (Yan et al., 2020). Other means of phytoremediation includes phytofiltration, phytotransformation, and phytovolatilization. Phytofiltration is a process of absorption or adsorption of contaminants from surface wastewater by plant roots which prevents them from leaching to the underground water. During phytotransformation the pollutants are chemically modified through plant metabolism into inactive forms and metabolized. In phytovolatilization, the pollutants are released into the air in volatile form (Usman et al., 2022).

Halophytes use varied mechanisms discussed in this chapter for their growth and remediate the soils during this process. Some characteristics like root morphology, compartmentalization of ions, and presence of salt glands are species specific. The genetic expression and tolerance to increasing concentration of different ions also vary from one plant type to another. However, the basic characteristics of all halophytes are similar. Figure 15.1 represents the mechanisms adapted by halophytes for phytoremediation processes (Adapted from Palliyath et al. (2017)). Typically, halophytes have a strong and deep-rooted system, similar to agricultural weeds, which enable them to thrive over a large area and persist in a habitat for a long time. These roots improve the physical properties of soil by acting as a biological drilling and tillage tool. Proportional to the biomass of halophytes in a particular area (i.e., extent of growth), they increase soil porosity and hence reduce bulk density in a short period of time. These factors stabilize the soil structure and enable optimum nutrient reserve by replacing salt ions with organic matter (from fallen leaves and dead seedlings), enzymes, and root exudates. Over the time, the amount of organic matter increases in soil and the salts (along with other pollutants) begin to deplete. At the same time, the microbiome of soil also changes supporting a diversity of microorganisms. These micro-organisms play an important role in improving root respiration and intake of nutrients from soil. Besides root respiration, the improved soil structure enhances the concentration of carbon dioxide in the soil atmosphere where it reacts with water to form carbonic acid. It is suggested that carbonic acid dissociates calcium carbonate to calcium, water, and carbon dioxide. The calcium ions facilitate the removal of Na^+ from the cation exchange sites of soil colloid which is either absorbed by roots (phytoextraction) or leached from soil through irrigation. Thus, calcium ions too play an important role in phytoremediation of saline soils (Caparrós et al., 2022; Manousaki and Kalogerakis, 2011; Chaudhri et al., 1964; Yan et al., 2020; Usman et al., 2022; Palliyath et al., 2017).

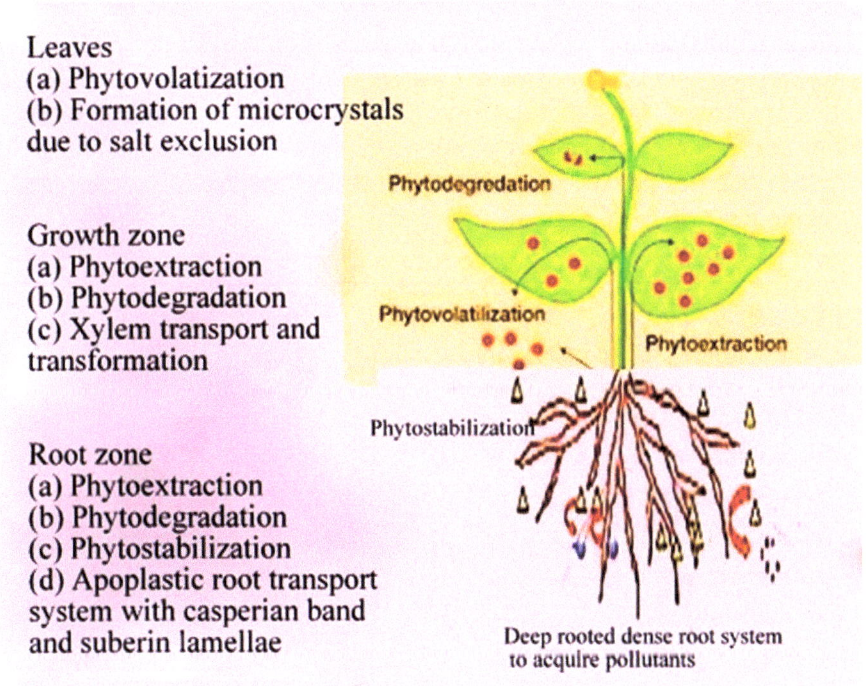

FIGURE 15.1 Mechanisms of phytoremediation by halophytes.

Source: Adapted from Palliyath et al. (2017).

15.6 CONCLUSIONS

The increase in soil salinity is among the major environmental concerns that we face currently. Soil salinity has affected the agricultural production; thus, threatening future food security. At the same time, decrease in plantation has caused a major climate change in the past few decades. Since halophytes are naturally adapted to grow in saline soil, they can become useful means to overcome these challenges. Besides, it is also a great tool for phytoremediation of polluted lands. The application of halophytes as a phytoremediation tool is particularly beneficial in semiarid desert regions with low precipitation rates. In such regions, the obligate halophytic plants can preferably reduce the salt burden in the rhizosphere region of soil to help salt-tolerant halophytes to thrive. Considering the potential of halophytes in different fields, a relatively new ecophysiological field of agroforestry has also emerged that promotes the agriculture of edible species of halophytes for food and fodder along the coast line. The nonedible woody halophytes can be utilized for fuel and nonedible grasses and shrubs can be a great source of organic compost. Thus, utilization of halophytes can present us with a winning situation in times of despair. Despite their recognized importance, studies of halophytes for soil reclamation are under an exploratory stage and more field research is needed to understand the long-term ecophysiological impact of halophytes on environment.

KEYWORDS

- **bioaccumulation**
- **halophytes**
- **phytoremediation**
- **salinity**
- **salt stress**
- **salt tolerance**

REFERENCES

Abeysinghe, P. D. Antibacterial Activity of Some Medicinal Mangroves Against Antibiotic Resistant Pathogenic Bacteria. *Indian J. Pharma. Sci.* **2010,** *72*, 167–172.

Agarie, S.; Shimoda, T.; Shimizu, Y.; Baumann, K.; Sunagawa, H.; Kondo, A. Salt Tolerance, Salt Accumulation, and Ionic Homeostasis in an Epidermal Bladder-Cell-Less Mutant of the Common Ice Plant *Mesembryanthemum crystallinum*. *J. Exp. Bot.* **2007,** *58*, 1957–1967.

Alhaddad, F. A.; Abu-Dieyeh, M. H.; ElAzazi, E. S. M.; Ahmed, T. A. Salt Tolerance of Selected Halophytes at the Two Initial Growth Stages for Future Management Options. *Sci. Rep.* **2021**, *11*, 10194.

Arora, S.; Bhuva, C.; Solanki, R. B.; Rao, G. G. Halophytes for Bio-Saline Agro-Forestry and Phyto-Remediation of Coastal Saline Lands. *J. Soil Water Conservat.* **2013,** *12*, 252–259.

Ayyappan, D.; Sathiyaraj, G.; Ravindran, K. C. Phytoextraction of Heavy Metals by *Sesuvium portulacastrum* L. A Salt Marsh Halophyte From Tannery Effluent. *Int. J. Phytoremediat.* **2016**, *18*, 453–459.

Bamidele, J. F.; Malloch, A. J. C. Morphological and Utrastructural Studies of the Salt Gland of Tree Malow, *Lavatera arborea* L. (Malvaceae). *Int. J. Bot.* **2006,** *2*, 313–318.

Barkla, B. J.; Vera-Estrella, R.; Raymond, C. Single-Cell-Type Quantitative Proteomic and Ionomic Analysis of Epidermal Bladder Cells From the Halophyte Model Plant *Mesembryanthemum crystallinum* to Identify Salt-Responsive Proteins. *BMC. Plant Bio.* **2016,** *16*, 110.

Barkla, B. J.; Vera-Estrella, R. Single Cell-Type Comparative Metabolomics of Epidermal Bladder Cells From the Halophyte *Mesembryanthemum crystallinum*. *Fron. Plant Sci.* **2015,** *6*, 435.

Batanouny, K. H. Adaptation of Plants to Saline Conditions in Arid Regions. In *Plants in the Deserts of the Middle East. Adaptations of Desert Organisms*; Springer: Berlin, Heidelberg, 2001. https://doi.org/10.1007/978-3-662-04480-3_12

Bhatt, A.; Bhat, N.R.; Murru, V.; Santo, A. Eco-Physiological Studies on Desert Plants: Germination of *Halothamnus iraqensis* Botsch. Seeds Under Different Conditions. *J. Arid Land* **2019,** *11*, 75–85.

Bing, C.; Ranran, L.; Timothy, F.; Jie, S. Casparian Bands and Suberin Lamellae: Key Targets for Breeding Salt Tolerant Crops?. *Environ. Experiment. Bot.* **2021,** *191*, 104600.

Bomani, B. M. M.; Hendricks, R. C.; Elbluk, M.; Okon, M.; Lee, E.; Gigante, B. NASA's GreenLab Research Facility- A Guide for a Self-Sustainable Renewable Energy Ecosystem 2011. http://ntrs.nasa.gov/archive/nasa/casi.ntrs.nasa.gov/20120001794.pdf (accessed July 24, 2022).

Cao, J.; Wang, L.; Lan, H. Validation of Reference Genes for Quantitative RT-PCR Normalization in *Suaeda aralocaspica*, an Annual Halophyte with Heteromorphism and C^4 Pathway Without Kranz Anatomy. *Peer J.* **2016,** *4*, e1697.

Cao, D.; Baskin, C. C.; Baskin, J. M.; Yang, F.; Huang, Z. Dormancy Cycling and Persistence of Seeds in Soil of a Cold Desert Halophyte Shrub. *Annals Bot.* **2014,** *113*, 171–179.

Caparrós, P. G.; Ozturk, M.; Gul, A.; Batool, T. S.; Pirasteh-Anosheh, H.; Unal, B. T., Altay, V.; Toderich, K. N. Halophytes Have Potential as Heavy Metal Phytoremediators: A Comprehensive Review. *Environ. Exp. Bot.* **2022,** *193*, 104666.

Chaudhri, I. I.; Shah, B. H.; Naqvi, N.; Mallick, I. A. Investigations on the Role of *Suaeda fruticosa* Forsk in the Reclamation of Saline and Alkaline Soils in West Pakistan Plains. *Plant Soil* **1964,** *21*, 1–7.

Chen, M.; Yang, Z.; Liu, J.; Zhu, T.; Wei, X.; Fan, H.; Wang, B. Adaptation Mechanism of Salt Excluders Under Saline Conditions and its Applications. *Int. J. Mol. Sci.* **2018,** *19*, 3668.

Drought and Climate Change. https://www.c2es.org/content/drought-and-climate-change/ (accessed July 24, 2022).

Fan, C. Genetic Mechanisms of Salt Stress Responses in Halophytes. *Plant Signal Behav.* **2020,** *15*, 1704528.

Faustino, M. V.; Faustino, M. A. F.; Pinto, D. C. G. A. Halophytic Grasses, a New Source of Nutraceuticals? A Review on Their Secondary Metabolites and Biological Activities. *Int. J. Mol. Sci.* **2019,** *20*, 1067.

Food and Agriculture Organization (FAO) of United Nation, 2022. https://www.fao.org/global-soil-partnership/areas-of-work/soil-salinity/en/ (accessed July 24, 2022).

González, M. B. Adaptation of Halophytes to Different Habitats. In *Seed dormancy and germination* [Internet]; Jimenez-Lopez, J. C., Ed.; IntechOpen: London, 2019. https://www.intechopen.com/chapters/67884 (accessed July 24, 2022).

Hafeez, M. B.; Raza, A.; Zahra, N.; Shaukat, K.; Akram, M.Z.; Iqbal, S.; Basra, S. M. A. Gene Regulation in Halophytes in Conferring Salt Tolerance. Handbook of Bioremediation, 2021. DOI:10.1016/b978-0-12-819382-2.00022-3.

Hasanuzzaman, M.; Nahar, K.; Alam, M.; Bhowmik, P. C.; Hossain, M. A.; Rahman, M. M.; Prasad, M. N. V.; Ozturk, M.; Fujita, M. Potential Use of Halophytes to Remediate Saline Soils. *Bio. Med. Res. Int.* **2014,** 589341, 12.

Hassan, M.; Estrelles, E.; Soriano, P.; López-Gresa, M. P.; Bellés, J. M.; Boscaiu, M.; Vicente, O. Unraveling Salt Tolerance Mechanisms in Halophytes: A Comparative Study on Four Mediterranean *Limonium* Species With Different Geographic Distribution Patterns. *Front. Plant Sci.* **2017**. https://doi.org/10.3389/fpls.2017.01438

Herrera, A. Are Thick Leaves, Large Mesophyll Cells and Small Intercellular Air Spaces Requisites for CAM?. *Ann. Bot.* **2020,** *125*, 859–868.

Himabindu, Y.; Chakradhar, T.; Reddy, M. C.; Kanygin, A.; Redding, K. E.; Chandrasekhar, T. Salt-Tolerant Genes From Halophytes Are Potential Key Players of Salt Tolerance in Glycophytes. *Environ. Exp. Bot.* **2016,** *124*, 39–63.

Hussain, M. I.; Al-Dakheel, A. J.; Reigosa, M. J. Genotypic Differences in Agro-Physiological, Biochemical and Isotopic Responses to Salinity Stress in Quinoa (*Chenopodium quinoa* Willd.) Plants: Prospects for Salinity Tolerance and Yield Stability. *Plant Physiol. Biochem.* **2018,** *129*, 411–420.

Jithesh, M. N.; Prashanth, S. R.; Sivaprakash, K. R.; Parida, A. K. Antioxidative Response Mechanisms in Halophytes: Their Role in Stress Defence. *J. Genet.* **2006,** *85*, 237–254.

Kashyap, S. P.; Prasanna, H. C.; Kumari, N.; Kevin, K.; Redding, K. E.; Chandrasekharad, T. Understanding Salt Tolerance Mechanism Using Transcriptome Profiling and de Novo Assembly of Wild Tomato *Solanum chilense*. *Sci Rep.* **2020,** *10*, 15835.

Kaur, B.; Gupta, S. R.; Singh, G. Carbon Storage and Nitrogen Cycling in Silvi-Pastoral Systems on a Sodic Soil in Northwestern India. *Agroforestry Sys*. **2002,** *54*, 21–29.

Kumar, P.; Mishra, A. K.; Kumar, M.; Chaudhari, S. K.; Singh, R.; Singh, K.; Rai, P.; Sharma, D. K. Biomass Production and Carbon Sequestration of *Eucalyptus tereticornis* Plantation in Reclaimed Sodic Soils of North-West India. *Indian J. Agri. Sci.* **2019,** *89*, 1091–1095.

Kumar, P.; Sharma, P. K. Soil Salinity and Food Security in India. *Front. Sustain. Food Syst.* **2020**. 06 October 2020 Sec. Agroecology and Ecosystem Services https://doi.org/10.3389/fsufs.2020.533781

Liu, H. Q.; Lu, X. B.; Li, Z. X.; Tian, C. Y.; Song, J. The Role of Root-Associated Microbes in Growth Stimulation of Plants Under Saline Conditions. *Land Degrad. Develop.* **2021,** *32*, 3471–3486.

Lu, C.; Yuan, F.; Guo, J.; Han, G.; Wang, C.; Chen, M.; Wang, B. Current Understanding of Role of Vesicular Transport in Salt Secretion by Salt Glands in Recretohalophytes. *Int. J. Mol. Sci.* **2021,** *22*, 2203.

Lucia, Y. M.; Vivo, C. N.; Manuel, J. M. F.; Agatha, A.; Micaela, C.; Mcarmen, M. B. Analysis of Physiological Traits in the Response of *Chenopodiaceae*, *Amaranthaceae*, and *Brassicaceae* Plants to Salinity Stress. *Plant Physiol. Biochem.* **2018,** *132*.

Ma, Y.; Wang, J.; Zhang, J.; Zhang, S.; Liu, Y.; Lan, H. Seed Heteromorphism and Effects of Light and Abiotic Stress on Germination of a Typical Annual Halophyte *Salsola ferganica* in Cold Desert. *Front. Plant Sci.* **2018**. https://doi.org/10.3389/fpls.2017.02257.

Manousaki, E.; Kalogerakis, N. Halophytes Present New Opportunities in Phytoremediation of Heavy Metals and Saline Soils. *Ind. Eng. Chem. Res.* **2011,** *50*, 656–670.

Mishra, A.; Tanna, B. Halophytes: Potential Resources for Salt Stress Tolerance Genes and Promoters. *Front Plant Sci.* **2017,** *8*, 829.

Mishra, A.; Tanna, B. Halophytes: Potential Resources for Salt Stress Tolerance Genes and Promoters. *Front. Plant Sci.* **2017**. DOI: 10.3389/fpls.2017.00829.

Moir-Barnetson, L. A. B.; Veneklaas, E. J. A.; Colmer, T. D. A. Salinity Tolerances of Three Succulent Halophytes (*Tecticornia* spp.) Differentially Distributed Along a Salinity Gradient. *Function. Plant Biol.* **2016,** *43*, 739–750.

Morris, L.; Yun, K.; Rutter, A.; Zeeb, B. A. Characterization of Excreted Salt From the Recretohalophytes *Distichlis spicata* and *Spartina pectinata. J. Env. Qual.* **2019,** *48*, 1775–1780.

Oh, D. H.; Dassanayake, M.; Haas, J. S.; Kropornika, A.; Wright, C.; d'Urzo, M. P.; Hong, H.; Ali, S.; Hernandez, A.; Lambert, G. M. Genome Structures and Halophyte-Specific Gene Expression of the Extremophile *Thellungiella parvula* in Comparison with *Thellungiella salsuginea* (*Thellungiella halophila*) and Arabidopsis. *Plant Physiol.* **2010,** *154*, 1040–1052.

Oh, D. H.; Leidi, E.; Zhang, Q.; Hwang, S. M.; Li, Y.; Quintero, F. J. Loss of Halophytism by Interference with SOS1 Expression. *Plant Physiol.* **2009,** *151*, 210–222.

Otlewska, A.; Migliore, M.; Dybka-Stępień, K.; Manfredini, A.; Struszczyk-Świta, K.; Napoli, R.; Białkowska, A.; Canfora, L.; Pinzari, F. Plant Root Interaction with Associated Microbiomes to Improve Plant Resiliency and Crop Biodiversity. *Front. Plant Sci.* **2020**. DOI: org/10.3389/fpls.2020.553087.

Palliyath, S.; Am, S.; Puthur, J. Heavy Metal Detoxification Mechanisms in Halophytes: An Overview. *Wetlands Ecol. Manage.* **2017**. DOI: 10.1007/s11273-016-9513-z.

Pujol, J. A.; Calvo, J. F.; Ramírez-Díaz, L. Recovery of Germination From Different Osmotic Conditions by Four Halophytes from Southeastern Spain. *Annals Bot.* **2000,** *85*, 279–286.

Qu, X. X.; Huang, Z. Y.; Baskin, J. M.; Baskin, C. C. Effect of Temperature, Light and Salinity on Seed Germination and Radicle Growth of the Geographically Widespread Halophyte Shrub *Halocnemum strobilaceum*. *Ann. Bot.* **2008,** *101*, 293–299.

Rahman, M. M.; Mostofa, M. G.; Keya, S. S.; Siddiqui, M. N.; Ansary, M. M. U.; Das, A. K.; Rahman, M. A.; Son-Phan, L. T. Adaptive Mechanisms of Halophytes and Their Potential in Improving Salinity Tolerance in Plants. *Int. J. Mol. Sci.* **2021,** *22*, 10733.

Serag, M. S. Ecology of Four Succulent Halophytes in the Mediterranean Coast of Damietta Egypt. *Est. Coast. Shelf Sci.* **1999,** *49*, 29–36.

Shabala, S. Learning From Halophytes: Physiological Basis and Strategies to Improve Abiotic Stress Tolerance in Crops. *Ann. Bot.* **2013,** *112*, 1209–1221.

Song, J.; Shi, W. W.; Liu, R. R.; Xu, Y. G.; Sui, N.; Zhou, J. C.; Feng, C. The Role of the Seed Coat in Adaptation of Dimorphic Seeds of the Euhalophyte *Suaeda salsa* to Salinity. *Plant Species Biol.* **2017,** *32*, 107–114.

Soundararajan, P.; Manivannan, A.; Jeong, B. R. Different Antioxidant Defense Systems in Halophytes and Glycophytes to Overcome Salinity Stress. *Tasks Vegetat. Sci.* **2019**. DOI: 10.1007/978-3-030-04417-6_20.

Usman, K.; Al-Ghouti, M. A.; Abu-Dieyeh, M. H. Phytoremediation: Halophytes as Promising Heavy Metal Hyperaccumulators. In *Heavy Metals* [Internet]; Saleh, H. E. M., Aglan, R. F., Eds.; IntechOpen: London, 2018. https://www.intechopen.com/chapters/59441 (accessed July 26, 2022).

Vineeth, T.; Kumar, S.; Shukla, M.; Chinchmalatpure, A.; Sharma, P. C. Ecological and Economic Potential of Major Halophytes and Salt Tolerant Vegetation in India. In *Abiotic Stress in Plants* [Internet]; Fahad, S., Saud, S., Chen, Y., Wu, C., Wang, D., Eds.; IntechOpen: London, 2020. https://www.intechopen.com/chapters/73439 (accessed July 24, 2022).

Vuai, S. A. H. Characterization of Agar Extracted From *Gracilaria* Species Collected Along Tanzanian Coast. *Heliyon* **2022,** *8*, e09002.

Wang, J.; Li, B.; Meng, Y.; Ma, X.; Lai, Y.; Si, E.; Yang, K.; Ren, P.; Shang, X.; Wang, H. Transcriptomic Profiling of the Salt-Stress Response in the Halophyte *Halogeton glomeratus*. *BMC Genom.* **2015,** *16*, 169.

Wang, S.; Zhao, Z.; Ge1, S.; Peng, B.; Zhang, K.; Hu, M.; Mai, W.; Tian, C. Root Morphology and Rhizosphere Characteristics are Related to Salt Tolerance of *Suaeda salsa* and *Beta vulgaris* L. *Front. Plant Sci.* **2021**. https://doi.org/10.3389/fpls.2021.677767.

Winter, K. Ecophysiology of Constitutive and Facultative CAM Photosynthesis. *J. Exp. Bot.* **2019,** *70*, 6495–6508.

Wungrampha, S.; Joshi, R.; Rathore, R. S.; Singla-Pareek, S. L.; Govindjee, P. A. CO^2 Uptake and Chlorophyll a Fluorescence of *Suaeda fruticosa* Grown Under Diurnal Rhythm and After Transfer to Continuous Dark. *Photosynthesis Res.* **2019,** *142*, 211.

Xian-Zhao, L.; Chun-Zhi, W.; Qing, S.; Chao-Kui, L. The Potential Resource of Halophytes for Developing Bio-Energy in China Coastal Zone. *J. Agri. Food Sci. Res.* **2012,** *1*, 44–51.

Yan, A.; Wang, Y.; Tan, S. N.; Yusof, M. M. L.; Ghosh, S.; Chen, Z. Phytoremediation: A Promising Approach for Revegetation of Heavy Metal-Polluted Land. *Front. Plant Sci.* **2020**. https://doi.org/10.3389/fpls.2020.00359

Yuan, F.; Leng, B.; Wang, B. Progress in Studying Salt Secretion From the Salt Glands in Recretohalophytes: How do Plants Secrete Salt?. *Front Plant Sci.* **2016,** *7*, 977.

Zhang, H.; Zhang, G.; Lü, X.; Zhou, D.; Han, X. Salt Tolerance During Seed Germination and Early Seedlings Stages of 12 Halophytes. *Plant Soil.* **2015,** *388*, 229–241.

Zhao, C.; Zhang, H.; Song, C.; Zhu, J. K.; Shabala, S. Mechanisms of Plant Responses and Adaptation to Soil Salinity. *Innovation* **2020,** *1*, 100017.

Zvinavashe, A. T.; Lim, E.; Sun. H.; Marelli, B. A Bioinspired Approach to Engineer Seed Microenvironment to Boost Germination and Mitigate Soil Salinity. *PNAS* **2019,** *2019116*, 25555–25561.

CHAPTER 16

Adaptive Strategies and Role of Endophytes in Halophytes for Amelioration of Salt Stress

SHREYASI BISWAS and MAHUYA MUKHOPADHYAY

Department of Microbiology, Lady Brabourne College, Kolkata, West Bengal, India

ABSTRACT

Salinization is a worldwide problem and a significant abiotic stress that limits the growth of plants. High concentration of salts results in hyperosmotic stress, ionic imbalance and nutrient deficiency in plants. Halophytes are salt-tolerant plants that grow in soil or water with high salinity such as mangroves, marshes, seashores, sloughs, and so on. These plants have ability to complete their life cycle in a substrate rich in NaCl, which is usually detrimental to other plants. Halophytes that require a particular salt concentration to grow are referred to as "obligate halophytes" and some that may grow in the soil devoid of salt are called "facultative halophytes." There are various mechanisms of stress tolerance in halophytes like ion compartmentalization, succulence, selective transport, uptake of ions, and enzymatic and non-enzymatic antioxidant response to alleviate the salt stress. At the molecular level, halophytes survive under salt stress by regulating various stress-responsive genes. Apart from this, beneficial endophytes may help to protect their hosts (halophytes) under high salinity stress by influencing phytohormonal levels, enhancing photosynthesis, maintaining homeostasis, contributing biomass production. As, soil salinity disrupts the physiological and morphological processes of plants, they become more susceptible to phytopathogens. The plant growth promoting endophytes may inhibit the

Physiology of Halophytes: Signaling, Omics, and Tolerance Mechanisms.
Nivas Manohar Desai & Ganesh Chandrakant Nikalje (Eds.)

phytopathogens and protect their host plants under stress. In this chapter, different strategies adapted by the halophytes and the role of plant growth-promoting endophytes for amelioration of salt stress are discussed.

16.1 INTRODUCTION

Soil salinity is one of the major abiotic constraints for plants and crop production worldwide. Plant growth and yield is being affected severely due to salt stress. Soils are classified as saline when the electric conductivity of saturated paste extract is 4 dS m^{-1} or even more which is almost equivalent to 40 mM NaCl which is one of the biggest problems in coastal areas. Due to low precipitation, high evaporation, saline irrigation water, and poor water management, soil salinity increases. According to the findings of Flowers et al. (2010) about 6% land of the world has been plagued by increasing salinity. Most of the Earth surface is covered in a salt solution dominated by Na^+ and Cl^-. The concentration of salt in seawater is required for some halophytes but not for the others.

Plants have evolved on earth and have grown from tundra to tropical forests, from deserts to swamps, and in ocean as well. Based on the soil salinity, plants are classified as halophytes like euhalophytes, mesohalophytes, and mesoeuhalophyte (Chapman, 1942). A well-evolved group of halophytes have well-adapted morphological and anatomical characteristics, which reflect their ability to grow at high saline condition (Munns and Tester, 2008; Flowers and Colmer, 2008). Halophytes inhabit in land within tidal zone like mangrove forests, near-shore shallows and estuaries like seagrasses, inland salt lakes, and saline deserts. Halophytes may use the substrate present in saline soil which may not be possible for the salt-sensitive plants. The plant growth-promoting endophytes may help to survive their halophytic hosts in under salt stress by producing ACC deaminase enzyme which lowers the ethylene level in plants (Afridi et al., 2019). In this contest, several authors have reported that endophytic bacteria enhance plant productivity and salt tolerance (Ruppel et al., 2013; Belaouni et al., 2022). Endophytes may enhance plant growth either directly by producing phytohormones, fixing nitrogen, solubilizing phosphate, and so on, or indirectly by, for example, producing bioactive compounds and extracellular lytic enzymes to inhibit phytopathogens (Liotti et al., 2018; Herrera et al., 2016).

16.2 DIVERSITY OF HALOPHYTES

There are various types of saline habitats like aquatic-haline and terrestrohaline. According to Chapman (1942) halophytes are classified into euhalophytes, miohalophytes, mesohalophytes, mesoeuhalophytes, and so on. The hydrohalophytes have the ability to survive in areas subjected to flooding and fluctuations in salinity level, such as mangrove forest in the distinct saline woodland or shrub land grow in tropical and subtropical coastal areas across the world. Mangrove are tropical plants, well adapted to tidal habitat life or more specifically to high salinity coastal areas and often have been found in combination with seagrass beds and coral reefs (Kathiresan and Bingham, 2001; Sternberg et al., 2007). According to Tomlinson (1986), there are three types of mangrove namely major mangrove, minor mangrove, and mangrove associates. About 42%, which is the highest percentage, of worldwide mangrove area are found in Asia followed by Africa (20%), and North and Central America (15%). Mangroves are facultative halophytes. One such report by Dangremond et al. (2015) showed that *Avicennia* had a survival rate >90% under different salinity treatments like 0, 35, and 60 psu under low to medium light. *Arthrocnemum macrostachyum* is a perennial shrub, commonly found in the Mediterranean region, Middle East, and Asia, was reported with marked growth stimulation in a range between 171 mM and 510 mM NaCl for relative growth rate (Redondo-Gómez et al., 2010). Pssamophytes or the desert halophytes have the ability to grow in non-arable soil under extreme desert condition (Eshel et al., 2010). *Aeluropus lagopoides* (Linn.) Trin. ex Thw. is a perennial grass and grow easily in the high saline habitats especially in deserts (Gulzar and Khan, 2001). *Nitraria retusa* and *Zygophyllum album* L.F are very important desert halophytes or halo-xerophiles. They have been found mostly in El sokhna region adjacent to western coast Red sea, Egypt, with various morphological adaptation to thrive in harsh saline conditions (Abd Elhalim et al., 2016). *Cakile maritime* Scop. is the most widely studied psamophytes distributes in Mediterranean, Black sea and Atlantic areas, sandy coasts of New Zealand and Australia have maximum growth, seed production at moderate salinity level like 50–100 mM NaCl (Hmidi et al., 2018).

16.3 PHYSIOLOGICAL STRATEGIES OF HALOPHYTES AGAINST SALT STRESS

High salinity of soil restricts the uptake of essential macronutrients like Ca^{2+} and Mg^{2+} from soil which hinders the signal transduction pathways, membrane stability, and chlorophyll synthesis (Geilfus et al., 2018; Arif et al., 2020). Halophytes are the kind of plants which acquire several salt stress tolerance features like changes in shoot growth, Na^+ uptake (blocking Na^+ influx, enhancing Na^+ efflux from the roots), increased toxic ion exclusion or their compartmentalization into vacuoles, compatible solute accumulation, relative water content of leaf, ion concentration, lipid peroxidation, and so on, that ensure normal plant growth. Na^+ influx is a process through which Na^+ enters into the root via epidermal and cortical cells (Cheeseman, 1982). For example in *Salvadora persica,* the chlorophyll content, relative water content, and water-use efficiency remain unchanged in response to salt stress while photosynthesis, transpiration, and stomatal conductance have been reduced under high salinity (Rangani et al., 2016). Na^+ influx consist of Ca^{2+}-sensitive, Ca^{2+}-insensitive pathways and bypass flow (Tester and Davenport, 2003). In Ca^{2+}-sensitive process, about 10 mM Ca^{2+} in the external solution may reduce Na^+ influx into the roots and in Ca^{2+}-insensitive pathway nonselective cation channels are involved (Zhu, 2001; Munns, 2005). The third process in which Na^+ influx is blocked, halophytes excrete excess amount of salt by ultrafiltration—a root-induced mechanism. An apoplastic barrier in the roots can improve bypass flow resistance and results in salt exclusion from the roots as well as reduced accumulation of toxic ions in the aboveground shoots (Hussain et al., 2021). An ion homeostasis (high K^+ and low Na^+) in the cytoplasm is very important for both the halophytes and glycophytes under salt stress. Well, in this, SOS (salt overly system) pathway is very essential for Na^+ efflux. There are key proteins SOS3, SOS2, and SOS1 involved in this pathway and they play vital roles to efflux Na^+ from cytoplasm according to Liu and Zhu (1998). *SOS3* basically encodes a calcium-binding protein which upon binding to Ca^{2+} ion activates a serine threonine/kinase *SOS2* (Hrabak et al., 2003). Findings of Quan et al. (2007) show a SOS3-like protein called SOS3-like calcium-binding protein 8 (SCaBP8) acts as an alternative regulator of SOS2 in the shoots of Arabidopsis. *Halimione portulacoides* and *Sarcocornia fruticosa* are two halophytes that respond in a different manner to the salt-stress condition in terms of their leaf relative water content and concentration of leaf elements (Duarte et al., 2013); in case of *S. fruticosa*, it had lower water content while *H. portulacoides* showed higher water content subjected to higher salinities. The salinity stress tolerance is evolved by the halophytes easily than the glycolphytes, for example, *Cakile maritime,* a halophyte showed less pronounced results of early growth rate reduction, rapid reduction of H^2O^2 concentration and higher level of catalase, oxidase and superoxide dismutase activities in comparison to a glycophyte *Arabidopsis thaliana* (Ellouzi et al., 2011). Apart from this, some of the halophytes may excrete the excess amount of salt from tissues with the help of salt glands (diverse in structural and functional features) and can be classified into four groups in terms of their structural similarities namely multicellular salt glands, bi-cellular salt glands, and unicellular vacuolated secretary hairs and salt bladders (Yuan et al., 2016; Dassanayake and Larkin, 2017). Salt glands consists of multiple cells and differentiated into two basic cells namely basal collecting cells and distal secretory cells which results in active transport of salts through the collecting cells into secretory cells (Dassanayake and Larkin, 2017). Salt glands from different species of halophytes have various structural features. Generally dicotyledonous halophytes possess multicellular salt glands, like *Tamarix aphylla*, which contain total eight cells in which two collecting cells and six secreting cells make a symmetrical structure (Thomson and Platt-Aloia, 1985). *Limonium bicolor*, an extreme halophyte, has salt glands with a total of 16 cells (Feng et al., 2015; Yuan et al., 2015). Mangroves have multiple numbers of secretory cells like *Avicinnea officinalis*, which contain —six to eight secretory cells (Tan et al., 2010). *Avicinnea germinans* reported eight cells (Balsamo and Thomson, 1993). Autofluorescence and acidic nature of salt glands are very important features for salt secretion. The cuticles surrounding the salt glands have autofluorescence complexes which are acidic in nature as reported in many studies of mangroves and salt grasses (Tan et al., 2010; Deng et al., 2015). Bi-cellular salt glands are mostly found in monocotyledonous halophytes (Ramadan and Flowers, 2004; Semenova et al., 2010). Salinity has an effect on the development of xylem vessel and shows maximum hydraulic conductance when seedlings are grown in 75% seawater and minimum conductance observed without salt (Nguyen et al., 2015). The simple form of salt glands are salt bladders which are basically swollen epidermal cells found in various halophytes like *Atriplex lentiformis*, *Bienertia sinuspersici*, and *Chenopodium quinoa* (Adolf et al., 2013; Shabala et al., 2014). The halophytes use mechanisms of sequestering salt in their salt bladders which resemble the storage of salt in large vacuoles of mesophyll cells within the succulent leaves (Park et al., 2009). Recirculation of Na^+ into phloem actually decreases the accumulation of Na^+ into the leaves (Matsushita and Matoh, 1991). In case of halophytes, osmotic balance is achieved by the accumulation of Na^+, Cl^- ions, and low molecular weight organic molecules known as osmolytes (Glenn et al., 1999). Under enhanced stress tolerance, the level of osmolytes also increased and protect the cells through scavenging the free radicals (Szabados et al., 2011). Apart from these, phenotypic plasticity (ability of an organism to change its morphology in response to the changes in the environments) also protects the plants under several stress conditions (Arnold et al., 2019) In *Suaeda maritima*, variations in colors like green, greenish-red, and reddish traits are found in low or high salt marshes. Though their growth response was quite similar, but their growth forms are notable like those grow in low salt marshes are small in size and less branched (Rittirong-sakul et al., 2020).

16.4 MOLECULAR STRATEGIES OF HALOPHYTES AGAINST SALT STRESS

Salinity or salt stress results in various responses in the halophytes due to ion and osmotic stresses, and one of the most important responses is induction of reactive oxygen species (ROS) (Miller et al., 2010) which are really toxic and cause oxidative damage to the lipid, proteins, and DNA in the absence of protective mechanisms. Plants can control the accumulation of ROS with the help of enzymatic and non-enzymatic antioxidants. To cope with the salinity stress, plants decrease stomatal conductance to minimize the water loss and due to this, limitation occurs in CO^2 availability for fixation through Calvin cycle (Jithesh et al., 2006; Volpe et al., 2011). ROS is accumulated due to the limited gas exchange by photorespiration under salinity. In this carbon fixation process, ribulose-1,5-bisphosphate carboxylase/oxygenase (Rubisco) uses oxygen instead of CO^2 and produces phosphoglycolate. After that, in the detoxification and recycling step, when phosphoglycolate is back to 3-phosphoglycerate, hydrogen peroxide (H^2O^2) is produced in peroxisome (Bauwe et al., 2010). The enzymatic antioxidants include catalase (CAT), superoxide dismutase (SOD), ascorbate peroxidase (APX), and peroxidase (POX), and non-enzymatic antioxidants are ascorbic acid, β-carotene, α-tocopherol, are glutathione. Apart from this, there is an array of enzymes involved in the redox regulation like glutathione reductase (GR), glutathione peroxidase (GPX), and glutathion-S-transferases (GST) which are expressed as key components of antioxidant defence (Mittler, 2002). Many studies showed that halophyte *Hordeum marinum* and *Cakile maritima* had catalase, superoxide dismutase, and glutathione reductase as antioxidant defence system under salinity (Seckin et al., 2010; Ellouzi et al., 2011). The SOD is a very important scavenging enzyme and has also been referred to as the first line of defense against oxidative stress. This enzyme has several isoenzymes namely FeSOD (iron superoxide dismutase, generally localized in chloroplast), Cu/ZnSOD (copper/zinc superoxide dismutase in cytosol and chloroplast), and MnSOD (manganese superoxide dismutase mitochondria, and peroxisomes) which are usually distinguished by inhibitor treatment (Vitória et al., 2001; Alscher et al., 2002). But the regulation or expression of these antioxidant enzymes are different in various halophytes under stress. Like *Sesuvium portulacastrum*, a sea purslane, grows in the coastal areas had enhanced CAT, SOD, and APX enzymatic activities under in vitro salt treatment with 1M NaCl (Lokhande et al., 2011). An induction of H^2O^2 level was observed along with an increased peroxidase activity but no significant activity of CAT was reported in *Mesembryanthemum crystallinum* under salt stress (Shevyakova et al., 2006). On the other hand, *Crithmum maritimum* known as samphire or sea fennel showed no significant change in the activities of antioxidant enzymes (Ben Hamed et al., 2007). The rapid response of these antioxidant enzymes in halophytes is actually a vital trait for salt-stress tolerance and prevention of the irreversible cellular damages. It is evident in studies that these enzymatic activities are sometimes tissue specific like highest CAT activity was detected in roots and highest SOD activity was observed in shoots of *Crithmum maritimum* (Ben Hamed et al. 2007; Ben Amor et al. 2005). Halophytes have another set of antioxidants which are non-enzymatic by nature like carotenoids scavenges peroxyl and $^1O^2$ radicals reported in *Plantago coronopus* (Koyro, 2006). The polyphenol content is another important non-enzymatic antioxidant which may differ in halophytes, depending on the halophyte species, climate, and ecotypes. For example, *Mesembrythemum edule* has higher polyphenol content than *Mesembrythemum nodiflorum* and *Mesembrythemum crystallinum* (Falleh et al., 2009), on the other hand polyphenol content level is high in *Mesembrythemum edule* when it is collected from an arid area in comparison to the same plants collected from a semiarid region (Falleh et al., 2012). According to Motilva et al. (2013), polyphenols are classified in three gropus: flavonoids, non-flavonoids, and tannins. One of the most important characteristics of them is to scavenge ROS. In the halophyte *Prosopis strombulifera,* an increased production of polyphenols, specifically flavonoids and flavan-3-ols, was reported under salinity stress (Reginato et al., 2014). Polyphenols also have capability of chelation transition metals (Andjelkovic et al., 2008). Due to this, they protect DNA from free-radical damage as they chelate iron (III)/copper (II), iron (II)/Copper(I) involved in the conversion of H^2O^2 and $O^{2\circ-}$ into highly reactive hydroxyl radicals (Perron and Brumaghin, 2009; Leopoldini et al., 2011). Proline is another important factor or an osmo-protectant which may scavenge hydroxyl radicals, and an increased concentration of proline was also observed in halophytes under salt stress (Ben Hassine et al., 2008; Szabados and Savoure, 2010). Thioredoxins (TRX) are ubiquitous proteins that play an important role in the oxidative stress tolerance of the plants by preventing protein oxidation and lipid peroxidation. Though identification and characterization of trx gene needs more investigation, TRX-like homologue genes have been found in many halophytes (Chen et al., 2012; Baisakh et al., 2008). Peroxiredoxins (Prx), a well-known thioredoxin-dependent reductase, has various functions like H^2O^2 homeostasis, scavenging of hydroperoxides, and others. According to Guo et al. (2004), Prx Q gene was present and involved in H^2O^2-dependent signaling pathways in chloroplast in the halophyte *Suaeda salsa* under salt-stress condition. Glutaredoxins are also important in ROS signaling (Zhang et al., 2008), but the specific gene regulation under salinity has not been fully elucidated yet, which can be an additional mechanism to combat oxidative damages under salt stress. Halophytes have an interesting feature to survive under salt stress by vacuolar sequestration of Na^+ and that needs Na^+ could be pumped into the vacuole against the electrochemical gradient and prevent it from leaking back into the cytosol with the help of Na^+/H^+ antiporter, for example, *SsNHX1* gene may encode this particular antiporter as this gene reported an upregulated expression in a halophyte *Suaeda salsa* under in vitro salt treatment (Chen et al., 2010; Yang et al., 2010; Song et al., 2011).

The *SOS* genes, especially *SOS3*, *SOS2*, and *SOS1,* are involved in Na^+ efflux from cells under salinity stress. The elevated level of Ca^{2+} binds with a protein known as calcineurin B-like 4(CBLA/SOS3), and this activated protein binds and activates a serine/threonine protein kinase (CIPK24/SOS2). The SOS3/CBL4 and SOS2/CIPK complex phosphorylates the C-terminus of SOS1 and activates its protein activity which leads to Na^+ efflux with the help of plasma membrane ATPases (Halfter et al., 2000; Shi et al., 2000; Qiu et al., 2002; Guo et al., 2004). Various *SOS1*, *SOS2*, *NHX*, and other salt-tolerant-related genes have been isolated and characterized in halophytes (Table 16.1).

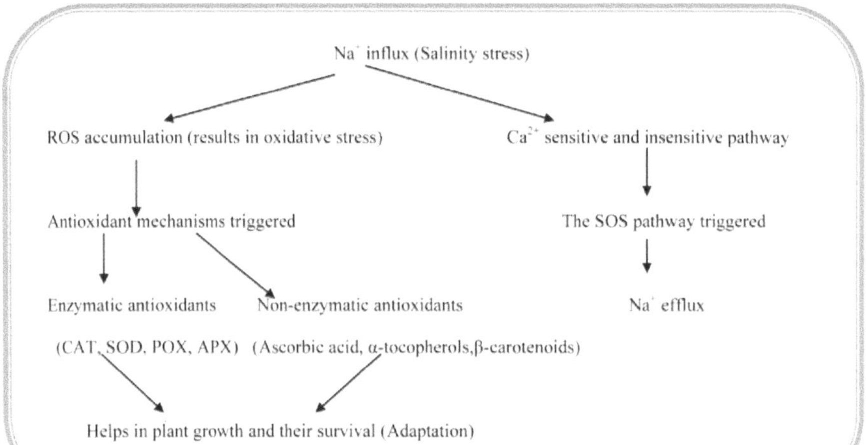

FIGURE 16.1 An overview of ROS signaling (antioxidants) and halophytes adaptation under salinity.

TABLE 16.1 Genes (Transporters, Antiporters, Antioxidants, and Others) from Halophytes and their Applications.

Name of gene	Host	Target	Result	References
ThSOS1	*Thellungiella halophila*	*Arabidopsis* and Yeast defective in Na^+ efflux transporter	Increased salt tolerance	Oh et al., 2009; Nah et al., 2009
PtNHA1 or PtSOS1	*Puccinellia tenuiflora*	*Arabidopsis*	Na^+ accumulation was lower and ROS-scavenging enzyme activity was enhanced	Kobayashi et al., 2012; Wang et al., 2011
SbSOS1	*Salicornia brachiata*	Tobacco	Helps in salt tolerance by decreasing cytosolic Na^+	Yadav et al., 2012
AgNHX1	*Atriplex gmelini*	Yeast and rice	Partial complementation function of the yeast antiporter and high activity in transgenic rice vacuolar antiporter activity	Hamada et al. 2001; Ohta et al. 2002
PtSOS2	*Populus trichocarpa*	Hybrid poplar	Both Na^+ exclusion of plasma membrane and ROS-scavenging activity enhanced	Yang et al., 2015
SaNHX1	*Spartina anglica*	Rice	It increased the salt tolerance in transgenic rice	Lan et al., 2011
SsNHX1	*Salsola soda*	*Alfalfa*	Toxic effect of Na^+ was reduced as Na^+ transported into vacuoles	Li et al., 2011
ScVP	*Suaeda corniculata*	*Arabidopsis*	Improved salt and drought stress tolerance along with accumulation of Na^+ in leaves and roots	Liu et al., 2011
AeNHX1	*Agropyron elongatum*	Yeast, *Arabidopsis*, and festuca	These genes complements antiporter function, improved salt stress tolerance, and photosynthesis in Arabidopsis and festuca	Qiao et al., 2007
LfNHX1	*Leptochloa fusca*	Tobacco	Improved germination and root growth was observed	Rauf et al., 2014
KfVP1	*Kalidium foliatum*	*Arabidopsis*	Helps to improve salt and drought tolerance	Yao et al., 2012
ThVHAc1	*Tamarix hispida*	Yeast	Helps in various abiotic stresses	Gao et al., 2011

TABLE 16.1 *(Continued)*

Name of gene	Host	Target	Result	References
SaVHAc1	*Spartina alterniflora*	Rice	Helps in salt tolerance and up-regulates gene for ABA signaling	Baisakh et al., 2012
PhaHAK2	*Phragmites australis*	Reed plant	Stable maintenance of high k^+/Na^+ ratio under slat stress	Takahashi et al., 2007
PutAKT1	*Puccinellia tenuiflora*	*Arabidopsis*	Improved salt tolerance	Ardie et al., 2010
KPutB1	*Puccinellia tenuiflora*	*Arabidopsis* and yeast	Na^+ concentration in roots decreased and shoot K^+ increased	Ardie et al., 2011
SaINO1	*Spartina alterniflora*	*Arabidopsis*	Improved salt stress tolerance and also become less sensitive to photoinhibition	Joshi et al., 2013
SbMT-2	*Salicornia brachiata*	Tobacco	Helps in salt tolerance and ROS scavenging	Chaturvedi et al., 2014
SbASR-1	*Salicornia brachiata*	Tobacco	Germination and seedling growth showed better results under salt stress	Jha et al., 2012
Sod1	*Avicennia marina*	Rice var. Pusa Basmati-1	Salt tolerance up to 150 mM NaCl by this rice variant	Prashanth et al., 2008
SaSce9	*Spartina alterniflora*	*Arabidopsis*	Results in an increased expression of stress responsive genes like *AtSOD, AtCAT, AtNHX1, AtSOS1, AtP5CS, AtRD22*	Karan and Subudhi, 2012
ZmVP1 (vacuolar H^+-pyrophosphatase)	*Zoysia matrella*	Salt-sensitive yeast and *Arabidopsis* (wild type and transgenic)	Restores salt tolerance in yeast, vigorous growth in wild type *Arabidopsis*, increased activities of V-ATPase, V-PPase, increased expression of stress-related genes in transgenic *Arabidopsis*	Chen et al., 2015
ZjZFN1	*Zoysia japonica*	*Arabidopsis thaliana*	Enhanced seed germination and salt adaptation	Teng et al., 2018

16.5 BENEFICIAL ENDOPHYTIC MICROBES AND THEIR IMPACT ON SALT TOLERANCE OF HALOPHYTES

Endophytes, either bacteria or fungi, thrive inside the plant tissues without causing any harm to their hosts. These endophytes have several plant-growth promoting (PGP) abilities like production of phytohormones, fixation of atmospheric nitrogen, solubilize phosphate, inhibition of phytopathogens, and so on (Shah et al., 2018; Kumar et al., 2020; Khalil et al., 2021). One of the most important phytohormones is ethylene after auxin and gibberellins. As this gaseous hormone ethylene is accumulated under various abiotic stress conditions, it may inhibit the plant growth (Yuan et al., 2015). There were evidences to support that plant-growth-promoting endophytes can reduce ethylene level by producing ACC deaminase as this enzyme cleaves ACC, the precursor of ethylene, into ammonia and α-ketobutyrate (Khan et al., 2019; Choudhury et al., 2021) which helps plants to survive under salt stress by lowering ethylene level and enhancing nutrient uptake capacity (Win et al., 2018); it also may supply a source of nitrogen (ammonia) to the plants (Hardoim et al., 2008). Endophytes help to promote plant growth in both host-specific and -non-specific manner. The endophytic bacteria help to survive non-host plants as well by increasing its non-enzymatic, enzymatic antioxidant defence, and photosynthetic pigments (Sofy et al., 2021). Salt-accumulating halophytes, like *Salicornia europaea, Salicornia bigelovii,* have been found in semi-arid saline area, and these have shown enhanced growth yield in terms of their increase in root and shoot length, fresh and dry weight in the presence of isolated plant growth promoting endophytic bacteria in a host-specific manner under varied saline conditions (Zhao et al., 2016; El-Tarabily et al., 2019). IAA-producing endophytes could increase salt tolerance by increasing root proliferation, as IAA is directly involved in root generation (Birnbaum, 2016). It was reported in a study that IAA-producing endophytic bacteria results in an enhanced growth of wheat seedlings under salt stress (Sorty et al. 2016). There are many reports on the isolation of beneficial endophytic microbes from halophytes and their effects on either host or non-host plants under salt stress (Table 16.2).

16.6 CONCLUSIONS

Stress, whether biotic or abiotic, is very harmful for plants. Plants have various mechanisms to avoid or tolerate the stress. Halophytes are able to survive at different salinity stress with the help of their own physiological, metabolic mechanisms, and also with the help of endophytes. Endophytes have been used in various studies to ameliorate the salt stress in halophytes and crops. But it is important to find out the genetic regulation of new pathways of these endophytes for their plant growth promotion on their hosts (halophytes) and other plants in saline condition, which may also help in agriculture because salinity is a worldwide problem. So the halophytes may be used for the remediation of salinity.

TABLE 16.2 Isolated Endophytic Microbes from Halophytes and Their Impacts under Salt Stress.

Halophytes	Endophytic microorganisms	Target	Results	References
Arthrocnemum macrostachyum, Spergularia marina	*Bacillus, Brevibacillus, Agrobacterium, and Paenibacillus*	Broad bean plants (non-host plant)	IAA and ammonia production, phosphate solubilization, antimicrobial effects, and helped in salt stress tolerance	Mahgoub et al., 2021
Salicornia europaea	*Bacillus endophyticus, Bacillus tequilensis, Planococcus rifietoensis, Variovorax paradoxus,* and *Arthrobacter agilis*	*Salicornia europaea* (host plant)	ACC deaminase activity increases, phosphate solubilization, and enhanced growth under salinity stress	Zhao et al., 2016
Limonium sinense	*Streptomyces* sp. KLBMP 5084	*Limonium sinense* (host plant)	Increased fresh weight, root length, chlorophyll content, and antioxidant content	Qin et al., 2017
Limonium sinense	*Bacillus, Pseudomonas, Klebsiella, Serratia, Arthrobacter, Streptomyces, Isoptericola, and Microbacterium*	*Limonium sinense* (host plant)	ACC deaminase production increased enhanced seedling growth	Qin et al., 2014
Arthrocnemum macrostachyum	*Bacillus sp.*	*Arthrocnemum macrostachyum* (host plant)	IAA and siderophore producers, phosphate solubioliozers, and help to accumulate Na^+ in the shoots of *A.macrostachyum*	Navarro-Torre et al., 2017
Suaeda salsa	*Curvularia* sp.	Poplar (non-host)	Promoted poplar growth, chlorophyll content, increased antioxidant enzymes like APX and SOD and proline content	Pan et al., 2018
Various *Suaeda* sp.	*Zhihengliuella halotolerans, Brachybacterium* sp.	*Suaeda maritime* (host plant)	N^2 fixation ability in medium supplemented with 6% NaCl and showed enhanced growth of the plant	Alishahi et al., 2020

TABLE 16.2 *(Continued)*

Halophytes	Endophytic microorganisms	Target	Results	References
Limonium sinense	*Glutamicibacter halophytocola* KLBMP 5180	Tomato (non-host)	An increased growth of root, fibrous roots, fresh weight, dry weight	Xiong et al., 2019
Tetragonia tetragonioides (Pall.) Kuntze	*Bacillus amyloliquefaciens*, *Stenotrophomonas maltophilia*	*Tetragonia tetragonioides* (Pall.) Kuntze (host plant)	Stimulated root and shoot growth under 200 mM NaCl concentration, ability to produce ACC deaminase, IAA, hydrogen cyanide, and had antifungal property as well	Egamberdieva et al., 2021
Haloxylon aphyllum Minkw.	*Bacillus amyloliquefaciens*, *Priestia endophytica*, *Bacillus subtilis* , *Bacillus toyonensis*, *Halomonas sulfidaeris*, *Isoptericola halotolerans*, *Planomicrobium soli*, *Pseudomonas kilonensis*	Cucumber plants (non-host)	High growth promotion ability under various NaCl concentration along with ACC deaminase, siderophore, IAA production activities	Shurigin et al., 2022

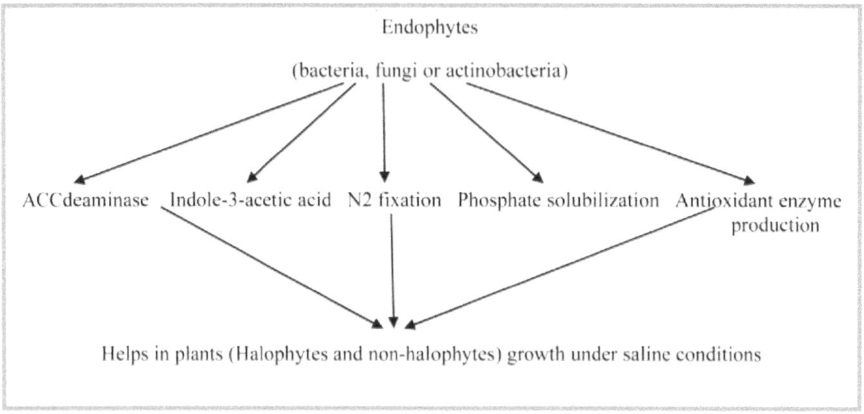

FIGURE 16.2 Beneficial effects of endophytes on plant growth.

ACKNOWLEDGMENT

The financial support for this work from University Grants Commission under NET-JRF scheme is duly acknowledged.

KEYWORDS

- **halophytes**
- **adaptations**
- **stress responsive genes**
- **beneficial endophytes**
- **sodium transport**
- **antioxidants**

REFERENCES

Abd Elhalim, M. E.; Abo-Alatta, O. K.; Habib, S. A.; Abd Elbar, O. H. The Anatomical Features of the Desert Halophytes *Zygophyllum album* LF and *Nitraria retusa* (Forssk.) *Asch. Ann. Agric. Sci.* **2016,** *61 (1)*, 97–104.

Adolf, V. I.; Jacobsen, S. E.; Shabala, S. Salt Tolerance Mechanisms in Quinoa (*Chenopodiu quinoa* Willd.). *Environ. Exp. Bot.* **2013,** *92*, 43–54.

Afridi, M. S.; Mahmood, T.; Salam, A.; Mukhtar, T.; Mehmood, S.; Ali, J.; Khatoon, Z.; Bibi, M.; Javed, M. T.; Sultan, T.; Chaudhary, H. J. Induction of Tolerance to Salinity in Wheat

Genotypes by Plant Growth Promoting Endophytes: Involvement of ACC Deaminase and Antioxidant Enzymes. *Plant Physiol. Biochem.* **2019,** *139*, 569–577.

Alishahi, F.; Alikhani, H. A.; Khoshkholgh-Sima, N. A.; Etesami, H. Mining the Roots of Various Species of the Halophyte *Suaeda* for Halotolerant Nitrogen-Fixing Endophytic Bacteria with the Potential for Promoting Plant Growth. *Int. Microbiol.* **2020,** *23*, 415–427.

Alscher, R. G.; Erturk, N.; Heath, L. S. Role of Superoxide Dismutases (SODs) in Controlling Oxidative Stress in Plants. *J. Exp. Bot.* **2002,** *53*, 1331–1341.

Andjelkovic, M.; van Camp, J.; de Meulenaer, B.; Depaemelaere, G.; Socaciu, C.; Verloo, M.; Verhe, R. Iron-Chelation Properties of Phenolic Acids Bearing Catechol and Galloyl Groups. *Food Chem.* **2008,** *98*, 23–31.

Ardie, S.; Liu, S.; Takano, T. Expression of the AKT1-type K+ Channel Gene from *Puccinellia tenuiflora, PutAKT1*, Enhances Salt Tolerance in *Arabidopsis*. *Plant Cell Rep.* **2010**, *29*, 865–874.

Ardie, S.W.; Nishiguchi, S.; Liu, S.; Takano, T. Ectopic Expression of the K+ Channel b Subunits from *Puccinellia tenuiflora* (*KPutB1*) and Rice (*KOB1*) alters K+ Homeostasis of Yeast and *Arabidopsis*. *Mol. Biotechnol.* **2011,** *48*, 76–86.

Arif, Y.; Singh, P.; Siddiqui, H.; Bajguz, A.; Hayat, S. Salinity Induced Physiological and Biochemical Changes in Plants: An Omic Approach Towards Salt Stress Tolerance. *Plant Physiol. Biochem.* **2020,** *156*, 64–77.

Arnold, P. A.; Kruuk, L. E.; Nicotra, A. B. How to Analyse Plant Phenotypic Plasticity in Response to a Changing Climate. *New Phytol.* **2019,** *222*, 1235–1241.

Baisakh, N.; RamanaRao, M. V.; Rajasekaran, K.; Subudhi, P.; Janda, J.; Galbraith, D.; Vanier, C.; Pereira, A. Enhanced Salt Stress Tolerance of Rice Plants Expressing a Vacuolar H+-ATPase Subunit c1 (*SaVHAc1*) Gene from the Halophyte Grass *Spartina alterniflora* Löisel. *Plant Biotech. J.* **2012,** *10*, 453–464.

Baisakh, N.; Subudhi, P. K.; Varadwaj, P. Primary Responses to Salt Stress in a Halophyte, Smooth Cordgrass (*Spartina alterniflora* Loisel.) *Funct. Integr. Genomics* **2008,** *8*, 287–300.

Balsamo, R. A.; Thomson, W. W. Ultrastructural Features Associated with Secretion in the Salt Glands of *Frankenia grandifolia* (Frankeniaceae) and *Avicennia germinans* (Avicenniaceae). *Am. J. Bot.* **1993,** *80*, 1276–1283.

Bauwe, H.; Hagemann, M.; Fernie, A. R. Photorespiration: Players, Partners, Origin. *Trends Plant Sci.* **2010,** *15*, 330–336.

Bauwe, H.; Hagemann, M.; Kern, R.; Timm, S. Photorespiration has a Dual Origin and Manifold Links to Central Metabolism. *Curr. Opin. Plant Biol.* **2012,** *15*, 269–275.

Belaouni, H. A.; Compant, S.; Antonielli, L.; Nikolic, B.; Zitouni, A.; Sessitsch, A. In-depth Genome Analysis of *Bacillus* sp. BH32, A Salt Stress-Tolerant Endophyte Obtained from a Halophyte in a Semiarid Region. *Appl. Microbiol. Biotechnol.* **2022**, *106*, 3113–3137.

Ben Amor, N.; Ben Hamed, K.; Debez, A.; Grignon, C.; Abdelly, C. Physiological and Antioxidant Responses of the Perennial Halophyte *Crithmum maritimum* to Salinity. *Plant Sci.* **2005,** *168*, 889–899.

Ben Hamed, K.; Castagna, A.; Salem, E.; Ranieri, A.; Abdelly, C. Sea Fennel (*Crithmum maritimum* L.) Under Salinity Conditions: A Comparison of Leaf and Root Antioxidant Responses. *Plant Growth Regul.* **2007,** *53*, 185–194.

Ben Hassine, A.; Ghanem, M. E.; Bouzid, S.; Lutts, S. An Inland and a Coastal Population of the Mediterranean Xero-halophyte Species *Atriplex halimus* L. Differ in their Ability to Accumulate Proline and Glycinebetaine in Response to Salinity and Water Stress. *J. Exp. Bot.* **2008,** *59*, 1315–1326.

Birnbaum, K. D. How Many Ways are there to make a Root? *Curr. Opin. Plant Biol.* **2016,** *34*, 61–67.
Bose, J.; Rodrigo-Moreno, A.; Shabala, S. ROS Homeostasis in Halophytes in the Context of Salinity Stress Tolerance. *J. Exp. Bot.* **2014,** *66*, 1241–1257.
Chapman, V. J. The New Perspective in the Halophytes. *Q. Rev. Biol.* **1942,** *17*, 291–311.
Chaturvedi, A. K.; Patel, M. K.; Mishra, A.; Tiwari, V.; Jha, B. The SbMT-2 Gene from a Halophyte Confers Abiotic Stress Tolerance and Modulates ROS Scavenging in Transgenic Tobacco. *PLoS One* **2014,** *9 (10)*, e111379.
Cheeseman, J. M. Pump-Leak Sodium Fluxes in Low Salt Corn Roots. *J. Membrane Biol.* **1982,** *70*, 157–164.
Chen, J.; Cheng, T.; Wang, P.; Liu, W.; Xiao, J.; Yang, Y.; Hu, X.; Jiang, Z.; Zhang, S.; Shi, J. Salinity-Induced Changes in Protein Expression in the Halophytic Plant *Nitraria sphaerocarpa. J. Proteom.* **2012,** *75*, 5226–5243.
Chen,Y.; Li, L.; Zong, J.; Chen, J.; Guo, H.; Guo, A.; Liu, J. Heterologous Expression of the Halophyte *Zoysia matrella* H^+-Pyrophosphatase Gene Improved Salt Tolerance in *Arabidopsis thaliana. Plant. Physiol. Biochem.* **2015,** *91*, 49–55.
Chen, M.; Song, J.; Wang, B. S. NaCl Increases the Activity of the Plasma Membrane H^+-ATPase in C-3 Halophyte *Suaeda salsa* Callus. *Acta Physiol. Plant.* **2010,** *32*, 27–36.
Choudhury, A. R.; Choi, J.; Walitang, D. I.; Trivedi, P.; Lee, Y.; Sa, T. ACC Deaminase and Indole Acetic Acid Producing Endophytic Bacterial Co-inoculation Improves Physiological Traits of red Pepper (*Capsicum annum* L.) under Salt Stress. *J. Plant Physiol.* **2021,** *267*, 153544.
Dangremond, E. M.; Feller, I. C.; Sousa, W. P. Environmental Tolerances of Rare and Common Mangroves Along Light and Salinity Gradients. *Oecologia* **2015,** *179*, 1187–1198.
Dassanayake, M.; Larkin, J. C. Making Plants Break a Sweat: The Structure, Function, and Evolution of Plant Salt Glands. *Front. Plant Sci.* **2017,** *8*, 406.
Deng, Y.; Feng, Z.; Yuan, F.; Guo, J.; Suo, S.; Wang, B. Identification and Functional Analysis of the Autofluorescent Substance in *Limonium bicolor* Salt Glands. *Plant Physiol. Biochem.* **2015,** *97*, 20–27.
Duarte, B.; Santos, D.; Marques, J. C.; Cašador, I. Ecophysiological Adaptations of Two Halophytes to Salt Stress: Photosynthesis, PS II Photochemistry and Anti-Oxidant Feedback–Implications for Resilience in Climate Change. *Plant Physiol. Biochem.* **2013,** *67*, 178–188.
Egamberdieva, D.; Alimov, J.; Shurigin, V.; Alaylar, B.; Wirth, S.; Bellingrath-Kimura, S. D. Diversity and Plant Growth-Promoting Ability of Endophytic, Halotolerant Bacteria Associated with *Tetragonia tetragonioides* (Pall.) Kuntze. *Plants* **2021,** *11*, 49.
Ellouzi, H.; Ben Hamed, K.; Cela, J.; Munné-Bosch, S.; Abdelly, C. Early Effects of Salt Stress on the Physiological and Oxidative Status of *Cakile maritima* (halophyte) and *Arabidopsis thaliana* (glycophyte). *Physiol. Plant* **2011,** *142*, 128–143.
El-Tarabily, K. A.; AlKhajeh, A. S.; Ayyash, M. M.; Alnuaimi, L. H.; Sham, A.; ElBaghdady, K. Z.; Tariq, S.; AbuQamar, S. F. Growth Promotion of *Salicornia bigelovii* by *Micromonospora chalcea* UAE1, an Endophytic 1-aminocyclopropane-1-carboxylic Acid Deaminase-Producing Actinobacterial Isolate. *Front. Microbiol.* **2019,** *10*, 1694.
Eshel, A.; Zilberstein, A.; Alekparov, C.; Eilam, T.; Oren, I.; Sasson, Y.; Valentini, R.; Waisel, Y Biomass Production by Desert Halophytes: Alleviating the Pressure on Food Production. Rosen, M. A., Perryman, R. S., Dodds et al., Ed.; 2010; vol *23*, pp 362–367.

Falleh, H.; Ksouri, R.; Oueslati, S.; Guyot, S.; Abdelly, C.; Magne, C. Phenolic Nature, Occurrence and Polymerization Degree as Marker of Environmental Adaptation in the Edible Halophyte *Mesembryanthemum edule*. *S. Afr. J. Bot.* **2012,** *79*, 117–124.

Falleh, H.; Ksouri, R.; Oueslati, S.; Guyot, S.; Magne, C.; Abdelly, C. Interspeci!c Variability of Antioxidant Activities and Phenolic Composition in *Mesembryanthemum* Genus. *Food Chem. Toxicol.* **2009,** *47*, 2308–2313.

Feng, Z. T.; Deng, Y. Q.; Zhang, S. C.; Liang, X.; Yuan, F.; Hao, J. L.; Zhang, J. C.; Sun, S. F.; Wang, B. S. K+ Accumulation in the Cytoplasm and Nucleus of the Salt Gland Cells of *Limonium bicolor* Accompanies Increased Rates of Salt Secretion under NaCl Treatment Using NanoSIMS. *Plant Sci.* **2015,** *238*, 286–296.

Flowers, T. J.; Colmer, T. D. Salinity Tolerance in Halophytes. *New Phytol.* **2008,** *179*, 945–963.

Flowers, T. J.; Galal, H. K.; Bromham, L. Evolution of Halophytes: Multiple Origins of Salt Tolerance in Land Plants. *Funct. Plant Biol.* **2010,** *37*, 604–612.

Gao, C.; Wang, Y.; Jiang, B.; Liu, G.; Yu, L.; Wei, Z.; Yang, C. A Novel Vacuolar Membrane H+-ATPase c Subunit Gene (*ThVHAc1*) from *Tamarix hispida* Confers Tolerance to Several Abiotic Stresses in *Saccharomyces cerevisiae*. *Mol. Biol. Rep.* **2011,** *38*, 957–963.

Geilfus, C. M. Chloride in Soil: From Nutrient to Soil Pollutant. *Environ. Exp. Bot,* **2018,** *157*, 299–309.

Glenn, E. P.; Brown, J. J.; Blumwald, E. Salt Tolerance and Crop Potential of Halophytes. *Crit. Rev. Plant Sci.* **1999,** *18*, 227–255.

Gulzar, S.; Khan, M. Seed Germination of Halophytic Grass *Aeluropus lagopoides*. *Ann. Bot.* **2001,** *87*, 319–324.

Guo, X. L.; Cao, Y. R.; Cao, Z. Y.; Zhao, Y. X.; Zhang, H. Molecular Cloning and Characterization of a Stress-Induced Peroxiredoxin Q Gene in Halophyte *Suaeda salsa*. *Plant Sci.* **2004,** *167*, 969–975.

Guo, Y.; Qiu, Q. S.; Quintero, F. J.; Pardo, J. M.; Ohta, M.; Zhang, C.; Schumaker, K. S.; Zhu, J. K. Transgenic Evaluation of Activated Mutant Alleles of SOS2 Reveals a Critical Requirement for its Kinase Activity and C-terminal Regulatory Domain for Salt Tolerance in *Arabidopsis thaliana*. *Plant Cell* **2004,** *16*, 435–449.

Halfter, U.; Ishitani, M.; Zhu, J. K. The Arabidopsis SOS2 Protein Kinase Physically Interacts with and is Activated by the Calcium-Binding Protein SOS3. *Proc. Natl. Acad. Sci. U. S. A.* **2000,** *97*, 3735–3740.

Hamada, A.; Shono, M.; Xia, T.; Ohta, M.; Hayashi, Y.; Tanaka, A.; Hayakawa, T. Isolation and Characterization of a Na+/H+ Antiporter Gene from the Halophyte *Atriplex gmelini*. *Plant Mol. Biol.* **2001,** *46*, 35–42.

Hardoim, P. R.; Overbeek, L. S. V.; Elsas, J. D. V. Properties of Bacterial Endophytes and their Proposed Role in Plant Growth. *Trends Microbiol.* **2008,** *16*, 463–471.

Herrera, S. D.; Grossi, C.; Zawoznik, M.; Groppa, M. D. Wheat Seeds Harbour Bacterial Endophytes with Potential as Plant Growth Promoters and Biocontrol Agents of *Fusarium graminearum*. *Microbiol. Res.* **2016,** *186*, 37–43.

Hmidi, D.; Abdelly, C.; Athar, H. R.; Ashraf, M.; Messedi, D. Effect of Salinity on Osmotic Adjustment, Proline Accumulation and Possible Role of Ornithine-δ-aminotransferase in Proline Biosynthesis in *Cakile maritima*. *Physiol. Mol. Biol. Plants* **2018,** *24*, 1017–1033.

Hrabak, E. M.; Chan, C. W.; Gribskov, M.; Harper, J. F.; Choi, J. H.; Halford, N.; Kudla, J.; Luan, S.; Nimmo, H. G.; Sussman, M. R.; Thomas, M.; Walker-Simmons, K.; Zhu, J.

K.; Harmon, A. C. The Arabidopsis CDPK-SnRK Superfamily of Protein Kinases. *Plant Physiol.* **2003,** 132, 666–680.

Hussain, S.; Khalid, M. F.; Sohail, M.; Anjum, M. A.; Ejaz, S.; Nafees, M.; Zakir, I.; Ahmad, M.; Ali, S.; Ahmad, S. Role of Transporters in Accumulating Salt Ions by Halophytes. In *Approaches to the Remediation of Inorganic Pollutants*, 2021; pp 11–40.

Jha, B.; Lal, S.; Tiwari, V.; Yadav, S.; Agarwal, P. The *SbASR-1* Gene Cloned from an Extreme Halophyte *Salicornia brachiata* Enhances Salt Tolerance in Transgenic Tobacco. *Mar. Biotech.* **2012,** *14*, 782–792.

Jithesh, M. N.; Prashanth, S. R.; Sivaprakash, K. R.; Parida, A. K. Antioxidative Response Mechanisms in Halophytes: Their Role in Stress Defence. *J. Genet.* **2006,** *85*, 237–254.

Joshi, R.; Ramanarao, M. V.; Baisakh, N. Arabidopsis Plants Constitutively Overexpressing a Myo-inositol 1-Phosphate Synthase Gene (*SaINO1*) from the Halophyte Smooth Cordgrass Exhibits Enhanced Level of Tolerance to Salt Stress. *Plant Physiol. Biochem.* **2013,** *65*, 61–66.

Karan, R.; Subudhi, P. Overexpression of an Adenosine Diphosphate-ribosylation Factor Gene from the Halophytic Grass *Spartina alterniflora* Confers Salinity and Drought Tolerance in Transgenic *Arabidopsis*. *Plant Cell Rep.* **2014,** *33*, 373–384.

Kathiresan, K.; Bingham, B. L. Biology of Mangroves and Mangrove Ecosystems. In*Advances in Marine Biology*; Southward, A. J., Tyler, P. A., Young, C. M., Fuiman, L. A., Eds.; 2001, pp 81–251.

Khalil, A. M.; Hassan, S. E.; Alsharif, S. M.; Eid, A. M.; Ewais, E. E.; Azab, E.; Gobouri, A. A.; Elkelish, A.; Fouda, A. Isolation and Characterization of Fungal Endophytes Isolated from Medicinal Plant *Ephedra pachyclada* as Plant Growth-Promoting. *Biomolecules* **2021,** *11*, 140.

Khan, M. A.; Asaf, S.; Khan, A. L.; Ullah, I., Ali, S.; Kang, S. M.; Lee, I. J. Alleviation of Salt Stress Response in Soybean Plants with the Endophytic Bacterial Isolate *Curtobacterium* sp. SAK1. *Ann. Microbiol.* **2019,** *69*, 797–808.

Kobayashi, S.; Abe, N.; Yoshida, K. T.; Liu, S.; Takano, T. Molecular Cloning and Characterization of Plasma Membrane and Vacuolar Type Na+/K+ Antiporters of an Alkaline Salt Tolerant Monocot *Puccinellia tenuiflora*. *J. Plant Res.* **2012,** *125*, 587– 594.

Koyro, H. W. Effect of Salinity on Growth, Photosynthesis, Water Relations and Solute Composition of the Potential Cash Crop Halophyte *Plantago coronopus* (L.). *Environ. Exp. Bot.* **2006,** *56*, 136–146.

Kumar, V.; Jain, L.; Jain, S. K.; Chaturvedi, S.; Kaushal, P. Bacterial Endophytes of Rice (*Oryza sativa* L.) and their Potential for Plant Growth Promotion and Antagonistic Activities. *S. Afr. J. Bot.* **2020,** *134*, 50–63.

Lan, T.; Duan, Y.; Wang, B.; Zhou, Y.; Wu, W. Molecular Cloning and Functional Characterization of a Na+/H+ Antiporter Gene from Halophyte *Spartina anglica*. *Turk. J.* **2011,** *35*, 535–543.

Leopoldini, M., Russo, N., Toscani, M. The Molecular Basis of Working Mechanism of Natural Polyphenolic Antioxidants. *Food Chem.* **2011,** *125*, 288–306.

Li, W.; Wang, D.; Jin, T.; Chang, Q.; Yin, D.; Xu, S.; Liu, B.; Liu, L. The Vacuolar Na+/H+ Antiporter Gene *SsNHX1* from the Halophyte *Salsola soda* Confers Salt Tolerance in Transgenic alfalfa (*Medicago sativa* L.). *Plant Mol. Biol. Rep.* **2011,** *29*, 229–278.

Liotti, R. G.; da Silva Figueiredo, M. I.; da Silva, G. F.; de Mendonça, E. A.; Soares, M. A. Diversity of Cultivable Bacterial Endophytes in *Paullinia cupana* and their Potential for Plant Growth Promotion and Phytopathogen Control. *Microbiol. Res.* **2018,** *207*, 8–18.

Liu, L.; Wang, Y.; Wang, N.; Dong, Y. Y.; Fan, X. D.; Liu, X. M.; Yang, J.; Li, H. Y. Cloning of a Vacuolar H+-Pyrophosphatase Gene from the Halophyte *Suaeda corniculata* whose Heterologous Overexpression Improves Salt, Saline–Alkali and Drought Tolerance in *Arabidopsis*. *J. Integ. Plant Biol.* **2011,** *53*, 731–742.

Liu, J.; Zhu, J. K. A Calcium Sensor Homolog Required for Plant Salt Tolerance. *Science* **1998,** *280*, 1943–1945.

Lokhande, V. H.; Nikam, T. D.; Patade, V. Y.; Ahire, M. L.; Suprasanna, P. Effects of Optimal and Supra-Optimal Salinity Stress on Antioxidative Defence, Osmolytes and in Vitro Growth Responses in *Sesuvium portulacastrum* L. *Plant Cell Tissue Organ Cult.* **2011,** *104*, 41–49.

Mahgoub, H. A.; Fouda, A.; Eid, A. M.; Ewais, E. E. D.; Hassan, S. E. D. Biotechnological Application of Plant Growth-Promoting Endophytic Bacteria Isolated from Halophytic Plants to Ameliorate Salinity Tolerance of *Vicia faba* L. *Plant Biotechnol. Rep.* **2021,** *15*, 819–843.

Matsushita, N.; Matoh, T.; Characterization of Na^+ Exclusion Mechanisms of Salt-Tolerant Reed Plants in Comparison with Salt-Sensitive Rice Plants. *Physiol. Plant* **1991,** *83*, 170–176.

Motilva, M. J.; Serra, A.; Macia, A. Analysis of Food Polyphenols by Ultra-High-Performance Liquid Chromatography Coupled to Mass Spectrometry: An Overview. *J. Chromatogr. A* **2013,** *1292*, 66–82.

Reginato, M. A.; Castagna, A.; Furlán,A.; Castro, S.; Ranieri, A.; Luna, V. Physiological Responses of a Halophytic Shrub to Salt Stress by Na^2SO^4 and NaCl: Oxidative Damage and the Role of Polyphenols in Antioxidant Protection. *AoB Plants* **2014,** *6*, plu042.

Rittirongsakul, K.; Srinual, A.; Vanijajiva O. Anatomical Features and SCoT Profiles Provide New Insight into Phenotypic Plasticity in the Halophyte *Suaeda maritima* in Thailand. *Biodivers. J. Biol. Divers.* **2020**, *21*.

Mittler. R. Oxidative Stress, Antioxidants and Stress Tolerance. *Trends Plant Sci.* **2002,** *7*, 405–410.

Miller, G.; Suzuki, N.;Ciftci-Yilmaz, S.; Mittler, R. Reactive Oxygen Species Homeostasis and Signalling during Drought and Salinity Stresses. *Plant Cell Environ.* **2010,** *33*, 453–467.

Munns, R. Genes and Salt Tolerance: Bringing them Together. *New Phytol.* **2005,** *167*, 645–663.

Munns, R.; Tester, M. Mechanisms of Salinity Tolerance. *Annu. Rev. Plant Biol.* **2008,** *59*, 651–681.

Nah, G.; Pagliarulo, C. L.; Mohr, P. G.; Luo, M.; Sisneros, N.; Yu, Y.; Collura, K.; Currie, J.; Goicoechea, J. L.; Wing, R. A.; Schumaker, K. S. Comparative Sequence Analysis of the Salt Overly Sensitive1 Orthologous Region in *Thellungiella halophila* and *Arabidopsis thaliana*. *Genomics* **2009,** *94*, 196–203.

Navarro-Torre, S.; Barcia-Piedras, J. M.; Mateos-Naranjo, E.; Redondo-Gómez, S.; Camacho, M.; Caviedes, M. A.; Pajuelo, E.; Rodríguez-Llorente, I. D. Assessing the Role of Endophytic Bacteria in the Halophyte *Arthrocnemum macrostachyum* Salt Tolerance. *Plant Biol.* **2017,** *19*, 249–256.

Nguyen, H. T.; Stanton, D. E.; Schmitz, N.; Farquhar, G. D.; Ball, M. C. Growth Responses of the Mangrove *Avicennia marina* to Salinity: Development and Function of Shoot Hydraulic Systems Require Saline Conditions. *Ann. Bot.* **2015,** *115*, 397–407.

Oh, D. H.; Leidi, E.; Zhang, Q.; Hwang, S. M.; Li, Y.; Quintero, F. J.; Jiang, X.; D'Urzo, M. P.; Lee, S. Y.; Zhao, Y.; Bahk, J. D.; Bressan, R. A.; Yun, D. J.; Pardo, J. M.; Bohnert, H. J. Loss of Halophytism by Interference with SOS1 Expression. *Plant Physiol.* **2009,** *151*, 210–222.

Ohta, M.; Hayashi, Y.; Nakashima, A.; Hamada, A.; Tanaka, A.; Nakamura, T.; Hayakawa, T. Introduction of a Na+/H+ Antiporter Gene from *Atriplex gmelini* Confers Salt Tolerance to Rice. *FEBS Lett.* **2002,** *532*, 279–282.

Pan, X.; Qin, Y.; Yuan, Z. Potential of a Halophyte-Associated Endophytic Fungus for Sustaining Chinese White Poplar Growth Under Salinity. *Symbiosis* **2018,** *76*, 109–116.

Park, J.; Okita, T. W.; Edwards, G. E. Salt Tolerant Mechanisms in Single-Cell C4 Species *Bienertia sinuspersici* and *Suaeda aralocaspica* (Chenopodiaceae). *Plant Sci.* **2009,** *176*, 616–626.

Perron, N. R.; Brumaghin, J. L. A Review of the Antioxidant Mechanisms of Polyphenol Compounds Related to Iron Binding. *Cell Biochem. Biophys.* **2009,** *53*, 75–100.

Prashanth, S. R.; Sadhasivam,V.; Parida, A. Over Expression of Cytosolic Copper/Zinc Superoxide Dismutase from a Mangrove Plant *Avicennia marina* in Indica Rice var Pusa Basmati-1 Confers Abiotic Stress Tolerance. *Trans. Res.* **2008,** *17*, 281–291.

Qiao, W. H.; Zhao, X. Y.; Li, W.; Luo, Y.; Zhang, X. S. Overexpression of *AeNHX1*, A Root-Specific Vacuolar Na+/H+ Antiporter from Agropyron elongatum, Confers Salt Tolerance to Arabidopsis and Festuca Plants. *Plant Cell Rep.* **2007,** *26*, 1663–1672.

Qin, S.; Feng, W. W.; Wang, T. T.; Ding, P.; Xing, K.; Jiang, J. H. Plant Growth-Promoting Effect and Genomic Analysis of the Beneficial Endophyte *Streptomyces* sp. KLBMP 5084 Isolated from Halophyte *Limonium sinense*. *Plant Soil* **2017,** *416*, 117–132.

Qiu, Q. S.; Guo, Y.; Dietrich, M. A.; Schumaker, K. S.; Zhu, J. K. Regulation of SOS1, A Plasma Membrane Na+/H+ Exchanger in *Arabidopsis thaliana*, by SOS2 and SOS3. *Proc. Natl. Acad. Sci. U. S. A.* **2002,** *99*, 8436–8441.

Qin, S.; Zhang, Y. J.; Yuan, B.; Xu, P. Y.; Xing, K.; Wang, J.; Jiang, J. H. Isolation of ACC Deaminase-Producing Habitat-Adapted Symbiotic Bacteria Associated with Halophyte *Limonium sinense* (Girard) Kuntze and Evaluating their Plant Growth-Promoting Activity under Salt Stress. *Plant Soil.* **2014,** *374*,753–766.

Quan, R.; Lin, H.; Mendoza, I.; Zhang, Y.; Cao, W.; Yang, Y.; Shang, M.; Chen, S.; Pardo, J. M.; Guo, Y. SCABP8/CBL10, A Putative Calcium Sensor, Interacts with the Protein Kinase SOS2 to Protect Arabidopsis Shoots from Salt Stress. *Plant Cell* **2007,** *19*, 1415–1431.

Ramadan, T.; Flowers, T. J. Effects of Salinity and Benzyl Adenine on Development and Function of Microhairs of *Zea mays* L. *Planta* **2004,** *219*, 639–648.

Rangani, J.; Parida, A. K.; Panda, A.; Kumari, A. Coordinated Changes in Antioxidative Enzymes Protect the Photosynthetic Machinery from Salinity Induced Oxidative Damage and Confer Salt Tolerance in an Extreme Halophyte *Salvadora persica* L. *Front. Plant Sci.* **2016,** *7*, 50.

Rauf, M.; Shahzad, K.; Ali, R.; Ahmad, M.; Habib, I.; Mansoor, S.; Berkowitz, G.; Saeed, N. Cloning and Characterization of Na+/H+ Antiporter (*LfNHX1*) Gene from a Halophyte Grass *Leptochloa fusca* for Drought and Salt Tolerance. *Mol. Biol. Rep.* **2014,** *41*, 1669–1682.

Redondo-Gómez, S.; Mateos-Naranjo, E.; Figueroa, E.; Davy, A. J. Salt Stimulation of Growth and Photosynthesis in an Extreme Halophyte, *Arthrocnemum macrostachyum*. *Plant Biol.* **2010,** *12*, 79–97.

Ruppel, S.; Franken, P.; Witzel, K. Properties of the Halophyte Microbiome and their Implications for Plant Salt Tolerance. *Funct. Plant. Biol.* **2013,** *40*, 940–951.

Shabala, S.; Bose, J.; Hedrich, R. Salt Bladders: Do they Matter? *Trends Plant Sci.* **2014,** *19*, 687– 691.

Shah, S.; Shrestha, R.; Maharjan, S.; Selosse, M. A.; Pant, B. Isolation and Characterization of Plant Growth-Promoting Endophytic Fungi from the Roots of *Dendrobium moniliforme*. *Plants* **2018,** *8*, 5.

Seckin, B.; Turkan, I.; Sekmen, A. H.; Ozfidan, C. The Role of Antioxidant Defense Systems at Differential Salt Tolerance of *Hordeum marinum* Huds. (sea barleygrass) and *Hordeum vulgare* L. (Cultivated Barley). *Environ. Exp. Bot.* **2010,** *69*, 76–85.

Semenova, G. A.; Fomina, I. R.; Biel, K. Y. Structural Features of the Salt Glands of the Leaf of *Distichlis spicata* 'Yensen 4a' (Poaceae). *Protoplasma* **2010,** *240*, 75–82.

Shevyakova, N. I.; Rakitin, V. Y.; Stetsenko, L. A.; Aronova, E. E.; Kuznetsov, V. V. Oxidative Stress and Fluctuations of Free and Conjugated Polyamines in the Halophyte *Mesembryanthemum crystallinum* L. under NaCl Salinity. *Plant Growth Regul.* **2006,** *50*, 69–78.

Shi, H.; Ishitani, M.; Kim, C.; Zhu, J. K. The Arabidopsis Thaliana Salt Tolerance Gene SOS1 Encodes a Putative Na+/H+ Antiporter. *Proc. Natl. Acad. Sci. U. S. A.* **2000,** *97*, 6896–6901.

Shurigin, V.; Alikulov, B.; Davranov, K.; Ismailov, Z. Bacterial Endophytes from Halophyte Black Saxaul (*Haloxylon aphyllum* Minkw.) and their Plant Growth-Promoting Properties. *J. Appl. Biol. Biotechnol.* **2022,** *10*, 45–53.

Sofy, M. R.; Aboseidah, A. A.; Heneidak, S. A.; Ahmed, H. R. ACC Deaminase Containing Endophytic Bacteria Ameliorate Salt Stress in *Pisum sativum* Through Reduced Oxidative Damage and Induction of Antioxidative Defense Systems. *Environ. Sci. Pollut. Res.* **2021,** *28*, 40971–40991.

Song, J.; Shi, G. W.; Gao, B.; Fan, H.; Wang, B. S. Waterlogging and Salinity Effects on Two *Suaeda salsa* Populations. *Physiol. Plant* **2011,** *141*, 343–351.

Sorty, A. M.; Meena, K. K.; Choudhary K.; Bitla, U. M.; Minhas, P. S.; Krishnani, K. K. Effect of Plant Growth Promoting Bacteria Associated with Halophytic Weed (*Psoralea corylifolia* L) on Germination and Seedling Growth of Wheat under Saline Conditions. *Appl. Biochem. Biotechnol.* **2016,** *180*, 872–882.

Sternberg, L. D. L.; The, S. Y.; Ewe, S. M. L.; Miralles-Wilhelm, F.; De Angelis, D. L. Competition
between Hardwood Hammocks and Mangroves. *Ecosystems* **2007,** *10*, 648–660.

Szabados, L.; Kovacs, H.; Zilberstein, A.; Bouchereau, A. Plants in Extreme Environments: Importance of Protective Compounds in Stress Tolerance. *Adv. Bot. Res.* **2011,** *57*, 105–150.

Szabados, L.; Savoure, A. Proline: A Multifunctional Amino Acid. *Trens. Plant Sci.* **2010,** *15*, 89–97.

Takahashi, R.; Nishio, T.; Ichizen, N.; Takano, T. Cloning and Functional Analysis of the K+ Transporter, *PhaHAK2*, from Salt-Sensitive and Salt-Tolerant Reed Plants. *Biotech. Lett.* **2007,** *29*, 501–506.

Tan, W. K.; Lim, T. M.; Loh, C. S. A Simple, Rapid Method to Isolate Salt Glands for Three-Dimensional Visualization, Fluorescence Imaging and Cytological Studies. *Plant Methods* **2010,** *6*, 24.

Teng, K.; Tan, P.; Guo, W.; Yeu, Y.; Fan, X.; Wu, J. Heterologous Expression of a Novel *Zoysia japonica* C^2H^2 Zinc Finger Gene, *ZjZFN1*, Improved Salt Tolerance in Arabidopsis. *Front. Plant. Sci.* **2018,** *9*, 1159.

Tester, M.; Davenport, R. Na+ Tolerance and Na+ Transport in Higher Plants. *Ann. Bot. London* **2003,** *91*, 503–527.

Thomson, W.; Platt-Aloia, K. The Ultrastructure of the Plasmodesmata of the Salt Glands of *Tamarix* as Revealed by Transmission and Freeze-Fracture Electron Microscopy. *Protoplasma* **1985,** *125*, 13–23.

Tomlinson, P. B. In *The Botany of Mangroves*; Cambridge University Press: Cambridge, 1986; p 413.

Vitória, A. P.; Lea, P. J.; Azevedo, R. A. Antioxidant Enzymes Responses to Cadmium in Radish Tissues. *Phytochemistry* **2001,** *57*, 701–710.

Volpe, V.; Manzoni, S.; Marani, M.; Katul, G. Leaf Conductance and Carbon Gain under Salt-Stressed Conditions. *J. Geophys. Res.* **2011,** *116*, G04035.

Wang, X.; Yang, R.; Wang, B.; Liu, G.; Yang, C.; Cheng, Y. Functional Characterization of a Plasma Membrane Na+/H+ Antiporter from Alkali Grass (*Puccinellia tenuiflora*). *Mol. Biol. Rep.* **2011,** *38*, 4122–4813.

Win, K. T.; Tanaka, F.; Okazaki, K.; Ohwaki, Y. The ACC Deaminase Expressing Endophyte *Pseudomonas* spp. Enhances NaCl Stress Tolerance by Reducing Stress-Related Ethylene Production, Resulting in Improved Growth, Photosynthetic Performance, and Ionic Balance in Tomato Plants. *Plant Physiol. Biochem.* **2018,** *127*, 599–607.

Xiong, Y. W.; Gong, Y.; Li, X. W.; Chen, P.; Ju, X. Y.; Zhang, C. M.; Yuan, B.; Lv, Z. P.; Xing, K.; Qin, S. Enhancement of Growth and Salt Tolerance of Tomato Seedlings by a Natural Halotolerant Actinobacterium *Glutamicibacter halophytocola* KLBMP 5180 Isolated from a Coastal Halophyte. *Plant Soil.* **2019,** *445*, 307–322.

Yadav, N.; Shukla, P.; Jha, A.; Agarwal, P.; Jha, B. The *SbSOS1* Gene from the Extreme Halophyte *Salicornia brachiata* Enhances Na+ Loading in Xylem and Confers Salt Tolerance in Transgenic Tobacco. *BMC Plant Biol.* **2012,** *12*, 1–18.

Yang, M. F.; Song, J.; Wang, B. S. Organ-Specific Responses of Vacuolar H^+-ATPase in the Shoots and Roots of C-3 Halophyte *Suaeda salsa* to NaCl. *J. Integr. Plant Biol.* **2010,** *52*, 308–314.

Yang, Y.; Tang, R. J.; Jiang, C. M.; Li, B.; Kang, T.; Liu, H.; Zhao, N.; Ma, X. J.; Yang, L.; Chen, S. L.; Zhang, H. X. Overexpression of the *PtSOS2* Gene Improves Tolerance to Salt Stress in Transgenic Poplar Plants. *Plant Biotech. J.* **2015,** *13*, 962–973.

Yao, M.; Zeng, Y.; Liu, L.; Huang, Y.; Zhao, E.; Zhang, F. Overexpression of the Halophyte *Kalidium foliatum* H+-pyrophosphatase Gene Confers Salt and Drought Tolerance in *Arabidopsis thaliana*. *Mol. Biol. Rep.* **2012,** *39*, 7989–7996.

Yuan, C. H. E. N.; Cothren, J. T.; Chen, D. H.; Ibrahim, A. M.; Lombardini, L. Ethylene-Inhibiting Compound 1-MCP Delays Leaf Senescence in Cotton Plants under Abiotic Stress Conditions. *J. Integr. Agric.* **2015,** *14*, 1321–1331.

Yuan, F.; Leng, B.; Wang, B. Progress in Studying Salt Secretion from the Salt Glands in Recretohalophytes: How do Plants Secrete Salt?. *Front. Plant. Sci.* **2016,** *7*, 977.

Yuan, F.; Lyu, M. J.; Leng, B. Y.; Zheng, G. Y.; Feng, Z. T.; Li, P. H.; Zhu, X. G.; Wang, B. S. Comparative Transcriptome Analysis of Developmental Stages of the *Limonium bicolor* Leaf Generates Insights into Salt Gland Differentiation. *Plant Cell Environ.* **2015,** *38*, 1637–1657.

Zhang, Y.; Lai, J.; Sun, S.; Li, Y.; Liang, L.; Chen, M.; Xie, Q. Comparison Analysis of Transcripts from the Halophyte *Thellungiella halophila*. *J. Integr. Plant Biol.* **2008,** *50*, 1327–1335.

Zhao, S.; Zhou, N.; Zhao, Z, Y.; Zhang, K.; Wu, G, H., Tian, C. Y. Isolation of Endophytic Growth-Promoting Bacteria Associated with the Halophytic *Salicornia europaea* and Evaluation of their Promoting Activity Under Salt Stress. *Curr. Microbiol.* **2016,** *73*, 574–581.

Zhu, J. K. Plant Salt Tolerance. *Trends Plant Sci.* **2001,** *6*, 66–71.

CHAPTER 17

Potential Role of Halophytes in Environmental Clean-Up

BHUSHAN P. BHUSARE[1], GANESH C. NIKALJE[2], RAMESH R. SANAP[3], and AMIT D. KALE[4]

[1]*Ideal Institute of Biology (IIB)-PCB, Nanded, Maharashtra, India*

[2]*PG Department of Botany, Seva Sadan's R. K. Talreja College of Arts, Science and Commerce, Ulhasnagar, Thane, Maharashtra, India*

[3]*Department of Botany, S.S.G.M. College, Kopargaon, Maharashtra, India*

[4]*Department of Botany, S.S.K.A.C.S. College, Kolpewadi, Maharashtra, India*

ABSTRACT

To fulfill the increasing demand for food, fiber, and energy, there is excessive irrigation of agricultural land, resulting in a continuous increase in soil salinity. Metal toxicity in land and water increases consistently which has resulted in deleterious effects on living organisms. For the removal of heavy metals, various conventional (physical and chemical) remediation methods are used. However, higher cost, huge manpower, and detrimental effect on soil and its microorganism have restricted their use. As an alternative, bioremediation (microorganism-assisted), and phytoremediation (plants-assisted) techniques were used for the removal or transformation of this toxic material into a nontoxic form. Researchers have used the phytoremediation technique in high metal polluted sites due to its role in biodiversity restoration, reduction in soil erosion, and removal of maximum contamination. Several halophytic plants such as *Hordeum brevisubulatum*, *Lycium humile*, *Salvadora persica*, *Scorzonera hieraciifolia*, and *Urochondra setulose* showed best

Physiology of Halophytes: Signaling, Omics, and Tolerance Mechanisms.
Nivas Manohar Desai & Ganesh Chandrakant Nikalje (Eds.)

growth in highly saline soil and their potential to transform heavy metals into nontoxic compounds, making them excellent candidates for phytoremediation process. The present chapter is useful for understanding the potential use of halophytes which can be beneficial for the environmental cleanup process.

17.1 INTRODUCTION

The human population is increasing at an alarming rate, and for feeding such a huge population, food production must be increased on an urgent basis (Duan et al., 2021). When we consider agricultural land for feeding, it is not sufficient as salt affected land area creates a problem with productivity (Syed et al., 2021). Due to salinity-affected soil, half of all cultivated land in the 21st century cannot be used. It is possible to resolve the above-mentioned global issue by developing stress-tolerant crops (a crop that can survive biotic and abiotic stress, as well as nutrient-poor soil), domesticating wild plant varieties (plants with dissolved characteristics), and reclaiming soil degraded by agriculture. Domestication of a halophyte is the best alternative for the above because of the growth potential of these plants in saline soil. Even in soil containing 40 mM NaCl salinity, most of our conventional plants cannot survive or start their life cycle (Liu and Wang, 2021). While, in soil containing 200 mM NaCl or more, halophytes showed growth and completed their life cycle (Nikalje et al., 2019a), including physiology, strong selection pressure, development of potential anatomical characteristics (salt hairs, salt glands, etc.), alteration in biochemical as well as metabolic profiling (Nikalje et al., 2018c; Shahid et al., 2020). In addition to salinity, halophytes are also tolerant to other abiotic stresses such as heavy metals, drought, nutrient deficiency, etc. (Nikalje et al., 2019c). All the above-mentioned inferences make these halophytes appropriate candidates to study stress adaptation mechanism in plants. Instead of this, some halophytes have high nutritive value such as sugar, proteins, carbohydrates, vitamins, iron, fibers, and different useful secondary metabolites (Rangani et al., 2019). Nikalje et al. (2018a) reported high nutritive value of *Quinoa and Salicornia.* The use of many halophytes was also reported in different industries such as cosmetics, food, pharmaceutical, and medicine (Lopes et al., 2021). With this, halophytes such as *Atriplex helimus, Sesuvium portulacastrum.*, *Medicago lupulina*, *Salicornia* sp., and others played important roles in environmental protection such as phyto-desalination, phytoremediation, and phytostabilization (Saddhe et al., 2020; Liu and Wang, 2021; Nikalje et al., 2022). The present chapter summarizes information such as the salt-tolerance mechanism of halophytes, heavy metal and its uptake by halophytes, phytoremediation, and molecular processes for improvement of halophytes utilization. This will be useful for understanding the potential use of halophytes which can be beneficial for the environmental cleanup process.

17.2 SALT TOLERANCE MECHANISMS OF HALOPHYTES

Salt tolerance is one of the complex processes in plants that are connected with multigenic traits at different levels (Nikalje et al., 2017; Shelake et al., 2022). To attain salinity resistance, five features of plant activity can be significant for biomass production under stressful conditions. These features are: (1) growth rate and plant morphology, (2) water relation dynamics, (3) leaf CO^2/H^2O exchange, (4) ion hemostasis, and (5) defense against oxidative stress (Zilaie et al., 2022). Figure 17.1 represents significant characteristics of plants for salinity stress.

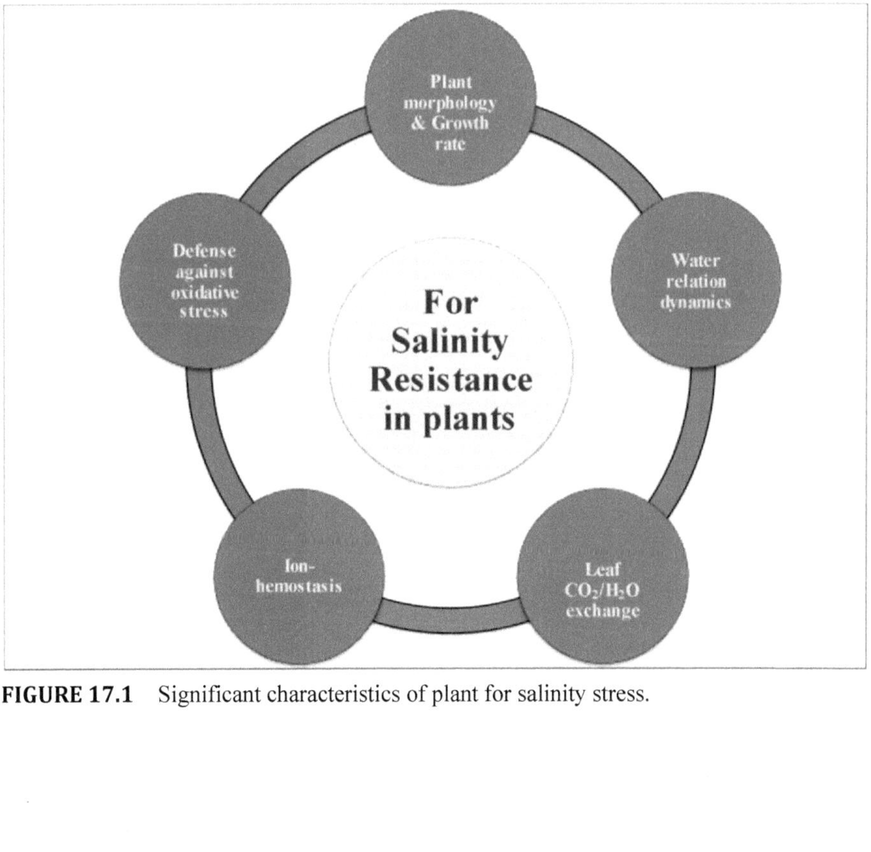

FIGURE 17.1 Significant characteristics of plant for salinity stress.

The growth responses are directly associated with the efficiency and salinity resistance of plants. Salinity causes decrease in plants' fresh and dry weight, leaf area expansion, shoot/root, plant height, leaf number, shoot length, and relative growth rate, etc. (Riffat et al., 2020). Various halophytes such as *Aeluropus lagopoides, Sporobolus ioclados, Urochondra setulosa,* and *Halopyrum mucronatum*) specify optimum growth under non-saline conditions (Atia et al., 2019). Nevertheless, moderate salinity stimulates growth parameters in *Phragmites karka, Phragmites australis, Phragmites communis, Spartina maritima,* and *Pennisetum clandestinum* but these parameters decline significantly in higher salinities (Shoukat et al., 2020). Improvement of biomass is accompanied by biomass distribution strategy between root/shoot and higher metabolic performance of leaf physiological activity (Zhang et al., 2020). For instance, the higher root development relates to the storage of Na^+ and Cl^- ions in the below-ground tissue that can avoid Na^+ loading in the above-ground foliage part (Javaid et al., 2019). Under high salinity, lesser root growth is also observed in halophytes that prevent net Na^+ influx and prevent Na^+ translocation to the shoot. Halophytes can minimize their leaf water and osmotic potential growing in harsh saline conditions. Some halophytes fluctuate their leaf succulence and water content for water uptake and osmotic adjustment to preserve leaf turgidity and water loss. Osmotic balance in halophytes can be accomplished by the accumulation of inorganic (Na^+, Cl^-, and K^+) and organic substances (sugars proline, glycine betaine, and sugar alcohol) for water uptake (Cruz et al., 2020). Organic osmotic content could be involved in support of other metabolic functions including osmoprotection and radical scavenging. Accumulation of inorganic substances in plants of saline soil could be cheaper for osmotic adjustment, as compared to the ones relying on organic compounds. Stomata closures are an instant response to uphold tissue water balance while reducing ion uptake under salinity stress. However, it comes at the cost of intercellular CO^2 (Ali et al., 2021). The subsequent deficiency of CO^2 leads to an over-reduction in the photosynthetic electron transport chain as well as photo-damage (Javaid et al., 2019). Plants can defend their photosystems from damage by using enzymatic and non-enzymatic antioxidant substances for optimum metabolic activity and defense against oxidative stress (Ali et al., 2021).

17.3 HEAVY METALS AND THEIR UPTAKE

The natural chemical element having the largest atomic number and high density is highly harmful to the consideration of environmental aspects known as heavy metal (Velusamy et al., 2021). Sodango et al. (2018) found that the rising industrialization globally caused in widespread advancement in the utilization and manufacturing of heavy metals in the environment. Li et al. (2019) and Kayode et al. (2021) reported that an extensive upsurge in fatal heavy metals including arsenic, cadmium, chromium, copper, lead, manganese, mercury, selenium, as well as uranium has polluted the agricultural land (soil as well as water) and hence it ultimately showed direct and indirect adverse effects on plants and animals. Figure 17.2 illustrate the effect of heavy metals on human health. Absorption of heavy metals in plants takes place either by the process of simple diffusion or facilitated diffusion, and additionally, metabolism and sequestration of the toxic elements are essential for detoxification (Nikalje et al., 2019b; Das et al., 2022). For salt and heavy metal stress, a general physiological tolerance mechanism has been seen in plants while phytostabilization and phytoextraction type of special mechanismsare also known that are comprised under the broad term phytoremediation (Sharma et al., 2019). An exception has been observed for halophytes, that is, presence of enzymes for pollutant degradation, one which performs a role in the natural defense system, serving the plant to become a natural cleaner of the ecosystem leading to phytoremediation (Nikalje et al., 2018b; Nikalje et al., 2020; Aileni, 2022).

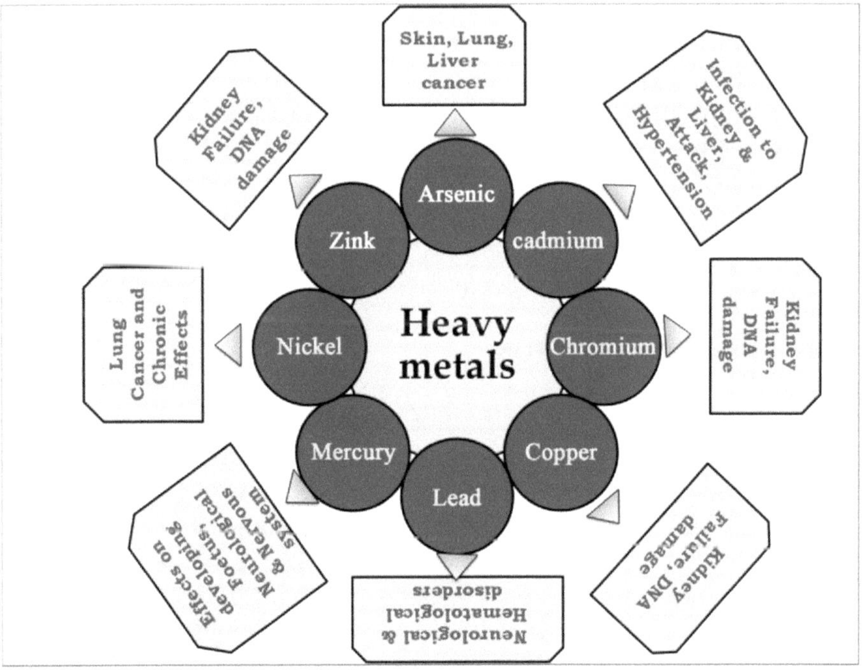

FIGURE 17.2 Toxicity of heavy metals on human health.

The capability to develop osmo-protectants was found to be beneficial as it helps to tolerate the metal stress. Therefore, to tolerate heavy metals, these plants (halophytes) are most appropriate (Shang et al., 2020). Phytochelatins are compounds that are produced from glutathione (tripeptide) and are used in metal detoxification by requisitioning metal in the storage organelle, that is, vacuole (Nikalje and Suprasanna, 2018; Dubey et al., 2022). Akhatar et al. (2022) have shown that metallophytes have not used this strategy for metal stress avoidance, whereas, for the avoidance of metal toxicity, phytochelatins synthesis takes place in non-metallophytes. Distinctive contemplation is taken for the elimination of noxious compounds from the environment and panting or utilization of those plants which can either reduce the maximum toxicity or tolerate such heavy toxicity (Hamidian et al., 2021). Abdullah (2020) states that phytoremediation comprises numerous mechanisms including the extraction of these toxic metals, degrade the same, and removal of these heavy metals from the ground. With this, phytoextraction and phytostabilization can be alternative options for remediation (Chen et al., 2022). As this method is interactive (i.e., Interaction of plant and soil) within environments so the chances of detrimental effects may not be observable. Also, the easy growing technique makes these environmental friendly and economical (Chen et al., 2022). Stabilization of these heavy metals in the plant's rhizosphere and inspection of the movement of these metals in water and soil consequently evade the nearby environment which is termed as phytostabilization (Shackira and Puthur, 2019). DalCorso et al. (2019) reported that phytoremediation is contemplated as a cluster of methods for metal toxicity removal and stabilizing these at the root surrounding area (rhizosphere). He also reported that plant selection depends on the site of contamination.

17.4 PHYTOREMEDIATION

Land degradation and desertification constitute chief constraints that affect agricultural areas and decrease crop productivity. They affect soil's biological and physicochemical properties, and decrease its fertility (Ferreira et al., 2022). But it is fascinating to note that some halophytes exhibit high effectiveness in soil fixation and landscaping (Sghaier-Hammami et al., 2020). Certainly, some Poaceae members such as *Imperata cylindrica, Leymus chinensis, Leymus racemosus, Puccinellia ciliata, Puccinellia tenuiflora, Spartina alterniflora, Spartina patens, Spartina townsendii, and*

Thinopyrum ponticum were used for soil fixation (Atia et al., 2019). The cultivation of these species enhances the biological and physicochemical properties of degraded land. This is advantageous to increase soil organic matter and nutrient content of soils (Yousaf et al., 2021). Some salt-tolerant species are found to be used in large-scale environmental cleanup efforts of metal-contaminated soils which increased with the increase of anthropogenic activity in the past few years (Orrego et al., 2020). Some other members of the Poaceae family such as *Sporobolus virginicus*, *Spartina alterniflora*, *Phragmites australis*, *Spartina anglica*, *Spartina densiflora*, *Puccinella scleroids*, *Spartina townsendii*, *Vetiveria zizanioides*, and *Dactylis glomerata* were used for phytoremediation of metal-contaminated soils (Sghaier-Hammami et al., 2020). There are binary key mechanisms by which halophytes remediate metal-contaminated soils: (1) Some halophytes have the capability to immobilize heavy metals through adsorption onto roots and/or precipitation within the rhizosphere (Shang et al., 2020). This mechanism is known as "metal phytostabilization." (2) Some other halophytes have the potential to permit metal accumulation in their above ground tissues, mainly shoots. Subsequently, metals are removed through shoot harvest. This process is called phytoextraction (Diarra et al., 2021). Numerous studies proved that some tolerance mechanisms operating at the whole plant level are not forever specific to sodium or chloride (Bakka and Challabathula, 2020). For example, copper, zinc, and cadmium may accumulate in salt glands or trichomes of *Tamarix aphylla* and *Armeria maritima* (Kumari et al., 2019). Hence, substantial capability for heavy metal removal has been well-known for halophytes that accumulate huge amounts of metals in their aerial parts (Caparrós et al., 2022).

17.5 MOLECULAR APPROACHES TO IMPROVE THE POTENTIAL USE OF HALOPHYTES

Halophytes can persist well under saline areas than glycophytes, which may be attributable to appropriate gene modulation which plays a vital role in promoting their salt tolerance mechanisms (Rahman et al., 2021). High-throughput transcriptome sequencing is progressively being used to know the plant responses to different abiotic stress by checking the expression of the gene (Snyman et al., 2017). Transcriptome analysis report of the model halophytic plant *Puccinellia tenuiflora* and some glycophytic Gramineae plants to divulge their basic response at the molecular level when there is extreme salt stress are explained in literature (Li et al., 2021). Greater expression of the gene families includes uptake of potassium, biosynthesis of sucrose, PPP (pentose phosphate pathway), and biosynthesis of flavonoid may be related to improvement in *P. tenuiflora* growth under high salt (Guo et al., 2021). In *Salicornia persica*, 1595 differentially expressed genes were reported in between the plants (shoot of the plants) which have not been given any treatments (control) and those treated with various salt concentrations, encoding transcription factors, enzyme (protein kinases) and transporters. Primary metabolite synthesis and higher energy homeostasis may have crucial roles in the salt tolerance of *S. persica* as specified in the analysis of functional annotation (Aliakbari et al., 2021). Gene network investigation discovered that Ca+ signaling, ABA signaling, and Na+ compartmentalization are vital constituents in conversing for *S. persica* salt tolerance (Hafeez et al., 2021). When different concentrations of NaCl (100 mM, 300 mM, and 500 mM), treatment was given to the *S. rigida* plant, in response to this 399 upregulated and 327 downregulated genes were reported (Rahman et al., 2021). Gene ontology (GO) enhancement analysis proposed that the downregulated genes are associated with the transport of intracellular protein, nitrogen compounds, and organic substances (Han et al., 2020). In comparison, the upregulated genes were intricate in the biogenesis of the cell wall, like the assembly of the cell wall, organization of the extracellular matrix, and transport in cell wall formation (Han et al., 2020).GO enrichment analysis displayed that salt stress intensely augmented the expression of genes associated with the transport of ions, signaling pathways depending on the ABA, (HB-7 and MYB78) pathway in transcriptional regulation and ROS-scavenging (Zhang et al., 2020). In *U. setulose,* recent *transcriptome* analysis showed genes upregulation which is associated with different functions including proteins of transport and cell membrane, enzymes used in photosynthesis, antioxidants, compatible solutes, transcription factors, mitogen-activated protein kinases (MAPKs) (Mann et al., 2021). In *Z. macrostachya*, for the treatment of NaCl (300 mM), many differentially expressed genes, especially 4903 upregulated and 3800 downregulated genes were observed. Most of the gene expression changes might be involved in plant growth improvements by modulating the ion homeostasis, the way of salt secretion, transduction of hormonal signaling, and accumulation of osmoregulatory substances, and by the activation of precise ROS-scavenging system and prevention of the oxidative damage (Rahman et al., 2021). Because they are genetically prepared with the potential of salt tolerance mechanism, halophytes can be used as an effective genetic resource for transgenic plant developments with great adaptability for salt stress. These genes play an important role in signaling and stabilization as they are associated with antioxidant, antiport, transport, and ROS scavengers (Nikalje et al., 2018d; Rahman et al., 2021).

17.6 CONCLUSIONS

Agricultural production is under high threat because of increased stress levels of drought and salinity. There have been no significant improvements made in agriculture for crop plants under various stress conditions due to complex responses and the limited capacity of crop plants for handling stress. Phytoremediation is one of the best available options for crop plants under stress conditions. Any biochemical or molecular improvements in halophytes would enable more benefits in terms of stress tolerance because no need to change the mechanisms or molecular structures of crop plants. Also, phytoremediation techniques could be incorporated with other biological and physical methods to increase the efficacy of the method. The present chapter tried to collect the most recent data to highlight both the basic and applied potentials of halophytes. This will be helpful to minimize the raising issue of the increasing salinization of lands which is one major challenge for developing countries. Fundamental and potential salt tolerance abilities of halophytes can be used as a model for crop improvement and salt stress response.

ACKNOWLEDGMENTS

The authors are thankful to Dr. T. D. Nikam (Professor and Emeritus Scientist, Department of Botany, SPPU, Pune) for constant guidance. The authors are also thankful to Dr. D. B. Shelke (Head, Department of Botany, Amruteshwar College, Pune) for sharing their expertise while writing this chapter.

KEYWORDS

- **halophytes**
- **salinity**
- **metal**
- **phytoremediation**
- **environmental cleanup**

REFERENCES

Abdullah, S. R.; Al-Baldawi, I. A.; Almansoory, A. F.; Purwanti, I. F.; Al-Sbani, N. H.; Sharuddin, S. S. Plant-Assisted Remediation of Hydrocarbons in Water and Soil: Application, Mechanisms, Challenges and Opportunities. *Chemosphere* **2020,** *247*, 125932.

Abideen, Z.; Koyro, H. W.; Huchzermeyer, B.; Ahmed, M. Z.; Gul, B.; Khan, M. A. Moderate Salinity Stimulates Growth and Photosynthesis of *Phragmite skarka* by Water Relations and Tissue-Specific Ion Regulation. *Environ. Exp. Bot.* **2014,** *105*, 70–76.

Aileni, M. Environment Sustainability and Role of. In *Innovations in Environmental Biotechnology*, 2022; pp 21–64.

Akhatar, O.; Sharma, H.; Zoomi, I.; Chaudhary, K. L.; Kumar, M. Metallophytes. In *Heavy Metals in Plants Physiological to Molecular Approach*; CRC Press, 2022; pp 354–368.

Ali, M.; Mustafa, A.; Abideen, Z.; Gul, B. Bioenergy Production from Halophytes Crops for Sustainable Development. In *Energy and Environmental Security in Developing Countries*; Springer: Cham, 2021; pp 571–586.

Aliakbari, M.; Razi, H.; Alemzadeh, A.; Tavakol, E. RNA-seq transcriptome Profiling of the Halophyte Salicornia persica in Response to Salinity. *J. Plant Growth Regul.* **2021,** *40* (2), 707–721.

Atia, A.; et al. Salt Tolerance and Potential Uses for Saline Agriculture of Halophytes from the Poaceae. In *Sabkha Ecosystems. Tasks for Vegetation Science*; Gul, B., Bӧer, B., Khan, M., Clüsener-Godt, M., Hameed, A., Eds.; Springer: Cham, 2019; vol 49. https://doi.org/10.1007/978-3-030-04417-6_14

Bakka, K.; Challabathula, D. Amelioration of Salt Stress Tolerance in Plants by Plant Growth-Promoting Rhizobacteria: Insights from “Omics” Approaches. In *Plant Microbe Symbiosis*; Springer: Cham, 2020; pp 303–330.

Caparrós, P. G.; Ozturk, M.; Gul, A.; Batool, T. S.; Pirasteh-Anosheh, H.; Unal, B. T.; Altay, V.; Toderich, K. N. Halophytes have Potential as Heavy Metal Phytoremediators: A Comprehensive Review. *Environ. Exp. Bot.* **2022,** *193*, 104666.

Chen, X.; Li, M.; Liang, Z.; Chen, C.; Chao, Y.; Qiu, R.; Wang, S. An AHP-Based Evaluation System Applied for Phytoremediation Method Selection in Heavy Metal Contaminated Farmland. *J. Hazard. Mater. Adv.* **2022,** 100138.

Cruz, J. L.; dos Santos, A. A.; Coelho, E. F.; Coelho Filho, M. A. Under Salinity the ‘PrataA nãGorutuba’clonerestrices Na+ Uptake and its Accumulation in Leaves. *Sci. Hortic.* **2020,** *270*, 109426.

DalCorso, G.; Fasani, E.; Manara, A.; Visioli, G.; Furini, A. Heavy Metal Pollutions: State of the Art and Innovation in Phytoremediation. *Int. J. Mol. Sci.* **2019,** *20* (14), 3412.

Das, S.; Das, S.; Ghangrekar, M. M. Efficacious Bioremediation of Heavy Metals and Radionuclides from Wastewater Employing Aquatic Macro-and Microphytes. *J. Basic Microbiol.* **2022,** *62* (3-4), 260–278.

Diarra, I.; Kotra, K. K.; Prasad, S. Assessment of Biodegradable Chelating Agents in the Phytoextraction of Heavy Metals from Multi–Metal Contaminated Soil. *Chemosphere* **2021,** *273*, 128483.

Duan, J.; Ren, C.; Wang, S.; Zhang, X.; Reis, S.; Xu, J.; Gu, B. Consolidation of Agricultural Land can Contribute to Agricultural Sustainability in China. *Nat. Food* **2021,** *2* (12), 1014–1022.

Dubey, S.; Shri, M.; Gupta, A.; Rani, V.; Chakrabarty, D. Toxicity and Detoxification of Heavy Metals during Plant Growth and Metabolism. *Environ. Chem. Lett.* **2018,** *16* (4), 1169–1192.

Ferreira, C. S.; Seifollahi-Aghmiuni, S.; Destouni, G.; Ghajarnia, N.; Kalantari, Z. Soil Degradation in the European Mediterranean Region: Processes, Status and Consequences. *Sci. Total Environ.* **2022,** *805*, 150106.

Guo, R.; Zhao, L.; Zhang, K.; Lu, H.; Bhanbhro, N.; Yang, C. Comparative Genomics and Transcriptomics of the Extreme Halophyte Puccinellia tenuiflora Provides Insights into Salinity Tolerance Differentiation between Halophytes and Glycophytes. *Front. Plant Sci.* **2021,** *12*, 649001.

Hafeez, M. B.; Raza, A.; Zahra, N.; Shaukat, K.; Akram, M. Z.; Iqbal, S.; Basra, S. M. Gene Regulation in Halophytes in Conferring Salt Tolerance. In *Handbook of Bioremediation* Academic Press, 2021; pp 341–370.

Hamidian, A. H.; Ozumchelouei, E. J.; Feizi, F.; Wu, C.; Zhang, Y.; Yang, M. A Review on the Characteristics of Microplastics in Wastewater Treatment Plants: A Source for Toxic Chemicals. *J. Clean. Prod.* **2021,** *295*, 126480.

Han, Z. J.; Sun, Y.; Zhang, M.; Zhai, J. T. Transcriptomic Profile Analysis of the Halophyte Suaedarigida Response and Tolerance under NaCl Stress. *Sci. Rep.* **2020,** *10* (1), 15148.

Javaid, T.; Farooq, M. A.; Akhtar, J.; Saqib, Z. A.; Anwar-ul-Haq, M. Silicon Nutrition Improves Growth of Salt-Stressed Wheat by Modulating Flows and Partitioning of Na+, Cl− and Mineral Ions. *Plant Physiol. Biochem.* **2019,** 141, 291–299.

Kayode, A. A. A.; Akram, M.; Laila, U.; Al-Khashman, O. A.; Kayode, O. T.; et al. Biological Implications of Atmospheric and Pedospheric Levels of Heavy Metals. *Adv. Toxicol. Toxic Effects* **2021,** *5* (1), 001–004. DOI: https://dx.doi.org/10.17352/atte.000010

Kumari, A.; Sheokand, S.; Kumar, A.; Mann, A.; Kumar, N.; Devi, S.; Rani, B.; Kumar, A.; Meena, B. L. Halophyte Growth and Physiology under Metal Toxicity. In *Ecophysiology, Abiotic Stress Responses and Utilization of Halophytes*; Springer: Singapore, 2019; pp 83–113.

Li, L.; Lu, H.; Wang, H.; Bhanbhro, N.; Yang, C. Genome-Wide DNA Methylation Analysis and Biochemical Responses Provide Insights into the Initial Domestication of Halophyte Puccinellia tenuiflora. *Plant Cell Rep.* **2021,** *40* (7), 1181–1197.

Li, Q.; Song, J. Analysis of Widely Targeted Metabolites of the Euhalophyte Suaeda Salsa under Saline Conditions Provides New Insights into Salt Tolerance and Nutritional Value in Halophytic Species. *BMC Plant Biol.* **2019,** *19* (1), 1–11.

Li, C.; Zhou, K.; Qin, W.; Tian, C.; Qi, M.; Yan, X.; Han, W. A Review on Heavy Metals Contamination in Soil: Effects, Sources, and Remediation Techniques. *Soil Sediment Contam. Int. J.* **2019,** *28* (4), 380–394.

Liu, L.; Wang, B. Protection of Halophytes and their uses for Cultivation of Saline-Alkali Soil in China. *Biology* **2021,** *10* (5), 353.

Lopes, M.; Sanches-Silva, A.; Castilho, M.; Cavaleiro, C.; Ramos, F. Halophytes as Source of Bioactive Phenolic Compounds and their Potential Applications. *Crit. Rev. Food Sci. Nutr.* **2021,** 1–24.

Mann, A.; Kumar, N.; Kumar, A.; Lata, C.; Kumar, A.; Meena, B. L.; Mishra, D.; Grover, M.; Gaba, S.; Parameswaran, C.; Mantri, N. de novo Transcriptomic Profiling of Differentially Expressed Genes in Grass Halophyte *Urochondra setulosa* Under High Salinity. *Sci. Rep.* **2021,** *11* (1), 1–4.

Nikalje, G. C.; Bhaskar, S. D.; Yadav, K.; Penna, S. Halophytes: Prospective Plants for Future. In *Ecophysiology, Abiotic Stress Responses and Utilization of Halophytes*; Springer: Singapore, 2019a; pp 221–234.

Nikalje, G. C.; Mirajkar, S. J.; Nikam, T.D.; Suprasanna, P. Multifarious Role of ROS in Halophytes: Signaling and Defense. In *Abiotic Stress-Mediated Sensing and Signaling in Plants: An Omics Perspective*; Zargar, S., et al., Eds.; Springer; pp 207–223.

Nikalje, G. C.; Nikam, T. D.; Suprasanna, P. Looking at Halophytic Adaptation to High Salinity through Genomics Landscape. *Curr. Genomics* **2017,** *18* (6), 542–552.

Nikalje, G. C.; Saini, N.; Suprasanna, P. Halophytes and Heavy Metals: Interesting Partnerships. In *Plant-Metal Interactions*; Srivastava, et al., Eds., Springer: Cham, 2019b; pp 99–118.

Nikalje, G. C.; Srivastava, A. K.; Pandey, G. K.; Suprasanna, P. Halophytes in Biosaline Agriculture: Mechanism, Utilization and Value-Added Products. *Land Degrad. Dev.* **2018a,** *29* (4): 1081–1095. DOI: 10.1002/ldr.2819.

Nikalje, G. C.; Srivastava, M.; Nikam, T. D.; Suprasanna, P. Physiological Responses and Tolerance of Halophyte *Sesuvium portulacastrum* L. to Cesium. *Adv. Agric.* **2022,** *2022*, 98630021–98630027.

Nikalje, G. C.; Suprasanna, P. Coping with Metal Toxicity – Cues from Halophytes. *Front. Plant Sci.* **2018b,** *9*, 777. DOI: 10.3389/fpls.2018.00777.

Nikalje, G. C.; Variyar, P. S.; Joshi, M. V.; Nikam, T. D.; Suprasanna, P. Temporal and Spatial Changes in Ion Homeostasis and Accumulation of Flavonoids and Glycolipid in a Halophyte *Sesuvium portulacastrum* (L.) L. *PLoS One* **2018c,** *13* (4), e0193394. DOI: 10.1371/journal.pone.0193394.

Nikalje, G. C.; Yadav, K.; Suprasanna, P. Halophyte Responses and Tolerance to Abiotic Stresses. In *Ecophysiology, Abiotic Stress Responses and Utilization of Halophytes*; Hasanuzzaman, M., et al., Eds.; Springer: Singapore, 2019c; pp 1–23.

Orrego, F. Use of Halophytes for the Remediation of Metal-Affected Soils in Arid Environments. In *Handbook of Halophytes: From Molecules to Ecosystems Towards Biosaline Agriculture*, 2020; pp 1–28.

Rahman, M. M.; Mostofa, M. G.; Keya, S. S.; Siddiqui, M. N.; Ansary, M. M.; Das, A. K.; Rahman, M. A.; Tran, L. S. Adaptive Mechanisms of Halophytes and their Potential in Improving Salinity Tolerance in Plants. *Int. J. Mol. Sci.* **2021,** *22* (19), 10733.

Rangani, J.; Kumari, A.; Patel, M.; Brahmbhatt, H.; Parida, A. K. Phytochemical Profiling, Polyphenol Composition, and Antioxidant Activity of the Leaf Extract from the Medicinal Halophyte *Thespesia populnea* Reveal a Potential Source of Bioactive Compounds and Nutraceuticals. *J. Food Biochem.* **2019,** *43* (2), e12731.

Riffat, A. L. I. A.; Sajid, M.; Ahmad, A. Alleviation of Adverse Effects of Salt Stress on Growth og Maize (Zea mays L.) by Sulfur Supplementation. *Pak. J. Bot.* **2020,** *52* (3), 763–773.

Saddhe, A.; Manuka, R.; Nikalje, G. C.; Suprasanna, P. Halophytes as a Potential Resource for Phytodesalination. In *Handbook of Halophytes: From Molecules to Ecosystems towards Biosaline Agriculture*; Grigore, M. N., Eds.; Springer: Cham, 2020; pp 1–21.

Sghaier-Hammami, B.; Baazaoui, N.; Hammami, S. B.; Abdelly, C. Priming-Mediated Tolerance to Biotic and Abiotic Stresses in Halophytes. In *Handbook of Halophytes: From Molecules to Ecosystems towards Biosaline Agriculture*; 2020; pp 1–26.

Shackira, A. M.; Puthur, J. T. Phytostabilization of Heavy Metals: Understanding of Principles and Practices. In *Plant-Metal Interactions*; Springer: Cham, 2019; pp 263–282.

Shahid, M. A.; Sarkhosh, A.; Khan, N.; Balal, R. M.; Ali, S.; Rossi, L.; Gómez, C.; Mattson, N.; Nasim, W.; Garcia-Sanchez, F. Insights into the Physiological and Biochemical Impacts of Salt Stress on Plant Growth and Development. *Agronomy* **2020,** *10* (7), 938. https://doi.org/10.3390/agronomy10070938

Shang, C.; Wang, L.; Tian, C.; Song, J. Heavy Metal Tolerance and Potential for Remediation of Heavy Metal-Contaminated Saline Soils for the euhalophyteSuaeda salsa. *Plant Signal. Behav*. **2020,** *15* (11), 1805902.

Sharma, A.; Devi, A.; Garg, C.; Kumari, A.; Mann, A.; Kumar, A. Behavior of Halophytes and Their Tolerance Mechanism Under Different Abiotic Stresses. In *Ecophysiology, Abiotic Stress Responses and Utilization of Halophytes*; Springer: Singapore, 2019; pp 25–38.

Shelake, R. M.; Kadam, U. S.; Kumar, R.; Pramanik, D.; Singh, A. K.; Kim, J. Y. Engineering Drought and Salinity Tolerance Traits in Crops through CRISPR-Mediated Genome Editing: Targets, Tools, Challenges, and Perspectives. *Plant Commun.* **2022,** 100417.

Shoukat, E.; Ahmed, M. Z.; Abideen, Z.; Azeem, M.; Ibrahim, M.; Gul, B.; Khan, M. A. Short and Long Term Salinity Induced Differences in Growth and Tissue Specific Ion Regulation of *Phragmiteskarka*. *Flora* **2020,** *263*, 1–10

Snyman, M. C.; Solofoharivelo, M. C.; Souza-Richards, R.; Stephan, D.; Murray, S.; Burger, J. T. The Use of High-Throughput Small RNA Sequencing Reveals Differentially Expressed MicroRNAs in Response to Aster Yellows Phytoplasma-Infection in Vitisviniferacv.'Chardonnay'. *PloS One* **2017,** *12* (8), e0182629.

Sodango, T. H.; Li, X.; Sha, J.; Bao, Z. Review of the Spatial Distribution, Source and Extent of Heavy Metal Pollution of Soil in China: Impacts and Mitigation Approaches. *J. Health Pollut*. **2018,** *8* (17), 53–70.

Syed, A.; Sarwar, G.; Shah, S. H.; Muhammad, S. Soil Salinity Research in 21st Century in Pakistan: Its Impact on Availability of Plant Nutrients, Growth and Yield of Crops. *Commun. Soil Sci. Plant Anal*. **2021,** *52* (3), 183–200.

Velusamy, S.; Roy, A.; Sundaram, S.; Kumar Mallick, T. A Review on Heavy Metal Ions and Containing Dyes Removal Through Graphene Oxide-Based Adsorption Strategies for Textile Wastewater Treatment. *Chem. Rec*. **2021,** *21* (7), 1570–1610.

Yao, Y.; Zhang, X.; Wang, N.; Cui, Y.; Zhang, L.; Fan, S. Transcriptome Analysis of Salt Stress Response in Halophyte Atriplexcentralasiatica Leaves. *Acta Physiol. Plant*. **2020,** *42* (1), 1–5.

Yousaf, M. T. B.; Nawaz, M. F.; Zia urRehman, M.; Gul, S.; Yasin, G.; Rizwan, M.; Ali, S. Effect of Three Different Types of Biochars on Eco-physiological Response of Important Agroforestry Tree Species under Salt Stress. *Int. J Phytoremediat.* **2021,** *23* (13), 1412–1422.

Zhang, Z.; Tariq, A.; Zeng, F.; Graciano, C.; Zhang, B. Nitrogen Application Mitigates Drought-Induced Metabolic Changes in Alhagisparsifolia Seedlings by Regulating Nutrient and Biomass Allocation Patterns. *Plant Physiol. Biochem.* **2020,** *155*, 828–841.

Zhang, X.; Yao, Y.; Li, X.; Zhang, L.; Fan, S. Transcriptomic Analysis Identifies Novel Genes and Pathways for Salt Stress Responses in Suaeda Salsa Leaves. *Sci. Rep*. **2020,** *10* (1), 1–2.

Zilaie, M. N.; Arani, A. M.; Etesami, H.; Dinarvand, M.; Dolati, A. Halotolerant Plant Growth-Promoting Rhizobacteria-Mediated Alleviation of Salinity and Dust Stress and Improvement of Forage Yield in the Desert Halophyte Seidlitzia rosmarinus. *Environ. Exp. Bot.* **2022,** *201*, 104952.

CHAPTER 18

Halophytes as Reservoirs of Stress-Mitigating Proteins and Osmolytes

SUNIL P. BHAGAT[1], JAYANT P. RATHOD[2], VIDYA R. KOTHARI[2], SATYAM D. PAWAR[1], and HARICHANDRA A. NIKULE[1]

[1]*Department of Botany, ADT's Shardabai Pawar Mahila Arts, Commerce and Science College Shardanagar, Malegaon Bk., Baramati, Maharashtra, India*

[2]*Department of Microbiology, ADT's Shardabai Pawar Mahila Arts, Commerce and Science College Shardanagar, Malegaon Bk., Baramati, Maharashtra, India*

ABSTRACT

Diverse anthropogenic activities restrict plant distribution, growth, and yield in agricultural ecosystems throughout the world. Specific group of plants like halophytes exhibit excellent tolerance to adverse environmental conditions and can withstand various stressful conditions such as salinity, drought, heavy metal stress, and so on. Halophytes have the ability to grow vigorously in saline soils, making them convenient for agriculture in salinity prone areas. Additionally, they are an excellent source of salt-responsive genes. Several halophytes show similar physiological and biochemical responses to salinity stress, whereas different mechanisms operate under multiple stresses. Halophytes have also developed numerous adaptations to tolerate high salinity disturbances and other stresses. The roles of stress responsive proteins that are potential determinants for stress adaptation are discussed in the present chapter. Most common adaptations of halophytes include accumulation of osmolytes such as proline, induction of antioxidant molecules, selectivity in potassium and nitrogen uptake, prohibition of

Physiology of Halophytes: Signaling, Omics, and Tolerance Mechanisms.
Nivas Manohar Desai & Ganesh Chandrakant Nikalje (Eds.)

sodium uptake and its vacuolar compartmentalization. Osmolytes protect membrane from injuries and stabilize proteins and enzymes under high salt or harmful solute concentrations. Also, they are a source of carbon, nitrogen, sulphur, and cellular energy. Under stress conditions, osmolytes control metabolic balance by interacting with various biomolecules. The connection between variations in proteome composition, abundance, and adjustments in morphophysiological and biochemical features associated with stress are also discussed. Thus, review of various mechanisms adopted by plants for abiotic stresses will help to recognize potential strategies and halophytic varieties for production of stress resistant crops for betterment of agriculture.

18.1 INTRODUCTION

18.1.1 HALOPHYTES

More than a century ago, halophytes were simply, but meaningfully, defined as species adapted to persistent salinity conditions (Von Marilaun, 1896). Plants that grow and complete their life processes in the high salt surroundings are called halophytes or salt plants. About 0.5% of the dry weight is the basic salinity level for plants to be classified as halophytes (Stocker, 1933). Depending on how well these organisms grow in salty environments, plants are classified as either halophytes or salt-sensitive plants called glycophytes. Most dicot halophytes can survive in more than 200 mM of salt, whereas glycophytes exhibit substantially suppressed growth and occasional death under 100–200 mM salt (Flowers and Colmer, 2008; Shabala, 2013). Halophytes are found in different ecosystems like pure limy semi-deserts, mangroves, semi-natural pastures; artificial habitats like tropical over-irrigated cultivable areas with inadequate drainage, and boreal-temperate city lawns around the globe contaminated with NaCl and $CaCl^2$ while de-icing (Hasanuzzaman, 2014; Kapler, 2019). Various response mechanisms have been developed by the plants to tackle high salinity conditions (Muchate et al., 2016a; Nikalje et al., 2018a; Shelke et al., 2019). Some plants prefer to be in a saline-free environment, while others fight or dodge it (Flowers et al., 1977). Avoidance of salinity is achieved by restricting germination, growth, and reproduction. Plants resist salinity by storing salt in cells and by secreting additional salts. In halophytes, the protoplasm tolerates high salt concentration without exhibiting apparent harm and functioning typically (Flowers, 2015).

With the exception of Antarctica, all continents of the planet are home to about 1500 different halophyte species. According to screening tests, about one-third of these might be resistant to salt well enough to grow in highly saline irrigation water, including saltwater (Epstein et al., 1980). Halophytes occur in around 37 of the total 65 plant orders (Flowers et al., 1977; 2010). According to phylogenetic analysis of angiosperms, halophily has independently evolved 59 times in various families (Saslis-Lagoudakis et al., 2014). General ecological conditions required for their growth are mud-abundant shallow water: water-logged saline soil, sandy or loose soil, or organic-matter rich heavy clays; high humidity and rainfall.

Regardless of the fact that some families have very high numbers of halophytic species for unknown causes, no family is totally halophytic (e.g., the Chenopodiaceae, now included in the Amaranthaceae). At the genus level, halophytes and non-halophytes regularly co-occur; well-known examples include *Aster*, *Glycine*, *Plantago*, and *Solanum* (Cheeseman, 2015).

18.1.2 CLASSIFICATION

Halophytes are characterized in several ways, including their distribution, ecological behavior, salinity-induced growth responses, and extent of salt (Eshel and Waisel, 1972). Albert and Popp (1977) classified the salt marsh halophytes into physiotypes. Halophytes can also be classified according to whether or not they have salt glands, such as black mangroves with well-developed salt glands, and red mangroves without such glands (Popp et al., 1995). Halophytes were categorized based on how they respond to internal salt content. According to him, there are three different sorts of salt marsh plants: (1) succulent halophytes (e.g., *Salicornia herbacea*), (2) non-succulent (e.g., *Spartina alterniflora*), and (3) accumulating halophytes (e.g., *Juncus gerardii*, *Suaeda fruticosa*). Based on the ecological and morphophysiological characteristics related to the tolerance to salinity, halophytes are classified as euhalophytes, recretohalophytes, and pseudohalophytes (Fan, 2020).

18.1.3 HALOPHYTES: MORPHOLOGY AND DIVERSITY

To facilitate water absorption within their habitats, the halophytes exhibit critical xerophytic mechanisms for water level management. Tropical and subtropical halophytes are majorly shrubs, but other herbaceous forms are also known (e.g., *Acanthus ilicifolius*) (Rao, 1974). Halophytic flora is absolutely herbaceous in temperate zones. The bushes have a dome-like appearance because of their cymose branching. Halophytes increase many shallow ordinary roots. To supplement conventional roots, the aerial branches develop tilt and/or prop roots for green attachment in loose sandy or muddy soil (Grigore and Toma, 2021). These downward-growing roots penetrate the hard soil strata of the soil to certain depth. Sometimes stilt roots are sturdy and have extensively evolved in some vegetation, including *Rhizophora mucronata*, but are no longer produced in others (*Rhizophora conjugata*) (Setyawan et al., 2014). Since in some plants, the stilt roots might not increase now and again; various types of root buttresses arise from the base of the trunk. The soil in the nearshore region includes meager O^2 levels due to water logging (Cao, 2012). To redress the absence of soil aeration, hydro-halophytes generate completely unique and negatively geotropic root gadgets known as pneumatophores, which develop from the underground roots and layout above the water and slush. On the outside, they have multiple lenticels or pneumathodes and on the inner side, prominent aerenchyma tissue encloses major air depressions through the lenticels, for gaseous trade (Kumari et al., 2015). The aerenchyma allows the inside air conduction to reach the subsurface or submerged roots. In few floras (e.g., *Bruguiera*), the vertical roots that grow above the slush, bend down, and go deeper into the slush (Lawton et al., 1981). In some *Rhizophora* species, pneumatophores no longer broaden. In those plants, the higher upstanding hall of descending stilt roots apparently soaks up the respiration effort. Stems of several halophytes showcase succulence (e.g., *Suaeda maritima, Salicornia herbacia*) which is dependent on the ratio of absorbed-to-lost ions inside plant cells (Arnold, 1955). Under saline conditions, ionic concentration inside cell is lowered while cell length is enhanced; two characteristics which are usual of succulents (Repp et al., 1959).

All tropical seashores with mangrove flowers, from East Africa to Australia, include nearly similar forms of plants. Likewise, there are numerous similarities between the mangroves of East Asia, East Africa, and West Asia (D'odorico et al., 2013). The tropical halophytes are particularly hairy and have dense cymose branching compared to the herbaceous temperate halophytes (Kumar et al., 2018). Submerged marine angiosperms are one of the few plants that do not expand succulent tissues. Most halophytes have thick, entire, succulent leaves which are generally tiny in length and often have a glassy appearance. There are also a few aphyllous species. Coastal aero halophytes' stems and leaves show off similar habitat-particular variations. Trichomes are abundantly present throughout their surfaces. Submerged marine halophytes have thin, incredibly underdeveloped vascular structures, and frequently inexperienced dermis on their leaves (Waisel, 2012). They can absorb vitamins and water from the media. Fruit partitions include various air chambers, seeds, and seedlings that can stay afloat for large periods of time are carried by water currents to some faraway places (Waisel, 2012). An unusual method of fruit dissemination is displayed by a coastal species of Spinifex (*S. quarrosus*), a Gramineae plant that is often observed growing on the sandy seashores of Andhra (Desai, 1995). The female inflorescence has a round shape and is made of several spikelets. The stiff bractiolar bristles are a useful resource in its dancing and flipping motions. The globular, hairy inflorescence separates from the plant and trails at the sand, while dropping its seeds sometimes. Subsequently, the final fruits and the inflorescence are buried within the mud. Halophytes and mangrove plant life developing in tidal marshes undergo "vivipary," which is the process of seed germination even as the fruits are attached to mother plants (Walsh, 1974). Excessive salinity inhibits seed germination, and consequently, viviparous germination is a significant model to tackle slow seed germination. *Rhizophora, Aegiceras, Avicennia, Cassula,* and *Ranansatia* species showcase vivipary (Kumar et al., 2018).

The current chapter outlines halophyte salt tolerance mechanisms by delineating recent studies focusing on morphophysiological and molecular adaptations, osmolytes produced, and listing important proteins-regulating halophily via proteome studies of diverse halophytes in response to salinity, with an emphasis on shortlisting potential gene/protein candidates for enhancing salt resistance in commonly cultivated crop plants.

18.2 GLOBAL IMPACT OF ABIOTIC STRESSES ON AGRICULTURE

Due to the anthropogenic activities, Earth's average temperature has increased progressively, directly affecting agriculture through various abiotic stresses. Major agroeconomic countries have faced severe production losses due to harsh climate. By 2050, the Indo-Gangetic Plains (IGP), producer of 15% world's wheat is expected to face drought and high temperature conditions (Gill et al., 2008). Increased salinity is a consequence of rising water scarcity, accumulation of chemical fertilizers in soil, and usage of water from underground sources for agriculture. Australia and Asia Pacific are the major salinity-affected regions with 27% (549.30 million ha) of total agricultural area of 2016.63 million ha being affected. About 72.2 million ha (6.40%) land in Africa, 130.5 million ha (10.6%) land across America, and 17.30% land in Europe is salinity-affected (World Bank Population Prospects, 2019). Almost 147 million ha of land is degraded in India due to water erosion (94 million ha), salinity/alkalinity/acidification (23 million ha), flooding/water logging (14 million ha), wind erosion (9 million ha), and combination of all factors (7 million ha) (Kumar and Sharma, 2020). According to climatic models, the Indian subcontinent, North America, and Central and East Asia are and most prone to heat stress among the major producers of growing the major crops maize, wheat, and rice (Teixeira et al., 2013). Rapid temperature rise has significantly reduced the yield of rice in China (Tian et al., 2009). By 2050, wheat yields are predicted to decrease by 0.45 ton/ha in India and by 5–7% in rainfed wheat-producing China as the average temperature rises by 0.5%. During April–May in Philippines, rice farms face significant heat stress (Manigbas and Sebastian, 2007).

The hot summer of 2003 caused significant yield reduction in Europe (Ciais et al., 2005). In Africa, various factors in combination are predicted to cause more than 2% yield reduction for maize as temperature rises by 1% (Lobell et al., 2011). During the period of 1976–2006, temperature changes have been predicted to cause 12% and 17% reduction in maize and soybean production in the USA (Kucharik and Serbin, 2008). Due to combined heat and drought stress, South and South-West USA are expected to face losses of $ 5 billion+ (National Climatic Data Center, 2011). Wheat-growing areas across the world like Australia (Zhang et al., 2015; Crimp et al., 2016), China (Xiao et al., 2018), Europe (Trnka et al., 2014), and the United States (Holman et al., 2011) often undergo low-temperature stress. In North America, the 1988 drought cost the US economy more than $40 billion (Riebsame et al., 2019). About 36% reduced yields and consequent losses of around AUD $3.5 billion were caused by drought in Australia (Wong et al., 2010). According to an IPCC (Intergovernmental Panel on Climate Change) study, in recent decades, wheat, maize, and rice production has declined in several Asian countries because of rising water stress caused by increasing temperatures, increasing El Nino events and reduced monsoon (Bates et al., 2008).

18.3 IMPACT OF SALINITY ON CROPS

For crop plants, seed germination regulates the plant growth, development, and yield. However, seed germination is regulated by numerous external (environmental) and internal factors (Wahid et al., 2016). The higher salinity level adversely affects seed germination, reduces water absorption capacity, which should increase in response to the osmotic or drought effect (Khan et al., 1999). Due to this, crop growth is generally reduced at elevated levels of salt. Metabolic changes occur in the plant due to osmotic stress and lead to water stress-induced wilting (Munns et al., 2002). Toxicity by accumulation of specific ions and nutritional imbalances are known to reduce growth (Lauchli and Epstein, 1990). Visual symptoms start to appear on plant morphology under salt stress. The signs include wilting, stunted growth, yellowing and necrosis of leaves, chlorosis, scorching, and burning of leaf tips, and so on (Shannon and Grieve, 1998). Salinity stress also decreases the marketable yield of roots, fruits, leaves, and tubers and overall productivity (De Pascale et al., 2015).

Salinity also affects photosynthesis due to decreasing CO^2 availability, which limits its diffusion (Flexas et al., 2007) and reduces the photosynthetic pigments (Ashraf and Harris, 2013). Reduction in leaf area expansion and stomatal conductance is also reported with their interference in light interception in radish (Marcelis and Van Hooijdonk, 1999). Salinity affects the plant's total photosynthetic activity by decreasing leaf growth and inhibiting growth ability (Yeo, 2007). For different crops, reduced growth, leaf area and plant biomass has also been observed under saline conditions (Giuffrida et al., 2012). Salt stress mainly affects through roots, but more focused studies are needed on root morphology under salinity to ascertain its role in halophytic adaptations. However, reports also show that, root biomass has been less affected under salt stress than aboveground parts of the plant (Munns and Tester, 2008). Plants illustrate differential responses to salinity which are significantly expressed through morphological, physiological, biochemical, and molecular attributes required for adapting to the shock from salinity stress (Kibria et al., 2017). In root zone, salinity stress expands osmotic stress and interrupts cellular ionic homeostasis by augmented Na^+ and Cl^- accumulation and consequently, inhibiting the uptake of essential nutrients like K^+ (Paranychianakis and Chartzoulakis, 2005), causing K^+ deficiency and decreased K^+/Na^+ ratio under salt stress (Kibria et al., 2017).

18.4 SALT TOLERANCE AND ADAPTATION MECHANISMS IN HALOPHYTES

18.4.1 SALT ADAPTATIONS

Halophytes possess three well-known mechanisms of salt adaptation such as uptake and accumulation of higher levels of ions in vacuoles, removal of salts absorbed through special cells, and restricting uptake of salts (Acosta-Motos et al., 2017). Euhalophytes can accumulate more salt and can tolerate up to 300 mM concentration of NaCl (Yuan et al., 2019). Genomic, transcriptomic, and proteomic studies have shown that ancestors of euhalophytes had lesser adaptations and growth under saline conditions. Some studies have shown that stems and leaves of certain euhalophyte species have tissues for water accumulation (Tang et al., 2015; Acosta-Motos et al., 2017). Through utilization of salt glands or salt bladders for secretion of salt ions from leaves, recretohalophytes form a distinct class (Flowers et al., 2015; Acosta-Motos et al., 2017).

18.4.2 SALT-TOLERANCE MECHANISM

Plant salt tolerance mechanisms are very complex; even though, it is studied and described at various levels including morphological (whole plant processes), cellular (ions distribution, synthesis of osmoprotectants and antioxidants and enzymatic activation), and molecular (gene expression) levels. Halophytes and glycophytes exhibit variations in defence mechanisms at the organismal and cellular levels (Himabindu et al., 2016). Halophytic plant cells are more competent at maintenance of cellular ionic homeostasis and exhibit greater biosynthesis and accumulation of osmolytes in cytoplasm (Himabindu et al., 2016). Salt-related studies have deciphered the important role of ion transport systems in maintaining ion homeostasis. Na^+ absorption at root zone-soil boundary is facilitated by aquaporins, High-affinity K^+ Transporters (HKTs), and Non-Selective Cationic Channels (NSCC). Ion transporters and ion channels are responsible for Na^+ diffusion to stem and leaf tips through xylem (Hanin et al., 2016; Byrt et al., 2017). The ionic absorption and compartmentalization regulate ion homeostasis in the cell and provide the proper growth and development opportunities for the plant under favorable conditions and salt stress. Na^+ compartmentalization into vacuoles is also one of the effective salt-adaptation strategies reported in some species (Wang et al., 2019). Maintenance of K^+/Na^+ ratio is a prerequisite for normal cellular functioning. Under salt stress, this ratio is disbalanced due to increase in the concentration of Na^+ (Shabala and Pottosin, 2014) and causes K^+ movement from the cell, thus resulting in decreased cellular ionic homeostasis (Cabot et al., 2014). Hence, among all physiological salt tolerance mechanisms, K^+/Na^+ selectivity is known to be one of the main strategies. While glycophytes exhibit similar mechanisms relating to K^+ and Na^+, their sensitivity to ion imbalance is significantly more as compared to halophytes (Assaha et al., 2017).

To restrict higher Na^+ accumulation inside cell, Na^+ sequestration and inhibition of its uptake are carried out by plants. At the molecular level, salinity stress modulation at molecular level is achieved through the combined action of certain enzymes like metabolic and Ca^{2+}/calmodulin activated Ser/Thr protein phosphatases, phospholipase D, transcription factors like C-repeat binding factor/dehydration-responsive element-binding protein (CBF/DREB), and heat shock factor (HSF) and secondary messengers. Certain kinases like two-component histidine kinases, mitogen-activated protein kinase (MAPK) cascades, Ca^{2+}/calmodulin-dependent protein kinases (CDPKs), and Ser/Thr protein kinases (e.g., SOS2) are also known to contribute to osmoregulation (Bahmani et al., 2015). Nowadays, transgenics is also used to mitigate salinity stress by overexpressing some essential genes viz. proline-, lysine-, and glutamic-rich protein gene *(PKE),* Na^+/H^+ antiporters *(NHX),* salt overly sensitive proteins *(SOS),* caffeic acid O-methyltransferases *(COMT),* myeloblastosis transcription factors *(MYB),* dehydration responsive element-binding protein *(DREB),* etc., to combat salt stress (Ibrahimova et al., 2021).

18.4.3 SALT RESPONSIVE METABOLITES/COMPOUNDS

Halophytes possess a variety of organic solutes such as quaternary ammonium compounds, amino acids, sugars, and sugar alcohols which help them to tolerate high salt concentrations and modulate essential physiological functions active inside the cell (Slama et al., 2015). The most abundant osmolyte in halophytes is glycine betaine, followed by proline. The reported mechanisms behind high salt tolerance include balancing the vacuolar Na^+ and Cl- osmotic potential by accumulating osmolytes in cytoplasm (Hasegawa et al., 2000). Another mechanism includes scavenging of excessive ROS and prevention of membrane damage (Szabados and Savoure, 2010). The third mechanism includes protecting the subcellular structures inside the cell. Osmolytes are highly soluble compounds and do not affect other metabolic processes. They are not harmful inside the cell even at higher concentrations (Slama et al., 2015).

Different osmolytes are present in plants with varying concentrations depends on the species chosen and conditions of its habitat (Szabados et al., 2011). Various abiotic stresses cause osmolyte accumulation in different plant species. One such stress is increased salt concentration in halophytes. Osmoprotectants accumulation, viz. proline, sugars, glycine betaine, and gamma-aminobutyric acid, is a common response found in various plants (Suprasanna et al., 2016). This leads to synthesis of different osmolyte in halophytes such as glycine betaine, ala betaine, ß-alanine betaine, sucrose, fructose, raffinose, myo-inositol, pinitol, maltose, ononitol, mannitol, dimethyl sulfoniopropionate, choline-O-sulfate, hydroxyproline betaine, proline, glycine, proline betaine, methylated onium compounds, pipecolate, and so on (Slama et al., 2015). The osmolytes present in halophytes are given in Figure 18.1 and details of osmolyte produced by respective genus/species of halophytes are given in Table 18.1. Under saline conditions, activities of different enzymes like CAT (catalase), GR (glutathione reductase), SOD (superoxide dismutase), APX (ascorbate peroxidase), and GPX (guaiacol peroxidase) were maintained through osmolytes accumulation along with maintaining Na^+/K^+ ratio (Muchate et al., 2016b).

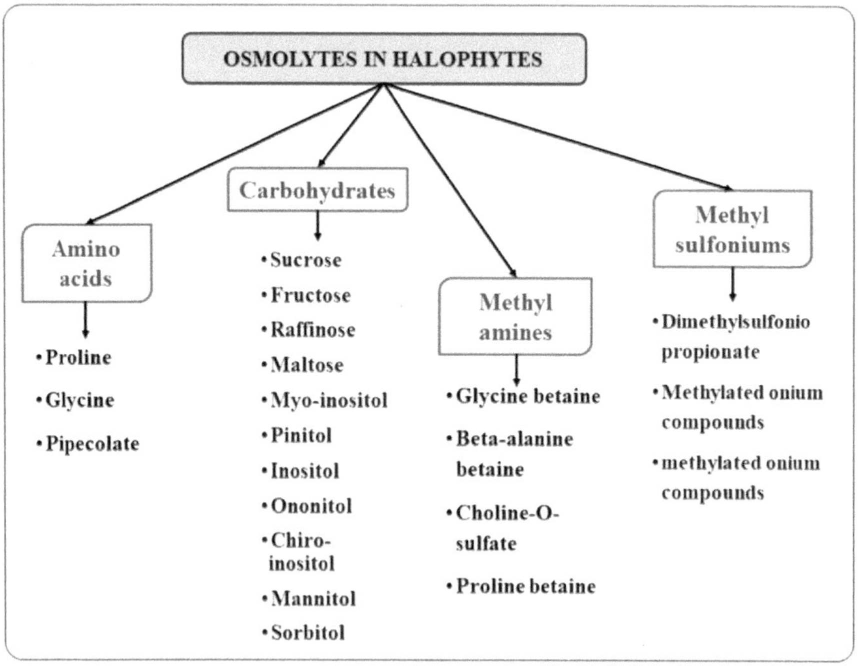

FIGURE 18.1 Osmolytes present in halophytes.

18.5 STRESS-RESPONSIVE PROTEINS FROM HALOPHYTES

Diverse processes like sugar and protein metabolism, redox homeostasis, and photosynthesis are affected by high salt concentrations (Nikalje et al., 2019).

TABLE 18.1 Stress-Responsive Osmolytes Found in Halophytes.

Sr. No.	Osmolyte	Name of the halophyte	Reference
1	Glycine betaine	*Anneurolepidium chinense*	Ochiai and Matoh, 2001
		Cynodon dactylon x C	Marcum and Murdoch, 1994
		Distichlis spicata	Marcum, 1999
		Distichlis stricta	Guy et al., 1984
		Halopyrum mucronatum	Khan et al., 1999
		Hordeum jubatum	Guy et al., 1984
		Hordeum marinum	Garthwaite et al., 2005
		Panicum antidotale	Ahmad et al., 2010
		Panicum hemitomon,	Hester et al., 2001
		Paspalum spp.	Lee et al., 2008
		Spartina alterniflora	Li et al., 2010
		Spartina anglica	Mulholland and Otte, 2000
		Spartina gracilis	Guy et al., 1984
		Spartina patens	Hester et al., 2001
		Spartina townsendii	Adrian-Romero et al., 1998
		Sporobolus airoides	Marcum, 1999
		Sporobolus virginicus	Marcum and Murdoch, 1992
		Stenotaphrum secundatum	Marcum and Murdoch, 1994
		Thinopyrum bessarabicum	Gorham et al., 1985
		Artemisia santonicum	Tipirdamaz et al., 2006
		Aster tripolium	Briens and Larher, 1982
		Cakile maritima Scott	Jdey et al., 2014
		Sesuvium portulacastrum	Kannan et al., 2013
		Atriplex barclayana	Nerd and Pasternak, 1992
		Atriplex gmelini	Matoh et al.,1987
		Atriplex griffithii	Khan et al., 2000
		Atriplex halimus	Hassine et al., 2008
		Atriplex lacinata	Adrian-Romero et al., 1998
		Atriplex lasiantha	Tipirdamaz et al., 2006
		Atriplex nuttalliana	Guy et al., 1984
		Atriplex patula	Guy et al., 1984
		Atriplex rosea	Tipirdamaz et al., 2006
		Atriplex littoralis	Adrian-Romero et al., 1998
		Atriplex nummularia	Silveira et al., 2009
		Atriplex prostrata	Wang and Showalter, 2004
		Atriplex tatarica	Tipirdamaz et al., 2006

TABLE 18.1 *(Continued)*

Sr. No.	Osmolyte	Name of the halophyte	Reference
		Bassia hysopifolia	Guy et al., 1984
		Beta vulgaris	Subbarao et al., 2001
		Bienertia sinuspersici	Park et al., 2009
		Borszczowia aralocaspica	Larher et al., 2009
		Camphorosma annua	Murakeozy et al., 2003
		Camphorosma monspeliaca	Tipirdamaz et al., 2006
		Chenopodium album	Tipirdamaz et al., 2006
		Chenopodium botrys	Tipirdamaz et al., 2006
		Chenopodium foliosum	Tipirdamaz et al., 2006
		Chenopodium glaucum	Guy et al., 1984
		Chenopodium rubrum	Adrian-Romero et al., 1998
		Halocnemum strobilaceum	Tipirdamaz et al., 2006
		Haloxylon recurvum	Khan et al., 1998; 2000
		Halopeplis perfoliata	Adrian-Romero et al., 1998
		Hammada elegans	Adrian-Romero et al., 1998
		Kochia scoparia	Kalinina et al., 2012
		Kochia prostrata	Karimi et al., 2005
		Kochia sieversiana	Yang et al., 2007
		Krasheninnikovia ceratoides	Tipirdamaz et al., 2006
		Microcnemum coralloides	Tipirdamaz et al., 2006
		Noaea mucronata	Tipirdamaz et al., 2006
		Petrosimonia brachiata	Tipirdamaz et al., 2006
		Petrosimonia nigdeensis	Tipirdamaz et al., 2006
		Salicornia dolichostachya	Katschnig et al., 2013
		Salicornia europaea	Guy et al., 1984
		Salicornia prostrata	Tipirdamaz et al., 2006
		Salicornia utahensis	Harris et al., 2001
		Salsola ikonnikovii	Xing et al., 2013
		Salsola rubescens	Adrian-Romero et al., 1998
		Salsola ruthenica	Tipirdamaz et al., 2006
		Sarcocornia fruticosa	Gil et al., 2011
		Seidlitzia rosmarinus	Adrian-Romero et al., 1998
		Suaeda altiissima	Tipirdamaz et al., 2006
		Suaeda aralocaspica	Park et al., 2009
		Suaeda confusa	Tipirdamaz et al., 2006

TABLE 18.1 *(Continued)*

Sr. No.	Osmolyte	Name of the halophyte	Reference
		Suaeda eltonica	Park et al., 2009
		Suaeda fruticosa	Hameed et al., 2012
		Suaeda glauca	Yang et al., 2008
		Suaeda heterophylla	Park et al., 2009
		Suaeda japonica	Tanimoto et al., 1997
		Suaeda maritima	Moghaieb et al., 2004
		Suaeda heterophylla	Park et al., 2009
		Suaeda salsa	Wu et al., 2012
		Tecticornia indica	English and Colmer, 2011
		Tecticornia mellaria	English and Colmer, 2011
		Tecticornia pergranulata	English and Colmer, 2011
		Frankenia hirsuta	Tipirdamaz et al., 2006
		Limonium perezii, L. sinuatum	Hanson et al., 1991
		Plumbago	Hanson et al., 1994
		Avicennia	Poljakoff-Mayber et al., 1987
		Plantago maritima	Tipirdamaz et al., 2006
		Lycium ferocissimum	Wyn Jones et al., 1981
2	Beta-alanine betaine	*Limonium aureum, L. latifolium*	Hanson et al., 1991
		Limonium gmelini	Murakeozy et al., 2003
3	Choline-O-sulfate	*Aegialitis*	Hanson et al.,1994
		Limoniastrum	Hanson et al., 1994
		Limonium aureum, L. latifolium	Hanson et al., 1991
		Limonium gmelini	Murakeozy et al., 2003
		Limonium perezii, L. sinuatum	Hanson et al., 1991
		Plumbago	Hanson et al., 1994
4	Pipecolate	*Limonium aureum, L. latifolium, L. vulgare*	Hanson et al., 1991
5	Proline betaine	*Limoniastrum*	Hanson et al., 1994
6	Proline	*Cynodon dactylon x C.*	Marcum and Murdoch, 1994
		Distichlis spicata	Marcum, 1999
		Hordeum marinum	Garthwaite et al., 2005
		Odyssea paucinervis	Naidoo et al., 2008
		Panicum antidotale	Ahmad et al., 2010
		Panicum hemitomon	Hester et al., 2001
		Paspalum spp.	Lee et al., 2008
		Puccinellia maritima	Ahmad et al., 1981

TABLE 18.1 *(Continued)*

Sr. No.	Osmolyte	Name of the halophyte	Reference
		Spartina alterniflora	Li et al., 2010
		Spartina patens	Hester et al., 2001
		Thinopyrum bessarabicum	Gorham et al., 1985
		Cymodocea nodosa	Sandoval-Gil et al., 2012
		Triglochin maritima	Briens and Larher, 1982
		Posidonia oceanica	Sandoval-Gil et al., 2012
		Zostera	Ye et al., 2003
		Aster tripolium	Briens and Larher, 1982
		Cakile maritima Scott	Jdey et al., 2014
		Cochlearia officinalis	Dauvergne et al., 2006
		Crambe maritima	de Vos et al., 2010
		Eutrema salsugineum	Ghars et al., 2008
		Lepidium crassifolium	Murakeozy et al., 2003
		Disphyma australe	Neales and Sharkey, 1981
		Mesembryanthemum crystallinum	Ishitani et al., 1996
		Sesuvium portulacastrum	Kannan et al., 2013
		Tetragonia tetragonioides	Yousif et al., 2010
		Trianthema portulacastrum	Mohammed and Sen, 1990
		Atriplex halimus	Hassine et al., 2008
		Atriplex hastata	Tipirdamaz et al., 2006
		Halocnemum strobilaceum	Tipirdamaz et al., 2006
		Kochia prostrata	Karimi et al., 2005
		Kochia sieversiana	Yang et al., 2007
		Noaea mucronata	Tipirdamaz et al., 2006
		Salsola ikonnikovii	Xing et al., 2013
		Suaeda fruticosa	Hameed et al., 2012
		Suaeda glauca	Yang et al., 2008
		Limoniastrum	Hanson et al., 1994
		Limonium aureum, L. latifolium	Hanson et al., 1991
		Limonium latifolium	Gagneul et al., 2007
		Portulaca oleracea	Yazici et al., 2007
		Casuarina equisetifolia	Tani et al., 2006
		Bruguiera parviflora	Parida et al., 2002
		Melaleuca cuticularis	Carter et al., 2005
		Plantago coronopus	Koyro, 2006

TABLE 18.1 *(Continued)*

Sr. No.	Osmolyte	Name of the halophyte	Reference
		Plantago maritima	Tipirdamaz et al., 2006
		Sesbania	Gorham et al., 1984
7	Glycine	*Cymodocea nodosa*	Sandoval-Gil et al., 2012
		Posidonia oceanica	Sandoval-Gil et al., 2012
8	Sucrose	*Scirpus maritimus*	Briens and Larher, 1982
		Juncus maritimus	Briens and Larher, 1982
		Distichlis spicata	Marcum, 1999
		Hordeum marinum	Garthwaite et al., 2005
		Phragmites australis	Briens and Larher, 1982
		Panicum hemitomon	Hester et al., 2001
		Spartina patens	Hester et al., 2001
		Thinopyrum bessarabicum	Gorham et al., 1985
		Triglochin maritima	Briens and Larher, 1982
		Posidonia oceanica	Sandoval-Gil et al., 2012
		Zostera	Ye et al., 2003
		Aster tripolium	Briens and Larher, 1982
		Cochlearia officinalis	Dauvergne et al., 2006
		Lepidium crassifolium	Murakeozy et al., 2003
		Limonium latifolium	Gagneul et al., 2007
		Bruguiera parviflora	Parida et al., 2002
9	Fructose	*Phragmites australis*	Briens and Larher, 1982
		Triglochin maritima	Briens and Larher, 1982
		Sarcocornia fruticosa	Gil et al., 2011
10	Raffinose	*Phragmites australis*	Briens and Larher, 1982
11	Maltose	*Triglochin maritima*	Briens and Larher, 1982
		Atriplex hastata	Tipirdamaz et al., 2006
12	Myo-inositol	*Porteresia coarctata*	Sengupta et al., 2008
		Aster tripolium	Briens and Larher, 1982
		Mesembryanthemum crystallinum	Ishitani et al., 1996
		Sesuvium portulacastrum	Kannan et al., 2013
13	Pinitol	*Porteresia coarctata*	Sengupta et al., 2008
		Mesembryanthemum crystallinum	Ishitani et al., 1996
		Sesuvium portulacastrum	Kannan et al., 2013
		Camphorosma annua	Murakeozy et al., 2003
		Limonium gmelini	Murakeozy et al., 2003
		Sesbania	Gorham et al., 1984

TABLE 18.1 *(Continued)*

Sr. No.	Osmolyte	Name of the halophyte	Reference
14	Inositol	*Cymodocea nodosa*	Sandoval-Gil et al., 2012
15	Ononitol	*Mesembryanthemum crystallinum*	Ishitani et al., 1996
		Sesuvium portulacastrum	Lokhande et al., 2010
16	Chiro-inositol	*Limonium latifolium*	Gagneul et al., 2007
17	Mannitol	*Laguncularia racemosa*	Polania, 1990
		Parachlorella kessleri	Rathod et al., 2022
18	Sorbitol	*Plantago coronopus*	Koyro, 2006
		Plantago maritima	Briens and Larher, 1982
		Plantago maritima	Tipirdamaz et al., 2006
19	Dimethylsulfonio-propionate	*Spartina alterniflora*	Li et al., 2010
		Spartina anglica	Mulholland and Otte, 2000
20	Methylated onium compounds	*Atriplex hastata*	Tipirdamaz et al., 2006
		Atriplex (Halimione) portulacoides	Adrian-Romero et al., 1998
		Beta vulgaris	Subbarao et al., 2001
		Salicornia europaea	Tipirdamaz et al., 2006
		Suaeda macrocarpa	Briens and Larher, 1982
21	Flavonoids	*Sesuvium portulacastrum* (L.) L.	Nikalje et al., 2018b
22	Glycolipids	*Sesuvium portulacastrum* (L.) L.	Nikalje et al., 2018b

Proteomics allows us to comprehend proteins and their changes in ways that gene expression analysis does not (Zhao et al., 2013; Edwards et al., 2014). Figure 18.2 and Table 18.2 highlight the major proteins involved in salt tolerance via diverse mechanisms discovered in research investigations in various plant species and functions of stress mitigating proteins, respectively.

Stomatal conductance reduces in response to salinity because of the salinity sensitive nature of stomatal aperture, causing reduced leaf intracellular CO^2 concentrations and photosynthesis (Centritto et al., 2003; Redondo-Gomez et al., 2007). Proteomic study indicated that under salt stress, numerous photosynthetic proteins differed in their expression. For light-induced water oxidation taking place in PSII, the significant role of oxygen-evolving enhancer proteins (OEE1-3) has been previously studied (Sugihara et al., 2000). Under salt stress, the abundance of proteins involved in Calvin cycle proteins like stomatal Rubisco involved in fixation of CO^2 and a related Rubisco-binding protein rose considerably. (Komatsu et al., 2009). Under salt stress, plants require a significant amount of energy, generated through metabolism of carbohydrates through glycolysis (EMP

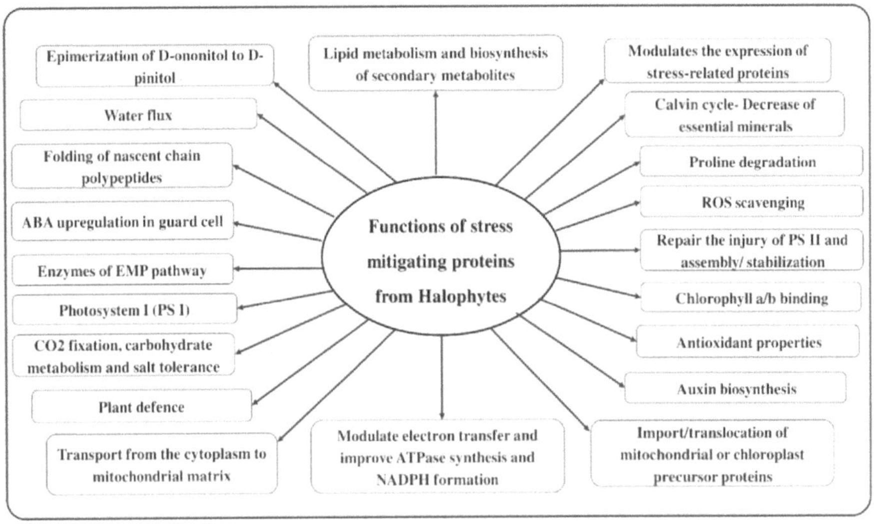

FIGURE 18.2 Various functions of stress mitigating proteins in halophytes.

pathway) and the Krebs cycle (Hoper et al., 2006). In *Kandelia candel*, EMP enzymes like phosphoglycerate mutase (PGAM) were elevated up to 600 mM concentration of NaCl-induced salt stress, whereas fructokinase-1 (FRK) was most abundant at 450 mM NaCl (Wang et al., 2013). Through catalysis of phosphate transfer to fructose, FRK is the most significant gatekeeper in the sugar inflow management into EMP pathway (Hoper et al., 2006). In addition, the levels of three TCA cycle enzyme proteins, malate dehydrogenase (MD), succinate dehydrogenase (SD), and aconitate hydratase (AH), were enhanced. An enhanced expression of several enzymes of EMP-TCA pathways, including FRK is responsible for greater energy production under salt stress through breakdown of glucose. Upregulation of these proteins under high salt in *Puccinellia tenuiflora* (Yu et al., 2011) and *K. candel* (Zhao et al., 2013) has been reported. For *A. lagopoides* (Sobhanian et al., 2010) and the gametophyte of *P. patens* (Wang et al., 2008), enolase, phosphoglycerate kinase (PGK), glyceraldehyde-3-phosphate dehydrogenase (GAPDH), and triosephosphate isomerase (TPI) have been found to be upregulated. For halophytes in salt stress, these results have shown that the coupling of upregulated EMP-TCA enzymes with ATPase activity for enhancing respiratory metabolism through energy remobilization. Together with the increase in photosynthetic light reactions, these mechanisms ensure that accessory molecules like NADPH and ATP are continuously created to regulate the salinity tolerance in halophytes (Zhao et al., 2013).

TABLE 18.2 List of Major Functional Proteins of Halophytes Involved in Salt Tolerance.

Sr. No.	Functional proteins	Plant species	Function	Reference
1	IMT1	*Mesembryanthemum crystallinum*	D-ononitol, which is further epimerized to D-pinitol	Bohnert et al., 1994
2	Major Intrinsic Proteins (MIPs) OEE	*M. crystallinum*	Involved in water flux, repair the injury of the PSII complex	Yamada et al., 1995
		Bruguiera gymnorrhiza		Tada and Kashimura, 2009
3	HSP70	*Aeluropus lagopoides*	Repair the injury of the PSII complex, folding of nascent chain polypeptides, Import/translocation of mitochondrial or chloroplast precursor proteins	Sobhanian et al., 2010
		Puccinellia tenuiflora		Yu et al., 2011
		B. gymnorrhiza		Tada and Kashimura, 2009
		Tangut nitraria		Cheng et al., 2015
		A. lagopoides		Sobhanian et al., 2010
4	HSP60	*Physcomitrella patens*	Transport of proteins from the cytoplasm to mitochondrial matrix, refolding of proteins	Wang et al., 2008
5	PGAM	*Kandelia candel*	Enzymes of EMP pathway	Wang et al., 2014
6	FRK	*Kandelia candel*	Enzymes of EMP pathway	Wang et al., 2014
7	Rubisco LSU and RCA	*Salicornia europaea*	ROS scavenging, Calvin cycle	Wang et al., 2009; Fan et al., 2011
		Porteresia coarctata	Calvin cycle	Sengupta and Majumder, 2009
		Porteresia tenuiflora	ROS scavenging, Calvin cycle	Yu et al., 2011
8	CAB	*A.lagopoides*	Photosystem I (PS I)	Sobhanian et al., 2010
		Cakilemaritima	Photosystem I (PS I)	Debeza et al., 2012
		Porteresia patens	Photosystem I (PS I)	Wang et al., 2008
		T. nitraria	Photosystem I (PS I)	Cheng et al., 2015
9	Proline dehydrogenase (PDH), D2 protein	*Thellungiella halophile*	Proline degradation	Wang et al., 2013
		Suaeda aegyptiaca	Photosystem II (PS II)	Askari et al., 2006
10	D1	*P. tenuiflora*	Photosystem II (PS II)	Sobhanian et al., 2010
		A. lagopoides	PSII assembly/ stabilization	Yu et al., 2011

TABLE 18.2 *(Continued)*

Sr. No.	Functional proteins	Plant species	Function	Reference
11	APX, Fe-SOD, PHGPX light harvesting complex (LHC) CABs	*Halogeton glomeratus*	Antioxidative	Wang et al., 2014
		K.candel	Photosystem I (PS I)	Wang et al., 2013
		P. tenuiflora	Photosystem I (PS I)	Yu et al., 2011
		P. coarctata	Chlorophyll a/b binding	Sengupta and Majumder, 2009
		P. patens	Chlorophyll a/b binding	Wang et al., 2008
		T. nitraria	Photosystem I (PS I)	Cheng et al., 2015
12	SOD	*S. europaea*	Conversion of H^2O^2 to H^2O and O^2	Wang et al., 2009
		K. candel	Antioxidant properties	Wang et al., 2013
		T. nitraria	Antioxidant properties	Cheng et al., 2015
		P. tenuiflora	Antioxidant properties	Yu et al., 2011
13	ATP synthase subunit α (*atpA*)	*Salicornia brachiata*	ATP synthesis	Benjamin et al., 2020
14	PFP (*PFP*)	*Hordeum vulgare*	CO2 fixation, carbohydrate metabolism and salt tolerance	Mostek et al., 2015
15	Ferredoxin-1 (FDX1)	*Halogeton glomeratus*	Electron transfer-related protein that modulates electron transfer effectively and improve ATPase synthesis and NADPH formation	Wang et al., 2016a, 2016b
16	Indole-3-acetaldehyde oxidase (*AAO1*)	*Salicornia brachiata*	Involve in auxin biosynthesis, and involve in salinity tolerance	Benjamin et al., 2020
17	Chaperonin CPN60-1 (CPN60-1)	*Salicornia brachiata*	Defense-related proteins	Benjamin et al., 2020
18	Desiccation-responsive protein 29B	*Chenopodium quinoa*	ABA upregulation in guard cell	Rasouli et al., 2021

TABLE 18.2 *(Continued)*

Sr. No.	Functional proteins	Plant species	Function	Reference
19	Osmotin-like protein OSML13	*Chenopodium quinoa*	ABA upregulation in guard cell	Rasouli et al., 2021
20	Polycystin-1, Lipoxygenase, Alpha-toxin, Triacylglycerol lipase (PLAT)	*Chenopodium quinoa*	ABA upregulation in guard cell	Rasouli et al., 2021
21	Dehydrin	*Chenopodium quinoa*	ABA upregulation in guard cell	Rasouli et al., 2021
22	Differentially accumulated proteins (DAPs)	*Sesuvium portulacastrum*	Involved in photosynthesis, carbohydrate and energy metabolism, lipid metabolism, and biosynthesis of secondary metabolites.	Ding et al., 2022
23	McHB7	*Mesembryanthemum crystallinum*	Modulates the expression of stress-related proteins	Zhang et al., 2022
24	RING-type E3 ligases	*Sesuvium verrucosum*	Possible involvement in salinity stress	Thayale Purayil et al., 2022

Salinity first produces osmotic stress and then affects nutritional balance (Munns and Tester, 2008). Reactive oxygen species (ROS) production is enhanced in both circumstances due to the breakdown of coordination across distinct metabolic pathways (Asada, 2006). ROS disrupts ion homeostasis within the cell, by directly activating non-selective cationic channels (NSCC), resulting in extra K^+ efflux in addition to sodium-induced depolarization of membrane (Demidchik et al., 2010). K^+ efflux due to NSCC activation causes a decrease in cytoplasmic K^+ concentration, which then activates endonucleases and caspase-like proteases causing apoptosis (Shabala, 2009; Demidchik et al., 2010). ROS oxidize proteins, lipids, and DNA affecting integrity of membrane, activity of enzymes, and normal operations of photosynthetic apparatus (Yu et al., 2011). ROS are also involved in boosting antioxidative enzymes in order to adapt to high salinity conditions (Abogadallah, 2010; Jaspers and Kangasjarvi, 2010). Plants have evolved a comprehensive antioxidant defence mechanism that include both non-enzymatic and enzymatic and components to deal with excessive build-up of these chemicals. SODs, GST, GPXs, class III peroxidases (PRXs), thioredoxins (Trxs), ascorbate peroxidase (APX), and dehydroascorbate reductase (DHAR) are some of the antioxidant enzymes that contribute to the positive effect of ROS (Yu et al., 2011; Bose et al., 2013; Wang et al., 2013). SODs are presumably among the first line of defence against the damage caused by free radicals during diverse stressors. SOD accumulation under salinity stress has been shown to protect wild species of tomato (*Solanum chilense*), *Salicornia europaea Kandelia candel,* and canola (Wang et al., 2009, 2013). Protein misfolding or unfolding can occur due to salt stress. During salt stress, chaperones like (HSPs), which aid in protein supramolecular assembly, native state folding, and restrict irreversible aggregation of protein by retaining native protein conformations (Sun et al., 2001; Pang et al., 2010). HSPs can alleviate stress in plants by restoring normal protein structure, refolding protein stores and maintaining homeostasis inside the cell. Reports provide insights of the roles of HSP70 in the folding of nascent chain polypeptides and the import or translocation of precursor proteins of chloroplast and mitochondria. *Physcomitrella patens, Aeluropus lagopoides,* and *Tangut nitraria* and have all been shown to use HSP70 under stress from excess salt (Wang et al., 2008; Cheng et al., 2015; Sobhanian et al., 2010). Also, HSP60, a mitochondrial chaperone, has been shown to be essential for protein transport into the mitochondrial matrix and folding proteins there, preventing protein aggregation when the mitochondria are stressed (Rospert et al., 1996). In *P. patens* under salt stress, the expression levels of HSP60 and a 14-3-3-like protein were found to increase (Wang et al., 2008). 14-3-3 proteins form a family of highly conserved proteins, which are involved in response to diverse stresses, including salinity (Zhao et al., 2013). Through modulation of multiple kinases, phosphatases and thus signaling pathways, 14-3-3 proteins are involved in salinity stress response in higher plants. 14-3-3 complexes are involved in regulating cytoplasmic pH, ion transport, cellular and subcellular defence mechanisms, signal transmission, and primary metabolism (Finnie et al., 1999). 14-3-3 proteins and phototropin in tandem positively regulate membrane-bound H+ ATPase activity and allow opening of ion channels and stomata; hence, are found to be upregulated in *P. patens* (Wang et al., 2008) and *Tangut nitraria* (Cheng et al., 2015) under saline conditions. CRT, a calcium-binding protein involved in Ca signaling in the ER under salty conditions is upregulated (Persson et al., 2003). Also, some studies suggest that in diverse halophytes, CRT upregulation may perform a key a role in signal transduction (Wang et al., 2008, 2013; Sobhanian et al., 2011; Cheng et al., 2015).

18.6 PERSPECTIVE, CONCLUSION, AND FUTURE PROSPECTS

Salinity is induced by natural processes and is augmented by anthropogenic activities, making soil water absorption difficult. Each year, close to 2% of irrigated agricultural land is lost to salinity (Hasegawa et al., 2000). Salinity is a global issue that costs billion-dollar losses each year. Salinity exerts its effects mainly by causing osmotic stress, ion specific toxicity, and oxidative stress, as well as nutrient imbalance. Tissue-specific toxicity leads to overproduction of ROS, resulting in oxidative damage. However, plants regulate salinity through, genetic- and hormonal controls as well as through the synthesis of stress-responsive proteins and osmolytes.

Salinity also exerts nutritional deficiency stress on the plant. Studies have shown that under adequate nutritional conditions, plants show better adaptations to salinity than nutrient deficient conditions. Genomic, proteomic, and transcriptomic studies have highlighted the role of ion transporters and antiporters in ionic homeostasis. Recent studies have expressed potential candidate genes in transgenic organisms to estimate their efficacy of salt tolerance in non-native conditions with respect to habitat and genetic environment. In general, transgenic organisms have shown better growth and development in higher concentrations of salinity as compared to their native counterparts. Another possible strategy evolved because of differential expression of several genes in halophytes and glycophytes, is the overexpression of such genes in glycophytes which also helps to enhance salt tolerance of these species. Upregulation of genes and enzymes involved in various metabolic pathways has also shown promising results with respect to salt tolerance. Plants are known to produce diverse osmolytes in various concentrations depending upon external levels of salinity. Studies overexpressing genes/enzymes involved in the synthesis of such osmolytes and external application of osmolyte through soil or nutrient medium also are known to supplement salt tolerance via mechanisms of ROS scavenging, protection of membrane, and vital organelles from oxidative damage. Wild varieties of crop plants are known to be more stress tolerant than their domestic counterparts. Inter-comparative studies among wild and domestic varieties for delineating morphophysiological, biochemical, and genetic determinants of salt tolerance are needed to be targeted. By comparison of transcriptomic differences and differential gene expression between halophytes and glycophytes, researchers can better understand mechanisms for salt tolerance. The promising results from such investigations may significantly enhance the chances of the developing novel varieties through mutagenic and transgenic approaches. For future perspectives, strategies may include targeting redox mechanisms, selective breeding for salt-resistant varieties, use of genetic transformations, application of molecular markers, nutrient management, as well as the application of PGRs.

KEYWORDS

- **halophytes**
- **salt stress**
- **tolerance**
- **osmolytes**
- **salt-responsive proteins**
- **salinity mitigation**

REFERENCES

Abogadallah, G. M. Insights into the Significance of Antioxidative Defense Under Salt Stress. *Plant Signal. Behav.* **2010,** *5* (4), 369–374.

Acosta-Motos, J. R.; Ortuño, M. F.; Bernal-Vicente, A.; Diaz-Vivancos, P.; Sanchez-Blanco, M. J.; Hernandez, J. A. Plant Responses to Salt Stress: Adaptive Mechanisms. *Agronomy* **2017,** *7* (1), 18.

Adrian-Romero, M.; Wilson, S. J.; Blunden, G.; Yang, M. H.; Carabot-Cuervo, A.; Bashir, A. K. Betaines in Coastal Plants. *Biochem. Syst. Ecol.* **1998,** *26* (5), 535–543.

Ahmad, M. S. A.; Ashraf, M.; Ali, Q. Soil Salinity as a Selection Pressure is a Key Determinant for the Evolution of Salt Tolerance in Blue Panicgrass (*Panicum antidotale* Retz.). *Flora Morphol. Distrib. Funct. Ecol. Plants* **2010,** *205* (1), 37–45.

Ahmad, I.; Larher, F.; Stewart, G. The Accumulation of Delta-acetylornithine and Other Solutes in the Salt Marsh Grass *Puccinellia maritima. Phytochemistry* **1981,** *20* (7), 1501–1504.

Albert, R.; Popp, M. Chemical Composition of Halophytes from the Neusiedler Lake Region in Austria. *Oecologia* **1977,** *27* (2), 157–170.

Arnold, A. The Importance of Chloride Ions for the Plant. Gustav Fischer Verlag: Jena, 1955; p 148.

Asada, K. Production and Scavenging of Reactive Oxygen Species in Chloroplasts and their Functions. *Plant Physiol.* **2006,** *141* (2), 391–396.

Ashraf, M. H. P. J. C.; Harris, P. J. Photosynthesis under Stressful Environments: An Overview. *Photosynthetica* **2013,** *51* (2), 163–190.

Askari, H.; Edqvist, J.; Hajheidari, M.; Kafi, M.; Salekdeh, G. H. Effects of Salinity Levels on Proteome of *Suaeda aegyptiaca* Leaves. *Proteomics* **2006,** *6* (8), 2542–2554.

Assaha, D. V.; Ueda, A.; Saneoka, H.; Al-Yahyai, R.; Yaish, M. W. The Role of Na+ and K+ Transporters in Salt Stress Adaptation in Glycophytes. *Front. Physiol.* **2017,** *8*, 509.

Bahmani, K.; Noori, S. A. S.; Darbandi, A. I.; Akbari, A. Molecular Mechanisms of Plant Salinity Tolerance: A Review. *Aust. J. Crop Sci.* **2015,** *9* (4), 321–336.

Bates, B.; Kundzewicz, Z.; Wu, S. Climate Change and Water. Technical Paper of the Intergovernmental Panel on Climate Change; IPCC Technical Paper VI; IPCC Secretariat: Geneva, 2008; p 210.

Benjamin, J. J.; Miras-Moreno, B.; Araniti, F.; Salehi, H.; Bernardo, L.; Parida, A.; Lucini, L. Proteomics Revealed Distinct Responses to Salinity between the Halophytes *Suaeda maritima* (L.) Dumort and *Salicornia brachiata* (Roxb). *Plants* **2020,** *9* (2), 227.

Bohnert, H. J.; Thomas, J. C.; DeRocher, E. J.; Michalowski, C. B.; Breiteneder, H.; Vernon, D. M.; ... Jensen, R. G. Responses to Salt Stress in the Halophyte *Mesembryanthemum crystallinum*. In *Biochemical and Cellular Mechanisms of Stress Tolerance in Plants*; Springer: Berlin, Heidelberg, 1994; pp 415–428.

Bose, J.; Rodrigo-Moreno, A.; Shabala, S. ROS Homeostasis in Halophytes in the Context of Salinity Stress Tolerance. *J. Exp. Bot.* **2014,** *65* (5), 1241–1257.

Briens, M.; Larher, F. Osmoregulation in Halophytic Higher Plants: A Comparative Study of Soluble Carbohydrates, Polyols, Betaines and Free Proline. *Plant Cell Environ.* **1982,** *5* (4), 287–292.

Byrt, C. S.; Zhao, M.; Kourghi, M.; Bose, J.; Henderson, S. W.; Qiu, J.; ... Tyerman, S. Non-selective Cation Channel Activity of Aquaporin AtPIP2; 1 Regulated by Ca2+ and pH. *Plant Cell Environ.* **2017,** *40* (6), 802–815.

Cabot, C.; Sibole, J. V.; Barceló, J.; Poschenrieder, C. Lessons from Crop Plants Struggling with Salinity. *Plant Sci.* **2014,** *226*, 2–13.

Cao, M.; Xin, P.; Jin, G.; Li, L. A Field Study on Groundwater Dynamics in a Salt Marsh–Chongming Dongtan wetland. *Ecol. Eng.* **2012,** *40*, 61–69.

Carter, J. L.; Colmer, T. D.; Veneklaas, E. J. Variable Tolerance of Wetland Tree Species to Combined Salinity and Waterlogging is Related to Regulation of Ion Uptake and Production of Organic Solutes. *New Phytol.* **2006,** *169* (1), 123–134.

Centritto, M.; Loreto, F.; Chartzoulakis, K. The Use of Low [CO2] to Estimate Diffusional and Non-diffusional Limitations of Photosynthetic Capacity of Salt-Stressed Olive Saplings. *Plant Cell Environ.* **2003,** *26* (4), 585–594.

Cheeseman, J. M. The Evolution of Halophytes, Glycophytes and Crops, and its Implications for Food Security Under Saline Conditions. *New Phytol.* **2015,** *206* (2), 557–570.

Cheng, T.; Chen, J.; Zhang, J.; Shi, S.; Zhou, Y.; Lu, L.; ... Shi, J. Physiological and Proteomic Analyses of Leaves from the Halophyte *Tangut Nitraria* Reveals Diverse Response Pathways Critical for High Salinity Tolerance. *Front. Plant Sci.* **2015,** *6*, 30.

Ciais, P.; Reichstein, M.; Viovy, N.; Granier, A.; Ogée, J.; Allard, V.; ... Valentini, R. Europe-Wide Reduction in Primary Productivity Caused by the Heat and Drought in 2003. *Nature* 2005, *437* (7058), 529–533.

Crimp, S. J.; Zheng, B.; Khimashia, N.; Gobbett, D. L.; Chapman, S.; Howden, M.; Nicholls, N. Recent Changes in Southern Australian Frost Occurrence: Implications for Wheat Production Risk. *Crop Pasture Sci.***2016,** *67* (8), 801–811.

Dauvergne, X. *Taxonomie du genre Cochlearia des cotes bretonnes: morphologie, ecologie, chimiotaxonomie et phylogenie moleculaire* (Doctoral dissertation, Brest), 2006.

De Pascale, S.; Maggio, A.; Orsini, F.; Stanghellini, C.; Heuvelink, E. Growth Response and Radiation use Efficiency in Tomato Exposed to Short-Term and Long-Term Salinized Soils. *Sci. Hortic.* **2015,** *189*, 139–149.

de Vos, A. C.; Broekman, R.; Groot, M. P.; Rozema, J. Ecophysiological Response of *Crambe maritima* to Airborne and Soil-Borne Salinity. *Ann. Bot.* **2010,** *105* (6), 925–937.

Debez, A.; Braun, H. P.; Pich, A.; Taamalli, W.; Koyro, H. W.; Abdelly, C.; Huchzermeyer, B. Proteomic and Physiological Responses of the Halophyte *Cakile maritima* to Moderate Salinity at the Germinative and Vegetative Stages. *J. Proteom.* **2012,** *75* (18), 5667–5694.

Demidchik, V.; Cuin, T. A.; Svistunenko, D.; Smith, S. J.; Miller, A. J.; Shabala, S.; ... Yurin, V. Arabidopsis Root K+-efflux Conductance Activated by Hydroxyl Radicals: Single-Channel Properties, Genetic Basis and Involvement in Stress-Induced Cell Death. *J. Cell Sci.* **2010,** *123* (9), 1468–1479.

Desai, K. N. The Structure and Functions of the Sand Dune Vegetation along the Goa Coast. Doctoral Dissertation, Goa University, 1995.

Ding, G.; Yang, Q.; Ruan, X.; Si, T.; Yuan, B.; Zheng, W.; ... Wang, X. Proteomics Analysis of the Effects for Different Salt Ions in Leaves of True Halophyte *Sesuvium portulacastrum*. *Plant Physiol. Biochem.* **2022,** *170*, 234–248.

D'odorico, P.; He, Y.; Collins, S.; De Wekker, S. F.; Engel, V.; Fuentes, J. D. Vegetation–Microclimate Feedbacks in Woodland–Grassland Ecotones. *Glob. Ecol. Biogeogr.* **2013,** *22* (4), 364–379.

Edwards, A. V.; Edwards, G. J.; Larsen, M. R. A Tool for Mapping and Displaying Spatial Patterns and Relationships between Post-Translational Modifications in Two and Three Dimensions. *J. Proteom. Bioinform.* **2014,** *7* (12), 385.

English, J. P.; Colmer, T. D. Salinity and Waterlogging Tolerances in Three Stem-Succulent Halophytes (*Tecticornia species*) from the Margins of Ephemeral Salt Lakes. *Plant Soil* **2011,** *348* (1), 379–396

Epstein, E.; Norlyn, J. D.; Rush, D. W.; Kingsbury, R. W.; Kelley, D. B.; Cunningham, G. A.; Wrona, A. F. Saline Culture of Crops: A Genetic Approach. *Science* **1980,** *210* (4468), 399–404.

Eshel, A.; Waisel, Y. Variations in Sodium Uptake along Primary Roots of Corn Seedlings. *Plant Physiol.* **1972,** *49* (4), 585–589.

Fan, C. Genetic Mechanisms of Salt Stress Responses in Halophytes. *Plant Signal. Behav.* **2020,** *15* (1), 1704528.
Fan, P.; Feng, J.; Jiang, P.; Chen, X.; Bao, H.; Nie, L.; ... Li, Y. Coordination of Carbon Fixation and Nitrogen Metabolism in *Salicornia europaea* under Salinity: Comparative Proteomic Analysis on Chloroplast Proteins. *Proteomics* **2011,** *11* (22), 4346–4367.
Finnie, C.; Borch, J.; Collinge, D. B. 14-3-3 Proteins: Eukaryotic Regulatory Proteins with many Functions. *Plant Mol. Biol.* **1999,** *40* (4), 545–554.
Flexas, J.; Diaz-Espejo, A.; Galmés, J.; Kaldenhoff, R.; Medrano, H.; Ribas-Carbo, M. Rapid Variations of Mesophyll Conductance in Response to Changes in CO2 Concentration around Leaves. *Plant Cell Environ.* **2007,** *30* (10), 1284–1298.
Flowers, T. J.; Colmer, T. D. Salinity Tolerance in Halophytes. *New Phytol.* **2008,** 945–963.
Flowers, T. J.; Galal, H. K.; Bromham, L. Evolution of Halophytes: Multiple Origins of Salt Tolerance in Land Plants. *Funct. Plant Biol.* **2010,** *37* (7), 604–612.
Flowers, T. J.; Munns, R.; Colmer, T. D. Sodium Chloride Toxicity and the Cellular Basis of Salt Tolerance in Halophytes. *Ann. Bot.* **2015,** *115* (3), 419–431.
Flowers, T. J.; Troke, P. F.; Yeo, A. R. The Mechanism of Salt Tolerance in Halophytes. *Ann. Rev. Plant Physiol.* **1977,** *28* (1), 89–121.
Gagneul, D.; Aïnouche, A.; Duhazé, C.; Lugan, R.; Larher, F. R.; Bouchereau, A. A Reassessment of the Function of the So-called Compatible Solutes in the Halophytic Plumbaginaceae *Limonium latifolium*. *Plant Physiol.* **2007,** *144* (3), 1598–1611.
Garthwaite, A. J.; von Bothmer, R.; Colmer, T. D. Salt Tolerance in Wild Hordeum Species is Associated with Restricted Entry of Na+ and Cl− into the Shoots. *J. Exp. Bot.* **2005,** *56* (419), 2365–2378.
Ghars, M. A.; Parre, E.; Debez, A.; Bordenave, M.; Richard, L.; Leport, L.; ... Abdelly, C. Comparative Salt Tolerance Analysis between *Arabidopsis thaliana* and *Thellungiella halophila*, with Special Emphasis on K+/Na+ Selectivity and Proline Accumulation. *J. Plant Physiol.* **2008,** *165* (6), 588–599.
Gil Ortiz, R.; Lull Noguera, C.; Boscaiu Neagu, M. T.; Bautista Carrascosa, I.; Lidón Cerezuela, A. L.; Vicente Meana, Ó. Soluble Carbohydrates as Osmolytes in Several Halophytes from a Mediterranean Salt Marsh. *Notulae Botanicae Horti Agrobotanici Cluj-Napoca* **2011,** *39* (2), 9–17.
Gill, M. S.; Pal, S. S.; Ahlawat, I. P. S. Approaches for Sustainability of Rice (*Oryza sativa*)-Weat (*Triticum aestivum*) Cropping System in Indo-Gangetic Plains of India- A Review. *Indian J. Agron.* **2008,** *53*, 81–96.
Giuffrida, F.; Scuderi, D.; Giurato, R.; Leonardi, C. In *Physiological Response of Broccoli and Cauliflower as Affected by NaCl Salinity*, VI International Symposium on Brassicas and XVIII Crucifer Genetics Workshop 1005, 2012; pp 435–441.
Gorham, J.; McDonnell, E.; Budrewicz, E.; Jones, R. W. Salt Tolerance in the Triticeae: Growth and Solute Accumulation in Leaves of *Thinopyrum bessarabicum*. *J. Exp. Bot.* **1985,** *36* (7), 1021–1031.
Grigore, M. N.; Toma, C. Morphological and Anatomical Adaptations of Halophytes: A Review. In *Handbook of Halophytes: From Molecules to Ecosystems towards Biosaline Agriculture*, 2021; pp 1079–1221.
Guy, R. D.; Warne, P. G.; Reid, D. M. Glycinebetaine Content of Halophytes: Improved Analysis by Liquid Chromatography and Interpretations of Results. *Physiol. Plant.* **1984,** *61* (2), 195–202.

Hameed, A.; Hussain, T.; Gulzar, S.; Aziz, I.; Gul, B.; Khan, M. A. Salt Tolerance of a Cash Crop Halophyte *Suaeda fruticosa*: Biochemical Responses to Salt and Exogenous Chemical Treatments. *Acta Physiol. Plant.* **2012,** *34* (6), 2331–2340.

Hanin, M.; Ebel, C.; Ngom, M.; Laplaze, L.; Masmoudi, K. New Insights on Plant Salt Tolerance Mechanisms and their Potential use for Breeding. *Front. Plant Sci.* **2016,** *7*, 1787.

Hanson, A. D.; Rathinasabapathi, B.; Chamberlin, B.; Gage, D. A. Comparative Physiological Evidence that β-alanine Betaine and Choline-O-sulfate Act as Compatible Osmolytes in Halophytic *Limonium species*. *Plant Physiol.* **1991,** *97* (3), 1199–1205.

Hanson, A. D.; Rathinasabapathi, B.; Rivoal, J.; Burnet, M.; Dillon, M. O.; Gage, D. A. Osmoprotective Compounds in the Plumbaginaceae: A Natural Experiment in Metabolic Engineering of Stress Tolerance. *Proc. Natl. Acad. Sci.* **1994,** *91* (1), 306–310.

Harris, L. C.; Khan, M. A.; Zou, J.; Smith, B. N.; Hansen, L. D. In *Effects of Salinity and Temperature on Respiratory Metabolism of Salicornia utahensis from a Great Basin Playa*, Shrubland Ecosystem Genetics and Biodiversity; McArthur, E. D., Fairbanks, D. J., comps; June 13–15, 2000; Provo, UT. Proc. RMRS-P-21; US Department of Agriculture, Forest Service, Rocky Mountain Research Station, Ogden, UT, 2001; pp 265–270.

Hasanuzzaman, M.; Nahar, K.; Alam, M.; Bhowmik, P. C.; Hossain, M.; Rahman, M. M.; ... Fujita, M. Potential use of Halophytes to Remediate Saline Soils. *BioMed Res. Int.* 2014, *2014*.

Hasegawa, P. M.; Bressan, R. A.; Zhu, J. K.; Bohnert, H. J. Plant Cellular and Molecular Responses to High Salinity. *Ann. Rev. Plant Biol.* **2000,** *51* (1), 463–499.

Hassine, A. B.; Ghanem, M. E.; Bouzid, S.; Lutts, S. An Inland and a Coastal Population of the Mediterranean Xero-Halophyte Species *Atriplex halimus* L. Differ in their Ability to Accumulate Proline and Glycinebetaine in Response to Salinity and Water Stress. *J. Exp. Bot.* **2008,** *59* (6), 1315–1326.

Hester, M. W.; Mendelssohn, I. A.; McKee, K. L. Species and Population Variation to Salinity Stress in *Panicum hemitomon*, *Spartina patens*, and *Spartina alterniflora*: Morphological and Physiological Constraints. *Environ. Exp. Bot.* **2001,** *46* (3), 277–297.

Himabindu, Y.; Chakradhar, T.; Reddy, M. C.; Kanygin, A.; Redding, K. E.; Chandrasekhar, T. Salt-Tolerant Genes from Halophytes are Potential Key Players of Salt Tolerance in Glycophytes. *Environ. Exp. Bot.* **2016,** *124*, 39–63.

Holman, J. D.; Schlegel, A. J.; Thompson, C. R.; Lingenfelser, J. E. Influence of Precipitation, Temperature, and 56 Years on Winter Wheat Yields in Western Kansas. *Crop Manag.* **2011,** *10* (1), 1–10.

Hoper, D.; Bernhardt, J.; Hecker, M. Salt Stress Adaptation of *Bacillus subtilis*: A Physiological Proteomics Approach. *Proteomics* **2006,** *6* (5), 1550–1562.

Ibrahimova, U.; Kumari, P.; Yadav, S.; Rastogi, A.; Antala, M.; Suleymanova, Z.; ... Brestic, M. Progress in Understanding Salt Stress Response in Plants using Biotechnological Tools. *J. Biotechnol.* **2021,** *329*, 180–191.

Ishitani, M.; Majumder, A. L.; Bornhouser, A.; Michalowski, C. B.; Jensen, R. G.; Bohnert, H. J. Coordinate Transcriptional Induction of Myo-inositol Metabolism during Environmental Stress. *Plant J.* **1996,** *9* (4), 537–548.

Jaspers, P.; Kangasjarvi, J. Reactive Oxygen Species in Abiotic Stress Signaling. *Physiol. Plant.* **2010,** *138* (4), 405–413.

Jdey, A.; Slama, I.; Rouached, A.; Abdelly, C. Growth, Na+, K+, Osmolyte Accumulation and Lipid Membrane Peroxidation of Two Provenances of *Cakile maritima* During Water Deficit Stress and Subsequent Recovery. *Flora Morphol. Distrib. Funct. Ecol. Plants* **2014,** *209* (1), 54–62.

Kalinina, E. B.; Keith, B. K.; Kern, A. J.; Dyer, W. E. Salt-and Osmotic Stress-Induced Choline Monooxygenase Expression in *Kochia scoparia* is ABA-Independent. *Biol. Plant.* **2012,** *56* (4), 699–704.
Kannan, P. R.; Deepa, S.; Kanth, S. V.; Rengasamy, R. Growth, Osmolyte Concentration and Antioxidant Enzymes in the Leaves of *Sesuvium portulacastrum* L. under Salinity Stress. *Appl. Biochem. Biotechnol.* **2013,** *171* (8), 1925–1932.
Kapler, A. Habitats of Halophytes. In *Halophytes and Climate Change: Adaptive Mechanisms and Potential Uses*; Fujita, M., Hasanuzzaman, M., Shabala, S., Eds.; 2019; pp 19–37.
Karimi, G.; Ghorbanli, M.; Heidari, H.; Khavari Nejad, R. A.; Assareh, M. H. The Effects of NaCl on Growth, Water Relations, Osmolytes and Ion Content in *Kochia prostrata. Biol. Plant.* **2005,** *49* (2), 301–304.
Katschnig, D.; Broekman, R.; Rozema, J. Salt Tolerance in the Halophyte *Salicornia dolichostachya* Moss: Growth, Morphology and Physiology. *Environ. Exp. Bot.* **2013,** *92*, 32–42.
Khan, M. A.; Ungar, I. A.; Showalter, A. M. Effects of Salinity on Growth, Ion Content, and Osmotic Relations in *Halopyrum mucronatum* (L.) Stapf. *J. Plant Nutr.* **1999,** *22* (1), 191–204.
Khan, M. A.; Ungar, I. A.; Showalter, A. M.; Dewald, H. D. NaCl-induced Accumulation of Glycine Betaine in Four Subtropical Halophytes from Pakistan. *Physiol. Plant.* **1998,** *102* (4), 487–492.
Khan, M. A.; Ungar, I. A.; Showalter, A. M. Effects of Salinity on Growth, Water Relations and Ion Accumulation of the Subtropical Perennial Halophyte, *Atriplex griffithii* var. *stocksii*. *Ann. Bot.* **2000,** *85* (2), 225–232.
Kibria, M. G.; Hossain, M.; Murata, Y.; Hoque, M. A. Antioxidant Defense Mechanisms of Salinity Tolerance in Rice Genotypes. *Rice Sci.* **2017,** *24* (3), 155–162.
Komatsu, S.; Yamada, E.; Furukawa, K. Cold Stress Changes the Concanavalin A-Positive Glycosylation Pattern of Proteins Expressed in the Basal Parts of Rice Leaf Sheaths. *Amino Acids* **2009,** *36* (1), 115–123.
Koyro, H. W. Effect of Salinity on Growth, Photosynthesis, Water Relations and Solute Composition of the Potential Cash Crop Halophyte *Plantago coronopus* (L.). *Environ. Exp. Bot.* **2006,** *56* (2), 136–146.
Kucharik, C. J.; Serbin, S. P. Impacts of Recent Climate Change on Wisconsin Corn and Soybean Yield Trends. *Environ. Res. Lett.* **2008,** *3* (3), 034003.
Kumar, A.; Mann, A.; Kumar, A.; Devi, S. Potential and Role of Halophyte Crops in Saline Environments. In *Engineering Practices for Management of Soil Salinity*; Apple Academic Press, 2018; pp 379–416.
Kumar, P.; Sharma, P. K. Soil Salinity and Food Security in India. *Front. Sustain. Food Syst.* **2020,** *4*, 533781. DOI: 10.3389/fsufs.2020.533781.
Kumari, A.; Das, P.; Parida, A. K.; Agarwal, P. K. Proteomics, Metabolomics, and Ionomics Perspectives of Salinity Tolerance in Halophytes. *Front. Plant Sci.* **2015,** *6*, 537.
Larher, F. R.; Lugan, R.; Gagneul, D.; Guyot, S.; Monnier, C.; Lespinasse, Y.; Bouchereau, A. A Reassessment of the Prevalent Organic Solutes Constitutively Accumulated and Potentially Involved in Osmotic Adjustment in Pear Leaves. *Environ. Exp. Bot.* **2009,** *66* (2), 230–241.
Lauchli, A.; Epstein, E. Plant Responses to Saline and Sodic Conditions. In *Agricultural Salinity Assessment and Management*, 1990; vol *71*, pp 113–137.

Lawton, J. R.; Todd, A. N. N.; Naidoo, D. K. Preliminary Investigations into the Structure of the Roots of the Mangroves, *Avicennia marina* and *Bruguiera gymnorrhiza*, in Relation to Ion Uptake. *New Phytol.* **1981,** *88* (4), 713–722.

Lee, G.; Carrow, R. N.; Duncan, R. R. Identification of New Soluble Sugars Accumulated in a Halophytic Seashore *Paspalum* Ecotype Under Salinity Stress. *Hortic. Environ. Biotechnol.* **2008,** *49* (1), 13–19.

Li, R.; Shi, F.; Fukuda, K. Interactive Effects of Various Salt and Alkali Stresses on Growth, Organic Solutes, and Cation Accumulation in a Halophyte *Spartina alterniflora* (Poaceae). *Environ. Exp. Bot.* **2010,** *68* (1), 66–74.

Lokhande, V. H.; Nikam, T. D.; Penna, S. Biochemical, Physiological and Growth Changes in Response to Salinity in Callus Cultures of *Sesuvium portulacastrum* L. *Plant Cell Tissue Organ Cult.* **2010,** *102* (1), 17–25.

Manigbas, N. L.; Sebastian, L. S. In *Breeding for High Temperature Tolerance in the Philippines*, Proceedings of the International Workshop on Cool Rice for a Warmer World; Huazhong Agricultural Univeristy: Wuhan, Hubei, China, 2007.

Marcelis, L. F. M.; Van Hooijdonk, J. Effect of Salinity on Growth, Water use and Nutrient use in Radish (*Raphanus sativus* L.). *Plant Soil* **1999,** *215* (1), 57–64.

Marcum, K. B. Salinity Tolerance Mechanisms of Grasses in the Subfamily Chloridoideae. *Crop Sci.* **1999,** *39* (4), 1153–1160.

Marcum, K. B.; Murdoch, C. L. Salt Tolerance of the Coastal Salt Marsh Grass, Sporobolus virginicus (L.) Kunth. *New Phytol.* **1992,** *120* (2), 281–288.

Marcum, K. B.; Murdoch, C. L. Salinity Tolerance Mechanisms of Six C4 Turfgrasses. *J. Am. Soc. Hortic. Sci.* 1994, *119* (4), 779–784.

Matoh, T.; Watanabe, J.; Takahashi, E. Sodium, Potassium, Chloride, and Betaine Concentrations in Isolated Vacuoles from Salt-Grown *Atriplex gmelini* Leaves. *Plant Physiol.* **1987,** *84* (1), 173–177.

Moghaieb, R. E.; Saneoka, H.; Fujita, K. Effect of Salinity on Osmotic Adjustment, Glycine Betaine Accumulation and the Betaine Aldehyde Dehydrogenase Gene Expression in Two Halophytic Plants, *Salicornia europaea* and *Suaeda maritima. Plant Sci.* **2004,** *166* (5), 1345–1349.

Mohammed, S.; Sen, D. N. Biology and ecophysiology of *Trianthema portulacastrum* L. (Molluginaceae) in arid ecosystem. In *Folia Geobotanica et Phytotaxonomica*, 1990; vol *25* (2), pp 145–157.

Mostek, A.; Börner, A.; Badowiec, A.; Weidner, S. Alterations in Root Proteome of Salt-Sensitive and Tolerant Barley Lines Under Salt Stress Conditions. *J. Plant Physiol.* **2015,** *174*, 166-–176.

Muchate, N. S.; Nikalje, G. C.; Rajurkar, N. S.; Suprasanna, P.; Nikam, T. D. Plant Salt Stress: Adaptive Responses, Tolerance Mechanism and Bioengineering for Salt Tolerance. *Bot. Rev.* **2016a,** *82* (4), 371–406.

Muchate, N. S.; Nikalje, G. C.; Rajurkar, N. S.; Suprasanna, P.; Nikam, T. D. Physiological Responses of the Halophyte *Sesuvium portulacastrum* to Salt Stress and their Relevance for Saline Soil Bio-Reclamation. *Flora* **2016b,** *224*, 96–105.

Mulholland, M. M.; Otte, M. L. Effects of Varying Sulphate and Nitrogen Supply on DMSP and Glycine Betaine Levels in *Spartina anglica. J. Sea Res.* **2000,** *43* (3-4), 199–207.

Munns, R.; Husain, S.; Rivelli, A. R.; James, R. A.; Condon, A. G.; Lindsay, M. P.; ... Hare, R. A. Avenues for Increasing Salt Tolerance of Crops, and the Role of Physiologically Based

Selection Traits. In *Progress in Plant Nutrition: Plenary Lectures of the XIV International Plant Nutrition Colloquium*; Springer: Dordrecht, 2002; pp 93–105.

Munns, R.; Tester, M. Mecanismos de tolerância à salinidade. *Ann. Rev. Plant Biol.* **2008,** *59*, 651–681.

Murakeozy, E. P.; Nagy, Z.; Duhaze, C.; Bouchereau, A.; Tuba, Z. Seasonal Changes in the Levels of Compatible Osmolytes in Three Halophytic Species of Inland Saline Vegetation in Hungary. *J. Plant Physiol.* **2003,** *160* (4), 395–401.

Naidoo, G.; Somaru, R.; Achar, P. Morphological and Physiological Responses of the Halophyte, *Odyssea paucinervis* (Staph)(Poaceae), to salinity. *Flora Morphol. Distrib. Funct. Ecol. Plants* **2008,** *203* (5), 437–447.

National Climatic Data Center (NCDC). Billion Dollar U.S. Weather/ Disasters. National Oceanic and Atmospheric Administration, National Climatic Data Center, Asheville, NC, 2011.

Neales, T. F.; Sharkey, P. J. Effect of Salinity on Growth and on Mineral and Organic Constituents of the Halophyte *Disphyma australe* (Soland.) JM Black. *Funct. Plant Biol.* **1981,** *8* (2), 165–179.

Nerd, A.; Pasternak, D. Growth, Ion Accumulation, and Nitrogen Fractioning in *Atriplex barclayana* Grown at Various Salinities. *Rangel. Ecol. Manag.J. Range Manag. Arch.* **1992,** *45* (2), 164–166.

Nikalje, G. C.; Bhaskar, S. D.; Yadav, K.; Penna, S. Halophytes: Prospective Plants for Future. In *Ecophysiology, Abiotic Stress Responses and Utilization of Halophytes*; Springer: Singapore, 2019; pp 221–234.

Nikalje, G. C.; Srivastava, A. K.; Pandey, G. K.; Suprasanna, P. Halophytes in Biosaline Agriculture: Mechanism, Utilization, and Value Addition. *Land Degrad. Dev.* **2018a,** *29* (4), 1081–1095.

Nikalje, G. C.; Variyar, P. S.; Joshi, M. V.; Nikam, T. D.; Suprasanna, P. Temporal and Spatial Changes in Ion Homeostasis, Antioxidant Defense and Accumulation of Flavonoids and Glycolipid in a Halophyte *Sesuvium portulacastrum* (L.) L. *PLoS One* **2018b,** *13* (4), e0193394.

Ochiai, K.; Matoh, T. Mechanism of Salt Tolerance in the Grass Species, *Anneurolepidium chinense*: I. Growth Response to Salinity and Osmotic Adjustment. *Soil Sci. Plant Nutr.* **2001,** *47* (3), 579–585.

Pang, Q.; Chen, S.; Dai, S.; Chen, Y.; Wang, Y.; Yan, X. Comparative Proteomics of Salt Tolerance in *Arabidopsis thaliana* and *Thellungiella Halophila*. *J. Proteome Res.* **2010,** *9* (5), 2584–2599.

Parida, A.; Das, A. B.; Das, P. NaCl Stress Causes Changes in Photosynthetic Pigments, Proteins, and Other Metabolic Components in the Leaves of a True Mangrove, *Bruguiera parviflora*, in Hydroponic Cultures. *J. Plant Biol.* **2002,** *45* (1), 28–36.

Park, J.; Okita, T. W.; Edwards, G. E. Salt Tolerant Mechanisms in Single-Cell C4 Species *Bienertia sinuspersici* and *Suaeda aralocaspica* (Chenopodiaceae). *Plant Sci.* **2009,** *176* (5), 616–626.

Persson, S.; Rosenquist, M.; Svensson, K.; Galvao, R.; Boss, W. F.; Sommarin, M. Phylogenetic Analyses and Expression Studies Reveal Two Distinct Groups of Calreticulin Isoforms in Higher Plants. *Plant Physiol.* **2003,** *133* (3), 1385–1396.

Polania, J. Physiological Adaptations in Some Species of Mangroves. *Acta Biologica Columbiana* **1990,** *2* (6), 23–36.

Poljakoff-Mayber, A.; Symon, D. E.; Jones, G. P.; Naidu, B. P.; Paleg, L. G. Nitrogenous Compatible Solutes in Native South Australian Plants. *Funct. Plant Biol.* **1987,** *14* (3), 341–350.

Popp, M. Salt Resistance in Herbaceous Halophytes and Mangroves. *Prog. Bot.* **1995,** 416–429.
Rao, T. A.; Sastry, A. R. K. An Outline of the Coastal Vegetation of India. *Nelumbo* **1974,** *16* (1-4), 101–115.
Rasouli, F.; Kiani-Pouya, A.; Shabala, L.; Li, L.; Tahir, A.; Yu, M.; ... Shabala, S. Salinity Effects on Guard Cell Proteome in *Chenopodium quinoa. Int. J. Mol. Sci.* **2021,** *22* (1), 428.
Rathod, J. P.; Vira, C.; Lali, A. M.; Prakash, G. Heterologous Mannitol-1-Phosphate Dehydrogenase Gene Over-Expression in Parachlorella kessleri for Enhanced Microalgal Biomass Productivity. *J. Genet. Eng. Biotechnol.* **2022,** *20* (1), 1–8.
Redondo-Gómez, S.; Mateos-Naranjo, E.; Davy, A. J.; Fernández-Muñoz, F.; Castellanos, E. M.; Luque, T.; Figueroa, M. E. Growth and Photosynthetic Responses to Salinity of the Salt-Marsh Shrub *Atriplex portulacoides. Ann. Bot.* **2007,** *100* (3), 555–563.
Repp, G. I.; McAllister, D. R.; Wiebe, H. H. Salt Resistance of Protoplasm as a Test for the Salt Tolerance of Agricultural Plants 1. *Agron. J.* **1959,** *51* (6), 311–314.
Riebsame, W. E.; Changnon, S. A.; Karl, T. R. In *Drought and Natural Resources Management in the United States: Impacts and Implications of the 1987-89 Drought*; Routledge, 2019.
Rospert, S.; Looser, R.; Dubaquié, Y.; Matouschek, A.; Glick, B. S.; Schatz, G. Hsp60-independent Protein Folding in the Matrix of Yeast Mitochondria. *EMBO J.* **1996,** *15* (4), 764–774.
Sandoval-Gil, J. M.; Marin-Guirao, L.; Ruiz, J. M. The Effect of Salinity Increase on the Photosynthesis, Growth and Survival of the Mediterranean Seagrass *Cymodocea nodosa. Estuar. Coast. Shelf Sci.* **2012,** *115*, 260–271.
Saslis-Lagoudakis, C. H.; Moray, C.; Bromham, L. Evolution of Salt Tolerance in Angiosperms: A Phylogenetic Approach. *Plant Ecol. Evol. Harsh Environ.* **2014,** 77–95.
Sengupta, S.; Majumder, A. L. Insight into the Salt Tolerance Factors of a Wild Halophytic Rice, *Porteresia coarctata*: A Physiological and Proteomic Approach. *Planta* **2009,** *229* (4), 911–929.
Sengupta, S.; Patra, B.; Ray, S.; Majumder, A. L. Inositol Methyl Tranferase from a Halophytic Wild Rice, *Porteresia coarctata* Roxb. (Tateoka): Regulation of Pinitol Synthesis under Abiotic Stress. *Plant Cell Environ.* **2008,** *31* (10), 1442–1459.
Setyawan, A. D.; Ulumuddin, Y. I.; Ragavan, P. Mangrove Hybrid of *Rhizophora* and its Parentals Pecies in Indo-Malayan Region. *Nus. Biosci.* **2014,** *6* (1).
Shabala, S. Salinity and Programmed Cell Death: Unravelling Mechanisms for Ion Specific Signalling. *J. Exp. Bot.* **2009,** *60* (3), 709–712.
Shabala, S. Learning from Halophytes: Physiological Basis and Strategies to Improve Abiotic Stress Tolerance in Crops. *Ann. Bot.* **2013,** *112* (7), 1209–1221.
Shabala, S.; Pottosin, I. Regulation of Potassium Transport in Plants under Hostile Conditions: Implications for Abiotic and Biotic Stress Tolerance. *Physiol. Plant.* **2014,** *151* (3), 257–279.
Shannon, M. C.; Grieve, C. M. Tolerance of Vegetable Crops to Salinity. *Sci. Hortic.* **1998,** *78* (1-4), 5–38.
Shelke, D. B.; Nikalje, G. C.; Nikam, T. D.; Maheshwari, P.; Punita, D. L.; Rao, K. R. S. S.; ... Suprasanna, P. Chloride (Cl−) Uptake, Transport, and Regulation in Plant Salt Tolerance. In *Molecular Plant Abiotic Stress: Biology and Biotechnology*, 2019; pp 241–268.
Silveira, J. A. G.; Araújo, S. A. M.; Lima, J. P. M. S.; Viégas, R. A. Roots and Leaves Display Contrasting Osmotic Adjustment Mechanisms in Response to NaCl-Salinity in *Atriplex nummularia. Environ. Exp. Bot.* **2009,** *66* (1), 1–8.
Slama, I.; Abdelly, C.; Bouchereau, A.; Flowers, T.; Savouré, A. Diversity, Distribution and Roles of Osmoprotective Compounds Accumulated in Halophytes under Abiotic Stress. *Ann. Bot.* **2015,** *115* (3), 433–447.

Sobhanian, H.; Motamed, N.; Jazii, F. R.; Nakamura, T.; Komatsu, S. Salt Stress Induced Differential Proteome and Metabolome Response in the Shoots of *Aeluropus lagopoides* (Poaceae), A Halophyte C4 Plant. *J. Proteome Res.* **2010,** *9* (6), 2882–2897.

Stocker, O. Salt Plants. In *Handbook of Science*, 1933; vol *8*, pp 699–712.

Subbarao, G. V.; Wheeler, R. M.; Levine, L. H.; Stutte, G. W. Glycine Betaine Accumulation, Ionic and Water Relations of Red-Beet at Contrasting Levels of Sodium Supply. *J. Plant Physiol.* **2001,** *158* (6), 767–776.

Sugihara, K.; Hanagata, N.; Dubinsky, Z.; Baba, S.; Karube, I. Molecular Characterization of cDNA Encoding Oxygen Evolving Enhancer Protein 1 Increased by Salt Treatment in the Mangrove *Bruguiera gymnorrhiza. Plant Cell Physiol.* **2000,** *41* (11), 1279–1285.

Sun, W.; Bernard, C.; Van De Cotte, B.; Van Montagu, M.; Verbruggen, N. At-HSP17. 6A, Encoding a Small Heat-Shock Protein in *Arabidopsis*, can Enhance Osmotolerance upon Overexpression. *Plant J.* **2001,** *27* (5), 407–415.

Suprasanna, P.; Nikalje, G. C.; Rai, A. N. Osmolyte Accumulation and Implications in Plant Abiotic Stress Tolerance. In *Osmolytes and Plants Acclimation to Changing Environment: Emerging Omics Technologies* Springer: New Delhi, 2016; pp 1–12.

Szabados, L.; Kovács, H.; Zilberstein, A.; Bouchereau, A. Plants in Extreme Environments: Importance of Protective Compounds in Stress Tolerance. In *Advances in Botanical Research*; Academic Press, 2011; vol 57, pp 105–150.

Szabados, L.; Savouré, A. Proline: A Multifunctional Amino Acid. *Trends Plant Sci.* **2010,** *15* (2), 89–97.

Tada, Y.; Kashimura, T. Proteomic Analysis of Salt-Responsive Proteins in the Mangrove Plant, *Bruguiera gymnorhiza. Plant Cell Physiol.* **2009,** *50* (3), 439–446.

Tang, X.; Mu, X.; Shao, H.; Wang, H.; Brestic, M. Global Plant-Responding Mechanisms to Salt Stress: Physiological and Molecular Levels and Implications in Biotechnology. *Crit. Rev. Biotechnol.* **2015,** *35* (4), 425–437.

Tani, C.; Sasakawa, H. Proline Accumulates in *Casuarina equisetifolia* Seedlings under Salt Stress. *Soil Sci. Plant Nutr.* **2006,** *52* (1), 21–25.

Tanimoto, S.; Itoh, Y.; Yokoishi, T. Possible Involvement of Greening in Cell Growth of *Suaeda japonica* under Salt Stress. *Plant Cell Physiol.* **1997,** *38* (2), 129–132.

Teixeira, E. I.; Fischer, G.; Van Velthuizen, H.; Walter, C.; Ewert, F. Global Hot-Spots of Heat Stress on Agricultural Crops due to Climate Change. *Agric. For. Meteorol.* 2013, *170*, 206–215.

Thayale Purayil, F.; Sudalaimuthuasari, N.; Li, L.; Aljneibi, R.; Al Shamsi, A. M. K.; David, N.; ... Amiri, K. M. Transcriptome Profiling and Functional Validation of RING-Type E3 Ligases in Halophyte *Sesuvium verrucosum* under Salinity Stress. *Int. J. Mol. Sci.* **2022,** *23* (5), 2821.

Tian, X.; Luo, H.; Zhou, H.; Wu, C. Research on Heat Stress of Rice in China: Progress and Prospect. *Chinese Agric. Sci. Bull.* **2009,** *25* (22), 166–168.

Tipirdamaz, R.; Gagneul, D.; Duhazé, C.; Aïnouche, A.; Monnier, C.; Ozkum, D.; Larher, F. Clustering of Halophytes from an Inland Salt Marsh in Turkey According to their Ability to Accumulate Sodium and Nitrogenous Osmolytes. *Environ. Exp. Bot.* **2006,** *57* (1-2), 139–153.

Trnka, M.; Rötter, R. P.; Ruiz-Ramos, M.; Kersebaum, K. C.; Olesen, J. E.; Žalud, Z.; Semenov, M. A. Adverse Weather Conditions for European Wheat Production will become more Frequent with Climate Change. *Nat. Clim. Change* **2014,** *4* (7), 637–643.

Von Marilaun, A. K. In *The Natural History of Plants, Their Forms, Growth, Reproduction, and Distribution: From the German of Anton Kerner Von Marilaun* Blackie and Son, 1895; vol 4.

Wahid, A.; Farooq, M.; Basra, S. M.; Rasul, E.; Siddique, K. H. Germination of Seeds and Propagules under Salt Stress. In *Handbook of Plant and Crop Stress, Third Edition*; CRC Press, 2016; pp 321–337.

Waisel, Y. In *Biology of Halophytes*, 1st Ed.; Elsevier: Amsterdam, The Netherlands, 2012; p 410.

Walsh, G. E. Mangroves: A Review. In *Ecology of Halophytes*, 1974; pp 51–174.

Wang, X.; Chang, L.; Wang, B.; Wang, D.; Li, P.; Wang, L.; ... Guo, A. Comparative Proteomics of *Thellungiella halophila* Leaves from Plants Subjected to Salinity Reveals the Importance of Chloroplastic Starch and Soluble Sugars in Halophyte Salt Tolerance. *Mol. Cell. Proteom.* **2013,** *12* (8), 2174–2195.

Wang, X.; Fan, P.; Song, H.; Chen, X.; Li, X.; Li, Y. Comparative Proteomic Analysis of Differentially Expressed Proteins in Shoots of *Salicornia europaea* under Different Salinity. *J. Proteome Res.* **2009,** *8* (7), 3331–3345.

Wang, L.; Liu, X.; Liang, M.; Tan, F.; Liang, W.; Chen, Y.; ... & Chen, W. Proteomic Analysis of Salt-Responsive Proteins in the Leaves of Mangrove *Kandelia candel* during Short-Term Stress. *PLoS One* **2014,** *9* (1), e83141.

Wang, T.; McFarlane, H. E.; Persson, S. The Impact of Abiotic Factors on Cellulose Synthesis. *J. Exp. Bot.* **2016b,** *67* (2), 543–552.

Wang, J.; Qiu, N.; Wang, P.; Zhang, W.; Yang, X.; Chen, M.; ... Sun, J. Na+ Compartmentation Strategy of Chinese Cabbage in Response to Salt Stress. *Plant Physiol. Biochem.* **2019,** *140*, 151–157.

Wang, L. W.; Showalter, A. M. Cloning and Salt Induced, ABA Independent Expression of Choline Monooxygenase in *Atriplex prostrata*. *Physiol. Plant.* **2004,** *120* (3), 405–412.

Wang, J.; Yao, L.; Li, B.; Meng, Y.; Ma, X.; Lai, Y.; ... Wang, H. Comparative Proteomic Analysis of Cultured Suspension Cells of the Halophyte *Halogeton glomeratus* by iTRAQ Provides Insights into Response Mechanisms to Salt Stress. *Front. Plant Sci.* **2016a,** *7*, 110.

Wang, X.; Yang, P.; Gao, Q.; Liu, X.; Kuang, T.; Shen, S.; He, Y. Proteomic Analysis of the Response to High-Salinity Stress in *Physcomitrella patens*. *Planta* **2008,** *228* (1), 167–177.

Wong, G.; Lambert, M. F.; Leonard, M.; Metcalfe, A. V. Drought Analysis using Trivariate Copulas Conditional on Climate States. *J. Hydrol. Eng.* **2010,** *15* (2), 129–141.

World Bank. World Population Prospects 2019: Highlights. [Online] (2019): https://data.worldbank.org/indicator/AG.LND.AGRI.ZS (accessed Aug 24, 2019)

Wu, H.; Liu, X.; You, L.; Zhang, L.; Zhou, D.; Feng, J.; ... Yu, J. Effects of Salinity on Metabolic Profiles, Gene Expressions, and Antioxidant Enzymes in Halophyte *Suaeda salsa*. *J. Plant Growth Regul.* **2012,** *31* (3), 332–341.

Wyn Jones, R. G.; Storey R. Betaines. In The *Physiology and Biochemistry of Drought Resistance in Plants*; Paleg, L. G., Aspinall, D., Eds.; Academic Press: Sydney, 1981; pp 171–204.

Xiao, L.; Liu, L.; Asseng, S.; Xia, Y.; Tang, L.; Liu, B.; ... Zhu, Y. Estimating Spring Frost and its Impact on Yield Across Winter Wheat in China. *Agric. For. Meteorol.* **2018,** *260*, 154–164.

Xing, J.; Cai, M.; Chen, S.; Chen, L.; Lan, H. Seed Germination, Plant Growth and Physiological Responses of *Salsola ikonnikovii* to Short-Term NaCl Stress. *Plant Biosyst.* **2013,** *147* (2), 285–297.

Yamada, S.; Katsuhara, M.; Kelly, W. B.; Michalowski, C. B.; Bohnert, H. J. A Family of Transcripts Encoding Water Channel Proteins: Tissue-Specific Expression in the Common Ice Plant. *Plant Cell* **1995,** *7* (8), 1129–1142.

Yang, C.; Chong, J.; Li, C.; Kim, C.; Shi, D.; Wang, D. Osmotic Adjustment and Ion Balance Traits of an Alkali Resistant Halophyte *Kochia sieversiana* during Adaptation to Salt and Alkali Conditions. *Plant Soil* **2007,** *294* (1), 263–276.

Yang, C.; Shi, D.; Wang, D. Comparative Effects of Salt and Alkali Stresses on Growth, Osmotic Adjustment and Ionic Balance of an Alkali-Resistant Halophyte *Suaeda glauca* (Bge.). *Plant Growth Regul.* **2008,** *56* (2), 179–190.

Yazici, I.; Türkan, I.; Sekmen, A. H.; Demiral, T. Salinity Tolerance of Purslane (*Portulaca oleracea* L.) is Achieved by Enhanced Antioxidative System, Lower Level of Lipid Peroxidation and Proline Accumulation. *Environ. Exp. Bot.* **2007,** *61* (1), 49–57.

Ye, C. J.; Zhao, K. F. Osmotically Active Compounds and their Localization in the Marine Halophyte Eelgrass. *Biol. Plant.* **2003,** *46* (1), 137–140.

Yeo, A. Salinity. In *Plant Solute Transport*, 2007; pp 340–370.

Yousif, B. S.; Liu, L. Y.; Nguyen, N. T.; Masaoka, Y.; Saneoka, H. Comparative Studies in Salinity Tolerance between New Zealand Spinach (*Tetragonia tetragonioides*) and Chard (*Beta vulgaris*) to Salt Stress. *Agric. J.* **2010,** *5* (1), 19–24.

Yu, J.; Chen, S.; Zhao, Q.; Wang, T.; Yang, C.; Diaz, C.; ... Dai, S. Physiological and Proteomic Analysis of Salinity Tolerance in *Puccinellia tenuiflora. J. Proteome Res.* **2011,** *10* (9), 3852–3870.

Yuan, F.; Xu, Y.; Leng, B.; Wang, B. Beneficial Effects of Salt on Halophyte Growth: Morphology, Cells, and Genes. *Open Life Sci.* **2019,** *14* (1), 191–200.

Zhang, X. X.; Sun, H. Y.; Shen, C. M.; Li, W.; Yu, H. S.; Chen, H. G. Survey of *Fusarium* spp. Causing Wheat Crown Rot in Major Winter Wheat Growing Regions of China. *Plant Dis.* **2015,** *99* (11), 1610–1615.

Zhang, X.; Tan, B.; Cheng, Z.; Zhu, D.; Jiang, T.; Chen, S. Overexpression of McHB7 Transcription Factor from *Mesembryanthemum crystallinum* Improves Plant Salt Tolerance. *Int. J. Mol. Sci.* **2022,** *23* (14), 7879.

Zhao, Q.; Zhang, H.; Wang, T.; Chen, S.; Dai, S. Proteomics-Based Investigation of Salt-Responsive Mechanisms in Plant Roots. *J. Proteom.* **2013,** *82*, 230-–253

CHAPTER 19

Salicornia brachiata: A Potential Halophyte of Coastal Ecosystem Having Ecological and Economic Significance

ASISH KUMAR PARIDA[1,2] and MONIKA PATEL[1,2]

[1]*Plant Omics Division, CSIR–Central Salt and Marine Chemicals Research Institute (CSIR–CSMCRI), Bhavnagar, Gujarat, India*

[2]*Academy of Scientific and Innovative Research (AcSIR), Ghaziabad, Uttar Pradesh, India*

ABSTRACT

Halophytes are important resources of coastal ecosystems possessing high economic value and industrial applications. However, in India, halophytes continue to be either under or unexplored. *Salicornia brachiata* is an important species of halophyte distributed on the Indian coast. It has plenty of ecological and industrial applications and can be cultivated as a sustainable crop in saline wastelands. This fleshy plant is frequently distributed in salt marshes, edges of wetlands, mudflats, and seashores. *S. brachiata* is a highly salt-tolerant halophytic species that inhabited the salt flats of the Indian coasts. This succulent halophyte can flourish in a highly saline environment, and seed germination could be inhibited by high salinity. The seeds of *S. brachiata* are rich sources of proteins along with high palmitic, stearic, oleic, and linoleic acids. The biomass and seed oil of *Salicornia sp.* have been reported as a good source of bioethanol and biodiesel production. The plant seeds oil contains high ester (538.32 mg/g) and saponification (547.52 mg/g), making it potential for industrial application. Its salty succulent stem contains unique oligosaccharides and antioxidants and can be consumed as

Physiology of Halophytes: Signaling, Omics, and Tolerance Mechanisms.
Nivas Manohar Desai & Ganesh Chandrakant Nikalje (Eds.)

a green salad. In addition, *Salicornia* shoots are a potent source of minerals (sodium, calcium, iodine, iron, and magnesium), vitamin C, β-carotene, fatty acids, amino acids, polyphenols, proteins, and fibers (Ventura et al., 2011). *S. brachiata* has excellent potential for the desalinization of saline soil and phytoremediation of salt-affected areas. *S. brachiata* can tolerate a moderate level of fluoride and bromide which is helpful to vegetate the marshy lands. The presence of various antioxidants and scavenging activity of ROS, amino acids, sulfur-rich proteins with special oligosaccharides in the seeds, high flavonoids, essential fatty acids, and PUFAs exhibit the nutritional potential of *S. brachiata* and are considered a promising candidate for dilatory supplements. Additionally, the bioactive metabolites of the plant suggest the medicinal potential of *S. brachiata*. Moreover, *S. brachiata* has been extensively explored for abiotic stress tolerance promoters and genes for developing stress-tolerant transgenic plants. This chapter comprises the uses and applications of *S. brachiata* in food, pharmacy, chemical, and bioenergy production, the restoration of saline land, and seawater agriculture to alleviate the poverty of the coastal region. This chapter also highlights the stress-responsive genes and promotes *S. brachiata*, which provides further insight for future research on the development of transgenic crops with high-stress tolerance potential.

19.1 INTRODUCTION

Halophytes are salt-tolerant plants having highly evolved and well-adapted morpho-physiological and anatomical features allowing them to complete their life cycle in a salt-rich environment. Halophytes grow in coastal habitats and are efficient in thriving under extreme salinity and hence are considered the best source of germplasm for saline agriculture. Several halophytic plant species, especially those belonging to the Chenopodiaceae family, can withstand high external salinity through osmotic adjustment by accumulating NaCl in the vacuole and organic osmolytes (proline and glycine betaine) in the cytoplasm. Most halophytes endure extreme saline habitat by osmoregulation, ion compartmentalization and accumulation, salt gland or bladder development, and induction of antioxidant components (Shabala et al., 2014; Slama et al., 2015; Himabindu et al., 2016). At a molecular level, salt tolerance efficiency is accomplished by regulating salt-responsive genes via ABA-dependent or ABA-independent regulation pathways. Salt stress adaptation and tolerance is a complex phenomenon that involves multiple physiological responses directed by numerous genes and proteins (Mishra and Tanna, 2017). Moreover, halophytes are widely explored for their salt adaptation and tolerance strategies at physio-biochemicals, molecular, and metabolomic levels. Numerous studies have shown that halophytic plants can accumulate and tolerate a high level of hazardous metals in the environment (Pedro et al., 2013; Kachout et al., 2015; Panda et al., 2017; Patel and Parida, 2021). Some halophytes' abilities make them ideal candidates for the phytoextraction/phytoremediation of heavy metals-contaminated saline soil.

The abiotic stress such as salinity, drought, cold, high light intensity, minerals deficiency, and extreme temperature negatively affect plant growth and productivity. To tolerate these challenges, plants follow various adaptation strategies at physiological, biochemical, and molecular levels through the coordinated action of the stress-responsive protein. In plants, the stress-tolerance mechanism is harmonized with ROS production and its detoxification pathways, signal transduction, osmoregulation, and regulation of ion levels through various osmoticum and expression of various stress-responsive genes and proteins (Mishra and Tanna, 2017). *Salicornia brachiata* belongs to the Chenopodiaceae family and is commonly known as sea asparagus. It is an annual, leafless fleshy shrub that grows luxuriantly in mangrove habitats of the estuarine regions of the tropical zones of the globe. *S. brachiata* is widely distributed in Eurasia, North America, and South Africa (Singh et al., 2018). This halophyte dominates in the salt marshes along the Indian coast from Bengal to Gujarat (Rathore et al., 2016). *S. brachiata* is an obligate halophyte that completes its life cycle within 8–9 months. These plants are highly branched and grow up to 25–35 cm in height; the branches have rather slender joints of 6–12 mm; flowers are sunk in cavities of the joints, three on each side; and the fruits are membranous. The progressive senescence in the plant is visibly shown by changes in shoot color from green to greenish-yellow at the initial to middle stage and red at the terminal stage (Parida et al., 2018). Due to its economic potential, *S. brachiata* is considered a potential alternative crop for seawater agriculture and wasteland rehabilitation (Stanley, 2008). The seeds are germinated in September–October and attain maximum growth and flowering phase in April–May. The biomass of *S. brachiata* is used as a source of high-valued low-sodium salt (herbal salt) suggested for high blood pressure patients (U.S. Patent 6929809). *S. brachiata* can accumulate 30–40% NaCl in its dry weight. (Sharma et al., 2010). This salt water-irrigated crop is an important species that provides vegetable, animal fodder, and seeds yielding oil. Due to the high saponin content in seeds, the oil is not preferable for human consumption. However, the oil can be leached of saponins by washing in running water and boiling at a higher temperature making it more suitable for edible purposes. The high content of glycerides from the seed oil can be an alternative in soap and detergent industries. *Salicornia* has been reported for higher total lipid and omega-3 content than other green leaves of mustard, spinach, and lettuce (Ventura et al., 2011). Therefore, *S. brachiata* may attract considerable interest as a substitute for polyunsaturated fatty acids for human consumption. *S. brachiata* seed oil contains unusual fatty acids like palmitic (16.48%), myristic (12.88%), oleic (32.79%), and 10-undecenoic (37.85%) and is commonly used as lubricants in industries (Eganathan et al., 2006). *S. brachiata* is a significant resource for salt stress-tolerance genes that can serve as a model plant for understanding salt adaptation which facilitates the development of abiotic stress-tolerant crops. The potential applications of *S. brachiata* have been depicted in Figure 19.1. This halophytic shrub of coastal mud lands has the potential to produce high biomass in the marine ecosystem and is a great source of plant growth-promoting halotolerant diazotrophic rhizobacteria (Jha et al., 2012a). *Salicornia* favors the growth of diverse microbiota or endophytes, which regulate plant growth by synthesizing different secondary metabolites under a stressed environment. Recently, diverse endophytic bacteria associated with *S. brachiata* have been reported by Singh et al. (2021), which showed great antimicrobial potential against various pathogens. The plant has also been reported to harbor plant growth-promoting endophytes (Karnwal, 2019). The *B. subtilis* 4NP-BL and *S. hyderabadensis* 7VPT5-5R isolates of *S. brachiata* protect the host against phytopathogens and survival in adverse environments through the production of melanin pigment (Ghadge et al., 2020, 2022). The extract of *S. brachiata* nanoparticles has powerful antibacterial activity against pathogenic bacteria (Behlol et al., 2014). The cultivation of *S. brachiata* using brackish RO reject water without any additional nutrients has been reported as an environmentally friendly option for reject water management (Singh et al., 2018). The cellulose obtained from the plant is comparable to Whatman filter paper No. 4 (Sanandiya et al., 2010). This halophyte has been extensively studied for wasteland reclamation, bioprospection, oils, stress-responsive genes, physio-biochemistry, and phytoremediation of heavy metals (Eganathan et al., 2006; Jha et al., 2009; Parida and Jha, 2010; Sharma et al., 2010; Agarwal et al., 2010; Tiwari et al., 2016). This halophytic species has high market value and ecological importance for future prospects and hence has gained the increased attention of scientists, farmers, and entrepreneurs.

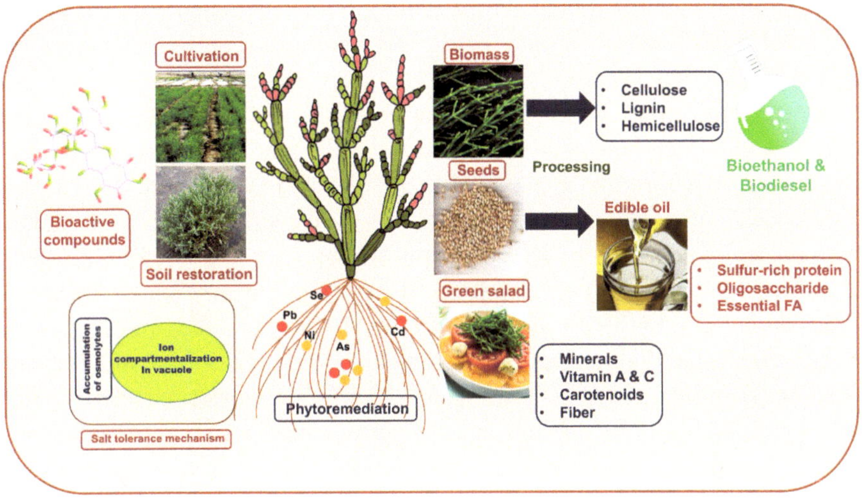

FIGURE 19.1 The schematic diagram showing various economic potential of *S. brachiata*.

19.2 *S. brachiata*: A POTENTIAL SOURCE OF NUTRITIVE SUPPLEMENTS

Salicornia has been traditionally used as a food material and for therapeutic uses such as oxidative stress treatment, source of vitamins A and C, minerals, fatty acids, and vegetables. In terms of nutritional properties, *Salicornia* has been reported for higher ascorbic acid and DHA (dehydroascorbic acids), and carotenoid content (Guil and Rodr, 1997). As an edible crop, *Salicornia* is used as a salt substitute since it has a positive effect on vascular dysfunction and hypertension due to trans-ferulic acid (Panth et al., 2016). This halophyte is known to accumulate a high amount of salt in the succulent stem used to prepare low-sodium vegetable salt (Saloni salt). Moreover, several species of *Salicornia* have medicinal properties in various diseases like immunomodulatory, osteoprotective, anti-proliferative, obesity, constipation, inflammation, diabetes, asthma, hepatitis, gastroenteritis, and cancer (Singh et al., 2018). Kumar et al. (2014) showed that the different dose of *S. brachiata* extract has antidiabetic and antioxidant effects and have better glucose utilization capacity. The proteome analysis of *S. brachiata* seeds showed the presence of inter-and intramolecular disulfide bonds in the glob ulins which makes this halophyte a potential source of dietary supplement (Jha et al., 2012b). The oligosaccharide mass profiling revealed the presence of a series of monomers which is similar to highly nutritious soybeans. This study revealed the nutritional potential of *S. brachiata* as a latent source of nutritional supplementation (Mishra et al., 2013). The plant content elevated the level of sulfur-rich amino acids like cysteine, methionine, tyrosine, valine, proline, leucine, and isoleucine under salt and heat stress, and are thus considered nutritious. The plant extract showed increased ROS-scavenging activity and reduced the capacity of H_2O_2, DPPH, and hydroxyl under stress conditions. Additionally, the total flavonoid content and total antioxidant activity are enhanced in stress conditions. However, the total polyphenol content declined under stress conditions (Mishra et al., 2015). The plant contains about 55% PUFAs and 38% SFAs, including linoleic acid (26%), alpha-linolenic acid (29%), and palmitic acid (26%), which is comparatively higher than other halophytes (Mishra et al., 2015). Nutritive supplements like benzyl sulfate and flavonoid myricetin provide nutritional value to this halophyte. Additionally, the presence of various bioactive compounds along with fungicidal, bactericidal, and insecticidal activity has been specified in this halophyte (Mishra et al., 2015). Several other metabolites such as oxomefruside are used in the treatment of hypertension. Clonidine is utilized in the treatment of high blood pressure, anxiety, and certain pain conditions. Carmustine is an anti-neoplastic compound used in cancer treatments. The compound cefazolin sodium acts on a broad range of bacteria that have been identified from the plant extract. Anti-tuberculosis activity has been reported in the plant extract (U.S. Patent No. 7,989,004). The presence of various metabolites with bioactivities, scavenging activities, nutritional anti-oxidants, flavonoids, essential PUFs and fatty acids, and amino acids makes this halophyte a promising functional food for dietary supplementation in nutraceutical industries.

19.3 *S. BRACHIATA* CAN COPE WITH ABIOTIC STRESS BY ALTERATION IN VARIOUS PHYSIOCHEMICAL ASPECTS

Soil salinity and drought stress are the major abiotic factors limiting growth and productivity worldwide. *S. brachiata* is a succulent stem halophyte that grows luxuriantly in coastal marshes and continuously encounters various environmental stresses. The reduction in plant height and total biomass of *S. brachiata* have been reported under high salt conditions (600 mM NaCl). However, under a low level of salt (200 mM NaCl), the plant expresses optimal growth potential (Parida and Jha, 2010). Higher seed germination under low salinity and drought (75% strength seawater, 200 mM NaCl, and 10% PEG) stress indicates the inherent potential of the plant to tolerate harsh environments (Jacob et al., 2020). In addition, the seed germination was inhibited in a harsh environment and maintained the potential to recover during favorable conditions. The recovery of stress-inhibited seed germination confirmed the recovery potential of seeds once the stress period is completed (Jacob et al., 2020). High plant biomass under saline conditions has been recorded by Jacob et al. (2020). However, high salinity and drought imposed growth reduction, which suggested the utilization of the plant's energy for the synthesis of compatible solutes to regulate ion toxicity (Jacob et al., 2020; Siddiqui et al., 2022). The fresh and dry biomass of *S. brachiata* reduced after seven days of PEG-mediated drought stress. However, the total plant water content and the shoot and root water content were adequately maintained in response to drought stress (Parida and Jha, 2012). A high salinity tolerance index and increased plant vigor under salinity indicate the obligation of NaCl for optimal growth (Haque et al., 2017; Rathore et al., 2019; Jacob et al., 2020). Fresh and dry biomass accumulation and increased relative water content have been reported in this halophyte under stress treatments. However, higher concentrations of fluoride and bromide declined both fresh and dry biomass accumulation (Reddy and Kaur, 2008; Reddy, 2009). A high level of salinity (1.0 M NaCl) and drought (20% PEG) caused stomatal closure and decreased stomatal conductance (g_s), which consequently decreased the net photosynthetic rate (P_N) in *S. brachiata.* However, under low salinity levels (100% seawater), g_s and P_N were maintained high in *S. brachiata* (Siddiqui et al., 2022). Salinity and drought (200% seawater salinity and 10% PEG) significantly lower the WUE of *S. brachiata* (Siddiqui et al., 2022). A significant decrease in the photosynthetic pigment level has been reported under high salinity, drought, fluoride, bromide, and heavy metals stress conditions. However, a better accumulation of photosynthetic pigments under seawater stress has been reported in the plant (Parida and Jha, 2012; Siddiqui et al., 2022). A significant increase in the anthocyanin content has been reported in fluoride treatment (Reddy and Kaur, 2008). In addition, a significant loss of photosynthetic pigments with the progress of senescence has been observed in *S. brachiata* by Parida et al. (2018).

19.4 CULTIVATION OF *S. BRACHIATA* ON SALINE LAND AND PHYTO-MANAGEMENT OF HEAVY METAL-CONTAMINATED SALINE SOIL

Soil salinity is one of the rising problems causing yield losses, especially in arid and semiarid regions in the world. A transitional region of ocean and land, coastal area, and sediments are typical zones contaminated with salinity and heavy metal. Halophytes have the potential to accumulate and excrete salt effectively and perform better growth compared to glycophytes in saline soil. Seawater agriculture is one of the important methods of agricultural production using saline water and soil. Thus, the cultivation of a halophytic plant that accumulates a high level of salt is a low-cost, highly effective, and economical solution to reduce soil salinity. Phytoremediation using halophytes has several unique advantages such as uniform salt removal, simultaneous absorption of other pollutants, and improved soil quality (Hasanuzzaman et al., 2014). Additionally, the cultivated halophytes provide food, fodder, and raw materials for industries. However, the process of phytoremediation is time-consuming and require several growing seasons to reduce soil salinity. *Salicornia* cultivation on inland saline soil and near coastal marshy areas provides additional sources of fodder components and offers a prospect to rehabilitate and remediate the coastal soil–water ecosystem to reduce salinity for traditional forage and crops. Moreover, commercial farming of *Salicornia* may create business opportunities and employment, support the local economy, and make a contribution to climate change mitigation. The different developmental stages of *S. brachiata* growing naturally in coastal saline areas have been depicted in Figure 19.2. Being a native of a high saline environment, *Salicornia* has attracted attention as a phytoremediator of heavy metal(loid)s contaminated saline soil. Heavy metals negatively affect seed germination, seedling growth, and biomass accumulation of plants (Sharma et al., 2010, 2011). This species has been reported to accumulate various heavy metals (Cd^{2+}, Ni^{2+}, and As^{3+}) and can serve as promising plant species for the phytoremediation of saline soils (Sharma et al., 2010, 2011). *S. brachiata* is C3 succulent halophyte that can efficiently accumulate a high concentration of salt in their tissue and is ideally suitable for restoration of salt-affected heavy metals polluted soil by removing salt from the soil. A plant with a high bioaccumulation factor (BCF) and translocation factor (TF) are potential remediators of saline soil. The higher BCF > 1 and TF > 1 value suggests *S. brachiata* as a suitable candidate for phytoremediation of saline soils (Chaudhary et al., 2018).

19.5 ACCUMULATION OF IONS AND BIOSYNTHESIS OF ORGANIC SOLUTES MAINTAIN CELLULAR HOMEOSTASIS IN *S. BRACHIATA* UNDER ADVERSE ENVIRONMENT

Halophytes can manage the excess salt in the surroundings using various physiochemical adaptation mechanisms such as maintaining water flux,

FIGURE 19.2 *Salicornia brachiata* naturally growing in the Gujarat coast of India at different developmental stages: (A) *S. brachiata* at early seedlings stage; (B) Young plants of *S. brachiata*; (C and D) *S. brachiata* at the middle stage of life cycle; (E and F) *S. brachiata* at matured stage.

regulation of ions, deposition of compatible solutes, use of gland tissue to sequester excess Na^+, and antioxidant defense system to scavenge ROS (Hasanuzzaman et al., 2014). It has been known that the plants of the Chenopodiaceae (e.g., *S. brachiata*) family can withstand high salinity through osmotic adjustment by accumulating NaCl and organic osmolytes. The excess Na^+ ions are stored in mesophyll tissue, which is morphologically suitable for the storage of salt and water. In addition, these cells also accumulate osmolytes like free proline, total amino acids, and soluble sugars. Moreover, increased ATPase activity has also been reported in the spongy mesophyll, which is responsible for increased ion influxes in the vacuoles in *S. brachiata* (Rathore et al., 2016). This halophytic plant species has a well-characterized nutrient uptake system that compartmentalizes excess NaCl and efficiently balances the essential nutrients for rapid growth and development. According to Parida et al. (2018), the major minerals (Na^+, K^+, Mg^{2+}, and Ca^{2+}) were preserved and the minor nutrients varied (Zn^{2+}, Mn^{2+}, Fe^{2+}, and Cu^{2+}) with the progress of senescence in *S. brachiata*. Higher accumulation of Na^+ and lower K^+ content under high salinity and drought have been reported in *S. brachiata* shoot (Parida and Jha, 2010; Haque et al., 2017; Jacob et al., 2020; Siddiqui et al., 2022). However, higher K^+ accumulation has been reported in shoots under various stress conditions (Parida and Jha, 2010; Jacob et al., 2020; Siddiqui et al., 2022). Drought stress increases the uptake and translocation of K^+, Mg^{2+}, and Ca^{2+} from root to shoot in the plant (Parida and Jha, 2012). The increased K^+ may have a vital role as inorganic osmolytes and increased Ca^{2+} protects the membrane during drought stress in *S. brachiata*. The higher Na^+/K^+ ratio suggests that most of the Na^+ may be compartmentalized in vacuoles and play an essential role in an osmotic adjustment under stress conditions (Parida and Jha, 2010; Haque et al., 2017; Jacob et al., 2020; Siddiqui et al., 2022). Moreover, the Na^+ content is significantly higher than proline in *S. brachiata* under salt treatment which indicates that Na^+ is a major inorganic solute in *S. brachiata* under salinity. The minor elements like Fe^{2+}, Mn^{2+}, Cu^{2+}, and Zn^{2+} showed differential responses under drought stress in *S. brachiata* (Parida and Jha, 2012). Bromide content significantly increased the Na^+ accumulation and reduced the K^+, Ca^{2+}, and Mg^{2+} content in this halophyte (Reddy, 2009). Additionally, F^-, Na^+, Mn^{2+}, and Fe^{2+} ions concentration increased, whereas K^+, Ca^{2+}, and Mg^{2+} contents decreased with fluoride treatment has been recorded in *S. brachiata* (Reddy and Kaur, 2008). Osmotic adjustment with inorganic solutes involved lower energy costs than organic solutes. In *S. brachiata,* a comparatively higher accumulation of inorganic solutes than organic solutes has been reported by

Haque et al. (2017). Storage of organic osmolytes is often considered the prime strategy for the protection and survival of halophytes in response to abiotic stresses. The major organic solutes such as proline, glycine betaine, polyols, and soluble sugars provide osmotic adjustment and protect from oxidative stress. Significantly higher storage of proline, total free amino acids, soluble sugars, and starch has been reported under various stress conditions in *S. brachiata* (Parida and Jha, 2010; Parida and Jha, 2012; Haque et al., 2017; Jacob et al., 2020; Siddiqui et al., 2022). In addition, proline and organic acids showed a higher contribution in osmotic adjustment than free amino acids and sugars under salinity (Haque et al., 2017). Heavy metals stress-induced higher proline accumulation in *S. brachiata* (Sharma et al., 2010). Reduced total free amino acid content has been observed with the progress of senescence in *S. brachiata,* indicating that free amino acid pools did not trigger senescence (Parida et al., 2018). However, the level of starch and total and reduced sugar significantly improved with senescence in *S. brachiata* (Parida et al., 2018).

19.6 *S. BRACHIATA* CAN THRIVE UNDER VARIOUS ABIOTIC STRESS CONDITIONS BY INDUCTION OF ANTIOXIDANTS DEFENSE MECHANISM AND MAINTAINING ROS HOMEOSTASIS

Various environmental constraints (salinity, drought, high temperature, cold, mineral starvation, and heavy metals) induce oxidative stress through the overproduction and accumulation of ROS. The excess ROS causes DNA damage, peroxidation of membrane lipids, denaturation of enzymes/protein, and degradation of pigments (Parida and Jha, 2010). This plant possesses a complex antioxidative defense system comprising enzymatic and non-enzymatic antioxidant components to scavenge ROS and reduce oxidative stress. Salinity, drought, and heavy metals stress have been reported to increase electrolyte leakage, MDA content, $O_2^{\bullet-}$, and H_2O_2 levels in *S. brachiata* (Sharma et al., 2010; Jacob et al., 2020; Parida and Jha, 2012, 2010; Siddiqui et al., 2022). Superoxide dismutase (SOD) is considered a first-line defense against oxidative stress and exhibits higher SOD activity in response to salinity. *S. brachiata* has been reported to induce an increase (135–149%) in SOD activity upon long-term exposure to salinity (Parida and Jha, 2012). Higher SOD activity has been reported under salinity, drought, and heavy metals stress conditions in *S. brachiata* (Jacob et al., 2020; Sharma et al., 2010; Siddiqui et al., 2022). Therefore, SOD plays a crucial role in ROS detoxification in halophytes *S. brachiata*. Moreover, long-term exposure to salinity and drought lowers catalase (CAT) activity (Parida and Jha, 2010; Siddiqui et al., 2022). Decreased CAT activity under stress conditions suggests the poor affinity of CAT for H_2O_2 and is not a potential H_2O_2 scavenger in *S. brachiata*. However, the non-significant changes and increases in CAT activity during germination and after-seedlings development have been observed in *S. brachiata* (Jacob et al., 2020). Ascorbate peroxidase (APX) inhibits the excess H_2O_2 accumulation in the cell by the ascorbate-glutathione or Halliwell-Asada pathway. Salinity-induced increase (19–24%) in APX activity has been observed in *S. brachiata,* demonstrating the key role of APX in H_2O_2 detoxification in response to salinity (Parida and Jha, 2010). Guaiacol-specific peroxidase (POX) has been used as a stress marker and guards the cell against the toxicity of H_2O_2. The increase in POX activity in *S. brachiata* indicates their function in abiotic stress tolerance (Parida and Jha, 2010; Siddiqui et al., 2022). However, a decrease in POX activity until the seedlings stage suggests that reduced POX activity might help preserve the threshold level of H_2O_2 as a signaling molecule in *S. brachiata* (Jacob et al., 2020). Glutathione reductase (GR) is a crucial enzyme to maintain the redox status of ascorbate and glutathione. GR regulates the endogenous level of H_2O_2 by the oxidoreduction cycle. The highest activity of GR (233%) in *S. brachiata* suggests that GR might plays an important role in the antioxidative defense mechanism in *S. brachiata* under salt stress. Furthermore, increased GR activity under saline and osmotic stress has been detected in *S. brachiata* (Jacob et al., 2020; Siddiqui et al., 2022). Ascorbate (AsA) and glutathione (GSH) are major non-enzymatic antioxidants involved in H^2O^2 detoxification through the AsA-GSH cycle. Salinity induced reduction in the total ascorbate and reduced ascorbate and AsA/DHA ratio in *S. brachiata* (Parida and Jha, 2010). Unlike the ascorbate, the glutathione content and the GSH/GSSG ratio increased with increased salinity in *S. brachiata* and protected the protein thiol groups from the singlet oxygen and hydroxyl radicals (Parida and Jha, 2010). Polyphenol is a non-enzymatic ROS scavenger and also serves as an osmoprotectant in harsh environments in plants. Higher accumulation of polyphenols are induced by salinity and drought conditions (Parida and Jha, 2010; Parida and Jha, 2012; Haque et al., 2017; Jacob et al., 2020; Siddiqui et al., 2022). A marginal increase in the polyphenols has also been observed in *S. brachiata* at a late stage of senescence (Parida et al., 2018). Jacob et al. (2020) reported higher flavonoid content during seed germination and seedling establishment, indicating its involvement in defense against stress conditions in *S. brachiata*. Carotenoids protect the photosystem from ROS-induced damage. The significant increase in carotenoid content indicates the role of carotenoids as a scavenger of singlet oxygen in *S. brachiata* in salinity and drought stress (Parida and Jha, 2012; Siddiqui et al., 2022). However, a noteworthy reduction in the carotenoid content has been reported with the progress of senescence in *S. brachiata* (Parida et al., 2018). Moreover, reduced carotenoid content in response to fluoride and bromide conditions has been reported in this halophyte (Reddy and Kaur, 2008; Reddy, 2009). A significant increase in betacyanin and betaxanthin content with the progress of senescence has been observed in the shoot of *S. brachiata,* which suggests its probable role in ROS detoxification and osmotic adjustment (Parida et al., 2018). According to Behlol et al. (2014), the level of polyphenols and the antioxidant enzymes like polyphenol oxidase (PPO), CAT, and SOD activity increased in *S. brachiata* exposed to various concentrations of tannery soaking water suggests that this halophyte has a potential role in remediation of NaCl from saline tannery wastewater. The SOD, POX, and ATPase activities increased; whereas, CAT activity decreased in bromide-treated *S. brachiata* (Reddy, 2009). Moreover, the activities of POX, SOD, acid phosphatase, and ATPase have been negatively regulated in this halophyte under fluoride treatment (Reddy and Kaur, 2008).

19.7 *S. BRACHIATA*: A POTENTIAL RESOURCE FOR SALT AND DROUGHT-TOLERANCE GENES AND PROMOTERS FOR ENGINEERING CROPS WITH ENHANCED STRESS TOLERANCE

Abiotic constraints like salinity and drought are significant plant growth limiting factors. Plant cellular and molecular responses against stress include the accumulation of stress-responsive proteins, up-regulation of genes for a transcription factor, osmolytes, antioxidants, phytohormones, and transporters. The molecular responses to stress play a crucial role in developing a strategy to engineer stress-resistant plants. Halophytic plant species are more resistant to abiotic stress than glycophytes due to the differential regulation of stress-responsive genes and higher expression of stress-inducible genes in non-stress conditions. The transcriptome database of *S. brachiata* has shown that it has more than 300 novel or unknown stress-responsive genes, which makes *S. brachiata* a model plant for unique genetic resources. Additionally, *S. brachiata* frequently encounters various abiotic stresses on salt marshes and tolerates adverse conditions and, hence, is considered a great source of stress-responsive promoters and genes. The overexpression genes of *S. brachiata* such as *SbGSTU* (Jha et al., 2011), *SbpAPX* (Singh et al., 2014), and *SbNHX1*(Jha et al., 2013) have high salinity tolerance (200–300 mM) in the transgenic glycophytes. Abiotic stress-responsive genes such as *SbSRP* (Udawat et al., 2017), *SbUSP* (Udawat et al., 2016), *SbSI-1* (Kumari et al., 2017), *SbMYB15* (Shukla et al., 2015), *SbpAPX*(Tiwari et al., 2014), *SbSLSP* (Singh et al., 2016), *SbNHX1* (Jha et al., 2011), *SbASR-1* (Tiwari et al., 2015), *SbMT-2* (Chaturvedi et al., 2014), and *SbSDR1* (Singh et al., 2016) from *S. brachiata* showed better stress tolerance in the transgenic plants. Moreover, *SbGSTU,* a promoter of *S. brachiata,* has been reported for the presence of several stress-responsive *cis*-regulatory motifs that regulate the *GSTU* gene expression (Tiwari et al., 2016).

19.8 CONCLUDING REMARKS

Salicornia is one of the most important genera of the halophyte promoted as a secondary vegetable and plant for the future. Presently, *S. brachiata* is being explored for its potential utilization as nutraceutical food, bioenergy and ecology, genetic resources, source of bioactive compounds, sustainable crop in seawater agriculture, and its application in bio-filtering and phytoremediation. *S. brachiata* is a good source of vitamins A and C, fatty acids, and minerals. Moreover, nutritional antioxidants, scavenging activities, flavonoids, polyphenols, carotenoids, and amino acids exhibit the nutritional potential of the halophyte. The seeds yield high-quality edible oil similar to safflower oil in high PUFA composition. Due to its nutraceutical value and medicinal properties, *S. brachiata* is the most significant candidate in the different fields of food and pharmaceutical industries. Additionally, the seeds and biomass of *S. brachiata* can be used to produce bioethanol and biodiesel. The high salt-tolerance potential of this halophyte makes it a potent candidate for the phytoremediation of saline soil. Cultivation of *S. brachiata* would hold the land's productivity and improve it in terms of both ecological and landscape value to create an attractive area for tourism. *S. brachiata* is also suitable for developing transgenic crops with enhanced stress tolerance, which increases its agricultural importance. This chapter promotes the further investigation of this halophytic species for deeper understanding to allow its application in land re-vegetation.

ACKNOWLEDGMENTS

MP is thankful to the Department of Science and Technology (DST), New Delhi, India, for providing financial support in the form of the INSPIRE

Fellowship. AKP duly acknowledges the financial support of CSIR, New Delhi, India, for the in-house grant (Grant No. MLP 066). This manuscript was assigned CSIR-CSMCRI Communication No. 135/2024.

KEYWORDS

- **abiotic stress**
- **antioxidative enzymes**
- **bioaccumulation**
- **bioprospection**
- **flavonoids**
- **ion homeostasis**
- **minerals**
- **nutraceuticals**
- **phytoremediation**
- **polyphenols**
- **Salicornia brachiata**
- **salinity**

REFERENCES

Agarwal, P. K.; Gupta, K.; Jha, B. Molecular Characterization of the *Salicornia brachiata* SbMAPKK Gene and Its Expression by Abiotic Stress. *Mol. Biol. Rep.* **2010,** *37*, 981–986. https://doi.org/10.1007/s11033-009-9774-1.

Behlol, K.; Ahmed, A.; Subramanian, S.; Sivasubramanian, A.; Veerappan, G.; Veerappan, A. Preparation of Gold Nanoparticles Using *Salicornia brachiata* Plant Extract and Evaluation of Catalytic and Antibacterial Activity. *Spectrochim. Acta Part A Mol. Biomol. Spectrosc.* **2014,** *130*, 54–58. https://doi.org/10.1016/j.saa.2014.03.070.

Chaturvedi, A. K.; Patel, M. K.; Mishra, A.; Tiwari, V.; Jha, B. The SbMT-2 Gene from a Halophyte Confers Abiotic Stress Tolerance and Modulates ROS Scavenging in Transgenic Tobacco. *PLoS One* **2014,** *9*. https://doi.org/10.1371/journal.pone.0111379.

Chaudhary, D. R.; Rathore, A. P.; Jha, B. Aboveground, Belowground Biomass and Nutrients Pool in *Salicornia brachiata* at Coastal Area of India: Interactive Effects of Soil Characteristics. *Ecol. Res.* **2018,** *33*, 1207–1218. https://doi.org/10.1007/s11284-018-1634-9.

Eganathan, P.; Subramanian, H. M.; Latha, R.; Rao, C. S. Oil Analysis in Seeds of *Salicornia brachiata*. *Ind. Crops Prod.* **2006,** *23*, 177–179. https://doi.org/10.1016/j.indcrop.2005.05.007.

Ghadge, V.; Kumar, P.; Maity, T. K.; Prasad, K.; Shinde, P. B. Facile Alternative Sustainable Process for the Selective Extraction of Microbial Melanin. *ACS Sustain. Chem. Eng.* **2022,** 2022. https://doi.org/10.1021/acssuschemeng.1c07135.

Ghadge, V.; Kumar, P.; Singh, S.; Mathew, D. E.; Bhattacharya, S.; Nimse, S. B.; Shinde, P. B. Natural Melanin Produced by the Endophytic *Bacillus subtilis* 4NP-BL Associated with the Halophyte *Salicornia brachiata*. *J. Agric. Food Chem.* **2020,** *68*, 6854–6863. https://doi.org/10.1021/acs.jafc.0c01997.

Guil, J. L.; Rodr, I. Nutritional and Toxic Factors in Selected Wild Edible Plants. *Plant Foods Hum. Nutr.* **1997,** *51*, 99–107.

Haque, M. I.; Rathore, M. S.; Gupta, H.; Jha, B. Inorganic Solutes Contribute More Than Organic Solutes to the Osmotic Adjustment in *Salicornia brachiata* (Roxb.) Under Natural Saline Conditions. *Aquat. Bot.* **2017,** *142*, 78–86. https://doi.org/10.1016/j.aquabot.2017.07.004.

Hasanuzzaman, M.; Nahar, K.; Alam, M. M.; Bhowmik, P. C.; Hossain, M. A.; Rahman, M. M.; Prasad, M. N. V.; Ozturk, M.; Fujita, M. Potential Use of Halophytes to Remediate Saline Soils. *Biomed Res. Int.* **2014,** *2014*. https://doi.org/10.1155/2014/589341.

Himabindu, Y.; Chakradhar, T.; Reddy, M. C.; Kanygin, A.; Redding, K. E.; Chandrasekhar, T. Salt-Tolerant Genes from Halophytes Are Potential Key Players of Salt Tolerance in Glycophytes. *Environ. Ex. Bot.* **2016,** *136*, 68–77. https://doi.org/10.1016/j.envexpbot.2015.11.010.

Jacob, P. T.; Siddiqui, S. A.; Rathore, M. S. Seed Germination, Seedling Growth and Seedling Development Associated Physiochemical Changes in *Salicornia brachiata* (Roxb.) Under Salinity and Osmotic Stress. *Aquat. Bot.* **2020,** *166*, 103272. https://doi.org/10.1016/j.aquabot.2020.103272.

Jha, A.; Joshi, M.; Yadav, N. S.; Agarwal, P. K.; Jha, B. Cloning and Characterization of the *Salicornia brachiata* Na^+/H^+ Antiporter Gene SbNHX1 and Its Expression by Abiotic Stress. *Mol. Biol. Rep.* **2011,** *38*, 1965–1973. https://doi.org/10.1007/s11033-010-0318-5.

Jha, B.; Agarwal, P. K.; Reddy, P. S.; Lal, S.; Sopory, S. K.; Reddy, M. K. Identification of Salt-Induced Genes from *Salicornia brachiata*, an Extreme Halophyte Through Expressed Sequence Tags Analysis. *Genes Genet. Syst.* **2009,** *84*, 111–120. https://doi.org/10.1266/ggs.84.111.

Jha, B.; Gontia, I.; Hartmann, A. The Roots of the Halophyte *Salicornia brachiata* are a Source of New Halotolerant Diazotrophic Bacteria with Plant Growth-Promoting Potential. *Plant Soil* 2012a, *356*, 265–277. https://doi.org/10.1007/s11104-011-0877-9.

Jha, B.; Singh, N. P.; Mishra, A. Proteome Pro Filing of Seed Storage Proteins Reveals the Nutritional Potential of *Salicornia brachiata* Roxb.; An Extreme Halophyte. *J. Agric. Food Chem.* 2012b, *60*, 4320–4326.

Jha, B.; Mishra, A.; Jha, A.; Joshi, M. Developing Transgenic Jatropha Using the SbNHX1 Gene from an Extreme Halophyte for Cultivation in Saline Wasteland. *PLoS One* **2013,** *8*. https://doi.org/10.1371/journal.pone.0071136.

Jha, B.; Sharma, A.; Mishra, A. Expression of SbGSTU (Tau Class Glutathione S-Transferase) Gene Isolated from *Salicornia brachiata* in Tobacco for Salt Tolerance. *Mol. Biol. Rep.* **2011,** *38*, 4823–4832. https://doi.org/10.1007/s11033-010-0625-x.

Karnwal, A. Screening, Isolation and Characterization of Culturable Stress-Tolerant Bacterial Endophytes Associated with *Salicornia brachiata* and Their Effect on Wheat (*Triticum aestivum* L.) and Maize (*Zea mays*) Growth. *J. Plant Prot. Res.* **2019,** *59*, 293–303. https://doi.org/10.24425/jppr.2019.129747.

Kumari, J.; Udawat, P.; Dubey, A. K.; Haque, M. I.; Rathore, M. S.; Jha, B. Overexpression of SbSI-1, a Nuclear Protein from *Salicornia brachiata* Confers Drought and Salt Stress Tolerance and Maintains Photosynthetic Efficiency in Transgenic Tobacco. *Front. Plant Sci.* **2017,** *8*. https://doi.org/10.3389/fpls.2017.01215.

Mishra, A.; Joshi, M.; Jha, B. Oligosaccharide Mass Profiling of Nutritionally Important *Salicornia brachiata*, an Extreme Halophyte. *Carbohydr. Polym.* **2013,** *92*, 1942–1945. https://doi.org/10.1016/j.carbpol.2012.11.055.

Mishra, A.; Patel, M. K.; Jha, B. Non-Targeted Metabolomics and Scavenging Activity of Reactive Oxygen Species Reveal the Potential of *Salicornia brachiata* as. *J. Funct. Foods* **2015,** *13*, 21–31. https://doi.org/10.1016/j.jff.2014.12.027.

Mishra, A.; Tanna, B. Halophytes: Potential Resources for Salt Stress Tolerance Genes and Promoters. *Front. Plant Sci.* **2017,** *8*, 1–10. https://doi.org/10.3389/fpls.2017.00829.

Panth, N.; Park, S.; Kim, H. J.; Kim, D.; Oak, M. Protective Effect of *Salicornia europaea* Extracts on High Salt Intake-Induced Vascular Dysfunction and Hypertension. *Int. J. Mol. Sci.* **2016,** *17*, 1–13. https://doi.org/10.3390/ijms17071176.

Parida, A. K.; Jha, B. Physiological and Biochemical Responses Reveal the Drought Tolerance Efficacy of the Halophyte *Salicornia brachiata*. *J. Plant Growth Regul.* **2012,** *32*, 342–352. https://doi.org/10.1007/s00344-012-9303-7.

Parida, A. K.; Jha, B. Antioxidative Defense Potential to Salinity in the Euhalophyte *Salicornia brachiata*. *J. Plant Growth Regul.* **2010,** *29*, 137–148. https://doi.org/10.1007/s00344-009-9129-0.

Parida, A. K.; Kumari, A.; Panda, A.; Rangani, J.; Agarwal, P. K. Photosynthetic Pigments, Betalains, Proteins, Sugars, and Minerals During *Salicornia brachiata* Senescence. *Biol. Plant.* **2018,** *62*, 343–352. https://doi.org/10.1007/s10535-017-0764-1.

Rathore, A. P.; Chaudhary, D. R.; Jha, B. Biomass Production, Nutrient Cycling, and Carbon Fixation by *Salicornia brachiata* Roxb.: A Promising Halophyte for Coastal Saline Soil Rehabilitation. *Inter. J. Phytoremediation* **2016,** *8*, 801–811. https://doi.org/10.1080/15226514.2016.1146228.

Rathore, M. S.; Balar, N.; Jha, B. Population Structure and Developmental Stage-Associated Ecophysiological Responses of *Salicornia brachiata*. *Ecol. Res.* **2019,** *34*, 644–658. https://doi.org/10.1111/1440-1703.12033.

Reddy, M. P. Bromide Tolerance in *Salicornia brachiata* Roxb, an Obligate Halophyte. *Water Air Soil Pollut.* **2009,** *196*, 151–160. https://doi.org/10.1007/s11270-008-9764-2.

Reddy, M. P.; Kaur, M. Sodium Fluoride Induced Growth and Metabolic Changes in *Salicornia brachiata* Roxb. *Water Air Soil Pollut.* **2008,** *188*, 171–179. https://doi.org/10.1007/s11270-007-9533-7.

Sanandiya, N. D.; Prasad, K.; Meena, R.; Siddhanta, A. K. Cellulose of *Salicornia braciata*. *Nat. Prod. Commun.* **2010,** *4*, 603–606.

Shabala, S.; Bose, J.; Hedrich, R. Salt Bladders: Do They Matter? *Trends Plant Sci.* **2014,** *19*, 687–691. https://doi.org/10.1016/j.tplants.2014.09.001.

Sharma, A.; Gontia, I.; Agarwal, P. K.; Jha, B. Accumulation of Heavy Metals and Its Biochemical Responses in *Salicornia brachiata*, an Extreme Halophyte. *Mar. Biol.* **2010,** *6*, 511–518. https://doi.org/10.1080/17451000903434064.

Shukla, P. S.; Gupta, K.; Agarwal, P.; Jha, B.; Agarwal, P. K. Overexpression of a Novel SbMYB15 from *Salicornia brachiata* Confers Salinity and Dehydration Tolerance by Reduced Oxidative Damage and Improved Photosynthesis in Transgenic Tobacco. Planta **2015,** *242*, 1291–1308. https://doi.org/10.1007/s00425-015-2366-5.

Siddiqui, S. A.; Khatri, K.; Patel, D.; Rathore, M. S. Photosynthetic Gas Exchange and *Chlorophyll a* Fluorescence in *Salicornia brachiata* (Roxb.) Under Osmotic Stress. *J. Plant Growth Regul.* **2022,** *41*, 429–444. https://doi.org/10.1007/s00344-021-10311-8.

Singh, A.; Sharma, S.; Shah, M. T. Successful Cultivation of *Salicornia brachiata* -A Sea Asparagus Utilizing RO of Reject Water: A Sustainable Solution. *Int. J. waste Resour.* **2018,** *8*, 1–5. https://doi.org/10.4172/2252-5211.1000322.

Singh, D.; Yadav, N. S.; Tiwari, V.; Agarwal, P. K.; Jha, B. A SNARE-Like Superfamily Protein SbSLSP from the Halophyte *Salicornia brachiata* Confers Salt and Drought Tolerance by Maintaining Membrane Stability, K^+/Na^+ Ratio, and Antioxidant Machinery. *Front. Plant Sci.* **2016,** *7*, 737. https://doi.org/10.3389/fpls.2016.00737.

Singh, N.; Mishra, A.; Jha, B. Ectopic Over-Expression of Peroxisomal Ascorbate Peroxidase (SbpAPX) Gene Confers Salt Stress Tolerance in Transgenic Peanut (*Arachis hypogaea*). *Gene* **2014,** *547*, 119–125. https://doi.org/10.1016/j.gene.2014.06.037.

Singh, S.; Ghadge, V. A.; Kumar, P.; Mathew, D. E.; Dhimmar, A.; Sahastrabudhe, H.; Nalli, Y.; Rathod, M. R.; Shinde, P. B. Biodiversity and Antimicrobial Potential of Bacterial Endophytes from Halophyte *Salicornia brachiata*. Antonie van Leeuwenhoek. *Int. J. Gen. Mol. Microbiol.* **2021,** *114*, 591–608. https://doi.org/10.1007/s10482-021-01544-4.

Singh, V. K.; Mishra, A.; Haque, I.; Jha, B. A Novel Transcription Factor-Like Gene SbSDR1 Acts as a Molecular Switch and Confers Salt and Osmotic Endurance to Transgenic Tobacco. *Sci. Rep.* **2016,** *6*, 1–15. https://doi.org/10.1038/srep31686.

Slama, I.; Abdelly, C.; Bouchereau, A.; Flowers, T.; Savoure, A. Diversity, Distribution and Roles of Osmoprotective Compounds Accumulated in Halophytes Under Abiotic Stress. *Ann. Bot.* **2015,** *115*, 433–447. https://doi.org/10.1093/aob/mcu239.

Stanley, O. D. Bio Prospecting Marine Halophyte *Salicornia brachiata* for Medical Importance and Salt Encrusted Land Development. *J. Coast. Dev.* **2008,** *11*, 62–69.

Tiwari, V.; Chaturvedi, A. K.; Mishra, A.; Jha, B. Introgression of the SbASR-1 Gene Cloned from a Halophyte *Salicornia brachiata* Enhances Salinity and Drought Endurance in Transgenic Groundnut (*Arachis hypogaea*) and Acts as a Transcription Factor. *PLoS One* **2015,** *10*, 1–29. https://doi.org/10.1371/journal.pone.0131567.

Tiwari, V.; Chaturvedi, A. K.; Mishra, A.; Jha, B. The Transcriptional Regulatory Mechanism of the Peroxisomal Ascorbate Peroxidase (pAPX) Gene Cloned from an Extreme Halophyte, *Salicornia brachiata. Plant Cell Physiol.* **2014,** *55*, 201–217. https://doi.org/10.1093/pcp/pct172.

Tiwari, V.; Patel, M. K.; Chaturvedi, A. K.; Mishra, A.; Jha, B. Functional Characterization of the Tau Class Glutathione-S-Transferases Gene (SbGSTU) Promoter of *Salicornia brachiata* Under Salinity and Osmotic Stress. *PLoS One*. https://doi.org/10.1371/journal.pone.0148494.

Udawat, P.; Jha, R. K.; Mishra, A.; Jha, B. Overexpression of a Plasma Membrane-Localized SbSRP-Like Protein Enhances Salinity and Osmotic Stress Tolerance in Transgenic Tobacco. *Front. Plant Sci.* **2017,** *8*, 1–19. https://doi.org/10.3389/fpls.2017.00582.

Udawat, P.; Jha, R. K.; Sinha, D.; Mishra, A.; Jha, B. Overexpression of a Cytosolic Abiotic Stress Responsive Universal Stress Protein (SbUSP) Mitigates Salt and Osmotic Stress in Transgenic Tobacco Plants. *Front. Plant Sci.* **2016,** *7*, 1–21. https://doi.org/10.3389/fpls.2016.00518.

Ventura, Y.; Wuddineh, W. A.; Myrzabayeva, M.; Alikulov, Z.; Khozin-Goldberg, I.; Shpigel, M.; Samocha, T. M.; Sagi, M. Effect of Seawater Concentration on the Productivity and Nutritional Value of Annual *Salicornia* and Perennial *Sarcocornia halophytes* as Leafy Vegetable Crops. *Sci. Hortic.* **2011,** *128*, 189–196. https://doi.org/10.1016/j.scienta.2011.02.001.

Index

F

G

H

I

L

M

N

T

V

W

X